Stochastic Mechanics
Random Media
Signal Processing
and Image Synthesis
Mathematical Economics
and Finance
Stochastic Optimization
Stochastic Control

Applications of Mathematics

Stochastic Modelling and Applied Probability

39

Edited by I. Karatzas
M. Yor

Advisory Board P. Brémaud
E. Carlen
W. Fleming
D. Geman
G. Grimmett
G. Papanicolaou
J. Scheinkman

Springer

New York
Berlin
Heidelberg
Barcelona
Budapest
Hong Kong
London
Milan
Paris
Singapore
Tokyo

Applications of Mathematics

(continued after index)

Ioannis Karatzas   Steven E. Shreve

Methods of
Mathematical Finance

 Springer

Ioannis Karatzas
Departments of Mathematics
   and Statistics
Columbia University
New York, NY 10027, USA

Steven E. Shreve
Department of Mathematical Sciences
Carnegie Mellon University
Pittsburgh, PA 15213, USA

Managing Editors

I. Karatzas
Departments of Mathematics and Statistics
Columbia University
New York, NY 10027, USA

M. Yor
CNRS, Laboratoire de Probabilités
Université Pierre et Marie Curie
4, Place Jussieu, Tour 56
F-75252 Paris Cedex 05, France

Mathematics Subject Classification (1991): 90A09, 60G44, 93E20

Library of Congress Cataloging-in-Publication Data
Karatzas, Ioannis.
   Methods of mathematical finance / Ioannis Karatzas, Steven E.
Shreve.
      p.   cm.—(Applications of mathematics ; 39)
   Includes bibliographical references and index.
   ISBN 0-387-94839-2 (alk. paper)
      1. Business mathematics.   2. Finance—Mathematical models.
   3. Brownian motion processes.   4. Contingent valuation.   I. Shreve,
   Steven E.   II. Title.   III. Series
   HF5691.K3382   1998
   650′.01′513—dc21   98-14284

Printed on acid-free paper.

Production managed by Steven Pisano; manufacturing supervised by Joe Quatela.
Typeset by Integre Technical Publishing Co., Albuquerque, NM.
Printed and bound by R.R. Donnelley and Sons, Harrisonburg, VA.
Printed in the United States of America.

9 8 7 6 5 4 3 2 1

ISBN 0-387-94839-2 Springer-Verlag New York Berlin Heidelberg   SPIN 10544022

To Eleni, for all her patience.

I.K.

And to Dot, whom I love.

S.E.S.

Preface

This book is intended for readers who are quite familiar with probability and stochastic processes but know little or nothing about finance. It is written in the definition/theorem/proof style of modern mathematics and attempts to explain as much of the finance motivation and terminology as possible.

A mathematical monograph on finance can be written today only because of two revolutions that have taken place on Wall Street in the latter half of the twentieth century. Both these revolutions began at universities, albeit in economics departments and business schools, not in departments of mathematics or statistics. They have led inexorably, however, to an escalation in the level of mathematics (including probability, statistics, partial differential equations and their numerical analysis) used in finance, to a point where genuine research problems in the former fields are now deeply intertwined with the theory and practice of the latter.

The first revolution in finance began with the 1952 publication of "Portfolio Selection," an early version of the doctoral dissertation of Harry Markowitz. This publication began a shift away from the concept of trying to identify the "best" stock for an investor, and towards the concept of trying to understand and quantify the trade-offs between *risk* and *return* inherent in an entire portfolio of stocks. The vehicle for this so-called *mean–variance analysis* of portfolios is *linear regression*; once this analysis is complete, one can then address the *optimization problem* of choosing the portfolio with the largest mean return, subject to keeping the risk (i.e., the variance) below a specified acceptable threshold. The implementation of Markowitz's ideas was aided tremendously by William Sharpe, who developed the concept of determining covariances not between every possible pair of stocks, but between each stock and the "market." For purposes of the above optimization problem each stock could then be characterized by its mean rate of return (its "α") and its correlation with the market (its "β"). For their pioneering work, Markowitz and Sharpe shared with Merton Miller the 1990 Nobel Prize in economics, the first ever awarded for work in finance.

The portfolio-selection work of Markowitz and Sharpe introduced mathematics to the "black art" of investment management. With time, the mathematics has become more sophisticated. Thanks to Robert Merton and Paul Samuelson, one-period models were replaced by continuous-time, Brownian-motion-driven models, and the quadratic utility function implicit in mean–variance optimization was replaced by more general increasing, concave utility functions. Model-based mutual funds have taken a permanent seat at the table of investment opportunities offered to the public. Perhaps more importantly, the paradigm for thinking about financial markets has become a mathematical model. This affects the way we now understand issues of corporate finance, taxation, exchange-rate fluctuations, and all manner of financial issues.

The second revolution in finance is connected with the explosion in the market for *derivative securities*. Already in 1992, this market was estimated by the Bank for International Settlements to be a \$4 trillion business worldwide, involving every sector of the finance industry. According to this estimate, the size of the derivative securities market had increased eight-fold in five years. The foundational work here was done by Fisher Black, Robert Merton, and Myron Scholes in the early 1970s. Black, Merton, and Scholes were seeking to understand the value of the option to buy one share of a stock at a future *date* and *price* specified in advance. This so-called *European call-option* derives its value from that of the underlying stock, whence the name *derivative security*. The basic idea of valuing a European call-option is to construct a *hedging portfolio*, i.e., a combination of shares from the stock on which the call is written and of shares from a money market, so that the resulting portfolio replicates the option. At any time, the option should be worth exactly as much as the hedging portfolio, for otherwise some astute trader ("arbitrageur") could make something for nothing by trading in the option, the stock, and the money market; such trading would bring the prices back into line. Based on this simple principle, called *absence of arbitrage,* Black and Scholes (1973) derived the now famous formula for the value of the European call-option, which bears their name and which was extended by Merton (1973) in a variety of very significant ways. For this foundational work, Robert Merton and Myron Scholes were awarded the 1997 Nobel Prize in economics.

While options and other derivative securities can be used for speculation, their primary appeal is to investors who want to remove some of the risk associated with their investments or businesses. The sellers of derivative securities are faced with the twin problems of *pricing* and *hedging* them, and to accomplish this, current practice is to use Brownian-motion-based models of asset prices. Without such models and the analytical tractability that they provide, the market for derivative securities could not have grown to its present mammoth proportions.

Before proceeding further in this brief description of modern finance, there are two myths about the mathematical theory of finance that we need to explode.

The first myth is that this research is only about how to "beat the market." It is true that much of the portfolio optimization work growing out of the first revolution in finance is about how to "beat the market," but a substantial component is about how to understand the market for other purposes, such as regulation. The second revolution in finance, the derivative securities explosion, is not about beating the market at all.

The second myth maintains that since the finance industry does not manufacture tangible commodities, such as refrigerators or automobiles, it can be engaged in nothing but a zero-sum game, "robbing Peter to pay Paul." In fact, the role of financial institutions in a decentralized economy is to facilitate the flow of capital to sectors of society engaged in production. An efficient finance industry will facilitate this flow at the least possible cost, making available to the manufacturing sector a wide variety of instruments for borrowing and investing.

Consider, for example, a manufacturer who contemplates expansion of his production facilities and who chooses to finance this expansion by borrowing capital, in effect taking a mortgage on the new facilities. The terms (e.g., fixed or variable interest rate, term, prepayment options, collateralization) under which the manufacturer is willing to borrow money may not neatly match the terms under which any particular lender is willing to provide it. The finance industry should take the investments that lenders are willing to make, restructuring and recombining them as necessary, so as to provide a loan the manufacturer is willing to accept. The finance industry should perform this function in a wide variety of settings and manage its affairs so as to be exposed to minimal risk.

Let us suppose that the manufacturer is unable to plan effectively if he takes out a variable-rate mortgage, and so insists on a fixed-rate mortgage. Imagine also that an investment bank makes the mortgage, using money invested with it by depositors expecting to receive payments at the current (variable) interest rate. The bank is obliged to make monthly payments to these investors; the amounts of these payments fluctuate with the prevailing interest rates, and may be larger or smaller than what the bank receives from the manufacturer. To remove the risk associated with this position, the bank constructs a *hedge*. It may, for example, choose to sell short a number of bonds, i.e., receive money now in exchange for a promise to deliver bonds that it does not presently own and will have to buy eventually. If interest rates rise, the bank will have to pay its investors more than it receives on the loan from the manufacturer, but the cost of buying the bonds it has promised to deliver will fall. If the bank chooses its position carefully, its additional liability to its investors will be exactly offset by the downward movement of bond prices, and it will thus be protected against increases in the interest rate. Of course, decreases in interest rates will cause bond prices to rise, and the bank should choose its hedging position so as to be protected against this eventuality as well.

As one can see from this overly simplistic example, *a proliferation of financial instruments can enhance the efficiency of an economy.* The bank in this example "synthesizes" a fixed-rate mortgage using variable-rate investments and a position in the bond market. Such synthetic securities are the "products" of investment banks; while no one would claim that every "product" of this type contributes to the well-being of the nation, there is no doubt that an economy that has available a large variety of such products has a comparative advantage over one with a more limited offering. The firm that "manufactures" such products can do so only if it has reliable *models* for pricing and hedging them. Current models are built using stochastic calculus, are fit to the data by careful statistical estimation procedures, and require accurate and fast real-time numerical analysis.

This book is about some of these models. It treats only a small part of the whole picture, leaving completely untouched the issues of estimation and numerical analysis. Even within the range of models used in finance, we have found it necessary to be selective. Our guide has been to write about what we know best, namely areas of research in which we have had some level of personal involvement. Through the inclusion of an extensive bibliography and of notes at the end of each chapter, we have tried to point the reader toward some of the topics not touched. The bibliography is necessarily incomplete. We apologize to those whose work should have been included but is not. Such omissions are unintentional, and due either to ignorance or oversight.

In order to read this book one should be familiar with the material contained in the first three chapters of our book *Brownian Motion and Stochastic Calculus* (Springer-Verlag, New York, 1991). There are many other good sources for this material, but we will refer to the source we know best when we cite specific results.

Here is a high-level overview of the contents of this monograph. In *Chapter 1* we set up the generally accepted, Brownian-motion-driven model for financial markets. Because the coefficient processes in this model are themselves stochastic process, this is nearly the most general continuous-time model conceivable among those in which prices move continuously. The model of Chapter 1 allows us to introduce notions and results about portfolio and consumption rules, arbitrage, equivalent martingale measures, and attainability of contingent claims; it divides naturally into two cases, called *complete* and *incomplete,* respectively.

Chapter 2 lays out the theory of pricing and hedging contingent claims (the "synthetic" or "derivative" securities described above) in the context of a *complete market*. To honor the origins of the subject and to acquaint the reader with some important special cases, we analyze in some detail the pricing and hedging of a number of different options. We have also included a section on "futures" contracts, derivative securities that are conceptually more difficult because their value is defined recursively.

Chapter 3 takes up the problem of a single agent faced with optimal consumption and investment decisions in the complete version of the market model in Chapter 1. Tools from stochastic calculus and partial differential equations of parabolic type permit a very general treatment of the associated optimization problem. This theory can be related to Markowitz's mean–variance analysis and is ostensibly about how to "beat the market," although another important use for it is as a first step toward understanding how markets operate. Its latter use is predicated on the principle that a good model of individual behavior is to postulate that individuals act in their own best interest.

Chapter 4 carries the notions and results of Chapter 3 to their logical conclusion. In particular, it is assumed that there are several individuals in the economy, each behaving as described in Chapter 3; through the *law of supply and demand*, their collective actions determine the so-called *equilibrium* prices of securities in the market. Characterization of this equilibrium permits the study of questions about the effect of interventions in the market.

In *Chapter 5* we turn to the more difficult issue of pricing and hedging contingent claims in markets with *incompleteness* or other *constraints* on individual investors' portfolio choices. An approach based on "fictitious completion" for such a market, coupled with notions and results from *convex analysis* and *duality theory*, permits again a very general solution to the hedging problem.

Finally, *Chapter 6* uses the approach developed in Chapter 5 to treat the optimal consumption/investment problem for such incomplete or constrained markets, and for markets with different interest rates for borrowing and investing.

Note to the Reader

We use a hierarchical numbering system for equations and statements. The k-th equation in Section j of Chapter i is labeled $(j.k)$ at the place where it occurs and is cited as $(j.k)$ within Chapter i, but as $(i.j.k)$ outside Chapter i. A definition, theorem, lemma, corollary, remark, or exercise is a "statement," and the k-th statement in Section j of Chapter i is labeled *j.k Statement* at the place where it occurs, and is cited as *Statement j.k* within Chapter i but as *Statement i.j.k* outside Chapter i.

Acknowledgments

We are grateful to our collaborators, who have made it possible for us to work productively in this exciting new field: Abel Cadenillas, Jaksa Cvitanić, Mark Davis, Nicole El Karoui, Monique Jeanblanc-Picqué, Steve Kou, Peter Lakner, John Lehoczky, Daniel Ocone, Igor Pikovsky, Suresh Sethi, Martin Shubik, Mete Soner, Bill Sudderth, Gan-Lin Xu, and Xing-Xiong Xue. Over the years we have benefited enormously from the knowledge, expertise, and friendly good advice of colleagues such as Ludwig Arnold, Philippe Artzner, Marco Avellaneda, Süleyman Başak, Kerry Back, Mark Broadie, Michael Camp, Peter Carr, George Constantinides, Domenico Cuoco, Freddie Delbaen, Jérôme Detemple, Darrell Duffie, Phil Dybvig, Wendell Fleming, Hans Föllmer, Sanford Grossman, Michael Harrison, David Heath, Rainer Höhnle, Robert Jarrow, Robert Kertz, Shigeo Kusuoka, Shlomo Levental, Marek Musiela, Stan Pliska, Marie-Claire Quenez, Chris Rogers, Martin Schweizer, Michael Selby, David Siegmund, Suresh Sundaresan, Christophe Stricker, and Walter Willinger; we are extending to them our most sincere appreciation. Special thanks go to Nicole El Karoui for instructing us on the subtleties of the general theory of optimal stopping; Appendix D reflects what we have managed to understand with her help from her St.-Flour lecture notes (El Karoui 1981).

Portions of this book grew out of series of lectures and minicourses delivered over the last ten years at Columbia and Carnegie-Mellon Universities, and elsewhere. We are deeply grateful for these opportunities to our two home institutions, and also to our colleagues in other places who arranged lectures: Professors John Baras, Yuan-Shih Chow, Nicole El Karoui, Hans-Jürgen Engelbert, Avner Friedman, Martin Goldstein, Raoul Le Page, Duong Phong, Heracles Polemarchakis, Boris Rozovsky, Wolfgang Runggaldier, Robert Shay, Michael Steele, and Luc Vinet. We also owe a debt to the audiences in these courses for their interest and enthusiasm, and for helping us correct some of our mistakes and misconceptions.

Over the years, our research and the writing of this book have both been supported generously by the National Science Foundation, under grants DMS-87-23078, DMS-90-22188, DMS-93-19816, DMS-90-02588, DMS-92-03360, and DMS-95-00626. A big part of the manuscript was typed by Mrs. Doodmatie Kalicharan in her usual efficient and tireless fashion; we could not even have started this project without her assistance, and we are grateful for it.

And we would never have succeeded in finishing this book without the support, the understanding, and the patience of our families. We dedicate it to them with love and gratitude.

Ioannis Karatzas
Steven E. Shreve

Contents

1

A Brownian Model of
Financial Markets

1.1   Stocks and a Money Market

Throughout this monograph we deal with a financial market consisting
of $N + 1$ financial assets. One of these assets is instantaneously risk-free,
and will be called a *money market*. Assets 1 through N are risky, and
will be called *stocks* (although in applications of this model they are often
commodities or currencies, rather than common stocks). These financial
assets have continuous prices evolving continuously in time and driven by
a D-dimensional Brownian motion. The continuity of the time parameter
and the accompanying capacity for continuous trading permit an elegance
of formulation and analysis not unlike that obtained when passing from
difference to differential equations. If asset prices do not vary continuously,
at least they vary frequently, and the model we propose to study has proved
its usefulness as an approximation to reality. Our assumption that asset
prices have no jumps is a significant one. It is tantamount to the assertion
that there are no "surprises" in the market: the price of a stock at time t
can be perfectly predicted from knowledge of its price at times strictly prior
to t. We adopt this assumption in order to simplify the mathematics; the
additional assumption that asset prices are driven by a Brownian motion is
little more than a convenient way of phrasing this condition. Some literature
on continuous-time markets with discontinuous asset prices is cited in the
notes at the end of this chapter. The extent to which the results of this
monograph can be extended to such models has not yet been fully explored.

Let us begin then with a complete probability space $(\Omega, \mathcal{F}, P)$ on which is given a standard, D-dimensional Brownian motion $W(t) = (W^{(1)}(t), \ldots, W^{(D)}(t))', 0 \leq t \leq T$. Here prime denotes transposition, so that $W(t)$ is a column vector. We assume that $W(0) = 0$ almost surely. All economic activity will be assumed to take place on a finite horizon $[0, T]$, where T is a positive constant.[†] Define

$$\mathcal{F}^W(t) \stackrel{\triangle}{=} \sigma\{W(s); \quad 0 \leq s \leq t\}, \quad \forall t \in [0, T], \tag{1.1}$$

to be the filtration generated by $W(\cdot)$, and let $\mathcal{N}$ denote the P-null subsets of $\mathcal{F}^W(T)$. We shall use the *augmented filtration*

$$\mathcal{F}(t) \stackrel{\triangle}{=} \sigma(\mathcal{F}^W(t) \cup \mathcal{N}), \quad \forall t \in [0, T]. \tag{1.2}$$

One should interpret the σ-algebra $\mathcal{F}(t)$ as the *information available to investors at time t*, in the sense that if $\omega \in \Omega$ is the true state of nature and if $A \in \mathcal{F}(t)$, then at time t all investors know whether $\omega \in A$. Note that $\mathcal{F}(0)$ contains only sets of measure one and sets of measure zero, so every $\mathcal{F}(0)$-measurable random variable is almost surely constant.

Remark 1.1: The difference between $\{\mathcal{F}^W(t)\}_{0 \leq t \leq T}$ and $\{\mathcal{F}(t)\}_{0 \leq t \leq T}$ is a purely technical one. The filtration $\{\mathcal{F}(t)\}_{0 \leq t \leq T}$ is *left-continuous*, in the sense that

$$\mathcal{F}(t) = \sigma\left(\bigcup_{0 \leq s < t} \mathcal{F}(s)\right), \quad \forall t \in (0, T], \tag{1.3}$$

and $\{\mathcal{F}^W(t)\}_{0 \leq t \leq T}$ is also left-continuous. The filtration $\{\mathcal{F}(t)\}_{0 \leq t \leq T}$ is *right-continuous* in the sense that

$$\mathcal{F}(t) = \bigcap_{t < s \leq T} \mathcal{F}(s), \quad \forall t \in [0, T), \tag{1.4}$$

but $\{\mathcal{F}^W(t)\}_{0 \leq t \leq T}$ is not right-continuous (see Karatzas and Shreve (1991), Section 2.7, for more details). Equations (1.3), (1.4) express the notion alluded to in the first paragraph of this section, that "there are no surprises in the flow of information" in this model.

Remark 1.2: Every local martingale relative to the filtration $\{\mathcal{F}(t)\}$ has a modification whose paths are *continuous* (Karatzas and Shreve (1991), Problem 3.4.16); we shall always use this continuous modification. We shall also encounter processes $Y(\cdot)$ that are right-continuous with left-hand limits and whose total variation $Y(t)$ is finite on each interval $[0, t], 0 \leq t \leq T$. We shall refer to these as *finite-variation RCLL* processes. In our context,

---

[†]There are a few places in this book, namely, Sections 1.7, 2.6, 3.9, and 3.10, where the planning horizon is $[0, \infty)$.

Si is Price

an $\{\mathcal{F}(t)\}$-*semimartingale* will be the sum of a (continuous) local martingale and a finite-variation RCLL process. The decomposition of such a semimartingale into a local martingale and a finite-variation RCLL process is unique, up to an additive constant.

We introduce now a *money market* and N *stocks*. The precise conditions on these assets are given in Definition 1.3; here we content ourselves with giving the main properties of these objects.

A share of the *money market* has price $S_0(t)$ at time t, with $S_0(0) = 1$. The price process $S_0(\cdot)$ is continuous, strictly positive, and $\{\mathcal{F}(t)\}$-adapted, with finite total variation on $[0, T]$. Being of finite variation, $S_0(\cdot)$ can be decomposed into absolutely continuous and singularly continuous parts $S_0^{ac}(\cdot)$ and $S_0^{sc}(\cdot)$, respectively. We can then define

$$r(t) \triangleq \frac{\frac{d}{dt} S_0^{ac}(t)}{S_0(t)}, \qquad A(t) \triangleq \int_0^t \frac{dS_0^{sc}(u)}{S_0(u)}, \tag{1.5}$$

so that

$$dS_0(t) = S_0(t)[r(t)\, dt + dA(t)], \qquad \forall\, t \in [0, T], \tag{1.6}$$

or equivalently,

$$S_0(t) = \exp\left\{ \int_0^t r(u)du + A(t) \right\}, \qquad \forall\, t \in [0, T]. \tag{1.7}$$

In the special case that $S_0(\cdot)$ is itself absolutely continuous, so that $A(\cdot) \equiv 0$, the price of the money market evolves like the value of a savings account whose instantaneous *(risk-free) interest rate* at time t is $r(t)$. This is the case the reader should keep in mind. The risk-free rate process $r(\cdot)$ is random and time-dependent, but $r(t)$ is $\mathcal{F}(t)$-measurable, so the current risk-free rate is known to all investors.

Next we introduce N *stocks* with prices-per-share $S_1(t), \ldots, S_N(t)$ at time t and with $S_1(0), \ldots, S_N(0)$ positive constants. The processes $S_1(\cdot), \ldots, S_N(\cdot)$ are continuous, strictly positive, and satisfy stochastic differential equations

$$dS_n(t) = S_n(t)\left[b_n(t)\, dt + dA(t) + \sum_{d=1}^{D} \sigma_{nd}(t)\, dW^{(d)}(t) \right], \tag{1.8}$$

$$\forall\, t \in [0, T], n = 1, \ldots, N.$$

We show in Appendix B that every continuous, strictly positive, and $\{\mathcal{F}(t)\}$-adapted semimartingale satisfies a stochastic differential equation of this form, where $A(\cdot)$ is some $\{\mathcal{F}(t)\}$-adapted, singularly continuous process. In (1.8), however, $A(\cdot)$ is not an arbitrary such process, but rather it is the one defined in (1.5). We also show in Appendix B that if the $(\{\mathcal{F}(t)\}$-

adapted, singularly continuous) process $A(\cdot)$ in (1.8) were not given by the expression of (1.5), then *an arbitrage opportunity would exist*. The notion of arbitrage is defined in Section 4.

The solution of the equation (1.8) is

$$S_n(t) = S_n(0) \exp \left\{ \int_0^t \sum_{d=1}^D \sigma_{nd}(s) \, dW^{(d)}(s) + \int_0^t \left[b_n(s) \right. \right.$$
$$\left. \left. - \frac{1}{2} \sum_{d=1}^D \sigma_{nd}^2(s) \right] ds + A(t) \right\}, \forall\, t \in [0, T], \quad n = 1, \ldots, N. \quad (1.9)$$

Consequently, the singularly continuous process $A(\cdot)$ does not enter the *discounted stock prices*

$$\frac{S_n(t)}{S_0(t)} = S_n(0) \exp \left\{ \int_0^t \sum_{d=1}^D \sigma_{nd}(s) \, dW^{(d)}(s) + \int_0^t \left[b_n(s) - r(s) \right. \right.$$
$$\left. \left. - \frac{1}{2} \sum_{d=1}^D \sigma_{nd}^2(s) \right] ds \right\}, \forall\, t \in [0, T], \quad n = 1, \ldots, N. \quad (1.10)$$

In some applications, the stocks have associated *dividend rate processes*. We model these as real-valued processes $\delta_n(\cdot)$, where $\delta_n(t)$ is the rate of dividend payment per dollar invested in the stock at time t. Adding the dividend rate process into (1.8), we can define the *yield* (per share) *processes* by $Y_n(0) = S_n(0)$ and

$$dY_n(t) = S_n(t) \left[b_n(t) \, dt + dA(t) \right.$$
$$\left. + \sum_{d=1}^D \sigma_{nd}(t) \, dW^{(d)}(t) + \delta_n(t) \, dt \right], \quad n = 1, \ldots, N, \quad (1.11)$$

or equivalently,

$$Y_n(t) = S_n(t) + \int_0^t S_n(u) \delta_n(u) \, du, \forall\, t \in [0, T], \quad n = 1, \ldots, N. \quad (1.12)$$

We set $Y_0(t) = S_0(t), 0 \le t \le T$.

We formalize this discussion with the following definition.

Definition 1.3:   A *financial market* consists of

(i) a probability space $(\Omega, \mathcal{F}, P)$;
(ii) a positive constant T, called the *terminal time*;
(iii) a D-dimensional Brownian motion $\{W(t), \mathcal{F}(t); 0 \le t \le T\}$ defined on $(\Omega, \mathcal{F}, P)$, where $\{\mathcal{F}(t)\}_{0 \le t \le T}$ is the augmentation (by the null sets in $\mathcal{F}^W(T)$) of the filtration $\{\mathcal{F}^W(t)\}_{0 \le t \le T}$ generated by $W(\cdot)$;

(iv) a progressively measurable *risk-free rate process* $r(\cdot)$ satisfying $\int_0^T |r(t)|\, dt < \infty$ almost surely (a.s.);

(v) a progressively measurable, N-dimensional *mean rate of return process* $b(\cdot)$ satisfying $\int_0^T \|b(t)\|\, dt < \infty$ a.s.;

(vi) a progressively measurable, N-dimensional *dividend rate process* $\delta(\cdot)$ satisfying $\int_0^T \|\delta(t)\|\, dt < \infty$ a.s.;

(vii) a progressively measurable, $(N \times D)$-matrix-valued *volatility process* $\sigma(\cdot)$ satisfying $\sum_{n=1}^N \sum_{d=1}^D \int_0^T \sigma_{nd}^2(t)\, dt < \infty$ a.s.;

(viii) a vector of positive, constant initial stock prices $S(0) = (S_1(0), \dots, S_N(0))'$;

(ix) a progressively measurable, singularly continuous, finite-variation process $A(\cdot)$ whose total variation on $[0,t]$ is denoted by $\check{A}(t)$.

We refer to this financial market as $\mathcal{M} = (r(\cdot), b(\cdot), \delta(\cdot), \sigma(\cdot), S(0), A(\cdot))$.

Given a financial market $\mathcal{M}$ as above, the money market and stock price processes are determined by (1.7), (1.9), and then (1.5), (1.6), (1.8), and (1.10) hold. The initial conditions of the asset prices are nearly irrelevant. For investment purposes, the essential feature of an asset is its rate of price change and dividend payment relative to the current price, and these relative rates are captured by $r(\cdot), b(\cdot), \sigma(\cdot), \delta(\cdot), A(\cdot)$. Thus, for notational simplicity we have taken the liberty of declaring $S_0(0) = 1$. We could also have set the initial prices of the stocks equal to one, but have chosen not to do so because some of the formulas developed later are more informative when the dependence on $S_1(0), \dots, S_N(0)$ is explicitly displayed.

Remark 1.4: Much of the existing finance literature is based on *Markov models*, and exploits the connections between such models and partial differential equations. Such a model typically has a K-dimensional *state-process* $\psi(\cdot)$ with a given initial condition $\psi(0)$, and is driven by a stochastic differential equation of the form

$$d\psi(t) = \mu(t, \psi(t))\, dt + \rho(t, \psi(t))\, dW(t), \tag{1.13}$$

where $\mu : [0,T] \times \mathbb{R}^K \to \mathbb{R}^K$ and $\rho : [0,T] \times \mathbb{R}^K \to L(\mathbb{R}^D; \mathbb{R}^K)$ (the set of $K \times D$ matrices) are jointly Borel measurable and satisfy conditions (e.g., Lipschitz continuity in their second argument) that guarantee the existence of a unique solution to (1.13). The coefficients of the market model are taken to be measurable functions $r : [0,T] \times \mathbb{R}^K \to \mathbb{R}$, $b : [0,T] \times \mathbb{R}^K \to \mathbb{R}^N$, $\delta : [0,T] \times \mathbb{R}^K \to \mathbb{R}^N$, and $\sigma : [0,T] \times \mathbb{R}^K \to L(\mathbb{R}^D; \mathbb{R}^K)$ with $A(\cdot) \equiv 0$, so that the dependence of $r(t, \psi(t)), b(t, \psi(t)), \delta(t, \psi(t))$, and $\sigma(t, \psi(t))$ on the sample point $\omega \in \Omega$ occurs only through the dependence of $\psi(t)$ on ω. The simplest Markov model is the one with *constant coefficients*; in this model r, b, δ, and σ are constants, $A(\cdot) \equiv 0$, and there is no need for a state-process. From time to time we shall specialize the results of our more general model to obtain various classical Markov model results.

1.2   Portfolio and Gains Processes

In this section we model portfolio decisions and their consequences for an investor faced with the market in Definition 2.1. We begin with an informal discussion of an investor who makes decisions in discrete time, and this leads us eventually to Definition 2.1, which pertains to continuous trading.

Let $0 = t_0 < t_1 < \cdots < t_M = T$ be a partition of $[0, T]$. For $n = 1, \ldots, N$ and $m = 0, \ldots, M - 1$, let $\eta_n(t_m)$ denote the number of shares of stock n held by the investor over the time interval $[t_m, t_{m+1})$. Let $\eta_0(t_m)$ denote the number of shares held in the money market. For $n = 0, 1, \ldots, N$ the random variable $\eta_n(t_m)$ must be $\mathcal{F}(t_m)$-measurable; in other words, *anticipation of the future* (insider trading) *is not permitted.*

Let us define the associated gains process by the stochastic *difference* equation

$$G(0) = 0, \tag{2.1}$$

$$G(t_{m+1}) - G(t_m) = \sum_{n=0}^{N} \eta_n(t_m)[Y_n(t_{m+1}) - Y_n(t_m)], \tag{2.2}$$

$$m = 0, \ldots, M - 1.$$

Then $G(t_m)$ is the amount earned by the investor over the time interval $[0, t_m]$. On the other hand, the value of the investor's holdings at time t_m is $\sum_{n=0}^{N} \eta_n(t_m)S_n(t_m)$. We have

$$G(t_m) = \sum_{n=0}^{N} \eta_n(t_m)S_n(t_m), \quad m = 0, \ldots, M,$$

if and only if there is no infusion or withdrawal of funds over $[0, T]$. In this case the trading is called "self-financed."

Now suppose that $\eta(\cdot) = (\eta_0(\cdot), \ldots, \eta_N(\cdot))'$ is an $\{\mathcal{F}(t)\}$-adapted process defined on all of $[0, T]$, not just the partition points $t_0, \ldots, t_M$. The associated gains process is now defined by the initial condition $G(0) = 0$ and the stochastic *differential* equation

$$dG(t) = \sum_{n=0}^{N} \eta_n(t) \, dY_n(t). \tag{2.2$'$}$$

We take this equation as an axiom; references related to this point are cited in the Notes, Section 1.8. Defining $\pi_n(t) \stackrel{\triangle}{=} \eta_n(t)S_n(t), \pi(\cdot) \stackrel{\triangle}{=} (\pi_1(\cdot), \ldots, \pi_N(\cdot))'$, and recalling (1.6), (1.11), we may rewrite (2.2)$'$ as

$$dG(t) = [\pi_0(t) + \pi'(t)\underline{1}](r(t) \, dt + dA(t)) + \pi'(t)[b(t) + \delta(t) - r(t)\underline{1}] \, dt$$
$$+ \pi'(t)\sigma(t) \, dW(t), \tag{2.3}$$

where $\underline{1}$ denotes the N-dimensional vector with every component equal to one. Note that $\pi_n(t)$ is the *dollar amount* invested in security n, not the number of shares held.

Definition 2.1: Consider a financial market $\mathcal{M} = (r(\cdot), b(\cdot), \delta(\cdot), \sigma(\cdot),$ $S(0), A(\cdot))$ as in Definition 1.3. A *portfolio process* $(\pi_0(\cdot), \pi(\cdot))$ for this market consists of an $\{\mathcal{F}(t)\}$-progressively measurable, real-valued process $\pi_0(\cdot)$ and an $\{\mathcal{F}(t)\}$-progressively measurable, $\mathbb{R}^N$-valued process $\pi(\cdot) = (\pi_1(\cdot), \ldots, \pi_N(\cdot))'$ such that

$$\int_0^T |\pi_0(t) + \pi'(t)\underline{1}|[|r(t)|\, dt + d\breve{A}(t)] < \infty, \tag{2.4}$$

$$\int_0^T |\pi'(t)(b(t) + \delta(t) - r(t)\underline{1})|\, dt < \infty, \tag{2.5}$$

$$\int_0^T \|\sigma'(t)\pi(t)\|^2\, dt < \infty \tag{2.6}$$

hold almost surely. The *gains process* $G(\cdot)$ associated with $(\pi_0(\cdot), \pi(\cdot))$ is

$$G(t) \triangleq \int_0^t [\pi_0(s) + \pi'(s)\underline{1}](r(s)\, ds + dA(s)) + \int_0^t \pi'(s)[b(s) + \delta(s)$$
$$- r(s)\underline{1}]\, ds + \int_0^t \pi'(s)\sigma(s)\, dW(s), \quad 0 \le t \le T. \tag{2.7}$$

The portfolio process $(\pi_0(\cdot), \pi(\cdot))$ is said to be *self-financed* if

$$G(t) = \pi_0(t) + \pi'(t)\underline{1}, \quad \forall t \in [0, T]. \tag{2.8}$$

In other words, *the value of the portfolio at every time is equal to the gains earned from investments up to that time.*

Remark 2.2: Define the N-dimensional vector of *excess yield* (over the interest rate) *processes*

$$R(t) \triangleq \int_0^t [b(u) + \delta(u) - r(u)\underline{1}]\, du + \int_0^t \sigma(u)\, dW(u), \quad 0 \le t \le T, \tag{2.9}$$

and simplify (2.7) as

$$G(t) = \int_0^t (\pi_0(s) + \pi'(s)\underline{1})[r(s)\, ds + dA(s)] + \int_0^t \pi'(s)\, dR(s), \quad 0 \le t \le T. \tag{2.10}$$

If $(\pi_0(\cdot), \pi(\cdot))$ is self-financed, then (2.10) reads in differential form

$$dG(t) = \frac{G(t)}{S_0(t)}\, dS_0(t) + \pi'(t)\, dR(t), \tag{2.10'}$$

and has the solution

$$G(t) = S_0(t)\int_0^t \frac{1}{S_0(u)}\pi'(u)\, dR(u); \quad 0 \le t \le T. \tag{2.11}$$

The integrand $b(\cdot)+\delta(\cdot)-r(\cdot)\underline{1}$ appearing in (2.9) is called the *risk premium process*; its nth component is regarded as the compensation, in terms of mean growth rate, received by an agent willing to incur the risk of investing in the nth stock.

If we are given only an $\{\mathcal{F}(t)\}$-progressively measurable, $\mathbb{R}^N$-valued process $\pi(\cdot)$ satisfying (2.5) and (2.6), we can consider the process $G(\cdot)$ of (2.11) and then *define* $\pi_0(\cdot)$ by the self-financing condition (2.8). Because $G(\cdot)$ defined by (2.11) is continuous, each of its paths is bounded on $[0,T]$, and (2.4) follows from Definition 1.3 (iv) and (ix). It develops that $(\pi_0(\cdot),\pi(\cdot))$ is a self-financed portfolio process. *Thus, in order to specify a self-financed portfolio, we need only specify $\pi(\cdot)$.* Slightly abusing terminology, we will refer sometimes to $\pi(\cdot)$ alone as a portfolio process.

As defined above, a portfolio process $(\pi_0(\cdot),\pi(\cdot))$ is subject to few restrictions. In particular, $\pi_0(\cdot)$ may take negative values, which corresponds to *borrowing* from the money market. The investor is subject to the same "interest rate" regardless of whether he is a borrower or a lender. Finally, the position $\pi_n(\cdot)$ in stock n may be negative, for $n=1,\ldots,N$; such *short-selling* of stocks is permitted in real markets, subject to some restrictions. In Chapters 5 and 6 we study models in which short-selling is either prohibited or constrained and/or the interest rate for borrowing exceeds $r(t)$. Other related work is cited in the Notes.

The definition of the gains process in (2.7) does not take into account any cost for trading. An idealized market in which there are no transaction costs is called *frictionless*, and most of the existing theory of finance pertains only to frictionless markets. Models with transaction costs are reviewed in the Notes to Chapters 2 and 3.

The conditions (2.4)–(2.6) are imposed on portfolio processes in Definition 2.1 in order to ensure the existence of the integrals in (2.7). If these were the only conditions imposed on portfolio processes, then "outrageous" behavior could occur, as the following example demonstrates.

Example 2.3 *(Doubling strategy):* In a discrete-time betting situation, a *doubling strategy* is to place an initial bet and then to double the size of the bet after each loss until a win is finally obtained. If the initial stake is \$1 and the first win occurs on the nth bet, then the accumulated losses $\sum_{k=0}^{n-1} 2^k = 2^n - 1$ (prior to the win) are more than offset by the win of 2^n. When the outcomes of successive bets are independent and identically distributed, and the probability of winning on any one bet is positive, then the probability of an eventual win is one. In such a situation, a doubling strategy offers a sure way to make money. Unfortunately, a gambler using a doubling strategy must be prepared to bet arbitrarily many times, and to incur arbitrarily large accumulated losses, while waiting for the eventual win.

In continuous time, the analogue of a doubling strategy can be implemented on a finite time interval. Consider a financial market with one stock driven by a one-dimensional Brownian motion ($N = D = 1$) and with $r(\cdot) \equiv 0, b(\cdot) \equiv 0, \sigma(\cdot) \equiv 1, \delta(\cdot) \equiv 0, A(\cdot) \equiv 0$. For a self-financed portfolio process $(\pi_0(\cdot), \pi(\cdot))$ the gains process is

$$G(t) = \int_0^t \pi(s) \, dW(s), \quad \forall t \in [0, T].$$

We show that for each $\alpha > 0$, it is possible to construct a self-financed portfolio $(\pi_0(\cdot), \pi(\cdot))$ such that $G(T) = \alpha$ a.s. Thus, by investing in a money market with constant price per share and a stock with zero mean rate of return and zero dividends, the investor can make arbitrarily large amounts of money almost surely!

To construct such a portfolio, we consider the stochastic integral $I(t) \triangleq \int_0^t \sqrt{\frac{1}{T-u}} \, dW(u)$, which is a martingale on $[0, T)$ with quadratic variation

$$\langle I \rangle(t) = \int_0^t \frac{du}{T-u} = \log\left(\frac{T}{T-t}\right), \quad \forall t \in [0, T).$$

The inverse of $\langle I \rangle(\cdot)$ is the mapping $s \mapsto T - Te^{-s}$ from $[0, \infty)$ to $[0, T)$. The time-changed stochastic integral $\tilde{I}(s) \triangleq I(T - Te^{-s})$ has quadratic variation $\langle \tilde{I} \rangle(s) = s$ and is thus a Brownian motion defined for $0 \leq s < \infty$ (Karatzas and Shreve (1991), Theorem 3.4.6). Consequently,

$$\overline{\lim}_{t \uparrow T} I(t) = \overline{\lim}_{s \uparrow \infty} \tilde{I}(s) = \infty, \quad \underline{\lim}_{t \uparrow T} I(t) = \underline{\lim}_{s \uparrow \infty} \tilde{I}(s) = -\infty,$$

and therefore

$$\tau_\alpha \triangleq \inf\{t \in [0, T); \quad I(t) = \alpha\} \wedge T$$

satisfies $0 < \tau_\alpha < T$ a.s. Define $\pi(t) = (\frac{1}{T-t})^{\frac{1}{2}} 1_{\{t \leq \tau_\alpha\}}$ and $\pi_0(t) = I(t \wedge \tau_\alpha) - \pi(t)$ for all $t \in [0, T]$. Then $(\pi_0(\cdot), \pi(\cdot))$ is a self-financed portfolio with corresponding gains process

$$G(t) = \int_0^{t \wedge \tau_\alpha} \frac{1}{(T-u)^{\frac{1}{2}}} \, dW(u) = I(t \wedge \tau_\alpha), \quad \forall t \in [0, T]. \tag{2.12}$$

In particular, we have $G(T) = \alpha$ almost surely.

The gains process of (2.12) is not bounded from below; indeed, we have $\lim_{t \uparrow T} G(t) = G(T)$ a.s. and $EG(t) = 0 \, \forall t \in [0, T)$, so that if $G(\cdot)$ were bounded from below, then Fatou's lemma would imply $EG(T) \leq 0$ (impossible, since $G(T) = \alpha > 0$ a.s.). To rule out the behavior evident in Example 2.3 we impose the following conditions on portfolio processes.

Definition 2.4: An $\{\mathcal{F}(t)\}$-adapted, $\mathbb{R}^N$-valued process $\pi(\cdot)$ satisfying (2.5) and (2.6) is said to be *tame* if the discounted gains semimartingale

$$\frac{G(t)}{S_0(t)} = M_0^\pi(t) \triangleq \int_0^t \frac{1}{S_0(u)} \pi'(u) \, dR(u), \quad 0 \leq t \leq T, \tag{2.13}$$

is almost surely bounded from below by a real constant that does not depend on t (but possibly depends on $\pi(\cdot)$). If $(\pi_0(\cdot), \pi(\cdot))$ is a portfolio process and $\pi(\cdot)$ is tame, we say that the portfolio process $(\pi_0(\cdot), \pi(\cdot))$ is tame.

As already noted, the portfolio process $\pi(\cdot)$ of Example 2.3 is not tame, since the corresponding gains process $G(\cdot) = M_0^\pi(\cdot)$ of (2.12) is not bounded from below. The assumption of tameness rules out doubling strategies such as the one encountered in Example 2.3; see Remark 5.7 for a stronger assertion along these lines. Observe, however, that even a tame portfolio can exhibit the "opposite" of the behavior encountered in Example 2.3. Indeed, for the same market as in that example and for any $\alpha > 0$, one can easily construct a tame, self-financed portfolio process for which $G(T) = -\alpha$ almost surely, for any given real number $\alpha > 0$. Such a *suicide strategy* need not be ruled out by model assumptions; it will be eliminated from consideration by "hedging" criteria or optimality criteria, to be formulated and imposed later.

1.3   Income and Wealth Processes

An investor may have sources of income and expense other than those due to investments in the assets discussed so far. In this section, we include this possibility in the model.

Definition 3.1: Let $\mathcal{M}$ be a financial market (Definition 1.3). A *cumulative income process* $\Gamma(t), 0 \le t \le T$, is a semimartingale, i.e., the sum of a finite-variation RCLL process and a local martingale.

We interpret $\Gamma(t)$ as the cumulative wealth received by an investor on the time interval $[0, t]$. In particular, the investor is given initial wealth $\Gamma(0)$. Consumption by the investor can be captured by a decrease in $\Gamma(\cdot)$.

Definition 3.2: Let $\mathcal{M}$ be a financial market, $\Gamma(\cdot)$ a cumulative income process, and $(\pi_0(\cdot), \pi(\cdot))$ a portfolio process. The *wealth process* associated with $(\Gamma(\cdot), \pi_0(\cdot), \pi(\cdot))$ is

$$X(t) \triangleq \Gamma(t) + G(t), \tag{3.1}$$

where $G(\cdot)$ is the gains process of (2.7). The portfolio $(\pi_0(\cdot), \pi(\cdot))$ is said to be $\Gamma(\cdot)$-*financed* if

$$X(t) = \pi_0(t) + \pi'(t)\underline{1}, \quad \forall\, t \in [0, T]. \tag{3.2}$$

Remark 3.3: For a $\Gamma(\cdot)$-financed portfolio $(\pi_0(\cdot), \pi(\cdot))$, using the vector of excess yield process $R(\cdot)$ of (2.9) we may write the wealth equation (3.2) in differential form as

$$dX(t) = d\Gamma(t) + \frac{X(t)}{S_0(t)} dS_0(t) + \pi'(t)\, dR(t)$$

$$= d\Gamma(t) + X(t)[r(t)\, dt + dA(t)] + \pi'(t)[b(t) + \delta(t) - r(t)\underline{1}]\, dt$$
$$+ \pi'(t)\sigma(t)\, dW(t) \tag{3.3}$$

by analogy with $(2.10)'$, and therefore the *discounted wealth process* is given by

$$\frac{X(t)}{S_0(t)} = \Gamma(0) + \int_{(0,t]} \frac{d\Gamma(u)}{S_0(u)} + \int_0^t \frac{1}{S_0(u)} \pi'(u)\, dR(u), \quad 0 \le t \le T. \tag{3.4}$$

This formula does not involve $\pi_0(\cdot)$, which can be recovered from $\pi(\cdot)$ and its corresponding wealth process $X(\cdot)$ of (3.4) via (3.2). As with self-financed portfolios (Remark 2.2), we will sometimes refer to $\pi(\cdot)$ alone as a portfolio process.

1.4   Arbitrage and Market Viability

Definition 4.1:   In a financial market $\mathcal{M}$ we say that a given tame, self-financed portfolio process $\pi(\cdot)$ is an *arbitrage opportunity* if the associated gains process $G(\cdot)$ of (2.11) satisfies $G(T) \ge 0$ almost surely and $G(T) > 0$ with positive probability. A financial market $\mathcal{M}$ in which no such arbitrage opportunities exist is said to be *viable*.

Here is an *example of a market $\mathcal{M}$ that is not viable*. Take $N = D = 1$, $T = 1, r(\cdot) \equiv 0, \delta(\cdot) \equiv 0, \sigma(\cdot) = 1$, and $b(\cdot) \equiv \frac{1}{Q(\cdot)}$, where $Q(\cdot)$ is the Bessel process with drift given by

$$dQ(t) = \left(\frac{1}{Q(t)} - 2 \right) dt + dW(t), \quad Q(0) = 1.$$

Just as for the classical Bessel process (e.g. Karatzas and Shreve (1991), pp. 158–163) we have $P[Q(t) > 0, \forall 0 \le t \le 1] = 1$. Indeed, $Q(\cdot)$ is the classical Bessel process with dimension $d = 3$ under the probability measure $\tilde{P}$ that makes $W(t) - 2t$ a Brownian motion and is equivalent to P. The gains process $G(\cdot)$ corresponding to the constant portfolio $\pi(\cdot) \equiv 1$ satisfies

$$dG(t) = \frac{1}{Q(t)} dt + dW(t), G(0) = 0,$$

and thus $G(t) = Q(t) + 2t - 1$. Now, this process satisfies $P[G(t) \ge -1, \forall 0 \le t \le 1] = 1$ (so that $\pi(\cdot)$ is a tame portfolio) as well as $G(1) = 1 + Q(1) \ge 1$ a.s. (so that $\pi(\cdot)$ is also an arbitrage opportunity).

A theory of mathematical finance must be restricted to viable markets. If an arbitrage opportunity exists in a market, the portfolio process that exploits it can be scaled to make $EG(T)$ arbitrarily large and still keep $G(T) \geq 0$ almost surely. An investment opportunity in which there is no possibility of loss and the expected gain is arbitrarily large does not correspond to the reality available to most of us. Furthermore, it makes optimization meaningless. This section explores the mathematical ramifications of the assumption of market viability.

Theorem 4.2: *If a financial market $\mathcal{M}$ is viable, then there exists a progressively measurable process $\theta(\cdot)$ with values in $\mathbb{R}^D$, called the* market price of risk, *such that for Lebesgue-almost-every $t \in [0,T]$ the risk premium $b(t) + \delta(t) - r(t)\underset{\sim}{1}$ is related to $\theta(t)$ by the equation*

$$b(t) + \delta(t) - r(t)\underset{\sim}{1} = \sigma(t)\theta(t) \quad a.s. \tag{4.1}$$

Conversely, suppose that there exists a process $\theta(\cdot)$ that satisfies the above requirements, as well as

$$\int_0^T \|\theta(s)\|^2\, ds < \infty \quad a.s. \tag{4.2i}$$

$$E\left[\exp\left\{-\int_0^T \theta'(s)\, dW(s) - \frac{1}{2}\int_0^T \|\theta(s)\|^2\, ds\right\}\right] = 1. \tag{4.2ii}$$

Then the market $\mathcal{M}$ is viable.

The idea behind Theorem 4.2 is the following. Suppose that for all (t,ω) in some subset of $[0,T] \times \Omega$ with positive product measure, one can find $\pi(t)$ such that $\pi'(t)\sigma(t) = 0$ but $\pi'(t)[b(t)+\delta(t)-r(t)\underset{\sim}{1}] \neq 0$. It is clear from (3.4) that this portfolio holds a combination of stocks that entails no risk but has a nonzero mean rate of return and hence exposes an arbitrage opportunity. Thus, for a viable market, every vector in the kernel $\mathcal{K}(\sigma'(t))$ of $\sigma'(t)$ should be orthogonal to $b(t) + \delta(t) - r(t)\underset{\sim}{1}$. But from linear algebra we know that the orthogonal complement of the kernel of $\sigma'(t)$ is the range of $\sigma(t)$. Except for the issue of progressive measurability, Theorem 4.2 is just the assertion that $b(t) + \delta(t) - r(t)\underset{\sim}{1}$ is in the range of $\sigma(t)$. The following lemmas make this argument rigorous by addressing the relevant measurability issues; the reader may wish to skip these on first reading and proceed directly to Corollary 4.8.

Notation 4.3: Let $L(\mathbb{R}^D; \mathbb{R}^N)$ denote the space of $N \times D$ matrices. For such a matrix σ, let $\mathcal{K}(\sigma)$ and $\mathcal{K}(\sigma')$ denote the kernels of σ and σ', respectively:

$$\mathcal{K}(\sigma) = \{x \in \mathbb{R}^D; \sigma x = 0\}, \quad \mathcal{K}(\sigma') = \{y \in \mathbb{R}^N; \sigma' y = 0\}.$$

Let $\mathcal{R}(\sigma)$ and $\mathcal{R}(\sigma')$ denote the range spaces of these matrices:

$$\mathcal{R}(\sigma) = \{\sigma x; x \in \mathbb{R}^D\}, \quad \mathcal{R}(\sigma') = \{\sigma' y; y \in \mathbb{R}^N\}.$$

Then $\mathcal{K}^\perp(\sigma) = \mathcal{R}(\sigma')$ and $\mathcal{K}^\perp(\sigma') = \mathcal{R}(\sigma)$, where the superscript $\perp$ denotes orthogonal complement. Let proj_M denote the orthogonal projection mapping onto a subspace M.

Lemma 4.4: *The mappings $(x, \sigma) \mapsto \operatorname{proj}_{\mathcal{K}(\sigma)}(x)$ and $(x, \sigma) \mapsto \operatorname{proj}_{\mathcal{K}^{\perp}(\sigma)}(x)$ from $\mathbb{R}^D \times L(\mathbb{R}^D; \mathbb{R}^N)$, and the mappings $(y, \sigma) \mapsto \operatorname{proj}_{\mathcal{K}(\sigma')}(y)$ and $(y, \sigma) \mapsto \operatorname{proj}_{\mathcal{K}^{\perp}(\sigma')}(y)$ from $\mathbb{R}^N \times L(\mathbb{R}^D; \mathbb{R}^N)$ are Borel measurable.*

PROOF.   We treat only the first of the four mappings under consideration. Let $L(\mathbb{R}^D; \mathbb{R}^N)$ be endowed with the operator norm and with the Borel σ-algebra generated by the associated topology. Let $\mathbb{R}^D$ and $\mathbb{R}^N$ have their Borel σ-algebras, and let all product spaces have the product Borel σ-algebras. Finally, let $\mathbb{Q}^N$ be the set of vectors in $\mathbb{R}^N$ with rational coordinates, so $\mathbb{Q}^N$ is a countable, dense subset of $\mathbb{R}^N$.

Define the Borel-measurable function $F : \mathbb{R}^D \times L(\mathbb{R}^D; \mathbb{R}^N) \to \mathbb{R}$ by

$$F(z, \sigma) \overset{\triangle}{=} \inf_{q \in \mathbb{Q}^N} \|z - \sigma' q\|, \qquad \forall z \in \mathbb{R}^D, \quad \sigma \in L(\mathbb{R}^D; \mathbb{R}^N).$$

Then $\{(z, \sigma); z \in \mathcal{R}(\sigma')\} \subset \{(z, \sigma); F(s, \sigma) = 0\}$. On the other hand, if $F(z, \sigma) = 0$, then there is a sequence $\{q_n\}_{n=1}^{\infty} \subset \mathbb{Q}^N$ such that $\lim_{n \to \infty} \|z - \sigma' q_n\| = 0$. Because $\mathbb{R}^N = \mathcal{K}(\sigma') \oplus \mathcal{K}^{\perp}(\sigma')$, we can decompose each q_n as $q_n = p_n + r_n$, where $p_n \in \mathcal{K}(\sigma'), r_n \in \mathcal{K}^{\perp}(\sigma')$. Restricted to $\mathcal{K}^{\perp}(\sigma')$, the linear mapping σ' is invertible. Since $\sigma' r_n \to z$, the sequence $\{r_n\}_{n=1}^{\infty}$ converges to some $r \in \mathcal{K}^{\perp}(\sigma')$ that satisfies $\sigma' r = z$. Therefore, $z \in \mathcal{R}(\sigma')$. We have shown that

$$\{(z, \sigma); z \in \mathcal{R}(\sigma')\} = \{(z, \sigma); F(z, \sigma) = 0\}, \tag{4.3}$$

and thus this is a Borel set. Consequently,

$$\begin{aligned}
\{(x, \sigma, \xi) \in \mathbb{R}^D \times L(\mathbb{R}^D; \mathbb{R}^N) \times \mathbb{R}^D; & \quad \xi = \operatorname{proj}_{\mathcal{K}(\sigma)}(x)\} \\
= \{(x, \sigma, \xi); & \quad \xi \in \mathcal{K}(\sigma), \quad (x - \xi) \perp \mathcal{K}(\sigma)\} \\
= \{(x, \sigma, \xi); & \quad \xi' \sigma = 0, \quad x - \xi \in \mathcal{R}(\sigma')\} \tag{4.4}
\end{aligned}$$

is also a Borel set. Define $Q : \mathbb{R}^D \times L(\mathbb{R}^D; \mathbb{R}^N) \to \mathbb{R}^D$ by

$$Q(x, \sigma) \overset{\triangle}{=} \operatorname{proj}_{K(\sigma)}(x), \qquad \forall x \in \mathbb{R}^D, \quad \sigma \in L(\mathbb{R}^D; \mathbb{R}^N).$$

The set in (4.4) is the graph

$$Gr(Q) \overset{\triangle}{=} \{(x, \sigma, \xi); \quad (x, \sigma) \in \mathbb{R}^D \times L(\mathbb{R}^D; \mathbb{R}^N), \quad \xi = Q(x, \sigma)\}$$

of Q. Having a Borel graph, Q must be a Borel-measurable function. Indeed, for any Borel set $B \subset \mathbb{R}^D$, we have

$$\{(x, \sigma); Q(x, \sigma) \in B\} = \operatorname{proj}_{\mathbb{R}^D \times L(\mathbb{R}^D; \mathbb{R}^N)} Gr(Q) \cap (\mathbb{R}^D \times L(\mathbb{R}^D; \mathbb{R}^N) \times B),$$

and the one-to-one projection of a Borel set is Borel (Parthasarathy (1967), Chapter I, Theorem 3.9). □

Corollary 4.5: *The process $\operatorname{proj}_{\mathcal{K}(\sigma'(t))}[b(t) + \delta(t) - r(t)\underline{1}], 0 \leq t \leq T$, is progressively measurable.*

Lemma 4.6: *If the financial market $\mathcal{M}$ is viable, then $b(t)+\delta(t)-r(t)\underline{1} \in \mathcal{R}(\sigma(t))$ for Lebesgue-almost-every $t \in [0,T]$ almost surely.*

PROOF.   We define for $0 \le t \le T$,

$$p(t) = \mathrm{proj}_{\mathcal{K}(\sigma'(t))}[b(t) + \delta(t) - r(t)\underline{1}],$$

$$\pi(t) = \begin{cases} \dfrac{p(t)}{\|p(t)\|} & \text{if } p(t) \ne 0, \\ 0 & \text{if } p(t) = 0, \end{cases}$$

so that $\pi(\cdot)$ is a bounded, progressively measurable process. Conditions (2.5), (2.6) are satisfied by $\pi(\cdot)$ because of conditions (iv) and (vi)–(viii) of Definition 1.3. Using Remark 2.2, we develop from $\pi(\cdot)$ a self-financed portfolio $(\pi_0(\cdot), \pi(\cdot))$ with associated gains process

$$G(T) = S_0(T) \int_0^T \frac{\|p(t)\|}{S_0(t)} 1_{\{p(t) \ne 0\}} \, dt.$$

Because $G(T) \ge 0$ almost surely, viability implies that $G(T)$ must be zero almost surely. It follows that $p(t) = 0$ for Lebesgue-almost-every t almost surely, and this is equivalent to the assertion that $b(t) + \delta(t) - r(t)\underline{1} \in \mathcal{K}^{\perp}(\sigma'(t)) = \mathcal{R}(\sigma(t))$ for Lebesgue-almost-every t almost surely.   □

Lemma 4.7: *Consider the mapping $\psi_1 : \{(y,\sigma) \in \mathbb{R}^N \times L(\mathbb{R}^D; \mathbb{R}^N); y \in \mathcal{R}(\sigma)\} \to \mathbb{R}^D$ defined by the prescription that $\psi_1(y,\sigma)$ is the unique $\xi \in \mathcal{K}^{\perp}(\sigma)$ such that $\sigma\xi = y$. Consider also the mapping $\psi_2 : \{(x,\sigma) \in \mathbb{R}^D \times L(\mathbb{R}^D; \mathbb{R}^N); x \in \mathcal{R}(\sigma)\} \to \mathbb{R}^N$ defined by the prescription that $\psi_2(x,\sigma)$ is the unique $\eta \in \mathcal{K}^{\perp}(\sigma')$ such that $\sigma'\eta = x$. Both ψ_1 and ψ_2 are Borel measurable.*

PROOF.   We establish the Borel measurability of ψ_1 only. Define

$$\Delta = \{(y,\sigma,\xi) \in \mathbb{R}^N \times L(\mathbb{R}^D; \mathbb{R}^N) \times \mathbb{R}^D; y \in \mathcal{R}(\sigma), \xi \in \mathcal{R}(\sigma'), \sigma\xi = y\}.$$

The set $\{(\sigma,\xi); \ \xi \in \mathcal{R}(\sigma')\}$ was shown in the proof of Lemma 4.4 to be Borel, and the same argument applies to $\{(y,\sigma); \ y \in \mathcal{R}(\sigma)\}$. Therefore, Δ is a Borel set. But Δ is the graph of ψ_1, and just as in Lemma 4.4 we conclude that ψ_1 is a Borel-measurable function.   □

PROOF OF THEOREM 4.2.   According to Lemmas 4.6 and 4.7, the progressively measurable process

$$\theta(t) \overset{\triangle}{=} \psi_1\Big(b(t) + \delta(t) - r(t)\underline{1}, \sigma(t)\Big) \tag{4.5}$$

is defined and satisfies (4.1) for Lebesgue-almost-every $t \in [0,T]$, almost surely. This proves the first part of the theorem.

On the other hand, suppose that the $\mathbb{R}^N$-valued process $\theta(\cdot)$ is progressively measurable and satisfies the conditions (4.1) and (4.2). For any tame portfolio $\pi(\cdot)$ the associated discounted gains process can be written as

$$M^\pi(t) \triangleq \frac{G(t)}{S_0(t)} = \int_0^T \frac{\pi'(u)}{S_0(u)} \, dR(u) = \int_0^t \frac{\pi'(u)}{S_0(u)} \sigma(u) \, dW_0(u), \quad 0 \le t \le T$$

(cf. (2.11)). Here $W_0(t) \triangleq W(t) + \int_0^t \theta(s) \, ds, 0 \le t \le T$, is Brownian motion under the probability measure

$$P_0(A) \triangleq E\left[1_A \cdot \exp\left\{-\int_0^T \theta'(s) \, dW(s) - \frac{1}{2} \int_0^T \|\theta(s)\|^2 \, ds\right\}\right], A \in \mathcal{F}(T)$$

(from the Girsanov theorem, §3.5 in Karatzas and Shreve (1991)). Thus, the lower-bounded process $M^\pi(\cdot)$ is a local martingale under P_0, hence also a supermartingale: $E_0\left(\frac{G(T)}{S_0(T)}\right) \le E_0 M^\pi(0) = 0$. This shows that it is impossible to have an arbitrage opportunity (i.e., we cannot have both $P[G(T) \ge 0] = 1$ and $P[G(T) > 0] > 0$).   □

Corollary 4.8: *A viable market can have only one money market, and hence only one risk-free rate. In other words, if the nth stock pays no dividends and entails no risk (i.e., the nth row of $\sigma(\cdot)$ is identically zero), then $b_n(\cdot) = r(\cdot)$ and $S_n(\cdot) = S_n(0)S_0(\cdot)$.*

PROOF.   If the nth stock were as described, Theorem 4.2 would then imply that $b_n(t) = r(t)$ for Lebesgue-almost-every t almost surely, so $S_n(\cdot) = S_n(0)S_0(\cdot)$.   □

Corollary 4.9: *For a viable market, the excess-yield process $R(\cdot)$ of (2.9) is given by*

$$R(t) = \int_0^t \sigma(u)[\theta(u) \, du + dW(u)], \quad 0 \le t \le T, \tag{4.6}$$

where $\theta(\cdot)$ satisfies (4.1).

Remark 4.10: Within the framework of a viable market, when characterizing the wealth processes that can be achieved through investment, one can assume—without loss of generality—that *the number N of stocks is not greater than the dimension D of the underlying Brownian motion.* The intuitive basis for this claim is that if there are more than D stocks, then some of them can be duplicated by forming *mutual funds* (i.e., (t, ω)-dependent linear combinations) of others, and thus the number of stocks in the model can be reduced.

To make this intuition precise, let $\mathcal{M}$ be a viable market with $N > D$, and let $\theta(\cdot)$ be as in Theorem 4.2. Define a progressively measurable, $(D \times D)$-matrix-valued process $\tilde\sigma(\cdot)$ by specifying that the rows of $\tilde\sigma(t)$ be obtained by deleting the first $N - D$ rows of $\sigma(t)$ that can be written as linear combinations of their predecessors. Then the subspace of $\mathbb{R}^D$ spanned by the rows of $\sigma(t)$ is the same as the subspace spanned by the rows of $\tilde\sigma(t)$, and thus for every $\Gamma(\cdot)$-financed portfolio process $(\pi_0(\cdot), \pi(\cdot))$ in the market $\mathcal{M}$, there is a D-dimensional, progressively measurable process $\tilde{\pi}(\cdot)$ such that $\pi'(t)\sigma(t) = \tilde{\pi}'(t)\tilde{\sigma}(t)$ for every $t \in [0, T]$. Therefore,

$$\pi'(t)\, dR(t) = \pi'(t)\sigma(t)[\theta(t)\, dt + dW(t)] = \tilde{\pi}'(t)\tilde{\sigma}(t)[\theta(t)\, dt + dW(t)],$$

and the discounted wealth process corresponding to a $\Gamma(\cdot)$-financed portfolio can be written as

$$\frac{X(t)}{S_0(t)} = \Gamma(0) + \int_{(0,t]} \frac{d\Gamma(u)}{S_0(u)} + \int_0^t \frac{1}{S_0(u)} \tilde{\pi}'(u)\tilde{\sigma}(u)\theta(u)\, du$$

$$+ \int_0^t \frac{1}{S_0(u)} \tilde{\pi}'(u)\tilde{\sigma}(u)\, dW(u) \tag{4.7}$$

(see (3.4)). This discounted wealth process can be achieved in the D-stock market $\tilde{\mathcal{M}} = (r(\cdot), r(\cdot)\mathbf{1}_D + \tilde{\sigma}(\cdot)\theta(\cdot), 0, \tilde{\sigma}(\cdot), \tilde{S}(0), A(\cdot))$, where $\tilde{S}(0)$ is a D-dimensional vector of initial stock prices and $\mathbf{1}_D$ is the D-dimensional vector with every component equal to 1. In particular, the vector of excess yield processes for the reduced model is

$$\tilde{R}(t) = \int_0^t \tilde{\sigma}(u)\theta(u)\, du + \int_0^t \tilde{\sigma}(u)\, dW(u)$$

(cf. (2.9)), and in terms of it the representation (4.7) for the discounted wealth process becomes

$$\frac{X(t)}{S_0(t)} = \Gamma(0) + \int_{(0,t]} \frac{d\Gamma(u)}{S_0(u)} + \int_{(0,t]} \frac{1}{S_0(u)} \tilde{\pi}'(u)\, d\tilde{R}(u), \quad 0 \le t \le T.$$

The D stocks in this reduced market form a subset of the N stocks available in the original market; the composition of this subset may depend on (t, ω), albeit in a progressively measurable fashion.

Remark 4.11: For a viable market, the progressively measurable process $\theta(\cdot)$ constructed in (4.5) satisfies both (4.1) and

$$\theta(t) \in \mathcal{K}^\perp(\sigma(t)) \quad a.s. \tag{4.8}$$

for Lebesgue-almost-every $t \in [0, T]$. Elementary linear algebra shows that $\theta(\cdot)$ is uniquely determined by these conditions. If $Rank(\sigma(t)) = N$, then

$$\theta(t) = \sigma'(t)(\sigma(t)\sigma'(t))^{-1}[b(t) + \delta(t) - r(t)\mathbf{1}]. \tag{4.9}$$

1.5   Standard Financial Markets

Motivated by the developments of the previous section, in particular Theorem 4.2 and Remark 4.10, we introduce here the notion of a *standard* financial market model. It is mostly with such models that we shall be dealing in the sequel.

Definition 5.1: A financial market model $\mathcal{M}$ is said to be *standard* if

 (i) it is viable;
 (ii) the number N of stocks is not greater than the dimension D of the underlying Brownian motion;
(iii) the D-dimensional, progressively measurable market price of risk process $\theta(\cdot)$ of (4.1), (4.8) satisfies

$$\int_0^T \|\theta(t)\|^2 \, dt < \infty \tag{5.1}$$

  almost surely; and
(iv) the positive local martingale

$$Z_0(t) \overset{\triangle}{=} \exp\left\{ -\int_0^t \theta'(s) \, dW(s) - \frac{1}{2}\int_0^t \|\theta(s)\|^2 \, ds \right\}, \quad 0 \le t \le T, \tag{5.2}$$

  is in fact a martingale.

For a standard market, we define the *standard martingale measure* P_0 on $\mathcal{F}(T)$ by

$$P_0(A) \overset{\triangle}{=} E[Z_0(T)1_A], \quad \forall A \in \mathcal{F}(T). \tag{5.3}$$

Note that a set in $\mathcal{F}(T)$ has P_0-measure zero if and only if it has P-measure zero. We say that P_0 and P are *equivalent* on $\mathcal{F}(T)$.

Remark 5.2: The process $Z_0(\cdot)$ of (5.2) is a local martingale, because

$$dZ_0(t) = -Z_0(t)\theta'(t) \, dW(t), \quad Z_0(0) = 1 \tag{5.4}$$

or equivalently,

$$Z_0(t) = 1 - \int_0^t Z_0(s)\theta'(s) \, dW(s), \quad \forall t \in [0, T]. \tag{5.5}$$

A well-known sufficient condition for $Z_0(\cdot)$ to be a martingale, due to Novikov, is that $E[\exp\{\frac{1}{2}\int_0^T \|\theta(t)\|^2 \, dt\}] < \infty$ (Karatzas and Shreve (1991), Section 3.5.D). In particular, if $\theta(\cdot)$ is bounded in t and ω, then $Z_0(\cdot)$ is a martingale.

Remark 5.3: According to Girsanov's theorem (Karatzas and Shreve (1991), Section 3.5) the process

$$W_0(t) \overset{\triangle}{=} W(t) + \int_0^t \theta(s) \, ds, \quad \forall t \in [0, T] \tag{5.6}$$

is a D-dimensional Brownian motion under P_0, relative to the filtration $\{\mathcal{F}(t)\}$ of (1.2). In terms of $W_0(\cdot)$, the excess yield process of (2.9) and (4.6) can be rewritten as $R(t) = \int_0^t \sigma(u) \, dW_0(u)$, the discounted gains process becomes

$$\frac{G(t)}{S_0(t)} = M_0^\pi(t) \triangleq \int_0^t \frac{1}{S_0(u)} \pi'(u)\sigma(u)\, dW_0(u) \tag{5.7}$$

(see (2.11) and (2.13)), and the discounted wealth process of (3.4) corresponding to a $\Gamma(\cdot)$-financed portfolio is

$$\frac{X(t)}{S_0(t)} = \Gamma(0) + \int_{(0,t]} \frac{d\Gamma(u)}{S_0(u)} + \int_0^t \frac{1}{S_0(u)} \pi'(u)\sigma(u)\, dW_0(u), \quad 0 \le t \le T. \tag{5.8}$$

We see that P_0 permits a presentation of the excess yield process in which the risk premium term $\int_0^t [b(u) + \delta(u) - r(u)\mathbb{1}]\, du$ has been absorbed into the stochastic integral. This term represents the difference in return, including dividends, between the stocks and the money market. See Remark 5.11 for an elaboration of this point.

Remark 5.4: By definition, a cumulative income process $\Gamma(\cdot)$ is a semimartingale under the original measure P; i.e., $\Gamma(t) = \Gamma(0) + \Gamma^{fv}(t) + \Gamma^{\ell m}(t)$, where $\Gamma^{fv}(\cdot)$ is a finite-variation RCLL process and $\Gamma^{\ell m}(\cdot)$ is a P-local martingale, both beginning at zero. The process $\Gamma(\cdot)$ is also a semimartingale under P_0, i.e., has the unique decomposition

$$\Gamma(t) = \Gamma(0) + \Gamma_0^{fv}(t) + \Gamma_0^{\ell m}(t), \quad 0 \le t \le T,$$

where $\Gamma_0^{fv}(\cdot)$ is a finite-variation RCLL process with total variation on $[0, t]$ denoted by $\check{\Gamma}_0^{fv}(t)$, and $\Gamma_0^{\ell m}(\cdot)$ is a P_0-local martingale, again with $\Gamma_0^{fv}(0) = \Gamma_0^{\ell m}(0) = 0$. Indeed, according to Theorem 3.5.4 in Karatzas and Shreve (1991),

$$\Gamma_0^{\ell m}(t) = \Gamma^{\ell m}(t) + \int_0^t \theta'(s)d\langle \Gamma^{\ell m}, W\rangle(s),$$

$$\Gamma_0^{fv}(t) = \Gamma^{fv}(t) - \int_0^t \theta'(s)d\langle \Gamma^{\ell m}, W\rangle(s).$$

In particular, $\Gamma^{\ell m}(\cdot) \equiv 0$ if and only if $\Gamma_0^{\ell m}(\cdot) \equiv 0$, in which case $\Gamma^{fv}(\cdot) = \Gamma_0^{fv}(\cdot)$.

Definition 5.5: A cumulative income process is said to be *integrable* if

$$E_0 \int_0^T \frac{d\check{\Gamma}_0^{fv}(u)}{S_0(u)} < \infty, \quad E_0 \int_0^T \frac{1}{S_0^2(u)} d\langle \Gamma_0^{\ell m}\rangle(u) < \infty, \tag{5.9}$$

where we use the notation of Remark 5.4 and E_0 denotes the expectation corresponding to P_0.

Theorem 5.6: *Under the standard martingale measure P_0, the process of discounted wealth minus discounted cumulative income*

$$\frac{X(t)}{S_0(t)} - \Gamma(0) - \int_{(0,t]} \frac{d\Gamma(u)}{S_0(u)}, \quad 0 \le t \le T, \tag{5.10}$$

corresponding to any tame $\Gamma(\cdot)$-financed portfolio is a local martingale and bounded from below, hence a supermartingale. In particular,

$$E_0\left[\frac{X(T)}{S_0(T)} - \int_{(0,T]} \frac{d\Gamma(u)}{S_0(u)}\right] \le \Gamma(0). \tag{5.11}$$

The process in (5.10) is a martingale under P_0 if and only if equality holds in (5.11).

PROOF. From (5.7), (5.8) we see that the process in (5.10) has the stochastic integral representation

$$M_0^\pi(t) = \int_0^t \frac{1}{S_0(u)} \pi'(u)\sigma(u)\, dW_0(u), \quad 0 \le t \le T.$$

This process is a local martingale, and because $\pi(\cdot)$ is tame, it is bounded from below. A local martingale which is bounded from below is a supermartingale because of Fatou's lemma. A supermartingale is a martingale if and only if it has constant expectation. $\square$

Remark 5.7: The proof of Theorem 5.6 shows that the expectation on the left-hand side of (5.11) is defined and finite. For an integrable cumulative income process, the P_0-expectations of the individual terms $\frac{X(T)}{S_0(T)}$ and $\int_{(0,T]} \frac{d\Gamma(u)}{S_0(u)} = \int_{(0,T]} \frac{d\Gamma^{fv}(u)}{S_0(u)} + \int_0^T \frac{d\Gamma^{lm}(u)}{S_0(u)}$ are also defined and finite.

Remark 5.8: The process

$$H_0(t) \triangleq \frac{Z_0(t)}{S_0(t)}, \quad 0 \le t \le T, \tag{5.12}$$

often called the *state price density process*, will play a key role in subsequent chapters (e.g., Remark 2.2.4). Using $H_0(\cdot)$, we can rewrite conditions involving the martingale measure P_0 in terms of the original probability measure P. For example, when $\Gamma^{\ell m}(\cdot) \equiv 0$, (5.9) can be rewritten as

$$E\int_0^T H_0(u)\, d\check{\Gamma}^{fv}(u) < \infty. \tag{5.13}$$

Applying Itô's rule to the product of $Z_0(t)$ and $\frac{X(t)}{S_0(t)}$, we obtain from (5.8) that

$$H_0(t)X(t) - \int_{(0,t]} H_0(u)\, d\Gamma(u) = \Gamma(0) + \int_0^t H_0(u)[\sigma'(u)\pi(u)$$
$$- X(u)\theta(u)]'\, dW(u), 0 \le t \le T \tag{5.14}$$

is a local martingale under the original measure P. If this local martingale is also a supermartingale, we obtain the restatement of (5.11) in terms of P:

$$E\left[H_0(T)X(T) - \int_{(0,T]} H_0(u)\,d\Gamma(u)\right] \leq \Gamma(0). \qquad (5.15)$$

We already know that this inequality holds under the conditions of Theorem 5.6; but because of Fatou's lemma it also holds whenever $\pi(\cdot)$ is $\Gamma(\cdot)$-financed and

$$H_0(t)X(t) - \int_{(0,t)} H_0(u)\,d\Gamma(u) \geq 0, \quad \forall\, t \in [0,T],$$

holds a.s., even if $\pi(\cdot)$ is not tame and even if $Z_0(\cdot)$ of (5.2) is not a martingale (so that the measure P_0 of (5.3) is not a probability).

The concept of a tame portfolio $(\pi_0(\cdot), \pi(\cdot))$ (Definition 2.4) was introduced to get some control on the semimartingale $M_0^\pi(\cdot)$. Under P_0, this semimartingale is in fact a local martingale, and this suggests a new, noncomparable concept.

Definition 5.9: An $\{\mathcal{F}(t)\}$-adapted, $\mathbb{R}^N$-valued process $\pi(\cdot)$ satisfying (2.5) and (2.6) is said to be *martingale-generating* if under the probability measure P_0 of (5.3), the local martingale $M_0^\pi(\cdot)$ of (5.7) is a martingale. If $(\pi_0(\cdot), \pi(\cdot))$ is a portfolio process and $\pi(\cdot)$ is martingale-generating, we say that the portfolio process $(\pi_0(\cdot), \pi(\cdot))$ is martingale-generating.

Remark 5.10: If $\Gamma(\cdot) \equiv 0$, then the wealth process is the gains process. For a tame portfolio process $\pi(\cdot)$, (5.11) shows that $E_0\left[\frac{G(T)}{S_0(T)}\right] \leq 0$. For a martingale-generating portfolio process $\pi(\cdot)$, $E_0\left[\frac{G(T)}{S_0(T)}\right] = 0$. In either case, arbitrage (Definition 4.1) is ruled out.

Remark 5.11: In the notation of (5.6) and (4.1), we may rewrite (1.10) in differential form as

$$
\begin{aligned}
d\left(\frac{S_n(t)}{S_0(t)}\right) &+ \frac{S_n(t)}{S_0(t)}\delta_n(t)\,dt \\
&= \frac{S_n(t)}{S_0(t)}\left\{\sum_{d=1}^{D} \sigma_{nd}(t)\,dW^{(d)}(t) + [b_n(t) + \delta_n(t) - r(t)]\,dt\right\} \\
&= \frac{S_n(t)}{S_0(t)}\sum_{d=1}^{D} \sigma_{nd}(t)\,dW_0^{(d)}(t).
\end{aligned}
$$

This shows that under the martingale measure P_0 of (5.3), the process

$$\frac{S_n(t)}{S_0(t)} \cdot \exp\left\{ \int_0^t \delta_n(u)\, du \right\} \tag{5.16}$$

$$= S_n(0) \exp\left\{ \sum_{d=1}^{D} \int_0^t \sigma_{nd}(u)\, dW_0^{(d)}(u) - \frac{1}{2} \sum_{d=1}^{D} \int_0^t \sigma_{nd}^2(u)\, du \right\},$$

$$0 \le t \le T,$$

is an exponential local martingale, and hence a supermartingale. If the entries of the volatility matrix $\sigma(\cdot)$ satisfy the Novikov condition

$$E_0\left[\exp\left\{ \int_0^T \frac{1}{2} \sum_{d=1}^{D} \sigma_{nd}^2(u)\, du \right\} \right] < \infty \tag{5.17}$$

(e.g., Karatzas and Shreve (1991), Section 3.5D), then the exponential local martingale (5.16) is in fact a martingale under P_0, and this justifies calling P_0 a "martingale measure."

1.6   Completeness of Financial Markets

An important purpose of a financial market, perhaps even the principal purpose, is to afford investors the opportunity to hedge risk inherent in their other activities. Consider an agent who knows, at time $t = 0$, that at some future time T he must make a payment $B(\omega)$, but the size of the payment depends on a number of factors that are still undetermined and not within his control. This agent would like to set aside a fixed amount of money x at time $t = 0$ and be assured that this will enable him to make the payment at time T. A conservative strategy would be to set aside an amount equal to the maximal possible payment size, $\sup_{\omega \in \Omega} B(\omega)$, if this maximal size is finite! A more reasonable strategy entails setting aside less money, but investing it in such a way that if the actual payment size turns out to be large, the capital has in the meantime grown to match it. This is the process of *hedging the risk inherent in the random payment*, which leads to the following definition.

Definition 6.1: Let $\mathcal{M}$ be a standard financial market, and let B be an $\mathcal{F}(T)$-measurable random variable such that $\frac{B}{S_0(T)}$ is almost surely bounded from below and

$$x \triangleq E_0\left[\frac{B}{S_0(T)} \right] < \infty. \tag{6.1}$$

(i) We say that B is *financeable*, if there is a tame, x-financed portfolio process $(\pi_0(\cdot), \pi(\cdot))$ whose associated wealth process satisfies

$X(T) = B$; i.e.,

$$\frac{B}{S_0(T)} = x + \int_0^T \frac{1}{S_0(u)} \pi'(u)\sigma(u)\, dW_0(u) \qquad (6.2)$$

almost surely.

(ii) We say that the financial market $\mathcal{M}$ is *complete* if every $\mathcal{F}(T)$-measurable random variable B, with $\frac{B}{S_0(T)}$ bounded from below and satisfying (6.1), is financeable. Otherwise, we say that the market is *incomplete*.

Proposition 6.2: *A standard financial market $\mathcal{M}$ is complete if and only if for every $\mathcal{F}(T)$-measurable random variable B satisfying*

$$E_0\left[\frac{|B|}{S_0(T)}\right] < \infty \qquad (6.3)$$

and with x defined by (6.1), there is a martingale-generating, x-financed portfolio process $(\pi_0(\cdot), \pi(\cdot))$ satisfying (6.2).

PROOF.   Suppose the market is complete and B is an $\mathcal{F}(T)$-measurable random variable that satisfies (6.3). Then there exist tame, $x_\pm$-financed portfolio processes $(\pi_0^\pm(\cdot), \pi^\pm(\cdot))$ with

$$\frac{B^\pm}{S_0(T)} = x_\pm + \int_0^T \frac{1}{S_0(u)} (\pi^\pm(u))'\sigma(u)\, dW_0(u) \qquad (6.4)$$

almost surely, where $B^\pm \triangleq \max\{\pm B, 0\}$ and $x_\pm \triangleq E_0\left[\frac{B^\pm}{S_0(T)}\right]$. Taking expectations in (6.4) with respect to P_0, we see that the lower-bounded local martingale (hence supermartingale) $\int_0^t \frac{1}{S_0(u)}(\pi^\pm(u))'\sigma(u)\, dW_0(u), 0 \le t \le T$, has constant expectation (equal to zero) under P_0. Hence, $\pi^\pm(\cdot)$ is martingale-generating. Subtracting one version of (6.4) from the other, we obtain (6.2), where $\pi(\cdot) \triangleq \pi^+(\cdot) - \pi^-(\cdot)$ is also martingale-generating.

Now suppose that for any $\mathcal{F}(T)$-measurable random variable B, such that $\frac{B}{S_0(T)}$ is almost surely bounded from below and satisfies (6.1), there exists a martingale-generating x-financed portfolio process $(\pi_0(\cdot), \pi(\cdot))$ satisfying (6.2). Taking conditional expectations in (6.2), we obtain that

$$\int_0^t \frac{1}{S_0(u)} \pi'(u)\sigma(u)\, dW_0(u) = -x + E_0\left[\frac{B}{S_0(T)} \,\Big|\, \mathcal{F}(t)\right], \quad 0 \le t \le T,$$

is bounded from below. It follows that $(\pi_0(\cdot), \pi(\cdot))$ is tame, $\frac{B}{S_0(T)}$ is financeable, and thus $\mathcal{M}$ is complete. $\qquad \square$

Remark 6.3: In the context of Definition 6.1, if an x-financed, tame portfolio $(\pi_0(\cdot), \pi(\cdot))$ can be found whose associated wealth process $X(\cdot)$ satisfies $X(T) = B$ almost surely, then $E_0'\left[\frac{X(T)}{S_0(T)}\right] = x$, and the last sentence in Theorem 5.6 asserts that $\frac{X(\cdot)}{S_0(\cdot)}$ is a martingale under P_0.

Consequently, $X(\cdot)$ is uniquely determined by the equation

$$\frac{X(t)}{S_0(t)} = E_0 \left[\frac{B}{S_0(T)} \,\bigg|\, \mathcal{F}(t) \right], \quad 0 \le t \le T. \tag{6.5}$$

Equation (6.5) also holds if $(\pi_0(\cdot), \pi(\cdot))$ is martingale-generating.

The remark and example that follow offer additional insights on the notion of financeability; they can be skipped on first reading.

Remark 6.4: The question arises why Definition 6.1 permits only x-financed portfolios with x defined by (6.1). We first argue that *for $y < x$, there can be no tame, y-financed portfolio whose associated wealth process satisfies $X(T) \ge B$ almost surely*. Indeed, if $\tilde{X}(\cdot)$ is the wealth associated with a tame, y-financed portfolio and $\tilde{X}(T) \ge B$, then (5.11) implies

$$x = E_0 \left[\frac{B}{S_0(t)} \right] \le E_0 \left[\frac{\tilde{X}(T)}{S_0(T)} \right] \le y. \tag{6.6}$$

It is sometimes possible to find a representation of $\frac{B}{S_0(T)}$ of the form

$$\frac{B}{S_0(T)} = y + \int_0^T \frac{1}{S_0(u)} \tilde{\pi}'(u)\sigma(u)\,dW_0(u), \tag{6.7}$$

where $y > x$ and $(\tilde{\pi}_0(\cdot), \tilde{\pi}(\cdot))$ is a tame, y-financed portfolio, even when B is not financeable in the sense of Definition 6.1 (see Example 6.5); but the associated discounted wealth process $\frac{\tilde{X}(t)}{S_0(t)} = y + \int_0^t \frac{1}{S_0(u)} \tilde{\pi}'(u)\sigma(u)\,dW_0(u)$ cannot then be a martingale, because

$$E_0 \left[\frac{\tilde{X}(0)}{S_0(0)} \right] = y > x = E_0 \left[\frac{\tilde{X}(T)}{S_0(T)} \right].$$

The properties of random variables B that permit a representation of the form (6.7) with $y > x$ are not well understood. Moreover, one could argue that y-financed portfolio processes leading to discounted wealth processes that are not martingales under P_0 are undesirable and should be excluded from consideration.

Example 6.5: Consider a financial market $\mathcal{M}$ with one stock, with an underlying two-dimensional Brownian motion $(N = 1, D = 2)$ and with $r(\cdot) \equiv 0, b(\cdot) \equiv 0, \delta \equiv 0, \sigma(\cdot) \equiv [1, 0], A(\cdot) \equiv 0$. For a portfolio process $\pi(\cdot)$ the discounted gains process is $G(t) = \int_0^t \pi(s)\,dW^{(1)}(s)$. As in Example 2.3, define

$$I(t) \triangleq \int_0^t \sqrt{\frac{1}{T - s}}\,dW^{(1)}(s), \quad 0 \le t < T, \tag{6.8}$$

so that $\overline{\lim}_{t\uparrow T} I(t) = \infty$ and $\underline{\lim}_{t\uparrow T} I(t) = -\infty$ almost surely. Set

$$\tau \stackrel{\triangle}{=} \inf\{t \in [0,T); 2 + I(t) = \exp(W^{(2)}(t) - t/2)\} \wedge T,$$

$$B \stackrel{\triangle}{=} 2 + I(\tau) = 2 + \int_0^T 1_{\{s \leq \tau\}} \sqrt{\frac{1}{T-s}} \, dW^{(1)}(s), \tag{6.9}$$

and notice that $P[\tau < T] = 1$ and that the portfolio $\pi(t) = (T-t)^{-\frac{1}{2}} 1_{\{t \leq \tau\}}$ is tame, because $G(\cdot) \geq -2$. Equation (6.9) provides a representation for B of the form (6.7) with $y = 2$. In this example, $P_0 = P$ and

$$E(B) = E\left[\exp\left(W^{(2)}(\tau) - \tau/2\right)\right] = 1 < y.$$

If B had a representation of the form (6.2), then from (6.5) there would exist a tame portfolio process $\pi(\cdot)$ satisfying

$$1 + \int_0^t \pi(s) \, dW^{(1)}(s)$$

$$= E[\exp(W^{(2)}(\tau) - \tau/2) \mid \mathcal{F}(t)]$$

$$= \exp\left(W^{(2)}(t \wedge \tau) - \frac{t \wedge \tau}{2}\right)$$

$$= 1 + \int_0^t 1_{\{s \leq \tau\}} \exp\left(W^{(2)}(s) - s/2\right) dW^{(2)}(s). \tag{6.10}$$

This is clearly impossible (e.g., the martingale on the left-hand side of (6.10) has zero cross-variation with $W^{(2)}(\cdot)$, but the martingale on the right-hand side of (6.10) has nonzero cross-variation with $W^{(2)}(\cdot)$).

The theory of complete markets is simpler and much better developed than the theory of incomplete markets. Chapters 2–4 are devoted to complete markets, while Chapters 5 and 6 explore incomplete markets. For a standard financial market, the two cases are easily distinguished by the following theorem.

Theorem 6.6: *A standard financial market $\mathcal{M}$ is complete if and only if the number of stocks N is equal to the dimension D of the underlying Brownian motion and the volatility matrix $\sigma(t)$ is nonsingular for Lebesgue-a.e. $t \in [0,T]$ almost surely.*

The remainder of this section is devoted to the proof of Theorem 6.6. A standard financial market $\mathcal{M}$ is fixed throughout, $(\Omega, \mathcal{F}, P)$ and $\{\mathcal{F}(t)\}_{0 \leq t \leq T}$ are as in Definition 1.3, the processes $\theta(\cdot), Z_0(\cdot)$ and the measure P_0 are as in Definition 5.1, and $W_0(\cdot)$ is given by (5.6).

Lemma 6.7 (Martingale representation property under P_0): *Let $\{M_0(t), \mathcal{F}(t); 0 \leq t \leq T\}$ be a martingale under P_0. Then there is a progressively* *measurable*, $\mathbb{R}^N$*-valued process* $\varphi(\cdot)$ *such that*

$$\int_0^T \|\varphi(s)\|^2 \, ds < \infty, \tag{6.11}$$

$$M_0(t) = M_0(0) + \int_0^t \varphi'(s) \, dW_0(s), \quad 0 \le t \le T, \tag{6.12}$$

hold almost surely.

PROOF.   This lemma is almost a restatement of the standard representation theorem for martingales as stochastic integrals (Karatzas and Shreve (1991), Theorem 3.4.15 and Problem 3.4.16). The only complication is that the filtration $\{\mathcal{F}(t)\}$ is the augmentation by null sets of the filtration generated by $W(\cdot)$, not $W_0(\cdot)$. Therefore, we revert to the original probability measure P and represent the Lévy P-martingale

$$N(t) \overset{\triangle}{=} E[Z_0(T)M_0(T)|\mathcal{F}(t)], \quad 0 \le t \le T,$$

as a stochastic integral with respect to $W(\cdot)$, namely

$$N(t) = N(0) + \int_0^t \psi'(s) \, dW(s), \quad 0 \le t \le T,$$

where $\psi(\cdot)$ is a progressively measurable, $\mathbb{R}^N$-valued process satisfying (6.11). A simple calculation known as "Bayes's rule" (e.g., Karatzas and Shreve (1991), Lemma 3.5.3) shows that $M_0(t) = E_0[M_0(T)|\mathcal{F}(t)] = N(t)/Z_0(t)$, and Itô's formula yields

$$dM_0(t) = \frac{1}{Z_0(t)}[\psi'(t) + N(t)\theta'(t)] \, dW_0(t),$$

so that (6.11), (6.12) hold with

$$\varphi(t) = \frac{1}{Z_0(t)}[\psi(t) + N(t)\theta(t)].$$

Condition (6.11) for $\varphi(\cdot)$ follows from the same condition for $\psi(\cdot)$, (5.1), and the fact that the paths of $\frac{1}{Z_0(\cdot)}$ and $N(\cdot)$ are continuous on $[0, T]$ almost surely.   □

Corollary 6.8 (Sufficiency in Theorem 6.6): *If* $N = D$ *and* $\sigma(t)$ *is nonsingular for Lebesgue-almost-every* $t \in [0, T]$ *almost surely, then the financial market is complete.*

PROOF.   We verify the condition of Proposition 6.2. Let B be an $\mathcal{F}(T)$-measurable random variable satisfying (6.3), and define the Lévy P_0-martingale

$$M_0(t) = E_0\left[\frac{B}{S_0(T)} \,\middle|\, \mathcal{F}(t)\right], \quad 0 \le t \le T.$$

This martingale has a representation as in (6.12), and with

$$\pi'(t) \stackrel{\triangle}{=} S_0(t)\varphi'(t)\sigma^{-1}(t), \quad 0 \le t \le T,$$

we have (6.2). Condition (2.6) follows from (6.11); condition (2.5) follows from the almost sure inequalities

$$\int_0^T |\pi'(s)(b(s) + \delta(s) - r(s)\underline{1})|\, ds$$

$$= \int_0^T S_0(u)\varphi'(u)\theta(u)\, du$$

$$\le \max_{0 \le u \le T} S_0(u) \cdot \int_0^T \|\varphi(u)\|^2\, du \cdot \int_0^T \|\theta(u)\|^2\, du < \infty,$$

where we have used (4.9) and the Cauchy–Schwarz inequality. We construct $\pi_0(\cdot)$ as in Remark 2.2.   □

Lemma 6.9: *There is a bounded, Borel-measurable mapping* ψ_3 : $L(\mathbb{R}^D; \mathbb{R}^N) \to \mathbb{R}^D$ *such that*

$$\psi_3(\sigma) \in \mathcal{K}(\sigma), \tag{6.13}$$

$$\psi_3(\sigma) \ne 0 \quad \text{if} \quad \mathcal{K}(\sigma) \ne \{0\} \tag{6.14}$$

for every $\sigma \in L(\mathbb{R}^D; \mathbb{R}^N)$.

PROOF.   Let $\{e_1, e_2, \ldots, e_D\}$ be a basis for $\mathbb{R}^D$, and define

$$n(\sigma) = \begin{cases} \min\{i; \text{proj}_{\mathcal{K}(\sigma)}(e_i) \ne 0\} & \text{if} \quad \mathcal{K}(\sigma) \ne \{0\}, \\ 1 & \text{if} \quad \mathcal{K}(\sigma) = \{0\}, \end{cases}$$

$$\psi_3(\sigma) = \text{proj}_{\mathcal{K}(\sigma)}(e_{n(\sigma)}).$$

The Borel-measurability of ψ_3 follows from Lemma 4.4.   □

PROOF OF NECESSITY IN THEOREM 6.6.   Using the function ψ_3 from Lemma 6.9, define the *bounded*, progressively measurable process $\varphi(t) = \psi_3(\sigma(t))$ that satisfies $\varphi(t) \in \mathcal{K}(\sigma(t))$ for all $t \in [0, T]$, and $\varphi(t) \ne 0$ whenever $\mathcal{K}(\sigma(t)) \ne \{0\}$. Next, define the $\mathcal{F}(T)$-measurable random variable

$$B \stackrel{\triangle}{=} S_0(T)\left[1 + \int_0^T \varphi'(u)\, dW_0(u)\right].$$

Clearly, $E_0[\frac{|B|}{S_0(T)}] < \infty$ and $E_0[\frac{B}{S_0(T)}] = 1$. Market completeness and Proposition 6.2 imply the existence of a martingale-generating portfolio process π for which

$$\int_0^T \frac{1}{S_0(u)}\pi'(u)\sigma(u)\, dW_0(u) = \frac{B}{S_0(T)} - 1 = \int_0^T \varphi'(u)\, dW_0(u). \tag{6.15}$$

The stochastic integrals $\int_0^t \frac{1}{S_0(u)} \pi'(u)\sigma(u)\,dW_0(u)$ and $\int_0^t \varphi'(u)\,dW_0(u)$ are both martingales under P_0. Conditioning both sides of (6.15) on $\mathcal{F}(t)$, we see that these stochastic integrals agree. This implies that the integrands agree, so $\sigma'(t)\pi(t) = S_0(t)\varphi(t)$ for Lebesgue-a.e. $t \in [0, T]$ almost surely. This shows that $\varphi(t) \in \mathcal{R}(\sigma'(t)) = \mathcal{K}^\perp(\sigma(t))$. By construction, $\varphi(t) \in \mathcal{K}(\sigma(t))$, so $\varphi(t) = 0$, which happens only if $\mathcal{K}(\sigma(t)) = \{0\}$. Thus, $N = D$ and $\sigma(t)$ is nonsingular for Lebesgue-a.e. $t \in [0, T]$ almost surely. $\qquad\square$

Remark 6.10: In a complete market $\mathcal{M}$, there is a unique market price of risk process $\theta(\cdot)$ satisfying (4.1), defined by

$$\theta(t) = (\sigma(t))^{-1}[b(t) + \delta(t) - r(t)\underline{1}], \quad 0 \le t \le T. \tag{6.16}$$

1.7 Financial Markets with an Infinite Planning Horizon[†]

The time parameter in the financial market of Definition 1.3 takes values in the interval $[0, T]$, where the planning horizon T is finite. In order to consider certain financial instruments such as perpetual American options, we introduce in this section the notion of a financial market on $[0, \infty)$.

For the construction of this section, we need to work with Wiener measure on the "canonical" space of continuous, $\mathbb{R}^D$-valued functions, rather than with a general probability space on which D-dimensional Brownian motion is defined. Let $\Omega = C([0, \infty))^D$ be the space of continuous functions $\omega : [0, \infty) \to \mathbb{R}^D$. On this space we define the *coordinate mapping process* $W(t, \omega) = \omega(t), 0 \le t < \infty, \omega \in \Omega$. As in Section 1, we denote by $\mathcal{F}^W(t) = \sigma\{W(s); 0 \le s \le t\}$ the σ-field generated by $W(\cdot)$ on $[0, t]$, i.e., the smallest σ-algebra containing all sets of the form $\{\omega \in \Omega; \omega(s) \in \Gamma\}$, where s ranges over $[0, t]$ and Γ ranges over the collection of Borel subsets of $\mathbb{R}^D$. We set $\mathcal{F}^W(\infty) \triangleq \sigma(\bigcup_{0 \le t < \infty} \mathcal{F}^W(t))$.

Let P be *Wiener measure* on $\mathcal{F}^W(\infty)$, i.e., the probability measure under which $\{W(t); 0 \le t < \infty\}$ is a D-dimensional Brownian motion. Let $\mathcal{F}$ be the completion of $\mathcal{F}^W(\infty)$ under P; i.e., $\mathcal{F} \triangleq \sigma(\mathcal{F}^W(\infty) \cup \mathcal{N})$, where

$$\mathcal{N} \triangleq \{N \subseteq \Omega; \ \exists B \in \mathcal{F}^W(\infty) \ \text{ with } \ N \subseteq B \ \text{ and } \ P(B) = 0\}$$

is the collection of P-null sets of $\mathcal{F}^W(\infty)$. We call $(\Omega, \mathcal{F}, P)$ the *canonical probability space for D-dimensional Brownian motion.*

For each $T \in [0, \infty)$, we define

$$\mathcal{N}^T = \{N \subseteq \Omega; \ \exists B \in \mathcal{F}^W(T) \ \text{ with } \ N \subseteq B \ \text{ and } \ P(B) = 0\}$$

[†]This section can be omitted on first reading; its results will be used only in Sections 2.6, 3.9, and 3.10.

to be the collection of P-null sets in $\mathcal{F}^W(T)$, and we define the augmented filtration

$$\mathcal{F}^{(T)}(t) \overset{\triangle}{=} \sigma(\mathcal{F}^W(t) \cup \mathcal{N}^T), \quad 0 \le t \le T.$$

This is the filtration we have called $\{\mathcal{F}(t); 0 \le t \le T\}$ heretofore; in this section we indicate explicitly its dependence on T.

Definition 7.1: A stochastic process $Y = \{Y(t); 0 \le t < \infty\}$ is said to be *restrictedly progressively measurable* or *restrictedly adapted* if for every $T \in [0, \infty)$, there exists $\tilde{T} \in [T, \infty)$ such that the restricted process $\{Y(t); 0 \le t \le T\}$ is $\{\mathcal{F}^{(\tilde{T})}(t); 0 \le t \le T\}$-progressively measurable or adapted, respectively.

Definition 7.2: A *financial market* $\mathcal{M} = (r(\cdot), b(\cdot), \delta(\cdot), \sigma(\cdot), S(0), A(\cdot))$ *on the infinite planning horizon* $[0, \infty)$ consists of

- **(i)** a D-dimensional Brownian motion $W = \{W(t); 0 \le t < \infty\}$ that is the coordinate mapping process on the canonical probability space $(\Omega, \mathcal{F}, P)$;
- **(ii)** restrictedly progressively measurable processes $r(\cdot), b(\cdot), \delta(\cdot), \sigma(\cdot)$, and $A(\cdot)$, as described in Definition 1.3, satisfying the integrability conditions of Definition 1.3 for every finite T;
- **(iii)** a vector of positive, constant initial stock prices $S(0) = (S_1(0), \dots, S_N(0))'$.

It is easily verified that the asset price processes in $\mathcal{M}$ are restrictedly progressively measurable. In order to simplify the presentation, we *define* a standard, complete financial market in terms of the conditions obtained in Theorem 6.6.

Definition 7.3: A financial market $\mathcal{M} = (r(\cdot), b(\cdot), \delta(\cdot), \sigma(\cdot), S(0), A(\cdot))$ on an infinite planning horizon is *standard and complete* if

- **(i)** the number of stocks N equals the dimension D of the driving Brownian motion;
- **(ii)** the volatility matrix $\sigma(t)$ is nonsingular for Lebesgue-a.e. $t \in [0, \infty)$ almost surely;
- **(iii)** the positive local martingale

$$Z_0(t) \overset{\triangle}{=} \exp\left\{ -\int_0^t \theta'(s)\, dW(s) - \frac{1}{2}\int_0^t \|\theta(s)\|^2\, ds \right\}, \quad 0 \le t < \infty, \tag{7.1}$$

is in fact a P-martingale, where

$$\theta(t) \overset{\triangle}{=} \sigma^{-1}(t)[b(t) + \delta(t) - r(t)\underline{1}], \quad 0 \le t < \infty. \tag{7.2}$$

Of course, a process is a martingale only relative to some filtration. In Definition 7.3 (iii) we mean that for each $T \in [0, \infty)$, there is a $\tilde{T} \in [T, \infty)$ such that the restricted process $\{Z_0(t); 0 \le t \le T\}$ is a P-martingale relative to $\{\mathcal{F}^{(\tilde{T})}(t); 0 \le t \le T\}$. (One should say, more precisely, that $Z_0(\cdot)$ is "restrictedly" a martingale; one could likewise speak of processes being "restricted" semimartingales, submartingales, etc.) As in Definition 7.3, we shall generally omit the modifier "restricted."

With $\theta(\cdot)$ defined by (7.2), we set

$$W_0(t) \overset{\triangle}{=} W(t) + \int_0^t \theta(s)\, ds, \quad 0 \le t < \infty. \tag{7.3}$$

By analogy with (5.3), we may define for each $T \in [0, \infty)$ the martingale measure P_0^T on $\mathcal{F}^W(T)$ by

$$P_0^T(A) \overset{\triangle}{=} E[Z_0(T)1_A], \quad \forall A \in \mathcal{F}^W(T). \tag{7.4}$$

Under P_0^T, the restricted process $\{W_0(t); 0 \le t \le T\}$ is a Brownian motion. Furthermore, for $0 \le t \le T$, the probability measure P_0^T is *equivalent* to P on $\mathcal{F}^{(T)}(t)$; i.e., a set in $\mathcal{F}^{(T)}(t)$ is a P_0^T-null set if and only if it is a P-null set.

Proposition 7.4:   *There is a unique probability measure P_0 on*

$$\mathcal{F}^W(\infty) \overset{\triangle}{=} \sigma(W(s); \quad 0 \le s < \infty)$$

such that P_0 agrees with each P_0^T on $\mathcal{F}^W(T)$, for any $T < \infty$. In particular, $\{W_0(t); 0 \le t < \infty\}$ is a D-dimensional Brownian motion under P_0.

PROOF.   The family $\{P_0^T\}_{0 \le T < \infty}$ given by (7.4) is consistent: if $0 \le T \le S < \infty$, then P_0^T and P_0^S agree on $\mathcal{F}^W(T)$. Thus, a (finitely additive) set function P_0, with $P_0(\emptyset) = 0, P_0(\Omega) = 1$, is well-defined on the algebra $\mathcal{G} = \bigcup_{0 \le T < \infty} \mathcal{F}^W(T)$ by the recipe

$$P_0(A) \overset{\triangle}{=} E[Z_0(T)1_A]; \quad A \in \mathcal{F}^W(T), \quad 0 \le T < \infty.$$

The question is whether this P_0 is also countably additive on $\mathcal{G}$, and thus, by the Carathéodory extension theorem, uniquely extendable to a probability measure on $\mathcal{F}^W(\infty) = \sigma(\mathcal{G})$.

The countable additivity of P_0 on $\mathcal{G}$ is a consequence of the extension theorems in Parthasarathy (1967), pp. 140–143, as we now explain. For each $T \in [0, \infty)$, the measurable space $(\Omega, \mathcal{F}^W(T))$ is σ-isomorphic to the complete, separable metric space $\Omega_T \overset{\triangle}{=} C([0, T])^D$ of $\mathbb{R}^D$-valued, continuous functions on $[0, T]$, equipped with the supremum norm and the Borel σ-algebra $\mathcal{B}_T$ generated by the collection of open subsets of Ω_T. Indeed, the "truncation mapping" $\pi_T : \Omega \to \Omega_T$ defined by

$$\pi_T(\omega)(t) \overset{\triangle}{=} \omega(t), \quad \forall\, t \in [0, T], \quad \omega \in \Omega,$$

places $\mathcal{F}^W(T)$ and $\mathcal{B}_T$ into a one-to-one correspondence. Therefore, $(\Omega, \mathcal{F}^W(T))$ is a "standard Borel space" (Parthasarathy (1967), pp. 133–134). Let $\{T_n\}_{n=1}^\infty$ be a strictly increasing sequence in $(0, \infty)$ and $\{A_n\}_{n=1}^\infty$

a decreasing sequence in $\mathcal{F}^W(\infty)$ such that each A_n is an atom of $\mathcal{F}^W(T_n)$. Then, for every n,

$$A_n = \{\omega \in \Omega; \quad \omega(t) = \omega_n(t), \quad \forall t \in [0, T_n]\}$$

for some $\omega_n \in \Omega_{T_n}$. Setting $T_0 = 0$ and

$$\omega_\infty(t) = \omega_n(t), \quad \forall t \in [T_{n-1}, T_n), \quad n = 1, 2, \dots,$$

we see from the inclusions $A_1 \supseteq A_2 \supseteq \cdots$ that $\omega_\infty \in \cap_{n=1}^\infty A_n$, so that $\cap_{n=1}^\infty A_n \neq \emptyset$. From Theorem 4.2, p. 143 of Parthasarathy (1967), we conclude that there is a unique probability measure P_0 on $(\Omega, \mathcal{F}^W(\infty))$ such that $P_0^T(A) = P_0(A)$ for all $A \in \mathcal{F}^W(T)$ and $0 \leq T < \infty$.   □

Remark 7.5: For each $T \in [0, \infty)$ and $t \in [0, T]$, the σ-algebra $\mathcal{F}^{(T)}(t)$ is a sub-σ-algebra of the completion of $\mathcal{F}^W(T)$ with respect to P_0, and so P_0 is defined for all sets in $\mathcal{F}^{(T)}(t)$. On $\mathcal{F}^{(T)}(t)$, the two measures P and P_0 are equivalent.

The infinite-horizon model is made difficult by the fact that P *and* P_0 *are not necessarily equivalent on* $\mathcal{F}^W(\infty)$. In fact, it is not hard to check that P and P_0 are equivalent on $\mathcal{F}^W(\infty)$ if and only if the P-martingale $Z_0(\cdot)$ of (7.1) is *uniformly integrable*. To see how things can go wrong if this condition fails, consider the following example.

Example 7.6: Suppose $N = D = 1$, $r(\cdot) \equiv r > 0$, $b(\cdot) \equiv b > r + \frac{1}{2}$, $\delta(\cdot) \equiv 0$, $\sigma(\cdot) \equiv 1$, $A(\cdot) \equiv 0$. Then $W_0(t) = W(t) + (b - r)t$ and

$$\frac{S_1(t)}{S_0(t)} = S_1(0) \exp\left[W(t) + \left(b - r - \frac{1}{2}\right)t\right]$$

$$= S_1(0) \exp\left[W_0(t) - \frac{1}{2}t\right], \quad 0 \leq t < \infty.$$

According to the law of large numbers for Brownian motion (Karatzas and Shreve (1991), Problem 2.9.3 and solution, p. 124), as $t \to \infty$ we have

$$\frac{W(t)}{t} \to 0 \quad P\text{-a.s.}; \qquad \frac{W_0(t)}{t} \to 0 \quad P_0\text{-a.s.}$$

This means that the event $A \overset{\triangle}{=} \{\lim_{t \to \infty} \frac{S_1(t)}{S_0(t)} = \infty\}$ satisfies $P(A) = 1$, $P_0(A) = 0$, whereas the event $B \overset{\triangle}{=} \{\lim_{t \to \infty} \frac{S_1(t)}{S_0(t)} = 0\}$ satisfies $P(B) = 0$, $P_0(B) = 1$. Notice also that the P-martingale $Z_0(t) = \exp[-(b - r)W(t) - (b - r)^2 t/2]$ of (7.1) is not uniformly integrable. Indeed, $\lim_{t \to \infty} Z_0(t) = 0$, P-almost surely, but $E Z_0(t) = 1$ for all $t \in [0, \infty)$.

Example 7.6 shows that if for $t \in [0, \infty)$ we were to augment $\mathcal{F}^W(t)$ by the P_0-null sets of $\mathcal{F}^W(\infty)$, we would obtain a σ-algebra $\mathcal{F}_0(t)$ on which P and P_0 would disagree. Indeed, there may exist sets $A \in \mathcal{F}_0(t)$ for which $P(A)$ is not defined and for which $B_1 \in \mathcal{F}^W(t)$, $B_2 \in \mathcal{F}^W(t)$ can be found satisfying $P_0(A \triangle B_1) = P_0(A \triangle B_2) = 0$ but $P(B_1) \neq P(B_2)$. For this reason, we choose to work with the family of filtrations $\{\mathcal{F}^{(T)}(t)\}_{0 \le t \le T}$, indexed by $T \in [0, \infty)$, rather than with the augmentation of $\{\mathcal{F}^W(t)\}_{0 \le t < \infty}$ by the P_0-null sets of $\mathcal{F}^W(\infty)$.

Definition 7.7: Consider a financial market $\mathcal{M} = (r(\cdot), b(\cdot), \delta(\cdot), \sigma(\cdot),$ $S(0), A(\cdot))$ on $[0, \infty)$. A *portfolio process* $(\pi_0(\cdot), \pi(\cdot))$ is as described in Definition 2.1, except that now we require $\pi_0(\cdot)$ and $\pi(\cdot)$ to be restrictedly progressively measurable and we require (2.4)–(2.6) to hold for every finite T. We say that $\pi(\cdot)$ is *tame* if the P_0-local martingale

$$M^\pi(t) \triangleq \int_0^t \frac{1}{S_0(u)} \pi'(u)\sigma(u) \, dW_0(u), \quad 0 \le t < \infty, \tag{7.5}$$

is almost surely bounded from below by a constant not depending on t. We say that $\pi(\cdot)$ is *martingale-generating* if $\{M^\pi(t); 0 \le t < \infty\}$ is a P_0-martingale.

A *cumulative income process* $\Gamma(\cdot) = \{\Gamma(t); 0 \le t < \infty\}$ is a P-semimartingale and hence a "restricted" P_0-semimartingale. We say that $\Gamma(\cdot)$ is *integrable* if (5.9) is satisfied for every $T \in [0, \infty)$. We say that $(\pi_0(\cdot), \pi(\cdot))$ is $\Gamma(\cdot)$-*financed* if (3.3) holds for every $T \in [0, \infty)$. In this case the *wealth process* is given by

$$\frac{X(t)}{S_0(t)} = \Gamma(0) + \int_{(0,t]} \frac{d\Gamma(u)}{S_0(u)} + \int_0^t \frac{1}{S_0(u)} \pi'(u)\sigma(u) \, dW_0(u), \quad 0 \le t < \infty. \tag{7.6}$$

1.8   Notes

Sections 1–3: Finance models that allow continuous trading constitute a burgeoning field of mathematical research. In the notes to Chapter 3 we discuss the origin of these models within the capital asset pricing context. Their application to the hedging of contingent claims is presented in Chapter 2, and the related history is summarized in the notes to that chapter. For broad and exhaustive surveys of the issues of finance, including continuous-time models, one may consult the books of Cox and Rubinstein (1985), Dothan (1990), Duffie (1988, 1992), Huang and Litzenberger (1988), Hull (1993), Ingersoll (1987), Jarrow (1988), Merton (1990), Jarrow and Turnbull (1995), Baxter and Rennie (1996), Pliska (1997), Musiela and Rutkowski (1997). The book by Malliaris and Brock (1982) surveys stochastic models used in economics and finance. Additional surveys and/or lecture notes include Malliaris (1983), Müller (1985), Karatzas (1989, 1996), Lamberton and Lapeyre (1991); this latter text, along with Duffie (1992) and Wilmott, Dewynne, and Howison (1993, 1995), can be consulted for numerical and computational aspects of the theory. Questions of convergence of discrete-time and/or discrete-state models to their continuous-time counterparts are discussed, among others, by Cox, Ross, and Rubinstein (1979),

Madan, Milne, and Shefrin (1989), Nelson and Ramaswamy (1990), He (1990, 1991), Amin (1991), Amin and Khanna (1994), Cutland, Kopp, and Willinger (1991, 1993a,b), Kind, Liptser, and Runggaldier (1991), Willinger and Taqqu (1991), Duffie and Protter (1992), Eberlein (1992), Dengler (1993), and Bick and Willinger (1994). Recent books on numerical methods of general applicability in this area are Kloeden and Platen (1992), Kushner and Dupuis (1992), Talay and Tubaro (1997).

For early work on the subject, it is instructive to see the articles in the volume edited by Cootner (1964), in particular the translation of the famous dissertation by Bachelier (1900); this work is the first instance of *both* a mathematical treatment of Brownian motion *and* its application to finance. The use of Brownian-motion-based models of stock prices derives from the *efficient market hypothesis*, which asserts that all public information useful for making investment decisions is already incorporated in market prices. According to the efficient market hypothesis, past stock prices may be useful for purposes of estimating parameters in the distribution of future prices, but do not provide information that permits an investor to outperform the market. In particular, if there is public information which implies that a stock price is certain to rise, then it would have already risen. The efficient market hypothesis is still subject to some debate, although a substantial amount of empirical and theoretical justification has accumulated in its favor; see, e.g., Kendall (1953), Osborne (1959), Sprenkle (1961), Boness (1964), Alexander (1961), and Fama (1965). Originally, the mathematical content of the efficient market hypothesis was expressed as the belief that returns on stock prices follow a discrete-time random walk. Samuelson (1965a) proposed a discrete-time martingale model of security prices, a mathematical concept also in keeping with the efficient market hypothesis. A more recent theoretical examination of this matter is given by Samuelson (1973); see also Black (1986). The discrete-time martingale model is criticized on the basis of empirical studies by LeRoy (1989). Nontechnical discussions can be found in Bernstein (1992), Chapters 5–7, and Malkiel (1996).

The model presented in this chapter is an outgrowth of the "geometric" Brownian motion model introduced by Samuelson (1965b) to capture the limited-liability nature of corporation ownership. It was formulated within the framework of Itô's stochastic calculus by Merton (1969, 1971) and of the calculus for more general stochastic processes by Harrison and Kreps (1979), Harrison and Pliska (1981, 1983). The independence between past and future increments of the driving Brownian motion enforces the efficient market hypothesis in this model, provided that the model is viable (Definition 4.1). *The efficient market hypothesis does not claim any particular distribution for stock prices*, although it is often confused with the assumption of stock prices modeled by geometric Brownian motion and hence having a log-normal distribution. This distribution provides a reasonable fit to the data, but other distributions are known to be better (see, e.g., Fama (1965), Officer (1972), Hsu et al. (1974), Hagerman (1978), Kon (1984), Madan and Seneta (1990), Lo (1991)). For the huge subjects of *statistical estimation* and *econometrics* in the context of financial markets, we send the reader to the April 1994 volume of the journal *Mathematical Finance*, and to the recent monograph by Campbell, Lo, and MacKinley (1997).

The model of this chapter allows for the stock-price coefficients to be themselves random processes, which affords much greater generality than the log-normal model. Various special cases of this model besides the constant-coefficient case of Samuelson (1965b) have been studied. Cox and Ross (1976) examined a *constant elasticity of variance* model, in which stock prices have the form

$$dS(t) = bS(t) \, dt + \sigma S^{\gamma/2}(t) \, dW(t),$$

where b, $\sigma > 0$ and $0 \leq \gamma < 2$ are constant and $W(\cdot)$ is a Brownian motion; see also Beckers (1980), Schroder (1989). Another alternative, due to Föllmer and Schweizer (1993), contains the geometric Ornstein–Uhlenbeck process as a special case. Some works on hedging and/or optimization in models that allow for *jumps* in the stochastic equations (1.8), and thus do not fall within the purview of this text, are Aase (1993), Back (1991), Bates (1988, 1992), Beinert and Trautman (1991), Dritschel and Protter (1997), Elliott and Kopp (1990), Jarrow and Madan (1991b,c), Jones (1984), Jarrow and Rosenfeld (1984), Jeanblanc-Picqué and Pontier (1990), Madan and Seneta (1990), Madan and Milne (1991), Mercurio and Runggaldier (1993), Merton (1976), Naik and Lee (1990), Pham (1995), Schweizer (1988, 1991, 1992a,b), Scott (1997), Shirakawa (1990, 1991), Xue (1992), Zhang (1993).

A nonnegativity constraint on wealth was used by Harrison and Pliska (1981) to rule out doubling strategies. The notion of a *tame* portfolio, used here for the same purpose, also appears implicitly in Karatzas, Lehoczky, and Shreve (1987) and Dybvig and Huang (1988). Heath and Jarrow (1987) achieve the same end by imposing margin requirements.

Sections 4–6: The example of a nonviable market, which is based on the three-dimensional Bessel process and appears right after Definition 4.1, is due to A.V. Skorohod (private communication by S. Levental). Related results can be found in Delbaen and Schachermeyer (1995b).

The absence of arbitrage opportunities is implied by the existence of an equivalent probability measure, under which discounted prices (plus discounted cumulative dividends, if dividends are present) become martingales. This is essentially a rephrasing of the classical principle behind the optional sampling and martingale systems theorems, according to which "one cannot win *for certain* by betting on a martingale" (e.g., Doob (1953), Chung (1974)).

To what extent is the converse true? In other words, if "one cannot win for certain by betting on a given process" (i.e., if the process does not support arbitrage opportunities), then is this process a martingale, perhaps under an equivalent probability measure? For *discrete-parameter processes,* affirmative answers to this question were provided by Harrison and Kreps (1979), Harrison and Pliska (1981), and Taqqu and Willinger (1987) for finite probability spaces (see also Ross (1976) and Cox and Ross (1976) for earlier work along similar lines), and by Dalang, Morton, and Willinger (1990) for general probability spaces and multidimensional processes. This last work uses arguments based on measurable selection results and on the separating hyperplane theorem of convex analysis; the same result has since been derived, using somewhat simpler arguments, by Kabanov and Kramkov (1994a) and by Rogers (1995a). Related papers are Willinger and Taqqu (1988), Back and Pliska (1991), Kusuoka (1992), Schachermayer (1992). All these results on the relation between "no arbitrage" and the existence of an equivalent martingale measure bear a striking similarity to de Finetti's (1937, 1974) theory of coherent subjective probabilities and inferences; see Heath and Sudderth (1978), Boykov (1996), and the survey papers of Sudderth (1994), Ellerman (1984).

For general *continuous-parameter processes,* the question at the beginning of the previous paragraph becomes significantly more complex, and the results much harder and deeper. Absence of arbitrage is in general *not* sufficient for the existence of an equivalent martingale measure, and stronger conditions are needed; cf. Kreps (1981), Schachermayer (1993, 1994). Work related to the results of Sections 4 and 5 has been done by Stricker (1990), Delbaen (1992), Lakner (1993), Delbaen and Schachermayer (1994a,b, 1995a–c, 1996b, 1997a–c), Fritelli and Lakner (1994, 1995) who use rather refined functional-analytic tools (see also the earlier work by Duffie and Huang (1986)), and Levental and Skorohod (1995).

In particular, Levental and Skorohod (1995) consider a market $\mathcal{M}$ as in Definition 1.3 with $N = D$ and invertible volatility matrix $\sigma(\cdot)$, and define $\theta(\cdot)$ via (6.16), $W_0(\cdot)$ via (5.6) whenever $\theta(\cdot)$ is integrable,

$$\zeta^{(r)} \triangleq \inf \left\{ t \in (r, T); \int_r^t \|\theta(s)\|^2 \, ds = \infty \right\} \wedge T, \qquad (8.1)$$

$$Z^{(r)}(t) \triangleq 1_{\{t \leq \zeta^{(r)}\}} \exp \left[- \int_0^t 1_{\{r \leq s\}} \theta'(s) \, dW(s) \right.$$

$$\left. - \frac{1}{2} \int_0^t 1_{\{r \leq s\}} \|\theta(s)\|^2 \, ds \right], \quad 0 \leq t \leq T, \qquad (8.2)$$

for $r \in [0, T]$, as well as

$$\alpha \triangleq \inf \left\{ t \in [0, T); \int_t^{t+h} \|\theta(s)\|^2 \, ds = \infty, \forall h \in (0, T - t] \right\} \wedge T. \qquad (8.3)$$

They show that $\mathcal{M}$ *is viable if and only if both*

$$P[\alpha = T] = 1 \quad and \quad EZ^{(r)}(T) = 1, \forall\, r \in [0, T] \qquad (8.4)$$

hold. In particular, under the condition (5.1), $\mathcal{M}$ is viable if and only if

$$\left\{ \begin{array}{l} \text{there exists a probability measure } P_0, \text{ equivalent to } P, \\ \text{under which } W_0(\cdot) \text{ becomes Brownian motion.} \end{array} \right\} \qquad (8.5)$$

In the absence of condition (5.1), they define *approximate arbitrage* as a sequence of tame portfolios $\{\pi_n(\cdot)\}_{n=1}^{\infty}$ with corresponding gains processes $\{G_n(\cdot)\}_{n=1}^{\infty}$ satisfying $\lim_{n\to\infty} P[G_n(T) \geq 0] = 1$ and $\lim_{n\to\infty} P[G_n(T) > \delta] \geq \delta, \forall\, n \in \mathbb{N}$ for some $\delta > 0$ (cf. Stricker (1990), Duffie (1992), Kabanov and Kramkov (1994b) for related notions); then they show that (8.5) is equivalent to the *absence of approximate arbitrage*. This latter condition is *stronger* than viability (absence of arbitrage), as these authors demonstrate by example. In a similar vein, see the recent results of Delbaen and Schachermayer (1997b) for general semimartingale price processes, which might well be the "last word" on this subject.

The relations between market completeness and *uniqueness* of the equivalent martingale measure were brought out in the fundamental papers of Harrison and Kreps (1979) and Harrison and Pliska (1981, 1983), and more recently in Ansel and Stricker (1992, 1994), Artzner and Heath (1995), Chatelain and Stricker (1992, 1994), Delbaen (1992), Jacka (1992), Jarrow and Madan (1991a,b), Müller (1989), and Taqqu and Willinger (1987).

In the context of a complete market as in Section 6, what happens if the agent starts at $t = 0$ with initial capital $y > 0$ strictly less than the quantity $x = E^0[B/S_0(T)]$ of (6.1)? Remark 6.4 shows that there can exist no tame, y-financed portfolio with corresponding wealth process $X^{y,\pi}(\cdot)$ satisfying $P[X^{y,\pi}(T) \geq B] = 1$ for $0 < y < x$. In other words, it is then *not possible to hedge the contingent claim B without risk*, and the agent might wish simply to maximize the probability $P[X^{y,\pi}(T) \geq B]$ of achieving a perfect hedge, over a suitable subclass of tame y-financed portfolios $\pi(\cdot)$. One is thus led to a stochastic control problem of the so-called "goal type." Such problems have been studied by Kulldorff (1993); see also Heath (1993), Karatzas (1996), pp. 55–59, Föllmer & Leukert (1998), and Spivak (1998).

2

Contingent Claim Valuation in a Complete Market

2.1 Introduction

A *derivative security* (also called *contingent claim*; cf. Definition 2.1 and discussion following it) is a financial contract whose value is *derived* from the value of another underlying, more basic, security, such as a stock or a bond. Common derivative securities are *put options, call options, forward contracts, futures contracts,* and *swaps.* These securities can be used for both speculation and hedging, but their creation and marketing are based much more on the latter use than the former. Some derivative securities are traded on exchanges, while others are arranged as private contracts between financial institutions and their clients. The world-wide market in derivative securities is in the trillions of dollars.

In order better to understand the concept of derivative securities, let us begin by considering one of the older derivative securities, a *forward contract.* Under this contract, one agent agrees to sell to another agent some commodity or financial asset at a specified future date at a specified *delivery price.* Corporations that need a certain amount of a commodity at a future date often buy a forward contract for delivery of that commodity. Corporations that expect to be paid at a future date in a foreign currency sell forward contracts on that currency. In the first case the corporation would promise to receive delivery of the commodity, whereas in the second case it would promise to deliver the currency. In both cases, the corporation is using the forward contract to *lock in a price in advance,* i.e., to hedge uncertainty. Forward contracts can also be used for speculation; a forward contract on an asset requires no initial cash outlay by either party, and for this reason allows speculators much higher leverage than can be obtained by purchasing and holding the underlying assets.

Futures contracts are like forward contracts, except that unlike forwards, futures are traded on exchanges and are consequently subject to a number of regulations. The principal rule is that an agent must deposit money into a *margin account* at the time the contract is entered, and this margin account is credited or debited daily to reflect movement in the futures' price. If the margin-account balance falls too low, the agent must replenish it. This whole process is called *marking-to-market*. Unlike the case of options discussed below, forward and futures contracts are designed so that their initial cost to both parties is zero.

Derivative securities that became popular more recently are the various options on stocks. Options have been traded on public exchanges since 1973. The holder of a *European call option* has the right, but not the obligation, to buy an underlying security at a specified date (*expiration date*) for a contractually specified amount (*strike price*), irrespective of the market value of the security on that date. The *European put option* is the same as the European call option, except that it entitles the holder to sell. The *American call option* and the *American put option* entitle the holder to buy or sell, respectively, at *any* time prior to a specified expiration date. Besides the European and American options (which, incidentally, are both traded worldwide, although exchange-traded stock options are typically American), there is a variety of more "exotic" options. *Options allow risk to be hedged in various ways.* The most obvious one is that an investor who owns a security but intends to sell it by a known future date can buy a put option on the same security and thus be guaranteed at least the strike price when the security is sold. If the security price rises above the strike price, the investor is still able to sell the security at its market value. By taking combinations of long and short positions in puts and calls, investors can create a variety of customized contingent claims. Because one of the likely possibilities available to the holder of an option is to "receive a positive amount and pay nothing," arbitrage-based considerations suggest that the option's present worth should be positive.

First appearing in 1981, a *swap* is a contract between two agents in which cash flows are traded. A common situation is that one agent has income due to variable-rate interest on some investment, and the other has income due to fixed-rate interest on a different investment. Because of their different financial situations, the agents may wish to trade some part of their income streams. This trade is usually arranged by a financial institution, which guarantees the contract.

The existence of derivative securities leads to two mathematical questions: *pricing* and *hedging*. The price of publicly traded derivative securities is set by the law of supply and demand, but many derivative securities are private contracts in which both parties would like to be assured that a "fair" price has been reached. Even for publicly traded derivative securities, the fact that they are described in terms of an underlying security, whose price and price history are known, suggests that there should be some way of theoretically determining a "fair" price. Arbitrageurs are vigilant for discrepancies between the market price of derivatives and their estimation of the fair price, and immediately take positions in the derivative security when they perceive such a discrepancy.

More recently, model-based pricing of derivative securities has become the basis of risk management. A typical risk-management question is how much a portfolio value will be affected by a certain movement in underlying asset prices. If the portfolio contains derivative securities, a mathematical model is needed to answer this question.

The hedging of a derivative security is the problem faced by a financial institution that sells to a client some contract designed to reduce the client's risk. This risk has been assumed by the financial institution, which would now like to take a position in the underlying security, and perhaps in other instruments as well, so as to minimize its own exposure to risk. Assumption of client risk is a principal service of financial institutions; managing this risk well is a necessary prerequisite for offering this service.

In a complete market model, as set forth in Definition 1.6.1, there are definitive solutions to both the problem of pricing and the problem of hedging derivative securities. This method of solution is known as *arbitrage pricing theory*, because it proceeds from the observation that an incorrectly priced derivative security in a complete market presents an arbitrage opportunity. A correctly priced derivative security in a complete market is redundant, in the sense that it can be duplicated by a portfolio in the nonderivative securities. The portfolio that achieves this duplication is the *hedging portfolio*, which the seller of the derivative security can use to remove the risk incurred by the sale. Of course, the existence of the hedging portfolio prompts one to ask why an agent would buy rather than duplicate a derivative security. It appears that this occurs because markets are not frictionless. In order to duplicate a derivative security by portfolio management, an agent would have to develop a mathematical model, pay brokerage fees, and gather information about the statistics of the market. Agents avoid these costs by striking a deal with another agent (intermediated by an exchange or a financial institution) or by paying a fee to a financial institution. A financial institution has modeling expertise and smaller transaction costs than its clients because of the volume of its transactions and the fact that many of these take place in-house.

We present the arbitrage pricing theory for European contingent claims in Section 2.2, and for American contingent claims in Section 2.5. Special cases of European contingent claims, such as forward and futures contracts, are treated in Section 2.3, whereas Section 2.4 computes the prices and hedging portfolios for European call and put options as well as for a certain type of path-dependent option (Example 4.5) in the context of a model with constant coefficients. In particular, Example 4.1 derives the celebrated *Black and Scholes (1973) formula* for the price of a European call option in a model with constant interest rate and volatility. The development of the arbitrage pricing theory for American contingent claims (Section 2.5) is based on the theory of optimal stopping, which we survey in Appendix D; special cases, such as the American call and put options, are studied in detail in Sections 2.6 and 2.7, respectively.

2.2 European Contingent Claims

Throughout this chapter we shall operate in the context of a *complete, standard financial market* $\mathcal{M}$ (Definitions 1.1.3, 1.5.1, 1.6.1 and Theorem 1.6.6). In particular, the price of the money market is governed by

$$dS_0(t) = S_0(t)[r(t)\, dt + dA(t)] \tag{2.1}$$

and the prices of the stocks satisfy

$$dS_n(t) = S_n(t)\left[b_n(t)\, dt + dA(t) + \sum_{d=1}^{N} \sigma_{nd}(t)\, dW^{(d)}(t) \right], n = 1, \ldots, N, \tag{2.2}$$

with $\sigma(t) = (\sigma_{nd}(t))_{1 \leq n, d \leq N}$ nonsingular for Lebesgue-almost-every $t \in [0, T]$ almost surely. Recall that $S_0(0) = 1$ and $S_1(0), \ldots, S_N(0)$ are positive constants.

Definition 2.1: A *European contingent claim (ECC)* is an integrable cumulative income process $C(\cdot)$. To simplify the notation, we shall always assume that $C(0) = 0$ almost surely.

European contingent claims are bought and sold. The *buyer*, who is said to assume a *long position* in the claim, pays some nonrandom amount $\Gamma(0)$ at time zero and is thereby entitled to the cumulative income process $C(\cdot)$. The *seller (writer, issuer)*, who is said to assume a *short position*, receives $\Gamma(0)$ at time zero and must provide $C(\cdot)$ to the buyer. Thus, the seller has cumulative income process

$$\Gamma(t) = \Gamma(0) - C(t), \quad 0 \leq t \leq T. \tag{2.3}$$

The *seller's* objective is to choose a martingale-generating, $\Gamma(\cdot)$-financed portfolio process $(\pi_0(\cdot), \pi(\cdot))$ such that his corresponding wealth satisfies $X(T) \geq 0$ almost surely. In other words, the seller wants to "hedge" the short position in the contingent claim by trading in the market in such a way as to make the necessary payments and still be solvent at the final time, almost surely. Recall from (1.5.8) that the seller's discounted wealth process satisfies

$$\frac{X(t)}{S_0(t)} = \Gamma(0) - \int_{(0,t]} \frac{dC(u)}{S_0(u)} + \int_0^t \frac{1}{S_0(u)} \pi'(u)\sigma(u)\, dW_0(u), \quad 0 \le t \le T.$$
(2.4)

If $\pi(\cdot)$ is martingale-generating and if $X(T) \ge 0$ holds almost surely, we have, upon taking P_0-expectations in (2.4), that

$$x \overset{\triangle}{=} E_0 \left[\int_{(0,T]} \frac{dC(u)}{S_0(u)} \right] \le \Gamma(0).$$
(2.5)

This provides a lower bound on the time-zero price $\Gamma(0)$ the seller must charge for the ECC $C(\cdot)$.

Suppose the seller charges the amount x. Because the market was assumed to be complete and the cumulative income process $C(\cdot)$ is integrable (Definition 1.5.5), the random variable $S_0(T) \left[\int_{(0,T]} \frac{dC(u)}{S_0(u)} \right]$ is financeable: there is a martingale-generating portfolio $\hat{\pi}(\cdot)$ satisfying

$$\int_{(0,T]} \frac{dC(u)}{S_0(u)} = x + \int_0^T \frac{1}{S_0(u)} \hat{\pi}'(u)\sigma(u)\, dW_0(u) \quad a.s.$$
(2.6)

(see Proposition 1.6.2). Define $\hat{X}(\cdot)$ by (2.4) with x replacing $\Gamma(0)$ and $\hat{\pi}(\cdot)$ replacing $\pi(\cdot)$, namely,

$$\frac{\hat{X}(t)}{S_0(t)} = x - \int_{(0,t]} \frac{dC(u)}{S_0(u)} + \int_0^t \frac{\hat{\pi}'(u)}{S_0(u)}\sigma(u)\, dW_0(u), \quad 0 \le t \le T. \quad (2.4')$$

Then (2.6) shows that $\hat{X}(T) = 0$ almost surely. With $\hat{\pi}_0(t) \overset{\triangle}{=} \hat{X}(t) - \hat{\pi}'(t)\mathbf{1}$, the seller has found a martingale-generating, $\Gamma(\cdot)$-financed "hedging" portfolio $(\hat{\pi}_0(\cdot), \hat{\pi}(\cdot))$ that results in nonnegative wealth at time T, with $\Gamma(0) = x$. Thus, the seller of the contingent claim can charge x (but no less) for the ECC $C(\cdot)$ and use the portfolio $(\hat{\pi}_0(\cdot), \hat{\pi}(\cdot))$ to hedge his position.

Now let us take conditional expectations with respect to $\mathcal{F}(t)$ under the probability measure P_0 in (2.6) to obtain

$$x + \int_0^t \frac{1}{S_0(u)} \hat{\pi}'(u)\sigma(u)\, dW_0(u) = E_0 \left[\int_{(0,T]} \frac{dC(u)}{S_0(u)} \,\Big|\, \mathcal{F}(t) \right], \quad 0 \le t \le T.$$

Substitution into (2.4′) yields

$$\frac{\hat{X}(t)}{S_0(t)} = E_0 \left[\int_{(t,T]} \frac{dC(u)}{S_0(u)} \,\Big|\, \mathcal{F}(t) \right], \quad 0 \le t \le T.$$
(2.7)

This provides a simple representation for the seller's wealth process corresponding to the hedging portfolio.

The *buyer* of the contingent claim can, if he wishes, hedge his position by the reverse strategy. He has income process $-\Gamma(\cdot)$, and $(-\hat{\pi}_0(\cdot), -\hat{\pi}(\cdot))$

is a $(-\Gamma(\cdot))$-financed portfolio process corresponding to the wealth process $-\hat{X}(\cdot)$ (and, in particular, final wealth $-\hat{X}(T) = 0$ almost surely).

The above discussion shows that the "fair price" at time zero for the ECC $C(\cdot)$ is x given by (2.5). To wit, if the ECC were traded at any other price, then either the seller or the buyer would have an arbitrage opportunity.

We are also interested in the price of the ECC at other times $t \in [0, T]$. Subtracting (2.4) evaluated first at T and then at t, we obtain

$$\frac{X(T)}{S_0(T)} = \frac{X(t)}{S_0(t)} - \int_{(t,T]} \frac{dC(u)}{S_0(u)} + \int_t^T \frac{1}{S_0(u)} \pi'(u)\sigma(u)\, dW_0(u), \quad 0 \le t \le T.$$
(2.8)

We imagine that at time t, an agent sells the remainder (excluding any payment at time t) of the ECC $C(\cdot)$ for a price $X(t)$ (measurable with respect to $\mathcal{F}(t)$), invests $X(t)$ in the market, pays out the contingent claim between times t and T, and wants to have $X(T) \ge 0$ almost surely. Prompted by these considerations, we make the following definition.

Definition 2.2: Let $C(\cdot)$ be a European contingent claim. For $t \in [0, T]$ the *value of* $C(\cdot)$ *at* t, denoted by $V^{ECC}(t)$, is the smallest (in the sense of a.s. domination) $\mathcal{F}(t)$-measurable random variable ξ such that if $X(t) = \xi$ in (2.8), then for some martingale-generating portfolio process $\pi(\cdot)$ we have $X(T) \ge 0$ almost surely. The value $V^{ECC}(0)$ at time $t = 0$ is called the *(arbitrage-based) price* for the ECC at $t = 0$.

Proposition 2.3: *The value at time t of a European contingent claim* $C(\cdot)$ *is*

$$V^{ECC}(t) = S_0(t) \cdot E_0 \left[\int_{(t,T]} \frac{dC(u)}{S_0(u)} \,\Big|\, \mathcal{F}(t) \right], \quad 0 \le t \le T.$$
(2.9)

In particular, $V^{ECC}(0) = x$ *as in (2.5).*

PROOF.   If $X(T) \ge 0$ in (2.8) and $\pi(\cdot)$ is martingale-generating, then

$$\frac{X(t)}{S_0(t)} \ge E_0 \left[\int_{(t,T]} \frac{dC(u)}{S_0(u)} \,\Big|\, \mathcal{F}(t) \right], \quad 0 \le t \le T,$$
(2.10)

which shows that $V^{ECC}(t)$ must be at least as large as the right-hand side of (2.9). But the wealth process $\hat{X}(\cdot)$ of (2.4'), corresponding to the pair $(\hat{\pi}_0(\cdot), \hat{\pi}(\cdot))$ whose existence was established above, satisfies (2.10) with equality (see (2.7)). It follows that $V^{ECC}(t) = \hat{X}(t)$, which is (2.9).   □

Remark 2.4: We may use "Bayes's rule" (Karatzas and Shreve (1991), Lemma 3.5.3) to convert (2.9) to a conditional expectation with respect to the original probability measure P, namely

$$V^{ECC}(t) = \frac{1}{H_0(t)} E \left[\int_{(t,T]} H_0(u)\, dC(u) \,\Big|\, \mathcal{F}(t) \right], \quad 0 \le t \le T,$$

where $H_0(u) \triangleq Z_0(u)/S_0(u)$ is the *state price density process*. In particular, the fair price of the claim at $t = 0$ is

$$V^{ECC}(0) = E \int_{(0,T]} H_0(u) \, dC(u).$$

Remark 2.5: Unlike much of the finance literature, our definition of $V^{ECC}(\cdot)$ is set up so that $V^{ECC}(T) = 0$ almost surely. This is consistent with our convention that processes have RCLL paths, a convention that requires the integral $\int_{(t,T]} \frac{dC(u)}{S_0(u)}$ in (2.9) to be over the half-open interval $(t, T]$. From (2.9) we have $V^{ECC}(T-) = C(T) - C(T-)$, so a final jump in $C(\cdot)$ causes the same final jump in $V^{ECC}(\cdot)$. In particular, if B is an $\mathcal{F}(T)$-measurable random variable satisfying

$$E_0 \left[\frac{|B|}{S_0(T)} \right] < \infty \tag{2.11}$$

and $C(\cdot)$ is given by

$$C(t) = \begin{cases} 0 & \text{if } 0 \leq t < T, \\ B & \text{if } -t = T, \end{cases} \tag{2.12}$$

then we have

$$V^{ECC}(t) = S_0(t) \cdot E_0 \left[\frac{B}{S_0(T)} \,\middle|\, \mathcal{F}(t) \right], \quad 0 \leq t \leq T, \tag{2.13}$$

$$V^{ECC}(T-) = B, \quad V^{ECC}(T) = 0. \tag{2.14}$$

Definition 2.6: Let $C(\cdot)$ be a European contingent claim, and define $\Gamma(\cdot) = V^{ECC}(0) - C(\cdot)$. The martingale-generating, $\Gamma(\cdot)$-financed portfolio process $(\hat{\pi}_0(\cdot), \hat{\pi}(\cdot))$ satisfying

$$\frac{V^{ECC}(t)}{S_0(t)} = \Gamma(0) - \int_{(0,t]} \frac{dC(u)}{S_0(u)}$$

$$+ \int_0^t \frac{1}{S_0(u)} \hat{\pi}'(u)\sigma(u) \, dW_0(u), \quad 0 \leq t \leq T, \tag{2.15}$$

$$\hat{\pi}_0(t) = V^{ECC}(t) - \hat{\pi}'(t)\mathbb{1}, \quad 0 \leq t \leq T, \tag{2.16}$$

is called the *hedging portfolio* for (a short position in) $C(\cdot)$.

Remark 2.7: The representation (2.9) allows us to write (2.15) as

$$\int_0^t \frac{1}{S_0(u)} \hat{\pi}'(u)\sigma(u) \, dW_0(u) = E_0 \left[\int_{(0,T]} \frac{dC(u)}{S_0(u)} \,\middle|\, \mathcal{F}(t) \right]$$

$$- E_0 \left[\int_{(0,T]} \frac{dC(u)}{S_0(u)} \right], \quad 0 \leq t \leq T,$$

or, in differential form,

$$\hat{\pi}'(t)\sigma(t)\,dW_0(t) = S_0(t) \cdot d_t \left(E_0 \left[\int_{(0,T]} \frac{dC(u)}{S_0(u)} \,\Big|\, \mathcal{F}(t) \right] \right). \qquad (2.17)$$

This formula sometimes enables us to determine $\hat{\pi}(\cdot)$ explicitly.

2.3   Forward and Futures Contracts

Consider an $\mathcal{F}(T)$-measurable random variable B satisfying (2.11). We regard B as the market value at time T of some asset, such as a stock or commodity. In this section, we define the forward and futures prices for this asset.

Suppose that at time zero an agent buys a contract *obligating* him to purchase the above asset for the nonrandom *delivery price* q on the *delivery date* T. The seller of this contract, who is taking the short position, agrees to deliver the asset at time T in exchange for q, even though the market value at time T is B. According to Remark 2.5, the value process for this contract is

$$V^{FC}(t;q) = S_0(t) \cdot E_0 \left[\frac{B-q}{S_0(T)} \,\Big|\, \mathcal{F}(t) \right], \quad 0 \le t \le T \qquad (3.1)$$

provided that

$$E_0 \left[\frac{1}{S_0(T)} \right] < \infty. \qquad (3.2)$$

Example 3.1 *(Forward contract to purchase a stock that pays no dividends):*   Suppose the contract is to purchase one share of the first stock, i.e., $B = S_1(T)$. If the first stock pays no dividends and $\sigma(\cdot)$ satisfies the Novikov condition (1.5.17) with $n = 1$, then Remark 1.5.11 implies

$$V^{FC}(t;q) = S_1(t) - qS_0(t) \cdot E_0[1/S_0(T)|\mathcal{F}(t)], \quad 0 \le t \le T. \qquad (3.3)$$

If in addition $S_0(T)$ is nonrandom, the hedging portfolio is particularly simple. The agent assuming the short position sells the contract for $V^{FC}(0;q) = S_1(0) - \frac{q}{S_0(T)}$ at time zero, buys one share of the stock at cost $S_1(0)$, and borrows $\frac{q}{S_0(T)}$ from the money market. At time T, his money-market debt has grown to q. He delivers the first stock, receives the delivery price q, and pays off his money-market debt.

The quantity of interest in finance is the *forward price* of an asset, defined as follows.

Definition 3.2:   Let B be an $\mathcal{F}(T)$-measurable random variable, and assume that (2.11), (3.2) hold. The *forward price process* $f(\cdot)$ for B is defined by

$$f(t) \triangleq \frac{E_0[B/S_0(T)|\mathcal{F}(t)]}{E_0[1/S_0(T)|\mathcal{F}(t)]}, \quad 0 \le t \le T. \tag{3.4}$$

From (3.1) we see immediately that $f(t)$ is the unique, $\mathcal{F}(t)$-measurable solution to the equation

$$V^{FC}(t; f(t)) = 0. \tag{3.5}$$

In other words, at time t the value is zero for the contract to buy at date T the asset at delivery price $f(t)$.

Example 3.3 *(Forward price of a stock that pays no dividends):* If $B = S_1(T)$, the first stock pays no dividends, and $\sigma(\cdot)$ satisfies the Novikov condition (1.5.17) with $n = 1$, then (3.3) and (3.5) yield

$$f(t) = \frac{S_1(t)/S_0(t)}{E_0[1/S_0(T)|\mathcal{F}(t)]}, \quad 0 \le t \le T. \tag{3.6}$$

Example 3.4 *(Forward price of a stock with nonrandom dividend rate):* If $B = S_1(T)$, the dividend rate process $\delta_1(\cdot)$ is nonrandom and the processes $\sigma_{11}(\cdot), \ldots, \sigma_{1N}(\cdot)$ are uniformly bounded, then the forward price is

$$f(t) = \frac{S_1(t)/S_0(t)}{E_0[1/S_0(T)|\mathcal{F}(t)]} \exp\left\{-\int_t^T \delta_1(u)\,du\right\}, \quad 0 \le t \le T. \tag{3.7}$$

To see this, observe from (1.5.16) that the process $\frac{S_1(t)}{S_0(t)} \exp\left\{\int_0^t \delta_1(u)du\right\}$ is a P_0-martingale, so the numerator of (3.4) is

$$E_0[S_1(T)/S_0(T)|\mathcal{F}(t)] = \frac{S_1(t)}{S_0(t)} \exp\left\{-\int_t^T \delta_1(u)\,du\right\}.$$

Example 3.5 *(Forward price of a stock with nonrandom dividend payments, when the money market is nonrandom):* If $B = S_1(T)$, if the dividend-payment $\rho(\cdot) \triangleq \delta_1(\cdot)S_1(\cdot)$ and money-market prices $S_0(\cdot)$ are nonrandom, and if $\sigma(\cdot)$ satisfies the Novikov condition (1.5.17) with $n = 1$, then according to Remark 1.5.11 the process

$$\frac{S_1(t)}{S_0(t)} + \int_0^t \frac{\rho(u)}{S_0(u)}\,du, \quad 0 \le t \le T,$$

is a martingale under P_0. From (3.4) we have

$$f(t) = S_0(T)\left[\frac{S_1(t)}{S_0(t)} - \int_t^T \frac{\rho(u)}{S_0(u)}du\right]. \tag{3.8}$$

When a commodity rather than a stock is being priced, there can be a storage cost, which corresponds to negative $\rho(\cdot)$ in this example. In that case, (3.8) is called the *cost-of-carry* formula.

Forward contracts on assets are not always available, and if they are, they are available only with delivery price equal to the forward price. Thus, at the time of purchase, the forward contract has value zero. After purchase, the value of the contract is typically nonzero. In particular, just before the delivery date T, a forward contract bought at time zero has value

$$V^{FC}(T-; f(0)) = B - f(0) = B - \frac{E_0[B/S_0(T)]}{E_0[1/S_0(T)]}.$$

As the value of a forward contract moves away from zero, one of the parties to the contract might become concerned about the possibility of default by the other party. The worried party might wish to see the other party deposit money into an escrow account. Any such stipulation would, of course, change the nature of the contract and render inappropriate the reasoning that led to the forward price formula (3.4).

This leads us to the concept of the *futures price* $\varphi(t)$ for an asset with market value B at time T. This futures price process is set such that at every time $t \in [0, T)$, the futures contract has value zero. Suppose one party sells (for \$0) such a contract to another party at time t. At time $t + dt$, the futures price has moved by an amount $\varphi(t + dt) - \varphi(t)$. According to the provisions of the contract, if $\varphi(t + dt) - \varphi(t)$ is positive, the party holding the short position must transfer this amount of money to the party holding the long position. If $\varphi(t + dt) - \varphi(t)$ is negative, the transfer of money is in the other direction. This way, the futures contract is "continuously resettled"[†] and the value of the contract is always zero. The futures price for an asset must agree with the market price on the delivery date. We are now ready for a precise definition.

Definition 3.6: Let B be an $\mathcal{F}(T)$-measurable random variable, and let $\varphi(\cdot)$ be a European contingent claim whose value is zero for all $t \in [0, T]$ almost surely, and that satisfies $\varphi(T) = B$. Then we say that $\varphi(\cdot)$ is a *futures price process* for B.

Theorem 3.7: *Let B be an $\mathcal{F}(T)$-measurable random variable satisfying $E_0(B^2) < \infty$, and assume that $S_0(\cdot)$ is bounded from above and away from zero, uniformly in $(t, \omega) \in [0, T] \times \Omega$. Then the futures price process for B exists, is unique, and is given by*

$$\varphi(t) = E_0[B|\mathcal{F}(t)], \quad 0 \le t \le T. \tag{3.9}$$

[†]The mechanism for the transfer of money is provided by the margin accounts set up by brokers dealing in futures. Money is credited or debited to these accounts daily, depending on the movement of futures prices. Investors can withdraw money when the balance exceeds a certain threshold, and are subject to a margin call if the balance falls too low.

PROOF. We first show that the square-integrable P_0-martingale $\varphi(\cdot)$ defined by (3.9) satisfies the conditions of Definition 3.6. Define

$$I(t) \triangleq \int_0^t \frac{1}{S_0(u)} \, d\varphi(u), \quad 0 \leq t \leq T, \tag{3.10}$$

which is also a square-integrable martingale under P_0. From Proposition 2.3, the value process for $\varphi(\cdot)$ is

$$V^\varphi(t) = S_0(t) \cdot E_0[I(T) - I(t)|\mathcal{F}(t)] = 0, \quad 0 \leq t \leq T. \tag{3.11}$$

It is obvious that $\varphi(T) = B$.

We next prove uniqueness. Suppose $\varphi(\cdot)$ is any ECC satisfying the conditions of Definition 3.6. With $I(\cdot)$ as in (3.10) we have (3.11), which implies that $I(\cdot)$ is a martingale under P_0. Hence $\varphi(t) = \int_0^t S_0(u) dI(u), 0 \leq t \leq T$ is also a local martingale, so $\varphi^{fv}(\cdot) \equiv 0, \varphi^{\ell c}(\cdot) = \varphi(\cdot)$, and

$$E_0 \langle I \rangle (T) = E_0 \int_0^T \frac{1}{S_0^2(u)} d\langle \varphi \rangle(u) < \infty$$

by (1.5.9). Since $I(\cdot)$ is a square-integrable martingale, $\varphi(\cdot)$ is also. In particular, $\varphi(t) = E_0[\varphi(T)|\mathcal{F}(t)] = E_0[B|\mathcal{F}(t)], 0 \leq t \leq T$. □

Remark 3.8: Because the value of a futures contract is always zero, an investor who holds a position in futures can "close out" that position at any time and at no cost. This is in fact the fate of most futures contracts; the position is closed out before maturity. If a long position in a futures contract is not closed out prior to maturity T, then in actual markets the investor must receive delivery of the asset at market price B. Since purchasing the futures at some time $t \in [0, T]$, the investor has received a total cash flow of $\int_t^T d\varphi(s) = B - \varphi(t)$, and so after the terminal payment of B the investor has paid the net amount $\varphi(t)$ between times t and T. In this sense, the investor has purchased the asset for the futures price $\varphi(t)$ prevailing at the time the futures contract was entered. However, the payment of $\varphi(t)$ occurs continuously prior to maturity, whereas for a forward contract the payment occurs at maturity.

It is common in finance to approximate futures prices by forward prices. The relationship between these two quantities is described in the following corollary.

Corollary 3.9 (Forward-futures spread):   *Under the conditions of Theorem 3.7, we have*

$$f(t) = \varphi(t) + \frac{\text{Cov}_0[B, 1/S_0(T)|\mathcal{F}(t)]}{E_0[1/S_0(T)|\mathcal{F}(t)]}, \quad 0 \leq t \leq T,$$

where

$$\text{Cov}_0[X, Y|\mathcal{F}(t)] \triangleq E_0 \left\{ [X - E_0(X|\mathcal{F}(t))] \cdot [Y - E_0(Y|\mathcal{F}(t))] \, \Big| \, \mathcal{F}(t) \right\}.$$

In particular, the forward price $f(0)$ agrees with the future price $\varphi(0)$ if and only if B and $1/S_0(T)$ are uncorrelated under P_0.

PROOF.   Because

$$\mathrm{Cov}_0[X, Y|\mathcal{F}(t)] = E_0[XY|\mathcal{F}(t)] - E_0[X|\mathcal{F}(t)] \cdot E_0[Y|\mathcal{F}(t)],$$

we may rewrite (3.4) as

$$f(t) = E_0[B|\mathcal{F}(t)] + \frac{\mathrm{Cov}_0[B, 1/S_0(T)|\mathcal{F}(t)]}{E_0[1/S_0(T)|\mathcal{F}(t)]}.$$

The result follows from Theorem 3.7.   □

Clearly, $f(t) = \varphi(t) \forall 0 \leq t \leq T$ if the money-market price $S_0(T)$ is deterministic (nonrandom).

2.4 European Options in a Constant-Coefficient Market

In this section we present examples of European options in the context of a complete, standard market with *constant* risk-free rate $r(\cdot) \equiv r$, dividend rate vector $\delta(\cdot) \equiv \delta = (\delta_1, \ldots, \delta_N)'$, volatility matrix $\sigma(\cdot) \equiv \sigma$, and with $A(\cdot) \equiv 0$:

$$dS_0(t) = S_0(t)r\, dt \tag{4.1}$$

$$dS_n(t) = S_n(t) \left[b_n(t)\, dt + \sum_{d=1}^{N} \sigma_{nd}\, dW^{(d)}(t) \right]$$

$$= S_n(t) \left[(r - \delta_n)\, dt + \sum_{d=1}^{N} \sigma_{nd}\, dW_0^{(d)}(t) \right], \quad n = 1, \ldots, N \tag{4.2}$$

(see (1.5.6) and (1.6.16)). One of these is the Black–Scholes formula. Completeness of the market is equivalent to nonsingularity of σ. For this analysis, it is *not* necessary to assume that the vector $b(\cdot)$ of mean rates of return is constant, and consequently, the market price of risk process

$$\theta(t) = \sigma^{-1}[b(t) + \delta - r\underline{1}] \tag{4.3}$$

is not necessarily constant either.

Solving (4.1), (4.2), we obtain $S_0(t) = e^{rt}$ and

$$S_n(u) = h_n\left(u - t, S(t), \sigma(W_0(u) - W_0(t))\right), \quad 0 \leq t \leq u \leq T,$$
$$n = 1, \ldots, N, \tag{4.4}$$

where $h : [0, \infty) \times \mathbb{R}_+^N \times \mathbb{R}^N \to \mathbb{R}_+^N$ is the function defined by

$$h_n(t, p, y) \overset{\triangle}{=} p_n \exp\left[\left(r - \delta_n - \frac{1}{2} a_{nn} \right) t + y_n \right], \quad n = 1, \ldots, N, \tag{4.5}$$

and $a \stackrel{\triangle}{=} (a_{n\ell})_{1 \le n \le \ell \le N} = \sigma\sigma'$. Here we denote by $S(t) = (S_1(t), \ldots, S_N(t))'$ the vector of stock prices.

Consider an ECC of the type $C(t) = 0, 0 \le t < T$, and $C(T) = \varphi(S(T))$; here $\varphi : \mathbb{R}_+^N \to \mathbb{R}$ is a continuous function satisfying $E_0|\varphi(S(T))| < \infty$. According to Proposition 2.3, the value process for this claim is

$$
\begin{aligned}
V^{ECC}(t) &= e^{-r(T-t)} E_0[\varphi(S(T))|\mathcal{F}(t)] \\
&= e^{-r(T-t)} E_0 \left[\varphi\left(h(T-t, S(t), \sigma(W_0(T) - W_0(t)))\right) \Big| \mathcal{F}(t) \right] \\
&= e^{-r(T-t)} \int_{\mathbb{R}^N} \varphi(h(T-t, S(t), \sigma z)) \frac{1}{(2\pi(T-t))^{N/2}} \\
&\qquad \exp\left\{ -\frac{\|z\|^2}{2(T-t)} \right\} dz
\end{aligned}
$$

because $W_0(\cdot)$ is a Brownian motion under P_0, relative to the filtration $\{\mathcal{F}(t)\}_{0 \le t \le T}$. From these considerations, we see that with

$$
u(s, x) \stackrel{\triangle}{=}
\begin{cases}
e^{-rt} \int_{\mathbb{R}^N} \varphi(h(s, x, \sigma z)) \frac{1}{(2\pi s)^{N/2}} \exp\{-\frac{\|z\|^2}{2s}\} dz, & s > 0, x \in \mathbb{R}_+^N, \\
\varphi(x), & s = 0, x \in \mathbb{R}_+^N,
\end{cases}
$$
(4.6)

the value process of the ECC is

$$
V^{ECC}(t) = e^{-r(T-t)} E_0[\varphi(S(T))|\mathcal{F}(t)] = u(T-t, S(t)), \quad 0 \le t \le T. \quad (4.7)
$$

Using Remark 2.7, it is possible to compute the hedging portfolio. Indeed, under appropriate growth conditions on the function φ (e.g., polynomial growth in both $\|x\|$ and $1/\|x\|$), the function u of (4.6) is the unique classical solution of the Cauchy problem

$$
\frac{1}{2} \sum_{n=1}^N \sum_{\ell=1}^N a_{n\ell} x_n x_\ell \frac{\partial^2 u}{\partial x_n \partial x_\ell}
$$
$$
+ \sum_{n=1}^N (r - \delta_n) x_n \frac{\partial u}{\partial x_n} - ru = \frac{\partial u}{\partial s}, \quad \text{on } (0, T] \times \mathbb{R}_+^N,
$$
$$
u(0, x) = \varphi(x), \quad \forall x \in \mathbb{R}_+^N, \quad (4.8)
$$

by the Feynman–Kac theorem (e.g., Karatzas and Shreve (1991), Theorem 5.7.6 and Remark 5.7.8). Applying Itô's rule and invoking the second representation of $dS_n(t)$ in (4.2), we obtain

$$
du(T-t, S(t)) = ru(T-t, S(t)) \, dt + \sum_{n=1}^N \sum_{\ell=1}^N \sigma_{n\ell} S_n(t) \frac{\partial u}{\partial x_n}(T-t, S(t)) \, dW_0^{(\ell)}(t)
$$

or equivalently,

$$
S_0(t) \cdot d_t \left(\frac{u(T-t, S(t))}{S_0(t)} \right) = \hat{\pi}'(t) \sigma \, dW_0(t)
$$

with $\hat{\pi}(\cdot) = (\hat{\pi}_1(\cdot), \ldots, \hat{\pi}_N(\cdot))'$ given by

$$\hat{\pi}_n(t) = S_n(t) \frac{\partial u}{\partial x_n}(T - t, S(t)), \quad 0 \le t < T, \tag{4.9}$$

for $n = 1, \ldots, N$. Recalling (4.7) and Remark 2.7, we conclude that $\hat{\pi}(\cdot)$ is indeed the hedging portfolio of Definition 2.6. At any time t, this portfolio holds $\frac{\partial u}{\partial x_n}(T - t, S(t))$ shares of the nth stock, $n = 1, \ldots, N$. The hedging portfolio also has a component

$$\hat{\pi}_0(t) = u(T - t, S(t)) - \sum_{n=1}^{N} \hat{\pi}_n(t), \quad 0 \le t < T, \tag{4.10}$$

recording holdings in the money market.

It should be noted in (4.6), (4.7) and (4.9), (4.10) that the value of the ECC and the hedging portfolio depend on r, δ, and σ, but not on the vector $b(\cdot)$ of mean rates of return of the stocks. This fact makes the formulae particularly attractive, because the mean rates of return can be difficult to estimate in practice.

Example 4.1 *(European call option):*   A European call option on the first stock in our market is the ECC given by $C(t) = 0$, $\quad 0 \le t < T$ and $C(T) = (S_1(T) - q)^+$. The nonrandom constant $q > 0$ is called the *strike price*, and T is the *expiration date*. The random variable $(S_1(T) - q)^+$ is the value at time T of the option to buy one share of the first stock at the (contractually specified) price q. If $S_1(T) > q$, this option should be exercised by its holder; the stock can be resold immediately at the market price, at a profit of $S_1(T) - q$. If $S_1(T) < q$, the option should not be exercised; it is worthless to its holder.

With $\varphi(x) \stackrel{\triangle}{=} (x_1 - q)^+$, the Gaussian integration in (4.6) can be carried out explicitly, to yield

$$u^{ECC}(s, x_1; q) = \begin{cases} x_1 e^{-\delta_1 s} \Phi(\rho_+(s, x_1; q)) \\ \quad - q e^{-rs} \Phi(\rho_-(s, x_1; q)), & 0 < s \le T, \quad x_1 \in (0, \infty), \\ (x_1 - q)^+, & s = 0, \quad x_1 \in (0, \infty), \end{cases} \tag{4.11}$$

where

$$\rho_\pm(s, x_1; q) \stackrel{\triangle}{=} \frac{1}{\sqrt{sa_{11}}} \left[\log\left(\frac{x_1}{q}\right) + \left(r - \delta_1 \pm \frac{1}{2}a_{11}\right)s \right],$$

$$\Phi(z) \stackrel{\triangle}{=} \frac{1}{\sqrt{2\pi}} \int_{-\infty}^{z} e^{-u^2/2}\, du.$$

This is the celebrated *Black and Scholes (1973) option pricing formula*. The hedging portfolio is (see (4.9), (4.10))

$$\pi_1^{ECC}(t) = S_1(t)\frac{\partial V^{ECC}}{\partial x_1}(T - t, S_1(t); q), \quad \pi_2^{ECC}(t) = \cdots = \pi_N^{ECC}(t) = 0,$$

$$\pi_0^{ECC}(t) = V^{ECC}(T - t, S_1(t); q) - S_1(t)\frac{\partial V^{ECC}}{\partial x_1}(T - t, S_1(t)q), 0 \leq t < T.$$

$$(4.12)$$

Exercise 4.2: Show that the function $u(s, x_1) \equiv u^{ECC}(s, x_1; q)$ of (4.11) satisfies

$$x_1 \cdot \frac{\partial u}{\partial x_1}(s, x_1) \geq u(s, x_1), \quad 0 < s \leq T, \quad x_1 \in (0, \infty),$$

and hence that we have

$$\pi_0^{ECC}(t) \leq 0, \quad 0 \leq t < T, \tag{4.12'}$$

from (4.12). In other words, *the hedging portfolio for a (short position in a) European call option always borrows.*

Example 4.3 *(European put option):*   The European put option confers to its holder the right to sell a stock at a future time at a prespecified price. We model a put on the first stock as the ECC with $C(t) = 0$ for $0 \leq t < T$ and $C(T) = (q - S_1(T))^+$. Because $(q - S_1(T))^+ = -(S_1(T) - q) + (S_1(T) - q)^+$, holding a long position in a European put is equivalent to holding simultaneously a short position in a forward contract and a long position in a European call. This is the so-called *put–call parity* relationship. We have already priced the European call; the forward contract is easily priced, as we now describe.

First note from Remark 1.5.11 that

$$e^{-(r-\delta_1)t}S_1(t) = S_1(0) + \int_0^t e^{-(r-\delta_1)u}[dS_1(u) - (r - \delta_1)S_1(u)\,du]$$

$$= S_1(0) + \int_0^t e^{\delta_1 u}\left[d\left(\frac{S_1(u)}{S_0(u)}\right) + \frac{S_1(u)}{S_0(u)}\delta_1\,du\right]$$

is a martingale under P_0. Consequently, the forward contract value process (see (3.1)) is

$$S_0(t) \cdot E_0\left[\frac{S_1(T) - q}{S_0(T)} \,\middle|\, \mathcal{F}(t)\right] = e^{-\delta_1(T-t)}S_1(t) - e^{-r(T-t)}q$$

$$= u^{FC}(T - t, S_1(t); q),$$

where

$$u^{FC}(s, x_1; q) \stackrel{\triangle}{=} e^{-\delta_1 s}x_1 - e^{-rs}q. \tag{4.13}$$

The hedging portfolio is (see (4.9), (4.10))

$$\pi_1^{FC}(t) = e^{-\delta_1(T-t)}S_1(t), \quad \pi_2^{FC}(t) = \cdots = \pi_N^{FC}(t) = 0,$$

$$\pi_0^{FC}(t) = -e^{-r(T-t)}q, \quad 0 \leq t < T. \tag{4.14}$$

In other words, at time $t = 0$ the contract is sold for $e^{-\delta_1 T} S_1(0) - e^{-rT} q$. The seller invests the proceeds in the first stock and in the money market, respectively, buying $e^{-\delta_1 T}$ shares of stock and borrowing $e^{-rT} q$ from the money market. Dividends are reinvested in the stock. The yield per share of the stock is $Y_1(\cdot)$ defined by (1.11), and the number of shares owned is $\frac{\pi_1^{FC}(\cdot)}{S_1(\cdot)}$, so we have the stochastic differential equation

$$d\pi_1^{FC}(t) = \frac{\pi_1^{FC}(t)}{S_1(t)} dY_1(t) = \pi_1^{FC}(t) \left[(b_1 + \delta_1) dt + \sum_{d=1}^{N} \sigma_{nd} \, dW^{(d)}(t) \right],$$

whose solution is

$$\pi_1^{FC}(t) = \pi_1^{FC}(0) e^{\delta_1 t} \frac{S_1(t)}{S_1(0)} = e^{-\delta_1(T-t)} S_1(t).$$

At time T, the agent owns one share of the stock, which he delivers at price q. This q is the amount of money needed to pay off the debt to the money market.

Returning to the European put option, we define

$$u^{EP}(s, x_1; q) \triangleq -u^{FC}(s, x_1; q) + u^{EC}(s, x_1; q). \qquad (4.15)$$

The value process for the European put is $u^{EP}(T-t, S_1(t); q)$. The hedging portfolio is

$$\pi_n^{EP}(t) = -\pi_n^{FC}(t) + \pi_n^{EC}(t), \quad 0 \leq t < T, \quad n = 0, 1, \ldots, N. \qquad (4.16)$$

Example 4.4: Consider an ECC of the form $C(t) = 0, \quad 0 \leq t < T$ and $C(T) = \varphi(S_1(T))$, where $\varphi : [0, \infty) \to \mathbb{R}$ is a convex function. Such a function has a nondecreasing, right-continuous derivative $D^+\varphi$ satisfying

$$\varphi(x) = \varphi(0) + \int_0^x D^+\varphi(q) \, dq, \quad x \geq 0 \qquad (4.17)$$

(e.g., Karatzas and Shreve (1991), Problems 3.6.20 and 3.6.21).

We establish an integration-by-parts formula for $D^+\varphi$. There is a unique measure μ on the Borel subsets of $[0, \infty)$ characterized by

$$\mu((a, b]) = D^+\varphi(b) - D^+\varphi(a), \quad 0 \leq a \leq b.$$

Let $\rho : \mathbb{R} \to [0, \infty)$ be a function of class C^∞, with support in $[0, 1]$ and satisfying $\int_0^1 \rho(z) dz = 1$. Define a sequence of mollifications of φ by

$$\varphi_n(q) = \int_0^1 \varphi \left(q + \frac{z}{n} \right) \rho(z) \, dz = n \int_{-\infty}^{\infty} \varphi(y) \rho(ny - nq) \, dy.$$

Then each φ_n is of class C^∞, and $\lim_{n\to\infty} \varphi_n(q) = \varphi(q), \lim_{n\to\infty} \varphi_n'(q) = D^+\varphi(q)$ for all $q \geq 0$. Furthermore, for any bounded, Borel-measurable function $g : [0, \infty) \to \mathbb{R}$, we have

$$\lim_{n \to \infty} \int_a^b g(q) \, \varphi_n''(q) \, dq = \int_{(a,b]} g(q) \mu(dq), \quad 0 \le a \le b,$$

a fact that can be proved by first considering functions g that are indicators of intervals. If g is a $C^1(\mathbb{R})$-function, we may let $n \to \infty$ in the integration-by-parts formula

$$g(b)\varphi_n'(b) - g(a)\varphi_n'(a) - \int_a^b g'(q)\varphi_n'(q) \, dq = \int_a^b g(q)\varphi_n''(q) \, dq$$

to obtain

$$g(b)D^+\varphi(b) - g(a)D^+\varphi(a) - \int_a^b g'(q)D^+\varphi(q) \, dq = \int_a^b g(q)\mu \, (dq). \quad (4.18)$$

We now fix $x_1 \ge 0$ and apply (4.18) with $g(q) = (x_1 - q)^+, a = 0$ and $b = x_1$, to obtain

$$\varphi(x_1) - \varphi(0) - x_1 D^+\varphi(0) = \int_{(0,\infty)} (x_1 - q)^+ \mu \, (dq) \quad (4.19)$$

in conjunction with (4.17). Formula (4.19) allows us to compute the value process for the contingent claim $C(T) = \varphi(S_1(T))$ at the beginning of this example. Indeed, this contingent claim has the value process

$$e^{-r(T-t)} E_0[\varphi(S(T))|\mathcal{F}(t)] = e^{-r(T-t)} E_0\left[\varphi(0) + D^+\varphi(0) \cdot S_1(T)\right.$$

$$\left. + \int_{(0,\infty)} (S_1(T) - q)^+ \mu(dq) \middle| \mathcal{F}(t)\right]$$

$$= e^{-r(T-t)}\varphi(0) + D^+\varphi(0) \cdot u^{FC}(T - t, S_1(t); 0)$$

$$+ \int_{(0,\infty)} u^{EC}(T - t, S_1(t); q)\mu(dq)$$

(cf. (4.7)), where u^{FC} is defined by (4.13) and u^{EC} by (4.11).

Example 4.5 *(A path-dependent option):*   Let us take $N = 1$, $\sigma = \sigma_{11} > 0$ and assume that $b_1(\cdot)$ is deterministic, so the one-dimensional Brownian motions $W(\cdot)$ and $W_0(\cdot)$ both generate the filtration $\{\mathcal{F}(t)\}_{0 \le t \le T}$. We assume without loss of generality that we are on the canonical space $\Omega = C([0, T])$, the space of real-valued continuous functions on $[0, T]$, and that W_0 is the coordinate mapping process $W_0(t) = \omega(t), 0 \le t \le T$, for all $\omega \in \Omega$.

Consider an ECC of the form $C(t) = 0$ for $0 \le t < T$ and $C(T) = G(\omega)$, where $G : C([0, T]) \to \mathbb{R}$ is a functional satisfying under P_0 the conditions (E.4)–(E.6) of Appendix E. Then from the Clark formula (E.7), the value process (2.9) for the above ECC is

$$V^{ECC}(t) = e^{-r(T-t)}E_0[G(W_0)|\mathcal{F}(t)]$$
$$= e^{-r(T-t)}E_0G(W_0)$$
$$+ e^{-r(T-t)}\int_0^t E_0[\partial G(W_0;(s,T])|\mathcal{F}(s)]dW_0(s).$$

From Remark 2.7 we see that the hedging portfolio is

$$\pi_1(t) = \frac{1}{\sigma}e^{-r(T-t)}E_0[\partial G(W_0;(t,T])|\mathcal{F}(t)],$$
$$\pi_0(t) = V^{ECC}(t) - \pi_1(t).$$

In particular, if

$$C(T) = \max_{0\leq t\leq T} S_1(t) = S_1(0) \cdot \max_{0\leq t\leq T}\left\{\exp\left[\left(r - \frac{\sigma^2}{2}\right)t + \sigma W_0(t)\right]\right\},$$
$$(4.20)$$

we have a so-called *look-back option (LBO)*.

For the look-back option of (4.20), we have $G(W_0) = S_1(0) \cdot \exp[\sigma\max_{0\leq t\leq T}(W_0(t) + \nu t)]$, where $\nu \overset{\triangle}{=} \frac{r}{\sigma} - \frac{\sigma}{2}$. With $\tilde{W}(t) \overset{\triangle}{=} W_0(t) + \nu t$, $\tilde{M}(t) \overset{\triangle}{=} \max_{0\leq s\leq t}\tilde{W}(s)$, we obtain from Example E.5 of Appendix E:

$$\pi_1(t) = e^{-r(T-t)}S_1(0)\left[e^{\sigma\tilde{M}(t)}f(T-t,\tilde{M}(t)-\tilde{W}(t))\right.$$

$$\left. + \sigma e^{\sigma\tilde{W}(t)}\int_{\tilde{M}(t)-\tilde{W}(t)}^{\infty}f(T-t,\xi)e^{\sigma\xi}d\xi\right],$$
$$V^{LBO}(t) = e^{-r(T-t)}E_0[G(W_0)|\mathcal{F}(t)]$$

$$= e^{-r(T-t)}S_1(0)\left[e^{\sigma\tilde{M}(t)} + \sigma e^{\sigma\tilde{W}(t)}\int_{\tilde{M}(t)-\tilde{W}(t)}^{\infty}f(T-t,\xi)e^{\sigma\xi}d\xi\right],$$

where f is defined in (E.11). We deduce from this last formula that the look-back option of (4.20) has value

$$V^{LBO}(0) = e^{-rT}S_1(0)\left\{1 + \sigma\int_0^{\infty}e^{\sigma b}f(T,b)db\right\}$$

$$= e^{-rT}S_1(0)\left\{1 + \sigma\int_0^{\infty}e^{\sigma b}\left[1 - \Phi\left(\frac{b - \nu T}{\sqrt{T}}\right)\right]db\right.$$

$$\left. + \sigma\int_0^{\infty}e^{b(2\nu+\sigma)}\left[1 - \Phi\left(\frac{b + \nu T}{\sqrt{T}}\right)\right]db\right\}$$

at time $t = 0$.

2.5 American Contingent Claims

American contingent claims differ from European contingent claims in that the buyer of an American contingent claim can opt, *at any time of his choice*, for a lump-sum settlement of the claim. The amount of this settlement is specified by a stochastic process, which is part of the claim's description. In this section, we define the value of an American contingent claim and characterize it in terms of the Snell envelope of the discounted payoff for the claim.

Definition 5.1: An *American contingent claim (ACC)* consists of a *cumulative income process* $C(\cdot)$ satisfying $C(0) = 0$ almost surely, and of an $\{\mathcal{F}(t)\}$-adapted, RCLL *lump-sum settlement process* $L(\cdot)$. We assume that the *discounted payoff process*

$$Y(t) \triangleq \int_{(0,t]} \frac{dC(u)}{S_0(u)} + \frac{L(t)}{S_0(t)}, \quad 0 \le t \le T, \tag{5.1}$$

is *bounded from below*, uniformly in $t \in [0,T]$ and $\omega \in \Omega$, and *continuous* (jumps in $C(\cdot)$ and $L(\cdot)$ occur at the same time and offset one another, i.e., they are of equal size and opposite direction), and satisfies

$$E_0 \left[\sup_{0 \le t \le T} Y(t) \right] < \infty. \tag{5.2}$$

Just as with a European contingent claim, the *buyer* (or *holder*) of an ACC, who is said to assume a *long position* in the claim, pays some nonrandom amount γ at the initial time and is thereby entitled to the cumulative income process $C(\cdot)$. The *seller*, who is said to assume a *short position*, receives γ at time zero and provides $C(\cdot)$ to the buyer. Here, however, the buyer also gets to choose an $\{\mathcal{F}(t)\}$-stopping time $\tau : \Omega \to [0,T]$, called the *exercise time*. At time τ, the buyer forgoes all future income from $C(\cdot)$ and receives instead the lump-sum settlement $L(\tau)$. Thus, once τ is chosen, the cumulative income process to the seller is

$$\Gamma(t) = \gamma - C(t \wedge \tau) - L(\tau)1_{\{t \ge \tau\}}, \quad 0 \le t \le T. \tag{5.3}$$

In particular, on the event $\{\tau = 0\}$, the seller receives $\gamma(0) = \gamma - \Gamma(0)$ at the initial time and nothing more. The buyer has cumulative income process $-\Gamma(\cdot)$.

For many American contingent claims, $C(\cdot) \equiv 0$. This is the case, for example, with American options. An *American call option* entitles its holder to buy one share of a stock, say the first one, at any time prior to T at a specified strike price $q > 0$; it is modeled by setting $C(\cdot) \equiv 0$, $L(t) = (S_1(t) - q)^+$.

A *prepayable mortgage* is a more complex American contingent claim. We model this by assigning the long position to the borrower, since the borrower gets to choose the time of prepayment. The original principal is $-\gamma$, $-C(t)$ is the cumulative mortgage payment made up to time t, $-L(t)$ is the remaining principal after any payment at time t, and T is the term of the mortgage. Every upward jump in $-C(\cdot)$ corresponds to a regularly scheduled payment and creates a downward jump of equal magnitude in $-L(\cdot)$, so the discounted payoff process $Y(\cdot)$ is continuous.

As with a European contingent claim, the seller of an ACC must choose a portfolio to hedge the risk associated with his short position. This hedging is complicated by his uncertainty about the exercise time τ appearing in (5.3). The simplest case is when $\tau = T$, because then the hedging and pricing are like those of a European claim. In this case, the seller's cumulative income process is given by $\gamma - C(t) - L(T)1_{\{t=T\}}$ for $0 \le t \le T$, and if $\pi(\cdot)$ is a martingale-generating portfolio, the wealth process is given by

$$\frac{X(t)}{S_0(t)} = \gamma - \int_{(0,t]} \frac{dC(u)}{S_0(u)} - \frac{L(T)}{S_0(T)}1_{\{t=T\}}$$

$$+ \int_0^t \frac{1}{S_0(u)}\pi'(u)\sigma(u)\,dW_0(u), \quad 0 \le t \le T \qquad (5.4)$$

(see (1.5.8)). The seller wants $X(T) \ge 0$, or equivalently,

$$Y(T) = \int_{(0,T]} \frac{dC(u)}{S_0(u)} + \frac{L(T)}{S_0(T)} \le \gamma + \int_0^T \frac{1}{S_0(u)}\pi'(u)\sigma(u)\,dW_0(u) \quad (5.5)$$

almost surely. To ensure that he can make the lump-sum payment if the buyer should stop prematurely, the seller also wants $X(t) \ge L(t)$ a.s. for $0 \le t < T$. This condition, coupled with (5.5), yields

$$Y(t) = \int_{(0,t]} \frac{dC(u)}{S_0(u)} + \frac{L(t)}{S_0(t)}$$

$$\le \gamma + \int_0^t \frac{1}{S_0(u)}\pi'(u)\sigma(u)\,dW_0(u) \quad a.s. \quad \forall\, t \in [0,T]. \qquad (5.6)$$

Suppose (5.6) holds. Since both sides are continuous in t, the probability-zero event on which the inequality is violated can be chosen not to depend on t. Therefore, the inequality holds if t is replaced by any random time τ taking values in $[0, T]$:

$$Y(\tau) = \int_{(0,\tau]} \frac{dC(u)}{S_0(u)} + \frac{L(\tau)}{S_0(\tau)} \le \gamma + \int_0^\tau \frac{1}{S_0(u)}\pi'(u)\sigma(u)\,dW_0(u) \quad a.s.$$

$$(5.6')$$

This is just the statement that the seller, with cumulative income process (5.3), has nonnegative wealth after settling the ACC at any exercise time τ chosen by the buyer.

Definition 5.2: Let $(C(\cdot), L(\cdot))$ be an American contingent claim. The *value of the claim at time zero is*

$$V^{ACC}(0) \stackrel{\triangle}{=} \inf\{\gamma \in \mathbb{R}; \text{ there exists a martingale-generating}$$

portfolio process $\pi(\cdot)$ satisfying (5.6)\}.

A *hedging portfolio* for $(C(\cdot), L(\cdot))$ is a martingale-generating portfolio process $\hat{\pi}(\cdot)$ satisfying (5.6) when $\gamma = V^{ACC}(0)$.

Theorem 5.3: *We have*

$$V^{ACC}(0) = \sup_{\tau \in \mathcal{S}_{0,T}} E_0 Y(\tau), \tag{5.7}$$

where $\mathcal{S}_{0,T}$ is the set of stopping times taking values in $[0, T]$. Furthermore, there is a stopping time τ^* attaining this supremum and there is a hedging portfolio $\hat{\pi}(\cdot)$ such that

$$Y(\tau^*) = V^{ACC}(0) + \int_0^{\tau^*} \frac{1}{S_0(u)} \hat{\pi}'(u)\sigma(u) \, dW_0(u) \quad a.s. \tag{5.8}$$

Remark 5.4: Equation (5.8) asserts that if the seller uses the hedging portfolio $\hat{\pi}(\cdot)$ and the buyer chooses the stopping time τ^*, then after settlement the seller has wealth $X(\tau^*) = 0$. The buyer, whose cumulative income process is the negative of that of the seller, can hedge his long position with $-\hat{\pi}(\cdot)$ and after settlement have wealth $-X(\tau^*) = 0$. The stopping time τ^* is an *optimal exercise time* for the buyer of the ACC.

PROOF OF THEOREM 5.3. By the addition of a constant, if necessary, the process $Y(\cdot)$ can be assumed nonnegative, and we can bring the results of Appendix D to bear. According to Theorem D.7, there is a P_0-supermartingale $\{\xi(t), \mathcal{F}(t); \ 0 \le t \le T\}$ with RCLL paths, called the *Snell envelope* of $Y(\cdot)$, such that

$$\xi(t) \ge Y(t) \text{ for all } t \in [0, T]$$

almost surely, and

$$\xi(v) = \text{ess sup}_{\tau \in \mathcal{S}_{v,T}} E_0[Y(\tau)|\mathcal{F}(v)] \quad a.s., \quad \forall v \in \mathcal{S}_{0,T}, \tag{5.9}$$

where $\mathcal{S}_{v,T} \stackrel{\triangle}{=} \{\tau \in \mathcal{S}_{0,T}; v \le \tau \le T \ a.s.\}$. In particular, $\xi(0) = \sup_{\tau \in \mathcal{S}_{0,T}} E_0 Y(\tau)$. According to Theorem D.12, the stopping time

$$\tau^* \stackrel{\triangle}{=} \inf\{t \in [0, T); \ \xi(t) = Y(t)\} \wedge T$$

satisfies $\xi(0) = E_0 Y(\tau^*)$.

Theorem D.13 asserts that $\xi(\cdot) = M(\cdot) - \Lambda(\cdot)$, where $M(\cdot)$ is a uniformly integrable RCLL martingale under P_0, and $\Lambda(\cdot)$ is an adapted, continuous, nondecreasing process with $\Lambda(0) = \Lambda(\tau^*) = 0$ a.s. Because of our assumption of market completeness, the $\mathcal{F}(T)$-measurable random variable $B \stackrel{\triangle}{=} S_0(T)M(T)$ is financeable; i.e., there is a martingale-generating portfolio process $\hat{\pi}(\cdot)$ satisfying

$$M(T) = \xi(0) + \int_0^T \frac{1}{S_0(u)} \hat{\pi}'(u)\sigma(u) \, dW_0(u) \tag{5.10}$$

(cf. Proposition 1.6.2). Taking conditional expectations with respect to $\mathcal{F}(t)$ in (5.10), we obtain

$$Y(t) \le \xi(t) = M(t) - \Lambda(t)$$

$$= \xi(0) - \Lambda(t) + \int_0^t \frac{1}{S_0(u)} \hat{\pi}'(u)\sigma(u) \, dW_0(u)$$

$$\le \xi(0) + \int_0^t \frac{1}{S_0(u)} \hat{\pi}'(u)\sigma(u) \, dW_0(u), \quad 0 \le t \le T. \tag{5.11}$$

It follows immediately that $V^{ACC}(0) \le \xi(0)$.

Now suppose that (5.6) is satisfied for some $\gamma \in \mathbb{R}$ and some martingale-generating portfolio process $\pi(\cdot)$; take expectations in (5.6') to obtain $E_0 Y(\tau) \le \gamma, \forall \tau \in \mathcal{S}_{0,T}$. It develops that $\xi(0) \le \gamma$, and thus $\xi(0) \le V^{ACC}(0)$.

Having thus established $\xi(0) = V^{ACC}(0)$, we see from (5.11) that $\hat{\pi}(\cdot)$ is a hedging portfolio and (5.8) holds. $\quad\square$

Remark 5.5: The martingale $M(\cdot)$ in the proof of Theorem 5.3 is actually continuous, as one can see by considering the second equality in (5.11). Hence, the Snell envelope $\xi(\cdot)$ is also continuous.

Remark 5.6: If $\gamma = V^{ACC}(0)$, and $\pi(\cdot) = \hat{\pi}(\cdot)$ is the hedging portfolio of Theorem 5.3, then (5.6) holds and implies (5.6') *for any random time* τ taking values in $[0, T]$. Thus, even if the purchaser of the contingent claim is allowed to choose τ with knowledge of future prices, the seller of the claim is not exposed to any risk if he uses the hedging portfolio of Theorem 5.3. $\quad\square$

We wish to extend the notion of the value of an ACC to times other than zero. Suppose that an ACC is given, and consider that at time $s \in [0, T]$ a buyer pays the amount $\gamma(s)$ (an $\mathcal{F}(s)$-measurable random variable) to receive the remaining income process $\{C(t) - C(s); \quad t \in [s, \tau]\}$ up to a stopping time $\tau \in \mathcal{S}_{s,T}$, at which time he receives the lump sum $L(\tau)$. The argument that led to (5.6) now leads to the condition for the seller's desired hedging:

$$Y(t) - \int_{(0,s]} \frac{dC(u)}{S_0(u)} = \int_{(s,t]} \frac{dC(u)}{S_0(u)} + \frac{L(t)}{S_0(t)}$$

$$\le \frac{\gamma(s)}{S_0(s)} + \int_s^t \frac{1}{S_0(u)} \pi'(u)\sigma(u) \, dW_0(u) \quad a.s.,$$

$$\forall t \in [s, T]. \tag{5.12}$$

Definition 5.7:   Let $(C(\cdot), L(\cdot))$ be an American contingent claim. The *value of the claim at time* $s \in [0, T]$, denoted by $V^{ACC}(s)$, is the smallest $\mathcal{F}(s)$-measurable random variable $\gamma(s)$ such that (5.12) is satisfied by some martingale-generating portfolio process $\pi(\cdot)$.

Theorem 5.8:   *For $s \in [0, T]$, we have*

$$V^{ACC}(s) = S_0(s) \left[\xi(s) - \int_{(0,s]} \frac{1}{S_0(u)} \, dC(u) \right], \tag{5.13}$$

where $\xi(\cdot)$ is the Snell envelope of $Y(\cdot)$ and satisfies (5.9). Furthermore, the stopping time

$$\tau_s^* \triangleq \inf\{t \in [s, T); \xi(t) = Y(t)\} \wedge T$$

satisfies $\xi(s) = E[Y(\tau_s^*)|\mathcal{F}(s)]$ a.s., and with $\hat{\pi}(\cdot)$ the hedging portfolio of Theorem 5.3, equality holds in (5.12) at τ_s^*:

$$Y(\tau_s^*) - \int_{(0,s]} \frac{dC(u)}{S_0(u)} = \frac{V^{ACC}(s)}{S_0(s)} + \int_s^{\tau_s^*} \frac{1}{S_0(u)} \hat{\pi}'(u)\sigma(u) \, dW_0(u) \quad a.s. \tag{5.14}$$

PROOF.   Replace t in (5.12) by an arbitrary $\tau \in \mathcal{S}_{s,T}$ and take conditional expectations, to obtain

$$E_0[Y(\tau)|\mathcal{F}(s)] - \int_{(0,s]} \frac{dC(u)}{S_0(u)} \leq \frac{\gamma(s)}{S_0(s)} \quad a.s.$$

and thus

$$\xi(s) - \int_{(0,s]} \frac{dC(u)}{S_0(u)} \leq \frac{V^{ACC}(s)}{S_0(s)} \quad a.s. \tag{5.15}$$

For the reverse inequality, let $t \in [s, T]$ be given and observe from (5.11) that

$$\xi(t) - \xi(s) = \int_s^t \frac{1}{S_0(u)} \hat{\pi}'(u)\sigma(u) \, dW_0(u) - [\Lambda(t) - \Lambda(s)].$$

Because $Y(t) \leq \xi(t)$ and $\Lambda(t) - \Lambda(s) \geq 0$, we have

$$Y(t) - \int_{(0,s]} \frac{dC(u)}{S_0(u)} \leq \xi(t) - \int_{(0,s]} \frac{dC(u)}{S_0(u)}$$

$$\leq \xi(s) - \int_{(0,s]} \frac{dC(u)}{S_0(u)} + \int_s^t \frac{1}{S_0(u)} \hat{\pi}'(u)\sigma(u) \, dW_0(u) \ a.s. \tag{5.16}$$

This shows that (5.12) is satisfied with

$$\frac{\gamma(s)}{S_0(s)} = \xi(s) - \int_{(0,s]} \frac{dC(u)}{S_0(u)},$$

whence

$$\frac{V^{ACC}(s)}{S_0(s)} \leq \xi(s) - \int_{(0,s]} \frac{dC(u)}{S_o(u)}.$$

Replacing t by τ_s^* in (5.16), we obtain equality because $Y(\tau_s^*) = \xi(\tau_s^*)$ and $\Lambda(\tau_s^*) - \Lambda(s) = 0$ (Theorem D.13, especially (D.34)). $\qquad\square$

Remark 5.9: Just as in Remark 5.4, we see here that the buyer of the ACC can hedge his position with the portfolio process $-\hat{\pi}(\cdot)$ if he calls for settlement at time τ_s^*. Equation (5.14) is the statement that after settlement, both buyer and seller will have zero wealth. The stopping time τ_s^* is an *optimal exercise time in $\mathcal{S}_{s,T}$* for the buyer of the ACC.

Remark 5.10: Let $(C(\cdot), L(\cdot))$ be an American contingent claim. If the buyer is forced to choose the exercise time $\tau = T$, then the value of the claim at time s is determined by formula (2.9) for European claims, namely

$$V^{ECC}(s) = S_0(s)E_0 \left[\int_{(s,T]} \frac{dC(u)}{S_0(u)} + \frac{L(T)}{S_0(T)} \,\Big|\, \mathcal{F}(s) \right], \quad 0 \leq s \leq T.$$

The difference between the "American value"

$$V^{ACC}(s) = S_0(s) \left[\xi(s) - \int_{(0,s]} \frac{dC(u)}{S_0(u)} \right], \quad 0 \leq s \leq T,$$

and this "European value" is called the *early exercise premium*

$$e(s) \stackrel{\triangle}{=} V^{ACC}(s) - V^{ECC}(s), \quad 0 \leq s \leq T. \tag{5.17}$$

Because

$$\int_0^T \frac{dC(u)}{S_0(u)} + \frac{L(T)}{S_0(T)} = Y(T) = \xi(T) = M(T) - \Lambda(T),$$

we have

$$\frac{e(s)}{S_0(s)} = \xi(s) - E_0[Y(T)|\mathcal{F}(s)] = E_0[\Lambda(T)|\mathcal{F}(s)] - \Lambda(s)$$

$$= E_0 \left[\int_s^T \frac{S_0(u)d\Lambda(u)}{S_0(u)} \,\Big|\, \mathcal{F}(s) \right], \quad 0 \leq s \leq T. \tag{5.18}$$

Thus, from (2.9) of Proposition 2.3, the early exercise premium is itself the value process for an ECC, namely, the cumulative income process $\int_0^t S_0(u)d\Lambda(u)$, $0 \leq t \leq T$. We shall offer some precise computations for $e(\cdot)$ and $\Lambda(\cdot)$ in the case of the American put option in Section 2.7.

Remark 5.11: Setting $s = 0$ in (5.18), we see that

$$e(0) = \sup_{\tau \in \mathcal{S}_{0,T}} E_0 Y(\tau) - E_0 Y(T).$$

It is an easy consequence of the optional sampling theorem (e.g., Karatzas and Shreve (1991), pp. 19–20) that *if the discounted payoff process* $Y(\cdot)$ *is a* P_0-*submartingale, then* $e(0) = 0$, *and the ACC of Definition 5.1 is equivalent to its ECC counterpart.* Then, of course, the exercise time $\tau = T$ is optimal for the holder of the ACC (although earlier exercise might also be optimal).

2.6   The American Call Option

This section develops a variety of results for American call options. We show that the value of an American call does not exceed the price of the underlying stock. If the call is *perpetual*, i.e., the expiration time is $T = \infty$, and the stock pays no dividends, then, regardless of the exercise price, the value of the call agrees with the price of the stock, but there is no optimal exercise time. If $T < \infty$ and the stock pays no dividends, the American call need not be exercised before maturity, and therefore its value is the same as that of a European call. All these facts are simple consequences of the optional sampling theorem.

The latter part of this section is devoted to a perpetual American call on a dividend-paying stock, when most of the coefficient processes in the model are constant. In this case, it is optimal to wait to exercise the call until the underlying stock price rises to a threshold, although this may never happen. The threshold is characterized in Theorem 6.7.

In this section we are concerned with an American call on a single stock, and we simplify notation by assuming that this is the only stock. Thus, we set $N = D = 1$ and we write $S(\cdot)$, $\sigma(\cdot)$, $b(\cdot)$, and $\delta(\cdot)$ in place of $S_1(\cdot)$, $\sigma_{11}(\cdot)$, $b_1(\cdot)$, and $\delta_1(\cdot)$.

To cast an American call option on the stock, with exercise price $q > 0$, into the framework of American contingent claims (Definition 5.1), we set $C(\cdot) \equiv 0$, $L(t) = (S(t) - q)^+$, and thus the discounted payoff process is

$$Y(t) = \frac{(S(t) - q)^+}{S_0(t)}. \tag{6.1}$$

For a finite planning horizon (*expiration date*) T, let us denote by $V^{AC}(t;T)$ the value of the American call at time $t \in [0, T]$. According to Theorem 5.8, we have

$$\frac{V^{AC}(t;T)}{S_0(t)} = \text{ess sup}_{\tau \in \mathcal{S}_{t,T}} E_0[Y(\tau)|\mathcal{F}(t)], \quad 0 \le t \le T, \tag{6.2}$$

where $\mathcal{S}_{t,T}$ is the set of $\{\mathcal{F}(t)\}$-stopping times taking values in $[t, T]$. We note from Remark 1.5.11 that when $\delta(\cdot) \ge 0$, then

$$\frac{S(t)}{S_0(t)} = S(0) \cdot \exp\left\{ \int_0^t \sigma(u) \, dW_0(u) - \int_0^t \left(\frac{1}{2}\sigma^2(u) + \delta(u) \right) du \right\},$$
$$0 \le t \le T, \tag{6.3}$$

is a nonnegative P_0-supermartingale, and the optional sampling theorem (Karatzas and Shreve (1991), Theorem 1.3.22) implies

$$V^{AC}(t;T) = S_0(t) \cdot \operatorname{ess\,sup}_{\tau \in \mathcal{S}_{t,T}} E_0\left[\left(\frac{S(\tau)}{S_0(\tau)} - \frac{q}{S_0(\tau)}\right)^+ \Bigg| \mathcal{F}(t)\right]$$

$$\leq S_0(t) \cdot \operatorname{ess\,sup}_{\tau \in \mathcal{S}_{t,T}} E_0\left[\frac{S(\tau)}{S_0(\tau)} \Bigg| \mathcal{F}(t)\right]$$

$$\leq S(t), \quad 0 \leq t \leq T.$$

In other words, *the value of an American call with finite expiration date never exceeds the price of the underlying stock.*

Theorem 6.1: *With $T < \infty$, assume that the stock pays no dividends, $S_0(\cdot)$ is almost surely nondecreasing, and $\sigma(\cdot)$ satisfies the Novikov condition (cf. (1.5.13))*

$$E_0\left[\exp\left\{\frac{1}{2}\int_0^T \sigma^2(u)\, du\right\}\right] < \infty. \tag{6.4}$$

Then an American call on the first stock need not be exercised before maturity, and its value is the same as that of a European call (cf. (2.13)):

$$V^{AC}(t;T) = S_0(t) E_0\left[\left(\frac{S(T)}{S_0(T)} - \frac{q}{S_0(T)}\right)^+ \Bigg| \mathcal{F}(t)\right], \quad 0 \leq t \leq T.$$

PROOF.   Jensen's inequality and the fact that $S_0(\cdot)$ is nondecreasing can be used to show that $Y(\cdot)$ of (6.1) is a P_0-submartingale. The result follows from Remark 5.11.   □

We now turn our attention to *perpetual American call options.* Let us assume that the financial market has an infinite planning horizon (Definition 1.7.2), and recall the discussion of such markets in Section 1.7.

Definition 6.2: Consider the discounted payoff process $Y(\cdot)$ of (6.1), defined for all $t \in [0,\infty)$. Following Definition 5.7, for $s \in [0,\infty)$ we define the *value at time s of the perpetual American call option on the stock*, denoted by $V^{AC}(s;\infty)$, to be the minimal random variable $\gamma(s)$ that is $\mathcal{F}^{(T)}(s)$-measurable for some $T \in [s,\infty)$ and for which there exists a martingale-generating portfolio process $\pi(\cdot)$ satisfying almost surely

$$Y(t) \leq \frac{\gamma(s)}{S_0(s)} + \int_s^t \frac{1}{S_0(u)}\pi'(u)\sigma(u)\, dW_0(u), \quad \forall t \in [s,\infty). \tag{6.5}$$

If (6.5) holds, then the continuity of both sides of (6.5) allows one to choose, for each $k \in \mathbb{N}$, a P-null and P_0-null set N_k in $\mathcal{F}^{(T_k)}(k)$ for some $T_k \in [k,\infty)$, such that (6.5) holds for all $t \in [s,k]$ and all $\omega \in \Omega\backslash N_k$. Consequently, (6.5) holds for all $t \in [s,\infty)$ and all $\omega \in \Omega\backslash\bigcup_{k=1}^{\infty} N_k$. Thus, if $\tau : \Omega \to [s, \infty]$ is any random time, we have (cf. (5.6'))

$$Y(\tau) \leq \frac{\gamma(s)}{S_0(s)} + \int_s^\tau \frac{1}{S_0(u)} \pi'(u)\sigma(u)\,dW_0(u), P \text{ and } P_0 \text{ a.s. on } \{\tau < \infty\}.$$

This is the assertion that an agent who sells the option for $\gamma(s)$ at time s and, having no further income, invests the amount $\pi(u)$ in the stock at all times $u \in [s, \infty)$, will have sufficient wealth to pay off the option if the buyer chooses to exercise it at any finite time.

Theorem 6.3: *Assume that the stock pays no dividends, that $S_0(\cdot)$ is almost surely nondecreasing with $P_0[\lim_{t\to\infty} S_0(t) = \infty] = 1$, and that $\sigma(\cdot)$ satisfies the Novikov condition (6.4). Then the value of the perpetual American call on the stock is equal to the current stock price:*

$$V^{AC}(s; \infty) = S(s), \quad 0 \leq s < \infty, \quad a.s.$$

PROOF. For $0 \leq s < t < \infty$, we have immediately from (6.1) and Remark 1.5.11 that

$$Y(t) \leq \frac{S(t)}{S_0(t)} = \frac{S(s)}{S_0(s)} + \int_s^t \frac{S(u)}{S_0(u)}\sigma(u)\,dW_0(u), \tag{6.6}$$

which suggests taking $\pi(u) = S(u)$, i.e., holding one share of stock at all times. This $\pi(\cdot)$ is martingale-generating, since $S(t)/S_0(t)$ is a martingale. Thus, from (6.6) and Definition 6.2, we obtain $\gamma(s) \leq S(s)$.

On the other hand, whenever (6.5) holds for *some* random variable $\gamma(s)$ and *some* martingale-generating portfolio $\pi(\cdot)$, we have for $t \in [s, \infty)$ and sufficiently large $T \in [t, \infty)$:

$$\frac{S(s)}{S_0(s)} - qE_0\left[\frac{1}{S_0(t)} \,\Big|\, \mathcal{F}^{(T)}(s)\right] = E_0\left[\frac{S(t)}{S_0(t)} - \frac{q}{S_0(t)} \,\Big|\, \mathcal{F}^{(T)}(s)\right]$$

$$\leq E_0[Y(t)|\mathcal{F}^{(T)}(s)] \leq \frac{\gamma(s)}{S_0(s)}, \quad s \leq t < \infty.$$

Letting $t \to \infty$ we obtain $S(s) \leq \gamma(s)$, and thus $S(s)$ is the minimal random variable that can replace $\gamma(s)$ in (6.5). $\qquad\square$

Remark 6.4: The proof of Theorem 6.3 reveals that the seller of the perpetual American call should hedge by holding one share of the stock, and when this hedging portfolio is used and $\gamma(s) = V^{AC}(s; T) = S(s)$, the inequality (6.5) becomes

$$\frac{S(t)}{S_0(t)} \geq Y(t) \triangleq \left(\frac{S(t)}{S_0(t)} - \frac{q}{S_0(t)}\right)^+.$$

This inequality is strict for all finite t, which means that *there is no optimal exercise time* (see Remarks 5.4, 5.9) for the purchaser of the option.

For the remainder of this section we shall assume that

$$\sigma(\cdot) \equiv \sigma > 0, \quad \delta(\cdot) \equiv \delta > 0, \quad r(\cdot) \equiv r > \delta \tag{6.7}$$

are constant, although $b(\cdot)$ may not be. Then the stock price process is

$$S(t) = S(0) \exp\left\{\sigma W(t) - \frac{1}{2}\sigma^2 t + \int_0^t b(u)\,du\right\}$$

$$= S(0) \exp\left\{\sigma W_0(t) + (r - \delta - \sigma^2/2)t\right\}. \tag{6.8}$$

We consider the *perpetual American call* on the stock with exercise price $q > 0$.

The proof of Theorem 5.3 cannot be easily adapted to the present situation, because here the expiration date is $T = \infty$; we do not have a filtration parametrized by $t \in [0, \infty]$ and satisfying the usual conditions of right-continuity and augmentation by null sets. Nonetheless, Theorem 5.3 suggests that the value $V^{AC}(t; \infty)$ of the perpetual call at time zero can be found by maximizing $E_0 Y(\tau)$ over "stopping times" τ, where

$$Y(t) \triangleq \begin{cases} e^{-rt}(S(t) - q)^+, & 0 \le t < \infty, \\ 0, & t = \infty. \end{cases} \tag{6.9}$$

It should be noted from (6.8) that $0 \le Y(t) \le e^{-rt}S(t) = S(0)\exp\{\sigma W_0(t) - (\delta + \frac{1}{2}\sigma^2)t\}$, $0 \le t < \infty$, whence

$$Y(\infty) = 0 = \lim_{t \to \infty} Y(t), \quad P_0\text{-}a.s. \tag{6.10}$$

From (6.8) we see that the process $S(\cdot)$ is Markovian under P_0, so we expect the maximizing "stopping time" to be a *hitting time*, i.e., to be of the form

$$H_a \triangleq \inf\{t \ge 0; S(t) \ge a\}$$

$$= \inf\left\{t \ge 0; W_0(t) + \left(\frac{r-\delta}{\sigma} - \frac{1}{2}\sigma\right)t \ge \frac{1}{\sigma}\log\left(\frac{a}{S(0)}\right)\right\} \tag{6.11}$$

for some $a \in (0, \infty)$. Because H_a is the first time the P_0-Brownian motion $W_0(t) + \nu t$ with drift $\nu \triangleq \frac{r-\delta}{\sigma} - \frac{1}{2}\sigma$ hits or exceeds the level $y \triangleq \frac{1}{\sigma}\log(\frac{a}{S(0)})$, we have $E_0 e^{-rH_a} = 1$ for $a \le S(0)$, and for $a > S(0)$ we have the following transform formula for the hitting time of (6.11) (Karatzas and Shreve (1991), Exercise 3.5.10):

$$E_0\left(e^{-rH_a}\right) = \exp\left[\nu y - y\sqrt{\nu^2 + 2r}\right] = \left(\frac{S(0)}{a}\right)^\gamma,$$

where $\gamma \triangleq \frac{1}{\sigma}[-\nu + \sqrt{\nu^2 + 2r}]$. Note that γ satisfies the quadratic equation

$$\frac{1}{2}\sigma^2\gamma^2 + \sigma\nu\gamma - r = 0 \tag{6.12}$$

and the inequalities

$$1 < \gamma < \frac{r}{r - \delta}. \tag{6.13}$$

Let us define

$$g_a(x) \triangleq \begin{cases} (a - q)^+(x/a)^\gamma, & 0 < x < a, \\ (x - q)^+, & 0 < a \le x. \end{cases}$$

It is now easily verified that

$$g_a(x) = E_0[e^{-rH_a}(S(H_a) - q)^+ | S(0) = x], \quad a > 0, \quad x > 0.$$

We want to maximize this quantity over $a > 0$.

Lemma 6.5: *We have*

$$g_a(x) \le g_b(x), \quad \forall x > 0, \quad a > 0,$$

where $b \overset{\triangle}{=} \gamma q/(\gamma - 1) \in (q, \infty)$.

PROOF.   The function $\phi(a) \overset{\triangle}{=} (a - q)/a^\gamma$ is increasing on $(0, b)$, decreasing on (b, ∞), and thus has its maximum on $(0, \infty)$ at b. Let $x > 0$, $a > 0$ be given. We have

$$a \le x, b \le x \Longrightarrow g_a(x) = (x - q)^+ = g_b(x);$$
$$a \le x < b \Longrightarrow g_a(x) = (x - q)^+ = (\phi(x))^+ x^\gamma \le \phi(b) x^\gamma = g_b(x);$$
$$a > x \ge b \Longrightarrow g_a(x) = (\phi(a))^+ x^\gamma \le (\phi(x))^+ x^\gamma = (x - q)^+ = g_b(x);$$
$$a > x, b > x \Longrightarrow g_a(x) = (\phi(a))^+ x^\gamma \le \phi(b) x^\gamma = g_b(x). \qquad \square$$

Lemma 6.6: *The function* $g = g_b$ *is convex, of class* $C^1((0, \infty)) \bigcap C^2$ *$((0, \infty) \backslash \{b\})$, and satisfies the variational inequality*

$$\max \left\{ \frac{1}{2} \sigma^2 x^2 g'' + (r - \delta) x g' - rg, f - g \right\} = 0 \quad \text{on} \quad (0, \infty) \backslash \{b\},$$

where $f(x) \overset{\triangle}{=} (x - q)^+$. *More precisely, we have*

$$\frac{1}{2} \sigma^2 x^2 g''(x) + (r - \delta) x g'(x) - rg(x) = \begin{cases} 0, & 0 < x < b, \\ -(\delta x - rq) < 0, & x > b, \end{cases}$$

$$\tag{6.14}$$

$$g(x) > f(x), \quad 0 < x < b, \tag{6.15}$$

$$g(x) = f(x), \quad x \ge b. \tag{6.16}$$

PROOF.   In order for g_a to be of class $C^1(0, \infty)$, we must have $a \ge q$. Even if $a \ge q$, we must also have equality between $g'_a(a-) = \frac{\gamma}{a}(a - q)$ and $g'_a(a+) = 1$. This equality is equivalent to $a = b$, so g_b is the *only* function in the family $\{g_a\}_{a>0}$ that is of class C^1. (Here we have an instance of the "*smooth fit*" condition, common in optimal stopping.)

The remainder of the lemma follows from straightforward computation, taking (6.12) into account. The positivity of $\delta x - rq$ when $x > b$ follows from (6.13). Inequality (6.15) follows from $\phi(b) > \phi(x)$, valid for $0 < x < b$, where ϕ is the function used in the proof of Lemma 6.5. $\square$

Theorem 6.7 (McKean (1965)): *Under the assumption (6.7), the value process for a perpetual American call option is given by*

$$V^{AC}(t; \infty) = g(S(t)), \quad 0 \le t < \infty,$$

where the function g is

$$g(x) = \begin{cases} (b-q)(\frac{x}{b})^\gamma, & 0 < x < b, \\ x - q, & x \geq b \end{cases} \tag{6.17}$$

and $\gamma \triangleq \frac{1}{\sigma}[\sqrt{\nu^2 + 2r} - \nu]$, $\nu \triangleq \frac{r-\delta}{\sigma} - \frac{1}{2}\sigma$, *and* $b \triangleq \frac{\gamma q}{\gamma - 1}$. *Furthermore,*

$$g(S(0)) = \sup_\tau E_0 Y(\tau) = \sup_\tau E_0[e^{-r\tau}(S(\tau) - q)^+], \tag{6.18}$$

where the supremum is over all random times τ satisfying

$$\forall t \in [0, \infty), \quad \exists T \in [t, \infty) \quad \text{such that} \quad \{\tau \leq t\} \in \mathcal{F}^{(T)}(t). \tag{6.19}$$

The random time

$$H_b \triangleq \inf\{t \geq 0; S(t) \geq b\}$$

satisfies (6.19) and attains the supremum in (6.18).

PROOF.   Itô's rule for convex functions (e.g., Karatzas and Shreve (1991), Theorem 3.6.22 and Problem 3.6.7(i)) implies

$$d(e^{-rt}g(S(t))) = e^{-rt}S(t)g'(S(t))\sigma\, dW_0(t) - e^{-rt}(\delta S(t) - rq)1_{\{S(t)>b\}}dt$$
$$= dM(t) - d\Lambda(t), \tag{6.20}$$

where

$$M(t) \triangleq \int_0^t e^{-ru}S(u)g'(S(u))\sigma\, dW_0(u) \tag{6.21}$$

is a P_0-*martingale (because g' is bounded), and*

$$\Lambda(t) \triangleq \int_0^t e^{-ru}(\delta S(u) - rq)1_{\{S(u)>b\}}du \tag{6.22}$$

is *nondecreasing* (because $x > b$ implies $\delta x - rq > 0$). For every random time τ satisfying (6.19) and every $t \in [0, \infty)$, we have then

$$g(S(0)) = E_0[e^{-r(\tau \wedge t)}g(S(\tau \wedge t))] + E_0\Lambda(\tau \wedge t)$$
$$\geq E_0[e^{-r(\tau \wedge t)}(S(\tau \wedge t) - q)^+], \tag{6.23}$$

where we have used (6.15) and (6.16).

To let $t \to \infty$ in (6.23), we need to dominate the right-hand side. But

$$\sup_{0 \leq t \leq \infty} Y(t) = \sup_{0 \leq t < \infty} [e^{-rt}(S(t) - q)^+]$$
$$\leq S(0)\exp\left[\sigma \cdot \sup_{0 \leq t < \infty}\left\{W_0(t) - \left(\frac{\delta}{\sigma} + \frac{\sigma}{2}\right)t\right\}\right]$$
$$\leq S(0)\exp\{\sigma W_*\},$$

where W_* is the maximum value attained by the P_0-Brownian motion $W_0(t) - \beta t$, $0 \leq t < \infty$, with negative drift, where $\beta = \frac{\delta}{\sigma} + \frac{\sigma}{2} > 0$. According to Karatzas and Shreve (1991), Exercise 3.5.9, W_* has the exponential distribution $P_0[W_* \in d\xi] = 2\beta e^{-2\beta\xi} d\xi$, $\xi > 0$, and consequently,

$$E_0 \left(\sup_{0 \le t \le \infty} Y(t) \right) = E_0 \left[\sup_{0 \le t < \infty} (e^{-rt}(S(t) - q)^+) \right]$$

$$\le S(0) \int_0^\infty 2\beta e^{(\sigma - 2\beta)\xi} \, d\xi = \frac{S(0)\beta\sigma}{\delta} < \infty. \quad (6.24)$$

Inequality (6.24) and the dominated convergence theorem allow us to let $t \to \infty$ in (6.23) and obtain, thanks to (6.10),

$$g(S(0)) \ge E_0[1_{\{\tau < \infty\}} e^{-r\tau}(S(\tau) - q)^+] = E_0 Y(\tau) \quad (6.25)$$

for every random time τ satisfying (6.19).

Let us now consider $\tau = H_b$. We have $\Lambda(H_b) = 0$ and $g(S(H_b)) = g(b) = (b - q)^+ = (S(H_b) - q)^+$ almost surely on $\{H_b < \infty\}$, so (6.23) becomes

$$g(S(0)) = E_0[1_{\{H_b \le t\}} e^{-rH_b}(S(H_b) - q)^+] + E_0[1_{\{H_b > t\}} e^{-rt} g(S(t))]. \quad (6.26)$$

But

$$E_0[1_{\{H_b > t\}} e^{-rt} g(S(t))] \le E_0 e^{-rt} S(t) = S(0)e^{-\delta t} \longrightarrow 0$$

as $t \to \infty$, so passage to the limit in (6.26) yields

$$g(S(0)) = E_0[1_{\{H_b < \infty\}} e^{-rH_b}(S(H_b) - q)^+] = E_0 Y(H_b), \quad (6.27)$$

thanks to (6.10). In conjunction with (6.25), this establishes (6.18) and shows that $\tau = H_b$ attains the supremum in (6.18).

The discounted payoff process for the American call is $Y(\cdot)$ of (6.9), and according to Definition 6.2 its *value at $t = 0$* is the minimal constant $\gamma(0)$ satisfying

$$e^{-rt}(S(t) - q)^+ \le \gamma(0) + \int_0^t e^{-ru}\pi(u)\sigma \, dW_0(u), \quad 0 \le t < \infty, \quad (6.28)$$

P_0-almost surely, for some martingale-generating portfolio $\pi(\cdot)$. But (6.28) implies

$$E_0[e^{-r(H_b \wedge t)}(S(H_b \wedge t) - q)^+] \le \gamma(0), \quad 0 \le t < \infty,$$

and letting $t \to \infty$ we obtain $g(S(0)) \le \dot\gamma(0)$. Therefore, $g(S(0)) \le V^{AC}(0; \infty)$. On the other hand, (6.15), (6.16), and (6.20)–(6.22) imply

$$e^{-rt}(S(t) - q)^+ \le e^{-rt} g(S(t))$$

$$\le g(S(0)) + \int_0^t e^{-ru} S(u)g'(S(u))\sigma \, dW_0(u), \quad 0 \le t < \infty,$$

$$(6.29)$$

P_0-almost surely, which shows that $V^{AC}(0; \infty) = g(S(0))$. It is an easy exercise in the Markov property for $S(\cdot)$ to extend this result and obtain $V^{AC}(t; \infty) = g(S(t))$, $0 \le t < \infty$. $\qquad\square$

Remark 6.8: From (6.29) we see that a hedging portfolio for the seller of the American call is

$$\pi(u) = S(u)g'(S(u)), \quad 0 \le u < \infty.$$

In particular, the seller should hold more of the stock as the price rises, holding one share whenever the price reaches or exceeds b. The buyer of the call should exercise it as soon as the price reaches or exceeds b. From (6.8), (6.11) we see that $P[H_b < \infty]$ can be either one or strictly less than one; in either case $P_0[H_b < \infty]$ can be either one or strictly less than one, depending on the model parameters.

Remark 6.9: In Theorem 6.7, the process

$$\xi(t) \triangleq \begin{cases} e^{-rt}g(S(t)), & 0 \le t < \infty, \\ 0, & t = \infty \end{cases}$$

is a continuous supermartingale with decomposition $\xi(t) = M(t) - \Lambda(t)$ as in (6.20)–(6.22). Furthermore, $\xi(t) \ge Y(t)$, $0 \le t \le \infty$, and it can be shown that $\xi(\cdot)$ is the minimal supermartingale that dominates the process $Y(\cdot)$ of (6.9) in this way. In other words, $\xi(\cdot)$ is the *Snell envelope* of $Y(\cdot)$. Recall, however, that all these processes are only known to be *restrictedly progressively measurable* in the sense of Definition 1.7.1, and terms like "supermartingale" are to be understood only in this restricted sense. For this reason, the theory of Appendix D is not directly applicable.

Remark 6.10: In the setting of Theorem 6.7, the early exercise premium for the perpetual American call can be defined as

$$e(0) \triangleq V^{AC}(0; \infty) - V^{EC}(0; \infty),$$

where the value of the "perpetual European call"

$$V^{EC}(0, \infty) \triangleq \lim_{T \to \infty} E_0[e^{-rT}(S(T) - q)^+]$$

is zero because of (6.8); hence

$$e(0) = V^{AC}(0, \infty) = g(S(0)). \tag{6.30}$$

A formal application of (5.18), (6.22) yields the formula

$$e(0) = E_0\Lambda(\infty) = \int_0^\infty e^{-rt}E_0[(\delta S(t) - rq)1_{\{S(t)>b\}}]\,dt, \tag{6.31}$$

which agrees with (6.30), although the computational verification is long and painful.

2.7   The American Put Option

We shall concentrate in this section on the *American put option* with exercise price $q > 0$ and finite expiration date $T \in (0, \infty)$. As in Section 2.6, we set $N = D = 1$ and suppress the indices on $S_1(\cdot), \sigma_{11}(\cdot), b_1(\cdot)$, and $\delta_1(\cdot)$. Furthermore, we assume that

$$\sigma(\cdot) \equiv \sigma > 0, \quad \delta(\cdot) \equiv \delta \geq 0, \quad r(\cdot) \equiv r > 0 \tag{7.1}$$

are constant, so that the stock price is given by

$$S(t) = S(0)H(t), \tag{7.2}$$

where

$$H(t) \stackrel{\triangle}{=} \exp\left\{\sigma W(t) - \frac{1}{2}\sigma^2 t + \int_0^t b(u)\, du\right\}$$
$$= \exp\{\sigma W_0(t) + (r - \delta - \sigma^2/2)t\}. \tag{7.3}$$

We shall characterize the value of the American put as the unique solution to a free boundary problem, and we shall obtain regularity results and some explicit formulae for this solution.

The *discounted payoff process* for the American put is

$$Y(t) = e^{-rt}(q - S(t))^+. \tag{7.4}$$

According to Theorem 5.3, the *value of the put* with expiration T, when $S(0) = x$, is

$$p(T, x) \stackrel{\triangle}{=} \sup_{\tau \in \mathcal{S}_{0,T}} E_0[e^{-r\tau}(q - xH(\tau))^+], \quad 0 \leq x < \infty, \quad 0 \leq T < \infty, \tag{7.5}$$

and from the strong Markov property of $S(\cdot)$ we can compute the Snell envelope of $Y(\cdot)$ as

$$\xi(t) \stackrel{\triangle}{=} \sup_{\tau \in \mathcal{S}_{t,T}} E_0[Y(\tau)|\mathcal{F}(t)] = e^{-rt}p(T - t, S(t)), \quad 0 \leq t \leq T. \tag{7.6}$$

The proof of Theorem 5.3 shows that the *optimal exercise time* for the American put option with initial stock price $S(0) = x$ is given by

$$\tau_x \stackrel{\triangle}{=} \inf\{t \in [0, T]; \quad p(T - t, S(t)) = (q - S(t))^+\}. \tag{7.7}$$

This stopping time takes values in $[0, T]$ because $p(0, S(T)) = (q - S(T))^+$, and it attains the supremum in (7.5). The proof of Theorem 5.3 also shows that the stopped process

$$\{e^{-r(t \wedge \tau_x)}p(T - (t \wedge \tau_x), S(t \wedge \tau_x)), \mathcal{F}(t); \quad 0 \leq t \leq T\} \tag{7.8}$$

is a P_0-martingale.

Proposition 7.1: *The optimal expected payoff function* $p : [0, \infty)^2 \to [0, \infty)$ *is continuous and dominates the option's "intrinsic value"*

$$\varphi(x) \stackrel{\triangle}{=} (q - x)^+, \quad 0 \leq x < \infty. \tag{7.9}$$

PROOF.   Fix $(T, x) \in [0, \infty)^2$ and let τ_x be given by (7.7). Because $z_1^+ - z_2^+ \leq (z_1 - z_2)^+$ for any $z_1, z_2 \in \mathbb{R}$, we have for any $y \in [0, \infty)$ that

$$p(T, x) - p(T, y) \leq E_0[e^{-r\tau_x}\{(q - xH(\tau_x))^+ - (q - yH(\tau_x))^+\}]$$
$$\leq (y - x)^+ E_0[e^{-r\tau_x}H(\tau_x)] \leq |x - y|,$$

because $e^{-rt}H(t)$ is a P_0-supermartingale and $H(0) = 1$. Interchanging the roles of x and y, we obtain $|p(T, x) - p(T, y)| \leq |x - y|$, so $p(T, x)$ is Lipschitz continuous in x. Now let us define

$$\psi(t) \triangleq E_0 \left[\max_{0 \leq s \leq t} \left(1 - e^{-rs}H(s)\right)^+ \right].$$

Because of the bounded convergence theorem, $\lim_{t\downarrow 0} \psi(t) = 0$. Let $0 \leq T_1 \leq T_2$ and $x \in [0, \infty)$ be given. Set

$$\tau_2 = \inf\{t \in [0, T_2); \quad p(T_2 - t, xH(t)) = (q - xH(t))^+\} \wedge T_2, \quad \tau_1 = \tau_2 \wedge T_1.$$

Then with $S(t) = xH(t)$, we have

$$0 \leq p(T_2, x) - p(T_1, x) \leq E_0[e^{-r\tau_2}(q - S(\tau_2))^+ - e^{-r\tau_1}(q - S(\tau_1))^+]$$
$$\leq E_0(e^{-r\tau_1}S(\tau_1) - e^{-r\tau_2}S(\tau_2))^+$$
$$\leq E_0\{e^{-r\tau_1}S(\tau_1) \cdot E_0[(1 - \min_{T_1 \leq t \leq T_2} \exp\{\sigma(W_0(t) - W_0(T_1))$$
$$- (\delta + \sigma^2/2)(t - T_1)\})^+ | \mathcal{F}(T_1)]\}$$
$$= E_0\{e^{-r\tau_1}S(\tau_1)\} \cdot \psi(T_2 - T_1) \leq x\psi(T_2 - T_1).$$

It follows that $p(T, x)$ is uniformly continuous in T. The function p dominates φ, because we can always take $\tau \equiv 0$ in (7.5).   □

Although we are primarily interested in $p(T, x)$ for finite T, we digress to consider the behavior of this function at $T = \infty$. Let us suppose that the financial market has infinite planning horizon (Definition 1.7.2). Following Definition 6.2, we define the *value at time s of the perpetual American put option on the stock*, denoted by $V^{AP}(s; \infty)$, to be the minimal random variable $\gamma(s)$ that is $\mathcal{F}^{(T)}(s)$-measurable for some $T \in [s, \infty)$ and for which there exists a martingale-generating portfolio process $\pi(\cdot)$ satisfying almost surely

$$e^{-rt}(q - S(t))^+ \leq e^{-rs}\gamma(s) + \int_s^t e^{-ru}\pi'(u)\sigma(u)\,dW_0(u), \quad s \leq t < \infty.$$

We have the following counterpart to Theorem 6.7.

Theorem 7.2:   *Assume (7.1). The value process for the perpetual American put option is* $V^{AP}(t; \infty) = p(S(t))$, $0 \leq t < \infty$, *where*

$$p(x) \triangleq \begin{cases} q - x, & 0 \leq x \leq c, \\ (q - c)(\frac{x}{c})^{\tilde{\gamma}}, & x > c. \end{cases} \tag{7.10}$$

Here $\tilde{\gamma} \stackrel{\triangle}{=} -\frac{1}{\sigma}[\nu + \sqrt{\nu^2 + 2r}], \nu = \frac{r-\delta}{\sigma} - \frac{1}{2}\sigma,$ *and* $c \stackrel{\triangle}{=} \tilde{\gamma}q/(\tilde{\gamma} - 1) < q.$
Furthermore,

$$p(S(0)) = \sup_{\tau} E_0[e^{-r\tau}(q - S(\tau))^+], \qquad (7.11)$$

where the supremum is taken over all random times τ *satisfying (6.19) and is achieved by the random time* $K_c \stackrel{\triangle}{=} \inf\{t \geq 0; S(t) \leq c\}.$

PROOF.   Much as in Section 6, we compute

$$E_0 e^{-rK_c} = \exp\left[\nu y - |y|\sqrt{\nu^2 + 2r}\right] = \left(\frac{S(0)}{c}\right)^{\tilde{\gamma}}$$

for $S(0) > c$, where now $y = \frac{1}{\sigma}\log(\frac{c}{S(0)})$ is negative. Note that $\tilde{\gamma}$ satisfies (6.12), but rather than (6.13) we now have $\tilde{\gamma} < 0$, $\tilde{\gamma}(r - \delta) < r$ (recall that $r - \delta$ is allowed to be negative). The function p is convex, of class $C^1((0, \infty)) \cap C^2((0, \infty)\backslash\{c\})$, and satisfies the variational inequality

$$\frac{1}{2}\sigma^2 x^2 p''(x) + (r - \delta)xp'(x) - rp(x) = \begin{cases} \delta x - rq < 0, & 0 \leq x < c, \\ 0, & x > c, \end{cases} \quad (7.12)$$

$$p(x) = (q - x)^+, \qquad 0 \leq x \leq c, \quad (7.13)$$

$$p(x) > (q - x)^+, \qquad x > c. \quad (7.14)$$

All the claimed results can now be obtained by simple modifications of the proof of Theorem 6.7. The convergence arguments are easier than before because $e^{-rt}(q - S(t))^+$ is bounded. In particular, the bounded convergence theorem implies

$$p(S(0)) = \lim_{T \to \infty} E_0[e^{-r(K_c \wedge T)}(q - S(K_c \wedge T))^+]. \qquad (7.15)$$

□

Corollary 7.3:   *With* $p(T, x)$ *defined by (7.5) and* $p(x)$ *defined by (7.10), we have* $p(T, x) \leq p(x)$ *for all* $T \in [0, \infty), x \in [0, \infty)$, *and* $\lim_{T \to \infty} p(T, x) = p(x).$

PROOF.   For $x = 0$, the definitions give $p(T, 0) = q = p(0)$ for all $T \in [0, \infty)$. For $x \in (0, \infty)$, (7.11) implies $p(T, x) \leq p(x)$ for all $T \in [0, \infty)$; but (7.15) shows $p(x) \leq \underline{\lim}_{T \to \infty} p(T, x)$, and the result follows.   □

Lemma 7.4:   *The mappings* $T \mapsto p(T, x)$, $x \mapsto p(T, x)$, *and* $x \mapsto x + p(T, x)$ *are nondecreasing, nonincreasing, and nondecreasing, respectively, and the latter two are convex.*

PROOF.   The first two monotonicity assertions are obvious, so let us establish the third. With $0 \leq x < y < \infty$, we have

$$p(T, y) - p(T, x) = p(T, y) - E_0[e^{-r\tau_x}(q - xH(\tau_x))^+]$$
$$\geq E_0[e^{-r\tau_x}\{(q - yH(\tau_x))^+ - (q - xH(\tau_x))^+\}]$$
$$\geq (x - y) \cdot E_0[e^{-r\tau_x}H(\tau_x)] \geq x - y$$

because $e^{-rt}H(t)$ is a supermartingale with $H(0) = 1$. The convexity of $x \mapsto p(T, x)$ follows easily from that of $x \mapsto (q - x)^+$, and leads to the convexity of $x \mapsto x + p(T, x)$.   □

Lemma 7.5: *For every* $(T, x) \in (0, \infty)^2$, *we have* $0 < p(T, x) < q$.

PROOF.   Only the positivity needs discussion. For $0 < x < q$, we have $p(T, x) \geq (q - x)^+ > 0$. For $x \geq q$, let $\tau = T \wedge \inf\{t \geq 0; xH(t) \leq \frac{q}{2}\}$ and observe that

$$p(T, x) \geq E_0[e^{-r\tau}(q - xH(\tau))^+] \geq \frac{q}{2}E_0[e^{-r\tau}1_{\{\tau < T\}}] > 0.$$ □

We define the *continuation region*

$$\mathcal{C} = \{(T, x) \in (0, \infty)^2; \quad p(T, x) > (q - x)^+\}$$

and consider its sections

$$\mathcal{C}_T = \{x \in (0, \infty); \quad p(T, x) > (q - x)^+\}, \quad T \in (0, \infty).$$

Because p is continuous, $\mathcal{C}$ is open in $(0, \infty)^2$ and each $\mathcal{C}_T$ is open in $(0, \infty)$.

Proposition 7.6: *For every* $T \in (0, \infty)$, *there is a number* $c(T) \in (0, q)$ *such that* $\mathcal{C}_T = (c(T), \infty)$. *The function* $T \mapsto c(T)$ *is nonincreasing, upper semicontinuous, and left continuous on* $(0, \infty)$; *thus it may be extended by the definitions* $c(0+) \stackrel{\triangle}{=} \lim_{T \downarrow 0} c(T)$, $c(\infty) \stackrel{\triangle}{=} \lim_{T \to \infty} c(T)$. *We have* $c(0+) \leq q$ *and* $c(\infty) = c$, *as defined in Theorem 7.2.*

PROOF.   For $T \in (0, \infty)$, suppose $x \in \mathcal{C}_T$ and $y > x$. From Lemmas 7.4, 7.5 we have

$$p(T, y) \geq p(T, x) + x - y > (q - x)^+ + x - y \geq q - y$$

and $p(T, y) > 0$, whence $p(T, y) > (q - y)^+$ and $y \in \mathcal{C}_T$. This shows that $\mathcal{C}_T$ has the form $(c(T), \infty)$. Since $T \mapsto p(T, x)$ is nondecreasing, we have for any $\varepsilon > 0, \delta > 0$ that $p(T + \varepsilon, c(T) + \delta) \geq p(T, c(T) + \delta) > (q - c(T) - \delta)^+$. Therefore, $c(T + \varepsilon) < c(T) + \delta$. Since $\delta > 0$ is arbitrary, $c(T + \varepsilon) \leq c(T)$, which shows that $c(\cdot)$ is nonincreasing.

Now take any sequence $\{T_n\}_{n=1}^\infty$ in $(0, \infty)$ with limit $T_0 \in (0, \infty)$ and $\lim_{n \to \infty} c(T_n) = c_0$. Because $\mathcal{C}$ is open and $(T_n, c(T_n)) \notin \mathcal{C}$ for every n, we have $(T_0, c_0) \notin \mathcal{C}$ and thus $c_0 \leq c(T_0)$. In other words, $\overline{\lim}_{T \to T_0} c(T) \leq c(T_0)$ for every $T_0 \in (0, \infty)$. This proves the upper semicontinuity of $c(\cdot)$, and since $c(\cdot)$ is nonincreasing, $c(T-) = c(T)$.

From Lemma 7.5 we have $p(T, x) > 0 = (q - x)^+$ for $x \geq q$, so $c(T) < q$ for all $T \in (0, \infty)$. It follows that $c(0+) \leq q$. From Corollary 7.3 we have $(q - c)^+ \leq p(T, c) \leq p(c) = (q - c)^+$, so $c(T) \geq c$ for all $T \in (0, \infty)$, and $c(\infty) \geq c$. But for $x > c$, $\lim_{T \to \infty} p(T, x) = p(x) > (q - x)^+$ (see (7.14)), so $c(\infty) \leq c$. Finally, we have $c(T) \geq c(\infty) = c > 0$ for all $T \in (0, \infty)$.   □

Theorem 7.7: *The optimal expected payoff function* p *of (7.5) is the unique solution on* $\bar{\mathcal{C}}$ *of the initial–boundary value problem*

$$\mathcal{L}f = 0 \quad in \quad \mathcal{C} = \{(t,x) \in (0,\infty)^2; \quad x > c(t)\}, \tag{7.16i}$$

$$f(t, c(t)) = q - c(t), \quad 0 \leq t < \infty, \tag{7.16ii}$$

$$f(0, x) = (q - x)^+, \quad c(0) \leq x < \infty, \tag{7.16iii}$$

$$\lim_{x \to \infty} \max_{0 \leq t \leq T} |f(t, x)| = 0 \quad \forall T \in (0, \infty), \tag{7.16iv}$$

where $\mathcal{L}f \triangleq \frac{1}{2}\sigma^2 x^2 f_{xx} + (r - \delta)x f_x - rf - f_t$. In particular, the partial derivatives p_{xx}, p_x, and p_t exist and are continuous in $\mathcal{C}$.

PROOF. Clearly, p satisfies the boundary conditions (7.16ii) and (7.16iii). In order to verify the equation (i) for p, let us take a point $(t, x) \in \mathcal{C}$ and a rectangle $\mathcal{R} = (t_1, t_2) \times (x_1, x_2)$ with $(t, x) \in \mathcal{R} \subset \mathcal{C}$. Denote by $\partial_0 \mathcal{R} \triangleq \partial \mathcal{R} \backslash [\{t_2\} \times (x_1, x_2)]$ the "parabolic boundary" of this rectangle, and consider the initial–boundary value problem

$$\mathcal{L}f = 0, \text{ in } \mathcal{R},$$
$$f = p, \text{ on } \partial_0 \mathcal{R}.$$

Because $x_1 \geq c(t) > 0$, the classical theory for parabolic equations (e.g., Friedman (1964), Chapter 3) guarantees the existence of a unique solution f with f_{xx}, f_x, and f_t continuous. We have to show that f and p agree on $\mathcal{R}$.

Let $(t_0, x_0) \in \mathcal{R}$ be given, and consider the stopping time in $\mathcal{S}_{0, t_0 - t_1}$ given by

$$\tau \triangleq \inf\{\theta \in [0, t_0 - t_1); \quad (t_0 - \theta, x_0 H(\theta)) \in \partial_0 \mathcal{R}\} \wedge (t_0 - t_1)$$

and the process

$$N(\theta) \triangleq e^{-r\theta} f(t_0 - \theta, x_0 H(\theta)), \quad 0 \leq \theta \leq t_0 - t_1.$$

From Itô's rule, it follows that $N(\cdot \wedge \tau)$ is a bounded P_0-martingale, and thus

$$f(t_0, x_0) = N(0) = E_0 N(\tau) = E_0[e^{-r\tau} p(t_0 - \tau, xH(\tau))].$$

But $(t_0 - \tau, x_0 H(\tau)) \in \mathcal{C}$ implies

$$\tau \leq \tau_x \triangleq \inf\{\theta \in [0, t_0); \quad p(t_0 - \theta, x_0 H(\theta)) = (q - x_0 H(\theta))^+\} \wedge t_0$$

(cf. (7.7)), and so the optional sampling theorem and (7.8) yield

$$E_0[e^{-r\tau} p(t_0 - \tau, x_0 H(\tau))] = p(t_0, x_0).$$

Thus f and p agree on $\mathcal{R}$, and hence p_{xx}, p_x, and p_t are defined, continuous, and satisfy (7.16i) at the arbitrary point $(t, x) \in \mathcal{C}$.

To check (7.16iv), let $T \in (0, \infty)$ be given. For $(t, x) \in [0, T] \times (0, \infty)$, define (cf. (7.7))

$$\tau_x \overset{\triangle}{=} \inf\{\theta \in [0, t); \quad p(t - \theta, xH(\theta)) = (q - xH(\theta))^+\} \wedge t$$

and notice $p(t, x) = E_0[e^{-r\tau_x}(q - xH(\tau_x))^+]$. Set $\rho_x \overset{\triangle}{=} \inf\{\theta \in [0, \infty); \quad xH(\theta) \leq q\}$, so that $\tau_x \geq \rho_x$ on $\{\rho_x \leq t\}$ and $\tau_x = t$ on $\{\rho_x > t\}$. Then

$$0 \leq p(t, x) \leq qE_0[1_{\{\rho_x \leq t\}}e^{-r\rho_x}] + E_0[1_{\{\rho_x > t\}}e^{-rt}(q - xH(t))^+]$$
$$\leq qP_0[\rho_x \leq T]$$

and $\lim_{x \to \infty} P_0[\rho_x \leq T] = 0$.

For uniqueness, let f defined on $\bar{\mathcal{C}}$ be a solution of (7.16). Note that for each $T \in (0, \infty)$, f is bounded on $\{(t, x) \in [0, T] \times [0, \infty); x \geq c(t)\}$. For $x > c(T)$ define $M(t) \overset{\triangle}{=} e^{-rt}f(T - t, xH(t)), 0 \leq t \leq T$, and $\tau_x \overset{\triangle}{=} T \wedge \inf\{t \in [0, T]; xH(t) \leq c(T - t)\}$. Itô's rule shows that $\{M(t \wedge \tau_x), 0 \leq t \leq T\}$ is a bounded P_0-martingale. Since τ_x attains the supremum in (7.5), we have from optional sampling that

$$f(T, x) = M(0) = E_0 M(\tau_x) = E_0[e^{-r\tau_x}f(T - \tau_x, xH(\tau_x))]$$
$$= E_0[e^{-r\tau_x}(q - xH(\tau_x))^+] = p(T, x). \qquad \square$$

Theorem 7.7 asserts that p is smooth enough to permit the application of Itô's rule to the Snell envelope $\xi(t) = e^{-rt}p(T - t, S(t))$ of (7.6), as long as $(T - t, S(t)) \in \mathcal{C}$, or equivalently, $S(t) > c(T - t)$. On the other hand, in the region $\{(t, x) \in [0, \infty)^2; x < c(t)\}$ we have $p(t, x) = q - x$, which is also smooth. At issue then is the smoothness of p across the boundary $x = c(t)$. We have the following result.

Lemma 7.8: *Fix* $T \in (0, \infty)$. *The convex function* $x \mapsto p(T, x)$ *is of class* C^1, *even at* $x = c(T)$. *In particular,* $p_x(T, c(T)) = -1$.

PROOF.   Because $p(T, x) = q - x$ for $0 \leq x < c(T)$, we have $p_x(T, c(T)-) = -1$. The convexity of $x \mapsto p_x(T, x)$, which was proved in Lemma 7.4, implies that $p_x(T, c(T)+)$ is defined and $p_x(T, c(T)+) \geq -1$.

Thus, it suffices to show $p_x(T, c(T)+) \leq -1$. To this end, set $x = c(T)$ and define

$$\tau_{x+\varepsilon} \overset{\triangle}{=} \inf\{t \in [0, T); (x + \varepsilon)H(t) \leq c(T - t)\} \wedge T$$

for $\varepsilon \geq 0$, so $\tau_{x+\varepsilon}$ is nondecreasing in ε and $\tau_x \equiv 0$. Because $c(\cdot)$ is nonincreasing, we have

$$\tau_{x+\varepsilon} \leq \inf\left\{t \in [0, T]; H(t) \leq \frac{x}{x + \varepsilon}\right\} \wedge T.$$

The law of the iterated logarithm for the P_0-Brownian motion $W_0(\cdot)$ implies that $P[\min_{0 \leq t \leq a} H(t) < 1] = 1$ for every $a > 0$, and therefore

$$\tau_{x+\varepsilon} \downarrow 0 \quad \text{as} \quad \varepsilon \downarrow 0 \tag{7.17}$$

almost surely. We also have

$$
\begin{aligned}
p(T, x + \varepsilon) &= E_0[e^{-r\tau_{x+\varepsilon}}(q - (x+\varepsilon)H(\tau_{x+\varepsilon}))^+] \\
&= E_0[e^{-r\tau_{x+\varepsilon}}(q - xH(\tau_{x+\varepsilon}))^+] \\
&\quad - E_0[e^{-r\tau_{x+\varepsilon}}((q - xH(\tau_{x+\varepsilon}))^+ - (q - (x+\varepsilon)H(\tau_{x+\varepsilon}))^+)] \\
&\leq p(T, x) - E_0[1_{\{\tau_{x+\varepsilon}<T\}}e^{-r\tau_{x+\varepsilon}}((q - xH(\tau_{x+\varepsilon})) \\
&\quad - (q - (x+\varepsilon)H(\tau_{x+\varepsilon})))] \\
&\quad - E_0[1_{\{\tau_{x+\varepsilon}=T\}}e^{-rT}((q - xH(T))^+ - (q - (x+\varepsilon)H(T))^+)] \\
&\leq p(T, x) - \varepsilon \cdot E_0[1_{\{\tau_{x+\varepsilon}<T\}}e^{-r\tau_{x+\varepsilon}}H(\tau_{x+\varepsilon})] \\
&= p(T, x) - \varepsilon \cdot E_0[e^{-r\tau_{x+\varepsilon}}H(\tau_{x+\varepsilon})] \\
&\quad + \varepsilon \cdot E_0[1_{\{\tau_{x+\varepsilon}=T\}}e^{-rT}H(T)],
\end{aligned}
$$

for $\varepsilon > 0$, from which it follows that

$$p_x(T, x+) \leq \lim_{\varepsilon \downarrow 0} E_0[1_{\{\tau_{x+\varepsilon}=T\}}e^{-rT}H(T)] - \lim_{\varepsilon \downarrow 0} E_0[e^{-r\tau_{x+\varepsilon}}H(\tau_{x+\varepsilon})] = -1.$$

For the last equality we have used (7.17) and the uniform integrability of the supermartingale $e^{-rt}H(t), 0 \leq t \leq T$, in (7.3). □

Theorem 7.9: *Fix $T \in (0, \infty)$. The Snell envelope $\xi(t) = e^{-rt}p(T - t, S(t))$ of (7.6) has the Doob–Meyer decomposition*

$$\xi(t) = M(t) - \Lambda(t), \quad 0 \leq t \leq T, \tag{7.18}$$

where

$$M(t) \triangleq p(T, S(0)) + \sigma \int_0^t e^{-ru}S(u)p_x(T - u, S(u)) \, dW_0(u) \tag{7.19}$$

is a P_0-martingale and

$$\Lambda(t) \triangleq \int_0^t e^{-ru}(rq - \delta S(u))1_{\{S(u)<c(T-u)\}} \, du \tag{7.20}$$

is nondecreasing. In particular,

$$\delta c(0+) \leq rq. \tag{7.21}$$

PROOF. We mollify the function $p(\cdot, \cdot)$ in order to apply Itô's rule. Let $\zeta : \mathbb{R}^2 \to [0, \infty)$ be a C^∞ function integrating to 1 and having support in $[0, 1]^2$. For $\varepsilon > 0$, define

$$
\begin{aligned}
p^{(\varepsilon)}(t, x) &\triangleq \int_0^\infty \int_0^\infty p(t + \varepsilon u, x + \varepsilon v)\zeta(u, v) \, du \, dv \\
&= \frac{1}{\varepsilon^2} \int_0^\infty \int_0^\infty p(s, y)\zeta\left(\frac{s-t}{\varepsilon}, \frac{y-x}{\varepsilon}\right) \, ds \, dy.
\end{aligned}
$$

Then $p^{(\varepsilon)}(\cdot)$ is of class C^{∞} on $(0, \infty)^2$. Because $p_x(t, \cdot)$ is continuous, for $(t, x) \in (0, \infty)^2$ we may integrate by parts to obtain

$$
\begin{aligned}
p_x^{(\varepsilon)}(t, x) &= -\frac{1}{\varepsilon^3} \int_0^\infty \int_0^\infty p(s, y) \zeta_2 \left(\frac{s - t}{\varepsilon}, \frac{y - x}{\varepsilon} \right) dy\, ds \\
&= \frac{1}{\varepsilon^2} \int_0^\infty \int_0^\infty p_x(s, y) \zeta \left(\frac{s - t}{\varepsilon}, \frac{y - x}{\varepsilon} \right) dy\, ds \\
&= \int_0^\infty \int_0^\infty p_x(t + \varepsilon u, x + \varepsilon v) \zeta(u, v)\, du\, dv,
\end{aligned}
$$

$$
\begin{aligned}
p_{xx}^{(\varepsilon)}(t, x) &= \frac{1}{\varepsilon^4} \int_0^\infty \int_0^\infty p(s, y) \zeta_{22} \left(\frac{s - t}{\varepsilon}, \frac{y - x}{\varepsilon} \right) dy\, ds \\
&= -\frac{1}{\varepsilon^3} \int_0^\infty \int_0^{c(s)} p_x(s, y) \zeta_2 \left(\frac{s - t}{\varepsilon}, \frac{y - x}{\varepsilon} \right) dy\, ds \\
&\quad -\frac{1}{\varepsilon^3} \int_0^\infty \int_{c(s)}^\infty p_x(s, y) \zeta_2 \left(\frac{s - t}{\varepsilon}, \frac{y - x}{\varepsilon} \right) dy\, ds \\
&= -\frac{1}{\varepsilon^2} \int_0^\infty \left[p_x(s, c(s)-) \zeta \left(\frac{s - t}{\varepsilon}, \frac{c(s) - x}{\varepsilon} \right) \right. \\
&\quad \left. - \int_0^{c(s)} p_{xx}(s, y) \zeta \left(\frac{s - t}{\varepsilon}, \frac{y - \varepsilon}{\varepsilon} \right) dy \right] ds \\
&\quad -\frac{1}{\varepsilon^2} \int_0^\infty \left[-p_x(s, c(s)+) \zeta \left(\frac{s - t}{\varepsilon}, \frac{c(s) - x}{\varepsilon} \right) \right. \\
&\quad \left. - \int_{c(s)}^\infty p_{xx}(s, y) \zeta \left(\frac{s - t}{\varepsilon}, \frac{y - \varepsilon}{\varepsilon} \right) dy \right] ds \\
&= \frac{1}{\varepsilon^2} \int_0^\infty \int_0^\infty p_{xx}(s, y) \zeta \left(\frac{s - t}{\varepsilon}, \frac{y - \varepsilon}{\varepsilon} \right) ds\, dy \\
&= \int_0^\infty \int_0^\infty p_{xx}(t + \varepsilon u, x + \varepsilon v) \zeta(u, v)\, du\, dv,
\end{aligned}
$$

and

$$
p_t^{(\varepsilon)}(t, x) = \int_0^\infty \int_0^\infty p_t(t + \varepsilon u, x + \varepsilon v) \zeta(u, v)\, du\, dv,
$$

where ζ_i denotes the partial derivative of ζ with respect to its ith variable. These formulas show that $p_x^{(\varepsilon)}$ and $\mathcal{L}p^{(\varepsilon)}$ are bounded on compact subsets of $(0, \infty)^2$ and

$$
p_x(t, x) = \lim_{\varepsilon \downarrow 0} p_x^{(\varepsilon)}(t, x),
$$

$$
\mathcal{L}p(t, x) = \lim_{\varepsilon \downarrow 0} \mathcal{L}p^{(\varepsilon)}(t, x), \quad \forall (t, x) \in (0, \infty)^2, \quad x \neq c(t).
$$

According to Itô's rule,

$$e^{-rt}p^{(\varepsilon)}(T-t,S(t)) = p^{\varepsilon}(T,S(0)) + \int_0^t e^{-ru}\mathcal{L}p^{(\varepsilon)}(T-u,S(u))du$$

$$+ \sigma \int_0^t e^{-ru}S(u)p_x^{(\varepsilon)}(T-u,S(u))dW_0(u), \quad 0 \le t \le T. \qquad (7.22)$$

For each $u \in (0,T)$, we have $P_0[S(u) = c(T-u)] = 0$, so $\int_0^t e^{-ru}\mathcal{L}p(T-u,S(u))du$ is defined and equal to $\int_0^t e^{-ru}(\delta S(u) - rq)1_{\{S(u)<c(T-u)\}}du$ a.s. Letting $\varepsilon \downarrow 0$ in (7.22) we obtain (7.18), first for $0 \le t < T$ and then, by letting $t \uparrow T$ in (7.18), for $t = T$ as well.

The process $M(\cdot)$ in (7.19) is a martingale because $-1 \le p_x \le 0$ and $E_0 \int_0^T S^2(u)du < \infty$. To see that $\Lambda(\cdot)$ is nondecreasing, we recall that a decomposition like (7.18) of a process into a continuous martingale $M(\cdot)$ and a continuous, bounded-variation process $\Lambda(\cdot)$ is unique (Karatzas and Shreve (1991), Problem 3.3.2). But the Snell envelope $\xi(\cdot)$, being a bounded supermartingale, has a Doob–Meyer decomposition as a continuous martingale minus a continuous, nondecreasing process (e.g., Theorem D.13 in Appendix D). Consequently, this Doob–Meyer decomposition *must be* the decomposition of (7.18), which shows that $\Lambda(\cdot)$ is nondecreasing. Since $P_0[S(u) < c(T-u)] > 0$ for every $u \in (0,T]$, this can be the case only if $rq - \delta c(T-u) \ge 0$ for Lebesgue-almost-every $u \in (0,T]$. In particular, (7.21) must hold. □

Proposition 7.10: *The free-boundary function* $c(\cdot) : [0,\infty) \to (0,q]$ *is continuous, with* $c(0+) = q$ *if* $r \ge \delta$, *and* $c(0+) = rq/\delta$ *if* $r < \delta$.

PROOF.   Let us define $c(0) \overset{\triangle}{=} q$ if $r \ge \delta$ and $c(0) \overset{\triangle}{=} rq/\delta$ if $r < \delta$. We know from Proposition 7.6 that $c(\cdot)$ is left continuous and nondecreasing. To prove right continuity we shall suppose $c(t_0) > c(t_0+)$ for some $t_0 \in [0,\infty)$ and obtain a contradiction.

Under the assumption $c(t_0) > c(t_0+)$, define $x_1 \overset{\triangle}{=} \frac{1}{2}[c(t_0) + c(t_0+)] < c(t_0) \le c(0)$. Let $t \in (t_0,\infty)$ and $x \in (c(t),x_1)$ be given. From (2.7.16i) and the fact that $p(\cdot,x)$ is nondecreasing, we have

$$\frac{1}{2}\sigma^2 x^2 p_{xx}(t,x) \ge (\delta - r)x p_x(t,x) + rp(t,x).$$

Now, $p(t,x) \ge (q-x)^+ \ge q - x_1 > 0$, so if $r \ge \delta$, we may use the inequalities $-1 \le p_x(t,x) \le 0$ to write $\frac{1}{2}\sigma^2 x^2 p_{xx}(t,x) \ge r(q - x_1) > 0$. On the other hand, if $r < \delta$, we have

$$\frac{1}{2}\sigma^2 x^2 p_{xx}(t,x) \ge -(\delta - r)x_1 + r(q - x_1) = rq - \delta x_1 > 0$$

because $\delta x_1 < \delta c(0+) = rq$ (see (7.2)). In either case,

$$p_{xx}(t,x) \ge \eta \overset{\triangle}{=} \frac{2[rq - (r \vee \delta)x_1]}{\sigma^2 x_1^2} > 0, \quad \forall\, t \in (t_0,\infty), \quad x \in (c(t),x_1).$$

With $\varphi(\xi) \triangleq (q - \xi)^+, t \in (t_0, \infty)$, and $x_0 \in (c(t_0+), x_1)$, we compute

$$p(t, x_0) - \varphi(x) = \int_{c(t)}^{x_0} \int_{c(t)}^{y} [p_{xx}(t, \xi) - \varphi''(\xi)] \, d\xi \, dy \geq \frac{1}{2}\eta(x_0 - c(t))^2,$$

where we have used the relations

$$p(t, c(t)) = \varphi(c(t)), \quad p_x(t, c(t)) = \varphi'(c(t)).$$

Letting $t \downarrow t_0$ and using the continuity of p, we obtain $p(t_0, x_0) \geq (q-x_0)^+ + \frac{1}{2}\eta(x_0 - c(t_0+))^2 > q - x_0$. If follows that $c(t_0) < x_0$, a contradiction to the definition of x_1. □

Corollary 7.11: *Fix $T \in (0, \infty)$. The Snell envelope $\xi(t) = e^{-rt}p(T - t, S(t))$ of (7.6) admits the representation*

$$\xi(t) = E_0[e^{-rT}(q - S(T))^+|\mathcal{F}(t)] + E_0[\Lambda(T) - \Lambda(t)|\mathcal{F}(t)], \quad 0 \leq t \leq T,$$
$$\text{(7.23)}$$

where $\Lambda(\cdot)$ is defined by (7.20).

PROOF. From Theorem 7.9 we have

$$\begin{aligned}
E_0[e^{-rT}(q - S(T))^+|\mathcal{F}(t)] &= E_0[\xi(T)|\mathcal{F}(t)] \\
&= E_0[M(T)|\mathcal{F}(t)] - E_0[\Lambda(T)|\mathcal{F}(t)] \\
&= M(t) - \Lambda(t) - E_0[\Lambda(T) - \Lambda(t)|\mathcal{F}(t)] \\
&= \xi(t) - E_0[\Lambda(T) - \Lambda(t)|\mathcal{F}(t)], \quad 0 \leq t \leq T. \ \square
\end{aligned}$$

Corollary 7.11 facilitates the explicit computation of the value $u^{AP}(T, x; q) = p(T, x)$ at time zero of the American put with expiration $T > 0$ and exercise price $q > 0$, when $S(0) = x$. Of course, the value at time $t \in [0, T]$ is just $u^{AP}(T - t, S(t); q)$. According to Corollary 7.11,

$$u^{AP}(T, x; q) = p(T, x) = u^{EP}(T, x; q) + e(0), \quad 0 < T < \infty, \quad \text{(7.24)}$$

where, in the notation of (4.11), (4.13), and (4.15),

$$\begin{aligned}
u^{EP}(T, x; q) &= E_0[e^{-rT}(q - S(T))^+] \\
&= -u^{FC}(T, x; q) + u^{EC}(T, x; q) \\
&= qe^{-rT}[1 - \Phi(\rho_-(T, x; q))] - xe^{-\delta T}[1 - \Phi(\rho_+(T, x; q))]
\end{aligned}$$
$$\text{(7.25)}$$

is the value of the associated European put, and

$$\begin{aligned}
e(0) &\triangleq \int_0^T E_0[e^{-ru}1_{\{S(u) < c(T-u)\}}(rq - \delta S(u))] \, du \\
&= \frac{1}{\sqrt{2\pi}} \int_0^T \int_{-\infty}^{\gamma(T, x; u)} e^{-ru - \frac{w^2}{2}} \left[rq - \delta x \exp\left\{ \sigma w \sqrt{u} \right. \right. \\
&\quad \left. \left. + (r - \delta - \sigma^2/2)u \right\} \right] dw \, du
\end{aligned}$$
$$\text{(7.26)}$$

with

$$\gamma(T, x; u) \triangleq \frac{1}{\sigma\sqrt{u}} \left[\log\left(\frac{c(T-u)}{x} \right) - (r - \delta - \sigma^2/2)u \right]. \tag{7.27}$$

In particular, $e(0)$ is the *early exercise premium* of Remark 5.10. In the special case of $\delta = 0$ (no dividends), the early exercise premium

$$e(0) = rq \cdot E_0 \int_0^T e^{-ru} 1_{\{S(u) < c(T-u)\}} du$$

$$= rq \int_0^T e^{-ru} \Phi\left(\frac{1}{\sigma\sqrt{u}} \left[\log\left(\frac{c(T-u)}{x} \right) - (r - \sigma^2/2)u \right] \right) du$$

is the value of an income process that pays at rate rq whenever the stock price is in the region where the put should be exercised.

The formula (7.24) for the value of the American put still involves the unknown free-boundary function $c(\cdot)$ via its presence in (7.27). To obtain information about this free boundary, we can use in (7.24) the fact that $p(T, x) = q - x$ for $0 \le x \le c(T)$. When $\delta = 0$, this leads to the equation

$$q - x = qe^{-rT}[1 - \Phi(\rho_-(T, x; q))] - x[1 - \Phi(\rho_+(T, x; q))]$$

$$+ rq \int_0^T e^{-ru} \Phi\left(\frac{1}{\sigma\sqrt{u}} \left[\log\left(\frac{c(T-u)}{x} \right) - (r - \sigma^2/2)u \right] \right) du,$$

$$\forall x \in [0, c(T)]. \tag{7.28}$$

It can be shown (cf. Jacka (1991)) that *the equation (7.28) characterizes* $c(\cdot)$ *uniquely, among all nonincreasing left-continuous functions with values in* $(0, q]$. In particular, setting $x = c(T)$ in (7.28) yields the integral equation

$$\frac{c(T)}{q} \Phi(\rho_+(T, c(T); q)) = 1 - e^{-rT}[1 - \Phi(\rho_-(T, c(T); q))]$$

$$- r \int_0^T e^{-ru} \Phi\left(\frac{1}{\sigma\sqrt{u}} \left[\log\left(\frac{c(T-u)}{c(T)} \right) \right. \right.$$

$$\left. \left. - (r - \sigma^2/2)u \right] \right) du \tag{7.29}$$

for $c(T), 0 < T < \infty$. The initial condition for $c(\cdot)$ when $\delta = 0$, given in Proposition 7.10, is $c(0) = q$. Little (1998) shows further that (7.29) can be reduced to a nonlinear Volterra integral equation for $c(\cdot)$.

Theorem 7.7 characterizes the optimal expected payoff function $p(\cdot, \cdot)$ in terms of the free-boundary function $c(\cdot)$, and this has finally led to formula (7.24) for $p(\cdot, \cdot)$, also in terms of $c(\cdot)$. We have seen that (7.24) indirectly provides information about $c(\cdot)$ (e.g., (7.28), (7.29)). We close this section with a direct characterization, along the lines of Theorem 7.7, of the *pair* of functions $(p(\cdot, \cdot), c(\cdot))$.

Consider the problem of finding a pair of functions $f : [0, \infty)^2 \to \mathbb{R}$ and $d : [0, \infty) \to (0, q]$ such that:

$$f \text{ is continuous on } [0, \infty)^2, \tag{7.30i}$$

$$d \text{ is nonincreasing and left-continuous on } [0, \infty), \tag{7.30ii}$$

$f_t, f_x,$ and f_{xx} are defined and continuous on the
open (because of (7.30ii)) set $\mathcal{D} \triangleq \{(t, x) \in (0, \infty)^2; x > d(t)\},$ (7.30iii)

$$d(0) \triangleq d(0+) \leq \frac{rq}{\delta} \quad \text{if} \quad \delta > r, \tag{7.30iv}$$

$$\mathcal{L}f = 0 \text{ in } \mathcal{D}, \text{where } \mathcal{L}f \triangleq \frac{1}{2}\sigma^2 x^2 f_{xx} + (r - \delta)x f_x - rf - f_t, \tag{7.30v}$$

$$f(t, x) \geq (q - x)^+, \quad \forall (t, x) \in [0, \infty)^2, \tag{7.30vi}$$

$$f(t, x) = (q - x), \quad \forall t \in [0, \infty), \quad 0 \leq x \leq d(t), \tag{7.30vii}$$

$$f(0, x) = (q - x)^+, \quad \forall x \in [d(0), \infty), \tag{7.30viii}$$

$$\lim_{x \to \infty} \max_{0 \leq t \leq T} |f(t, x)| = 0, \quad \forall T \in (0, \infty), \tag{7.30ix}$$

$$f_x(t, d(t)+) = -1, \quad \forall t \in (0, \infty). \tag{7.30x}$$

Recall the function $\varphi(x) = (q - x)^+$ of (7.9), and define the set

$$\mathcal{G} = \begin{cases} [0, \infty) \times [0, q) & \text{if } r \geq \delta, \\ [0, \infty) \times [0, \frac{rq}{\delta}) & \text{if } \delta > r. \end{cases}$$

On $\mathcal{G}$, the function φ is smooth, and

$$\mathcal{L}\varphi(t, x) = \delta x - rq < 0, \quad \forall (t, x) \in \mathcal{G}. \tag{7.31}$$

If (f, d) satisfies (7.30), then f agrees with φ on the set $\{(t, x) \in [0, \infty)^2; 0 \leq x < d(t)\} \subseteq \mathcal{G}$. Thus, the free boundary divides $(0, \infty)^2$ into two open regions, $\mathcal{D}$ and $(0, \infty)^2 \backslash \bar{\mathcal{D}}$, such that

$$\mathcal{L}f = 0, \quad f \geq \varphi; \quad \text{on} \quad \mathcal{D}, \tag{7.32i}$$

$$\mathcal{L}f < 0, \quad f = \varphi; \quad \text{on} \quad (0, \infty)^2 \backslash \bar{\mathcal{D}}. \tag{7.32ii}$$

Although we do not know that f is smooth across the boundary between these regions, (7.30x) guarantees that f_x is defined and continuous on $(0, \infty)^2$.

Theorem 7.12: *The pair of functions $(p(\cdot, \cdot), c(\cdot))$ is the unique solution to the free-boundary problem (7.30).*

PROOF. We know from Propositions 7.1, 7.6, the definition of $\mathcal{C}$, Theorem 7.7, Lemma 7.8, and relation (7.21) that $(p(\cdot, \cdot), c(\cdot))$ solves (7.30). Suppose $(f(\cdot, \cdot), d(\cdot))$ is any solution to (7.30). Fix $(T, x) \in [0, \infty)^2$ and use the mollification argument of Theorem 7.9 to obtain the formula

$$e^{-rt} f(T - t, S(t)) = M^f(t) - \Lambda^f(t), \quad 0 \le t \le T,$$

where $x = S(0)$, $M^f(\cdot)$ is the P_0-local-martingale $M^f(t) \stackrel{\triangle}{=} f(T, x) + \sigma \int_0^t e^{-ru} S(u) f_x(T - u, S(u)) dW_0(u)$, and $\Lambda^f(\cdot)$ is the nondecreasing (because of (7.31)) process $\Lambda^f(t) \stackrel{\triangle}{=} \int_0^t e^{-ru} 1_{\{S(u) < d(T-u)\}} (rq - \delta S(u)) du$. Let $\{\tau_n\}_{n=1}^\infty$ be a sequence of stopping times with $\tau_n \uparrow T$ almost surely and such that $\{M^f(t \wedge \tau_n); 0 \le t \le T\}$ is a P_0-martingale. For any stopping time $\tau \in \mathcal{S}_{0,T}$ we have

$$E_0[e^{-r(\tau \wedge \tau_n)} f(T - (\tau \wedge \tau_n), S(\tau \wedge \tau_n))] = f(T, x) - E_0 \Lambda^f(\tau \wedge \tau_n). \quad (7.33)$$

The function f is bounded on $[0, T] \times [0, \infty)$ (see (7.30i, ix), so passage to the limit in (7.33) yields

$$E_0[e^{-r\tau} f(T - \tau, S(\tau))] = f(T, x) - E_0 \Lambda^f(\tau), \quad \forall \tau \in \mathcal{S}_{0,T}. \quad (7.34)$$

From (7.30vi) and the nonnegativity of $\Lambda^f(\tau)$, we conclude that $E_0[e^{-r\tau}(q - S(\tau))^+] \le f(T, x)$ for all $\tau \in \mathcal{S}_{0,T}$, whence $p(T, x) \le f(T, x)$ (recall (7.5)). On the other hand, defining $\tau_x \stackrel{\triangle}{=} T \wedge \inf\{t \ge 0; S(t) \le d(T - t)\}$ to be the hitting time of the closed set $[0, \infty)^2 \backslash \mathcal{D}$, we have $f(T - \tau_x, S(\tau_x)) = (q - S(\tau_x))^+$ (from (7.30vii, viii), and $\Lambda^f(\tau_x) = 0$, so (7.34) implies $E_0[e^{-r\tau_x}(q - S(\tau_x))^+] = f(T, x)$. It follows that $p = f$ on $[0, \infty)^2$.

To show that the functions $c(\cdot)$ and $d(\cdot)$ are equal, it suffices to establish equality between the open sets

$$\mathcal{C} \stackrel{\triangle}{=} \{(t, x) \in (0, \infty)^2; \quad p(t, x) > (q - x)^+\} = \{(t, x) \in (0, \infty)^2; \quad x > c(t)\}$$

and

$$\mathcal{D} \stackrel{\triangle}{=} \{(t, x) \in (0, \infty)^2; x > d(t)\}.$$

For $(t, x) \in \mathcal{C}$, we have $\mathcal{L}p(t, x) = 0$, which means that $(t, x) \notin (0, \infty)^2 \backslash \bar{\mathcal{D}}$ (see (7.30ii). Therefore, $\mathcal{C} \subseteq \bar{\mathcal{D}}$, but since both $\mathcal{C}$ and $\mathcal{D}$ are open, we must have in fact $\mathcal{C} \subseteq \mathcal{D}$. The roles of $\mathcal{C}$ and $\mathcal{D}$ in this argument may be reversed to obtain $\mathcal{D} \subseteq \mathcal{C}$. $\square$

2.8 Notes

In order to implement the contingent-claim pricing theory presented in this chapter, it is necessary not only to know the current interest rate, but also to have a *model for the evolution of interest rates into the future*, so that the statistics of the process $S_0(\cdot)$ appearing in pricing formulas such as (2.9) can be computed. Such a model should be consistent with current observed yields of default-free bonds of various maturities, each of which is subject to the same pricing formula (2.9). Furthermore, the model should include randomness in a way that enables it to remain consistent with observed yields as time evolves. The construction of such models belongs to the study of the *term structure of interest rates*, a subject not developed in this monograph. The most general term-structure model is that of Heath, Jarrow, and Morton (1992, 1996), although many other models are also popular among practitioners. For example, we refer the reader to the exhaustive treatment of this subject in the recent monographs by Musiela and Rutkowski (1997a) and Bisière (1997), to Chapters 7 and 9 of Duffie (1992), to the papers by Artzner and Delbaen (1989), Björk, Di Masi, Kabanov, and Runggaldier (1997), Björk, Kabanov, and Runggaldier (1997), Black, Derman, and Toy (1990), Brace, Gatarek, and Musiela (1997), Brennan, and Schwartz (1979, 1982) (but see Hogan (1993)), Chan (1992), Cox, Ingersoll, and Ross (1985b), Dothan (1978), Duffie and Kan (1994, 1996), Dybvig (1997), El Karoui, Myneni, and Viswanathan (1992), El Karoui and Rochet (1989), Ho and Lee (1986), Hull and White (1990), Jamshidian (1990, 1997b), Litterman and Scheinkman (1988), Miltersen (1994), Miltersen, Sandmann, and Sondermann (1997), Musiela and Rutkowski (1997b), Richard (1978), Rogers (1997), Sandmann and Sondermann (1993), and Vasicek (1977), and to the survey papers by Rogers (1995b) and Björk (1997) for up-to-date overviews. A less mathematical introduction to the issues of interest-rate instruments is provided by Sundaresan (1997).

Section 3: The distinction between *forward* and *futures contracts* has been only relatively recently recognized (see Margrabe (1976), Black (1976a)) and even more recently understood. Cox, Ingersoll, and Ross (1981) and Jarrow and Oldfield (1981) provide a discrete-time, arbitrage-based analysis of the relationship between forwards and futures, whereas Richard and Sundaresan (1981) study these claims in a continuous-time, equilibrium setting. Myneni (1992b) has used stochastic calculus to revisit the formulas of Cox et al. (1981). Our presentation of this material is similar to that of Duffie and Stanton (1992), which also considers options on futures. In particular, our Corollary 3.9 on the forward-futures spread may be found in Duffie and Stanton (1992), and also in Chapter 7 of Duffie (1992). For additional reading on forward and futures contracts, one may consult Anderson (1984), Dubofsky (1992), Duffie (1989), Edwards and Ma (1992), Merrick (1990), Musiela and Rutkowski (1997), Sutcliffe (1993).

Sections 2 and 4: The modern theory of the pricing of options (or, more generally, contingent claims) in a complete market begins with the seminal articles of Samuelson (1965a), Samuelson and Merton (1969), Black and Scholes (1973), and Merton (1973a). The arbitrage-based approach of these sections, which is not restricted to markets with constant coefficients, has its origins in the articles of Ross (1976) and Cox and Ross (1976), and matures with Harrison and Kreps (1979); this latter paper, along with the seminal Harrison and Pliska (1981, 1983), clarified the mathematical stucture of the problem and worked out its connections with martingale theory. The early work on option pricing is nicely surveyed in Smith (1976) and Müller (1985). Nontechnical discussions can be found in Bernstein (1992), Chapter 11, and Malkiel (1996), Chapter 11. Grabbe (1983), Barron (1990), Barron and Jensen (1990, 1991), Korn (1992), and Bergman (1995) have derived option pricing formulas when the interest rate for borrowing is greater than the interest rate for investing. See also Section 6.8 of this text. Cox, Ross, and Rubinstein (1979) provide a simple yet powerful discrete-time model for option pricing. Ocone and Karatzas (1991) have used the Malliavin calculus to identify hedging portfolios; Colwell, Elliott, and Kopp (1991) have relied on the Markov property for the same purpose. Carr and Jarrow (1990) use Brownian local time to resolve the paradox of hedging an option using the stop–loss start–gain strategy. Lamberton and Lapeyre (1993) and Madan and Milne (1993) discuss contingent-claim valuation when one is investing in a "basis" of assets.

The Black–Scholes option pricing formula (4.11) heralded a revolution marked by the widespread creation of derivative securities whose prices are set as much by theoretical considerations as by market forces. One of the basic insights of this formula is that it singles out the *volatility* of the underlying stock-price process as the crucial parameter. The success of the Black–Scholes model has been such that prices are often quoted now in terms of the volatility parameters implied by it.

Empirical evidence has shown, however, that the constant volatility model that supports the pricing formula (4.11) should not be used injudiciously (see, for instance, Blattberg and Gonedes (1974), Black (1976b), MacBeth and Merville (1979), Christie (1982), Bhattacharya (1983), Rubinstein (1983, 1985), Scott (1987) and the references cited there). Several procedures for estimating stock volatilities have been proposed; for procedures based on the extreme values of stock prices, see Garman and Klass (1980), Parkinson (1980), Beckers (1983), Rogers and Satchell (1991). Procedures for estimating stochastic volatility are described by Andersen (1994), Taylor (1994).

The formulae of Example 4.5 on *path-dependent* (or "look-back") options appear in Goldman, Sosin, and Gatto (1979). Our treatment was inspired by the preprint of Föllmer (1991), who credits Martin Schweizer with the idea of using the Clark formula in this context. Shepp and Shiryaev (1993, 1994) introduced some perpetual American look-back options with discounted payoff of the type

$$Y(t) = e^{-rt} \max_{0 \le u \le t} S(u), \quad 0 \le t \le \infty$$

with $r > 0$, which were later also studied by Duffie and Harrison (1993). For the rather difficult study of *Asian options*, i.e., path-dependent options with (terminal) payoff of the type $C(\cdot) \equiv 0$ on $[0, T)$ and

$$C(T) = \left(\frac{1}{T} \int_0^T S(t) \, dt - q \right)^+$$

depending on the average stock price over a given time interval, we refer the reader to the papers of Carverhill and Clewlow (1990), Kemna and Vorst (1990), Conze and Viswanathan (1991), Turnbull and Wakeman (1991), Geman and Yor (1992, 1993), Bouaziz et al. (1994), Kramkov and Vishnyakov (1994), and Rogers and Shi (1995) or Karatzas (1996), p. 23.

Another example of a path-dependent option is the so-called *barrier option* of the type $C(\cdot) \equiv 0$ on $[0, T)$ and

$$C(T) = (S(T) - q)^+ \cdot 1_{\{\tau_h \le T\}} \quad \text{with} \quad \tau_h \overset{\triangle}{=} \inf\{t \ge 0; S(t) \ge h\} \quad (8.1)$$

and $0 < S(0), q < h < \infty$; i.e., a European call option that is activated only if a certain upper barrier is hit before the expiration date. See, for instance, Merton (1973), Cox and Rubinstein (1985), Rubinstein (1991), Karatzas (1996), pp. 21–22, and Broadie, Glasserman, and Kou (1996, 1997) for numerical methods. One can think of several types of barrier options, depending on whether the barrier lies above or below the initial stock price, and on the function of the barrier (i.e., whether hitting the barrier activates or deactivates the option); all of these have received attention, as has the "Parisian option" (cf. Chesney et al. (1997)), a barrier-type option that is activated only if the price process spends a sufficient, prespecified amount of time above the upper barrier h of (8.1). There are also the *double-barrier options*, which, for example, are deactivated if the stock price hits either an upper or a lower barrier before expiration; see Kunimoto and Ikeda (1992), Geman and Yor (1996), Rogers and Zane (1997), Jamshidian (1997a), and Pelsser (1997). Yet another example of a path-dependent option is the so-called *quantile option*, apparently first studied by Miura (1992); see also Akahori (1995), Dassios (1995), Yor (1995), and Fujita (1997). The reader can consult Rubinstein (1991) for definitions and pricing formulae for a number of exotic options, and Carr (1993) and Zhang (1997) for more comprehensive treatments of the topic.

The notion of *compound options*, or options on options, is not as far-fetched as one might think. Stock can be viewed as an option on the value of a firm; the value of a share of stock in a firm cannot become negative, because if the value of the firm's assets falls below its level of debt, the stockholders have the option of declaring bankruptcy. Thus, a stock option can be regarded as an option on an option. Analysis of compound options is contained in Geske (1979) and Selby and Hodges (1987). In theory, one can price a compound option by simply regarding the underlying option as generating a European contingent claim $C(\cdot)$ that is substituted into the pricing formula (2.9). Similarly interesting are the *exchange options* of the type $C(T) = (S_1(T) - S_2(T))^+$, which give their holder the right to exchange, on their expiration date $t = T$, one asset for another; in the context of a model with constant volatilities, the pricing of such options admits closed-form solutions of the Black–Scholes type, as was demonstrated by Margrabe (1978) (see also Davis (1996) or Karatzas (1996), p. 24).

The "arbitrage-based" approach of these sections is not directly applicable to markets that are *incomplete*, or subject to *portfolio constraints* or *frictions* (such as transaction costs or taxes); in such markets, contingent claims typically cannot be replicated exactly using self-financed portfolios. This problem arises, for example, when the volatility of the underlying stock is stochastic in a way that cannot be hedged by investment in the stock and the money market (e.g., Hull and White (1987, 1988a)). One possible approach, then, is to relax the requirement that portfolios be self-financed (Definition 1.2.1), by requiring that the difference of the two sides in (1.2.8) be a *martingale*—and not identically equal to zero, as postulated by (1.2.8). A portfolio with this property was called *mean-self-financed* by Föllmer and Sondermann (1986). These authors introduced this notion and then established that for any square-integrable contingent claim and discounted price processes that are martingales, there exists a portfolio whose value is almost surely equal to that of the contingent claim at time T and that possesses a certain "risk-minimizing" optimality property; such a portfolio is unique and mean-self-financing. The Föllmer–Sondermann approach was extended by Schweizer (1988, 1990, 1991, 1992a, 1995a) and Föllmer and Schweizer (1991) to semimartingale price processes; in these papers, mean-self-financing portfolios were characterized in terms of a stochastic functional equation, which was shown to have a solution under the assumption that a certain *minimal* martingale measure exists. This measure has the property that although it turns prices into martingales, it does not otherwise change the structure of the model. Using this minimal equivalent martingale measure, Schweizer (1992b) solved a *mean-variance hedging problem* in a particular incomplete market problem, generalizing earlier results of Richardson (1989) and Duffie and Richardson (1991); an excellent survey of these results is Schweizer (1993). Numerical algorithms based on this point of view are provided by Hofmann, Platen, and Schweizer (1992). For these and related quadratic optimization problems, see also Schäl (1994), Schweizer (1995b), Musiela and Rutkowski (1997), Section 4.2, for discrete-time results, and Schweizer (1994, 1996), Delbaen and Schachermayer (1996a), Delbaen, Monat, et al. (1997), Pham et al. (1997) for continuous-time results.

A different, stochastic-control-based, approach to the pricing and hedging problems in incomplete markets was pioneered by El Karoui and Quenez (1991, 1995), building on a duality construction developed by Xu (1990), who was in turn inspired by Bismut (1973) (see the Notes to Chapters 5 and 6 for a fuller discussion of this history). This approach insists on self-financeability of portfolios, but abandons the requirement of exact duplication of the contingent claim; instead, it requires that the value of the portfolio at the terminal time T *dominate the contingent claim almost surely.* The results of El Karoui and Quenez (1991, 1995) were extended by Cvitanić and Karatzas (1993) and Karatzas and Kou (1996, 1998) (see also Kramkov (1996a,b), Föllmer and Kramkov (1997)) to the case of general convex constraints on portfolio choice (of which market incompleteness is but a very special case), and to the case of different interest rates for borrowing and for lending. We take up this topic in Chapter 5.

The arbitrage pricing theory for contingent claims assumes that there are no *costs for transactions,* and typical hedging portfolios require an infinite amount of trading. Figlewski (1989) conducted a simulation study which shows that transaction costs have a significant impact on the cost of hedging options. Indeed, Soner, Shreve, and Cvitanić (1995) recently proved a conjecture, first formalized by Davis and Clark (1994), that in the presence of proportional transaction costs in the continuous-time model, the cheapest way to construct a hedge that dominates a European call option is the trivial and unsatisfactory method of buying one share of the stock on which the call option is written and holding it until expiration; this conjecture was also proved independently by Levental and Skorohod (1997). In anticipation of this negative result, Leland (1985) had considered a continuous-time model in which trades occur at discrete times, and thus a certain "hedge slippage" is unavoidable. However, the total cost of the transaction remains finite, and the Black–Scholes partial differential equation (the one-dimensional version of (4.8)) remains valid, except that the volatility σ must be replaced by a larger constant. As the time between trades approaches zero, the hedge slippage disappears, the adjusted volatility approaches infinity, and the value of the European call approaches the value obtained by Soner et al. (1995). See also Hoggard, Whalley, and Wilmott (1994), Avellaneda and Parás (1994), Kabanov and Safarian (1997), for work that builds on and extends Leland's approach.

Another approach to option pricing in the presence of transaction costs, initiated by Hodges and Neuberger (1989), is to assign utility to the discrepancy between the terminal value of a portfolio and the terminal value of the option, and to set up an optimal portfolio control problem. Some works along these lines are Constantinides (1986, 1993, 1997), Panas (1993), Davis, Panas, and Zariphopoulou (1993), Davis and Panas (1994), Davis and Zariphopoulou (1995), Cvitanić and Karatzas (1996), Constantinides and Zariphopoulou (1997). Barles and Soner (1998) use the utility-based approach to derive a nonlinear Black–Scholes-type equation for option pricing with transaction costs.

In the discrete-time binomial model of Cox, Ross, and Rubinstein (1979), there are replicating and dominating portfolios even in the presence of transaction costs; see Bensaid, Lesne, Pagès, and Scheinkman (1992), Boyle and Vorst (1992), Edirisinghe, Naik, and Uppal (1993), as well as Musiela and Rutkowski (1997), Chapter 2. For additional work on hedging under transaction costs, see Gilster and Lee (1984), Dermody and Rockafellar (1991, 1995), Henrotte (1993), Dewynne, Whalley, and Wilmott (1994), Shirakawa and Konno (1995), Grannan and Swindle (1996), Kusuoka (1995), Jouini and Kallal (1995b), Cvitanić and Karatzas (1996), Kabanov (1997), Cvitanić, Pham, and Touzi (1998). Work on optimal investment and/or consumption in the presence of transaction costs is cited in the notes to Chapter 3.

A specific kind of incompleteness arises when, due to additional sources of randomness that cannot be perfectly hedged, the *volatility is stochastic*. Early work on this subject is by Wiggins (1987), Hull and White (1987, 1988a), Johnson and Shanno (1987), Scott (1987), Chesnay and Scott (1989), Schroder (1989), and is nicely surveyed in Hull (1993), Merton

(1990), Musiela and Rutkowski (1997). El Karoui, Jeanblanc-Picqué, and Viswanathan (1992) construct dominating portfolios when the (stochastic) volatility is known to take values either inside or outside a given interval; see also El Karoui, Jeanblanc-Picqué and Shreve (1998), Hobson (1998), Eberlein and Jacod (1997), and Frey and Sin (1997). Stein and Stein (1991) discuss the distribution of asset prices in such models with stochastic volatility, while Hofman, Platen, and Schweizer (1992), Heston (1993), Dupire (1993, 1994), Platen and Schweizer (1994), (1998), Lyons (1995), Avellaneda, Levy, and Parás (1995), Avellaneda and Parás (1996), Renault and Touzi (1995), Pham and Touzi (1996), Cvitanić, Pham, and Touzi (1997), Romano and Touzi (1997), Scott (1997), Frey and Sin (1997), Sin (1996), Lazrak (1997a,b), deal with various aspects of hedging contingent claims in the framework of such models. Hobson and Rogers (1998) deal with similar questions within stochastic volatility models which are complete.

Randomness in the volatility can arise also as the result of "feedback effects" that relate stock price to the value or debt of a firm, or are due to the trading of large investors or to "finite elasticity" in the market. For option-pricing results in such models, which need not be incomplete, see for instance Cox and Ross (1976), Rubinstein (1983), and the more recent papers by Bensoussan et al. (1994, 1994/95), Cvitanić and Ma (1996), Frey and Stremme (1997), Frey (1998), Papanicolaou and Sircar (1997), Platen and Schweizer (1998) and Hobson and Rogers (1998).

For the very important and huge subject of *real options*—concerning whether or not a firm should invest in new technology or equipment, or hire additional workforce, or develop new products, etc.—which is not touched upon in this book at all, we refer the reader to the monograph by Dixit and Pindyck (1994).

Sections 5-7: A general arbitrage-based theory for the pricing of American contingent claims and options begins with the articles of Bensoussan (1984) and Karatzas (1988); see Myneni (1992a) for a survey and additional references, and Karatzas and Kou (1998) for the hedging of American contingent claims under portfolio constraints. Theorems 6.1, 6.3 are taken from Merton (1973a), which contains a wealth of information and still makes excellent reading. Theorem 6.7 is due to McKean (1965), a "companion" paper to Samuelson (1965a). Van Moerbeke (1976) formulates the free-boundary problem associated with that of Theorem 6.7 but on a finite-time horizon; no explicit solution seems possible in that case, but the author obtains results on the existence and smoothness of the optimal exercise boundary. Peter Carr pointed out that the right-hand sides of formulae (6.30) and (6.31) must agree, and he provided the key part of the "long and painful" computational verification. The results of Section 7 on the American put option are taken from Jacka (1991). For related work see the references in Myneni (1992a), in particular the paper by Carr, Jarrow, and Myneni (1992), who provide a finance explanation of the early exercise premium formula (7.26), as well as Beaghton (1988), Broadie and Detemple (1995, 1997), Carr and Jarrow (1990), El Karoui and Karatzas (1995), Elliott, Myneni, and Viswanathan (1990), Jamshidian (1992), Kim (1990), Zhang (1993), Pham (1995), Mastroeni and Matzeu (1995), Mulinacci and Pratelli (1996).

Approximations and/or numerical solutions for the valuation of American options have been proposed by several authors, including the early work of Black (1975) and Brennan and Schwartz (1977); see also Jaillet et al. (1990) for a treatment of the American option optimal stopping problem via variational inequalities, which leads to a justification of the Brennan–Schwartz algorithm. Kramkov and Vishnyakov (1994) work out formulas for hedging portfolios. Roughly speaking, the most popular methods currently in use are:

(i) *Binomial trees* and their extensions; see for example Cox, Ross, and Rubinstein (1979), and Lamberton (1993, 1995a) for the convergence of the associated binomial and/or finite difference schemes, as well as Boyle (1988), Hull and White (1988b), Rogers and Stapleton (1998), Figlewski and Gao (1997), Reimer and Sandmann (1995).

(ii) *Numerical solution of PDEs and Variational Inequalities;* see, for example, Carr and Faguet (1994), Carr (1998), Wilmott, Dewynne, and Howison (1993, 1995).

(iii) *Analytic approximations,* such as those in Parkinson (1977), Johnson (1983), Geske and Johnson (1984), MacMillan (1986), Omberg (1987), Barone-Adesi and Whaley (1987), Bunch and Johnson (1992), Meyer and Van der Hoek (1995), Broadie and Detemple (1996), Chesney, Elliott, and Gibson (1993), Gao, Huang, and Subrahmanyam (1996).

(iv) *Monte Carlo* and *quasi–Monte Carlo simulation,* as in Tilly (1993), Barraquand and Pudet (1996), who deal with path-dependent American contingent claims, and Broadie and Glasserman (1997). Other papers on option price calculation by Monte Carlo simulation are Boyle (1977), Boyle, Evnine, and Gibbs (1989), Duffie and Glynn (1995), Paskov and Traub (1995), Barraquand and Martineau (1995), Boyle, Broadie, and Glasserman (1997), Schoenmakers and Heemink (1997). Lehoczky (1997) surveys Monte Carlo variance reduction methods for finance applications.

Barles et al. (1995) and Lamberton (1995b) study asymptotic properties of the critical stock-price near expirations. For surveys of the extant numerical work on American and path-dependent contingent claims, the interested reader should consult Carverhill and Webber (1990), Hull (1993), and Wilmott, Dewynne, and Howison (1993) and Broadie and Detemple (1996).

3

Single-Agent Consumption and Investment

3.1 Introduction

This chapter solves the problem of an agent who begins with an initial endowment and who can consume while also investing in a standard, complete market as set forth in Chapter 1. The objective of this agent is to maximize the expected utility of consumption over the planning horizon, or to maximize the expected utility of wealth at the end of the planning horizon, or to maximize some combination of these two quantities. Except for the completeness assumption, the market model is quite general, allowing the coefficient processes to be stochastic processes that are not even assumed to be Markovian. Specializations of this model to the case of deterministic and even constant coefficients are provided in Sections 3.8 and 3.9. The problem of this chapter is revisited in the context of incomplete markets in Chapter 6.

The agent acting in this chapter is assumed to be a "small investor," in the sense that his actions do not influence market prices. Chapter 4 considers the *equilibrium problem* of several agents whose joint actions determine market prices through the law of supply and demand. The model of this section is a basic building block in such equilibrium models, and these, in turn, are the basis for a theory of financial markets. Such equilibrium models can be used, for example, to study possible effects of taxation and market regulation. The Notes to this chapter contain a further discussion of the relationship between this material and capital-asset pricing models.

In addition to its role in theoretical studies of financial markets, the model of this section can form the basis of portfolio management. This requires, of course, that the investment manager either assume a utility function or else elicit a utility function from the client whose money is being managed. We show in Section 3.10 that maximization of the logarithm of terminal wealth results in maximization of the growth rate of wealth, and this is consequently a frequently used utility function. The class of utility functions of the form $U^{(p)}(x) = x^p/p$ for $x \geq 0$, where $p \neq 0$ is a number strictly less than one, are also commonly used. For this one-parameter family of utility functions, the *Arrow–Pratt index of relative risk aversion,*

$$J(x) \triangleq -x \frac{\partial^2}{\partial x^2} U^{(p)}(x) / \frac{\partial}{\partial x} U^{(p)}(x),$$
$$= 1 - p,$$

decreases with increasing p. One can elicit from an investor some measure of risk aversion and attempt to determine the corresponding utility function from this one-parameter family.

Sections 3.2 and 3.3 describe the market model and the set of consumption and portfolio processes from which the investor in this market is free to choose. Section 3.4 introduces the notion of utility function. We allow these functions to take the value $-\infty$ on a half-line extending to $-\infty$, which effectively places a lower constraint on consumption and/or wealth. Section 3.6 solves the problem of an agent who seeks to maximize expected utility from consumption plus expected utility from terminal wealth. The method of solution uses the *convex dual* function (Legendre transform) of the utility function. Related to this concept, we introduce and study the convex dual of the value function for the problem of Section 3.6. This foreshadows a duality theory that plays a critical role in the analysis of incomplete markets in later chapters.

Section 3.7 considers the problem of maximization of expected utility from consumption only, and the antithetical problem of maximization of expected utility from terminal wealth only. These problems are related to that of Section 3.6 in the following way. Given an initial endowment x, an agent who wishes to maximize the expected utility of consumption plus the expected utility of terminal wealth can partition his initial endowment into two parts, x_1 and x_2, such that $x = x_1 + x_2$. Beginning with initial endowment x_1, the agent should solve the problem of maximizing expected utility from consumption only; with x_2, he should solve the problem of maximizing expected utility from terminal wealth only. The superposition of these two solutions is then the solution to the problem of maximizing expected utility from consumption plus expected utility from terminal wealth. The partition of wealth that accomplishes this decomposition of the problems is derived in Section 3.7.

The results of Sections 3.6 and 3.7 are specialized to models with deterministic coefficients in Section 3.8. For such models a Markov-based analysis is provided, including the development of the *Hamilton–Jacobi–Bellman equation* and the optimal consumption and portfolio processes as *feedback* functions of the agent's wealth. The Hamilton–Jacobi–Bellman equation of Section 3.8 is a second-order parabolic differential equation. When the model coefficients are constant and the planning horizon is infinite, the Hamilton–Jacobi–Bellman equation is a second-order ordinary differential equation, and lends itself to very explicit analysis; this is the subject of Section 3.9.

3.2 The Financial Market

As in Chapter 2, we shall work in this chapter in the context of a *complete, standard financial market* $\mathcal{M} = (r(\cdot), b(\cdot), \delta(\cdot), \sigma(\cdot), S(0), A(\cdot))$ (see Definitions 1.1.3, 1.5.1, 1.6.1 and Theorem 1.6.6). In particular, the price of the money market is governed by

$$dS_0(t) = S_0(t)\left[r(t)\,dt + dA(t)\right], \tag{2.1}$$

and the prices of the stocks satisfy

$$dS_n(t) = S_n(t)\left[b_n(t)\,dt + dA(t) + \sum_{d=1}^{N} \sigma_{nd}(t)\,dW^{(d)}(t)\right],$$
$$n = 1, \ldots, N, \tag{2.2}$$

with $\sigma(t) = (\sigma_{nd}(t))_{1\le n,d\le N}$ nonsingular for Lebesgue-almost-every $t \in [0, T]$ almost surely. In Sections 3.9 and 3.10, we place the financial market on the infinite planning horizon $[0, \infty)$ (Definitions 1.7.2, 1.7.3). Recall that $S_0(0) = 1$ and $S_1(0), \ldots, S_N(0)$ are positive constants, and recall from (1.6.16), (1.5.6), (1.5.2), and (1.5.12) the processes

$$\theta(t) \stackrel{\Delta}{=} \sigma^{-1}(t)[b(t) + \delta(t) - r(t)\mathbb{1}], \tag{2.3}$$

$$W_0(t) \stackrel{\Delta}{=} W(t) + \int_0^t \theta(s)\,ds, \tag{2.4}$$

$$Z_0(t) \stackrel{\Delta}{=} \exp\left\{-\int_0^t \theta'(s)\,dW(s) - \frac{1}{2}\int_0^t \|\theta(s)\|^2\,ds\right\}, \tag{2.5}$$

$$H_0(t) \stackrel{\Delta}{=} \frac{Z_0(t)}{S_0(t)}. \tag{2.6}$$

In a standard market, the exponential local martingale $Z_0(\cdot)$ is in fact a martingale, which permits the definition of the standard martingale measure P_0 (see Definition 1.5.1 and equation (1.5.3)). Except in Section 3.8, we shall present the analysis of this chapter in a way that uses only the *local martingale property* of $Z_0(\cdot)$ and avoids the use of P_0 altogether. This permits the present analysis also to be used in the study of constrained and incomplete markets in Chapter 6.

Until further notice, we take T to be finite and restrict attention to the finite-horizon model on $[0, T]$. For this model, one of the following conditions will be imposed.

Assumption 2.1: *The state price density process H_0 satisfies*

$$E\left[\int_0^T H_0(t)\, dt\right] < \infty.$$

Assumption 2.2: *The state price density process H_0 satisfies*

$$EH_0(T) < \infty.$$

Assumption 2.3: *The state price density process H_0 satisfies*

$$E\left[\int_0^T H_0(t)\, dt + H_0(T)\right] < \infty.$$

A sufficient condition for these assumptions is that $S_0(\cdot)$ be bounded away from zero on $[0, T]$, so that $H_0(\cdot)$ is bounded from above by a constant times the nonnegative supermartingale $Z_0(\cdot)$.

3.3 Consumption and Portfolio Processes

An agent will act in the financial market of the previous section by choosing a *consumption process* and a *portfolio process*. In this section, these entities are defined, and it is shown as a consequence of the *admissibility condition* (3.2) that these processes must satisfy the *budget constraint* (3.4). It is further shown that if one starts with a consumption process and a nonnegative random variable that satisfy the budget constraint, then there is a *hedging portfolio* that, together with the given consumption process, results in terminal wealth equal to the given nonnegative random variable.

Definition 3.1: A *consumption process* is an $\{\mathcal{F}(t)\}$-progressively measurable, nonnegative process $c(\cdot)$ satisfying $\int_0^T c(t)\, dt < \infty$, almost surely.

An agent with initial endowment $x \geq 0$ who chooses a consumption process $c(\cdot)$ will have a cumulative income process $\Gamma(t) \overset{\Delta}{=} x - \int_0^t c(u)\, du, 0 \leq t \leq T$. If this investor chooses a $\Gamma(\cdot)$-financed portfolio process $\pi(\cdot)$, then his corresponding wealth process $X^{x,c,\pi}(\cdot)$ will be governed by equation (1.5.8):

$$\frac{X^{x,c,\pi}(t)}{S_0(t)} = x - \int_0^t \frac{c(u)\, du}{S_0(u)} + \int_0^t \frac{1}{S_0(u)}\pi'(u)\sigma(u)\, dW_0(u),$$

$$0 \leq t \leq T. \tag{3.1}$$

Definition 3.2: Given $x \geq 0$, we say that a consumption and portfolio process pair (c, π) is *admissible at* x, and write $(c, \pi) \in \mathcal{A}(x)$, if the wealth process $X^{x,c,\pi}(\cdot)$ corresponding to x, c, π satisfies

$$X^{x,c,\pi}(t) \geq 0, \quad 0 \leq t \leq T, \tag{3.2}$$

almost surely. For $x < 0$, we set $\mathcal{A}(x) = \emptyset$.

Remark 3.3: From (1.5.14), we have

$$H_0(t)X^{x,c,\pi}(t) + \int_0^t H_0(u)c(u)\, du$$

$$= x + \int_0^t H_0(u)\left[\sigma'(u)\pi(u) - X^{x,c,\pi}(u)\theta(u)\right]'\, dW(u),$$

$$0 \leq t \leq T. \tag{3.3}$$

When $(c, \pi) \in \mathcal{A}(x)$, the left-hand side of (3.3) is nonnegative, and so the Itô integral on the right side is not only a local martingale under P, but also a supermartingale (Fatou's lemma). This implies that the so-called *budget constraint*

$$E\left[\int_0^T H_0(u)c(u)\, du + H_0(T)X^{x,c,\pi}(T)\right] \leq x \tag{3.4}$$

is satisfied for every $(c, \pi) \in \mathcal{A}(x)$. The budget constraint has the satisfying interpretation that the *expected "discounted" terminal wealth plus the expected "discounted" total consumption cannot exceed the initial endowment.* Here the "discounting" is accomplished by the state price density process H_0.

Remark 3.4: Bankruptcy is an absorbing state for the wealth process $X^{x,c,\pi}(\cdot)$ when $(c, \pi) \in \mathcal{A}(x)$; if wealth becomes zero before time T, it stays there, and no further consumption or investment takes place. To see this, note that because the left-hand side of (3.3) is a supermartingale, so is the process $H_0(t)X^{x,c,\pi}(t), 0 \leq t \leq T$, for every $(c, \pi) \in \mathcal{A}(x)$. With $\tau_0 \overset{\Delta}{=} T \wedge \inf\{t \in [0,T]; X(t) = 0\}$, we have then

$$X^{x,c,\pi}(t, \omega) = 0, \quad \forall t \in [\tau_0(\omega), T],$$

for P-almost-every $\omega \in \{\tau_0 < T\}$ (e.g., Karatzas and Shreve (1991), Problem 1.3.29). On the other hand, the optional sampling theorem applied to the left-hand side of (3.3) gives

$$E\left[\int_{\tau_0}^T H_0(t)c(t)\, dt + H_0(T)X^{x,c,\pi}(T)\,\middle|\, \mathcal{F}(\tau_0)\right] \leq H_0(\tau_0)X(\tau_0)$$

almost surely, and thus, for P-a.e. $\omega \in \{\tau_0 < T\}$,

$$c(t, \omega) = 0, \pi(t, \omega) = 0 \quad \text{for Lebesgue-a.e. } t \in [\tau_0(\omega), T].$$

The budget constraint (3.4) is not only a necessary condition for admissibility, but is also a sufficient condition, in a sense that we now explain.

Theorem 3.5: *Let $x \geq 0$ be given, let $c(\cdot)$ be a consumption process, and let ξ be a nonnegative, $\mathcal{F}(T)$-measurable random variable such that*

$$E\left[\int_0^T H_0(u)c(u)\, du + H_0(T)\xi\right] = x. \tag{3.5}$$

Then there exists a portfolio process $\pi(\cdot)$ such that the pair (c, π) is admissible at x and $\xi = X^{x,c,\pi}(T)$.

PROOF. Let us define $J(t) \triangleq \int_0^t H_0(u)c(u)\, du$ and consider the nonnegative martingale

$$M(t) \triangleq E\left[J(T) + H_0(T)\xi \mid \mathcal{F}(t)\right], \quad 0 \leq t \leq T.$$

According to the martingale representation theorem (e.g., Karatzas and Shreve (1991), Theorem 3.4.15 and Problem 3.4.16), there is a progressively measurable, $\mathbb{R}^d$-valued process $\psi(\cdot)$ satisfying

$$\|\psi\|_2^2 \triangleq \int_0^T \|\psi(u)\|^2\, du < \infty$$

almost surely and

$$M(t) = x + \int_0^t \psi'(u)\, dW(u), \quad 0 \leq t \leq T.$$

In particular, $M(\cdot)$ has continuous paths, and

$$\|M\|_\infty \triangleq \max_{0 \leq t \leq T} |M(t)| < \infty$$

almost surely. Similarly, $\|J\|_\infty \triangleq \max_{0 \leq t \leq T} J(t)$, $\|S_0\|_\infty \triangleq \max_{0 \leq t \leq T} S_0(t)$, and $\kappa \triangleq \max_{0 \leq t \leq T} 1/Z_0(t)$ are finite almost surely.

Define a nonnegative process $X(\cdot)$ by

$$\frac{X(t)}{S_0(t)} \triangleq \frac{1}{Z_0(t)} E\left[\int_t^T H_0(u)c(u)\, du + H_0(T)\xi \,\Big|\, \mathcal{F}(t)\right]$$

$$= \frac{1}{Z_0(t)}\left[M(t) - J(t)\right], \tag{3.6}$$

so that $X(0) = M(0) = x$. Itô's rule implies

$$d\left(\frac{X(t)}{S_0(t)}\right) = -\frac{c(t)}{S_0(t)}\, dt + \frac{1}{S_0(t)}\pi'(t)\sigma(t)\, dW_0(t),$$

where

$$\pi(t) \triangleq \frac{1}{H_0(t)} \left(\sigma'(t)\right)^{-1} \left[\psi(t) + (M(t) - J(t))\, \theta(t)\right]. \tag{3.7}$$

We check that $\pi(\cdot)$ satisfies (1.2.5), (1.2.6), and hence is a portfolio process. From Remark 1.6.10 and equation (1.5.1), we have that $\theta(t) = \sigma^{-1}(t) \left[b(t) + \delta(t) - r(t)\underline{1} \right]$ satisfies $\|\theta\|_2 \overset{\triangle}{=} (\int_0^T \|\theta(t)\|^2 \, dt)^{\frac{1}{2}} < \infty$ almost surely. Therefore,

$$\int_0^T \left| \pi'(t) \big(b(t) + \delta(t) - r(t)\underline{1} \big) \right| dt$$

$$= \int_0^T \frac{S_0(t)}{Z_0(t)} \left| \psi'(t)\theta(t) + \|\theta(t)\|^2 \left(M(t) - J(t) \right) \right| dt$$

$$\leq \kappa \|S_0\|_\infty \left[\|\psi\|_2 \|\theta\|_2 + \|\theta\|_2^2 \left(\|M\|_\infty + \|J\|_\infty \right) \right]$$

$$< \infty$$

almost surely, and (1.2.5) holds. Similarly,

$$\int_0^T \|\sigma'(t)\pi(t)\|^2 \, dt = \int_0^T \frac{S_0^2(t)}{Z_0^2(t)} \|\psi(t) + \theta(t) \left(M(t) - J(t) \right)\|^2 \, dt$$

$$\leq \kappa^2 \|S_0\|_\infty^2 \Big[\|\psi\|_2^2 + \|\psi\|_2 \|\theta\|_2 \left(\|M\|_\infty + \|J\|_\infty \right)$$

$$+ \|\theta\|_2^2 \left(\|M\|_\infty + \|J\|_\infty \right)^2 \Big] < \infty$$

almost surely, and (1.2.6) holds as well.

We conclude that

$$\frac{X(t)}{S_0(t)} = x - \int_0^t \frac{c(u) \, du}{S_0(u)} + \int_0^t \frac{1}{S_0(u)} \pi'(u)\sigma(u) \, dW_0(u), \ 0 \leq t \leq T,$$

and comparison with (3.1) shows that $X(\cdot) = X^{x,c,\pi}(\cdot)$. Since $X(t) \geq 0$ for $0 \leq t \leq T$, the pair (c,π) is admissible. Finally, $X(T) = \frac{S_0(T)}{Z_0(T)} E[H_0(T)\xi | \mathcal{F}(T)] = \xi$ almost surely.   □

3.4   Utility Functions

The agent in this chapter desires to maximize his utility. In this section we develop the properties of the utility functions we consider. We also introduce the *convex dual* of a utility function.

Definition 4.1: A *utility function* is a concave, nondecreasing, upper semicontinuous function $U:\mathbb{R} \to [-\infty, \infty)$ satisfying:

(i) the half-line $dom(U) \overset{\triangle}{=} \{x \in \mathbb{R}; U(x) > -\infty\}$ is a nonempty subset of $[0, \infty)$;

(ii) U' is continuous, positive, and strictly decreasing on the interior of $dom(U)$, and

$$U'(\infty) \overset{\triangle}{=} \lim_{x \to \infty} U'(x) = 0. \tag{4.1}$$

We set

$$\overline{x} \overset{\Delta}{=} \inf\{x \in \mathbb{R}; U(x) > -\infty\} \tag{4.2}$$

so that $\overline{x} \in [0, \infty)$ and either $dom(U) = [\overline{x}, \infty)$ or $dom(U) = (\overline{x}, \infty)$. We define

$$U'(\overline{x}+) \overset{\Delta}{=} \lim_{x \downarrow \overline{x}} U'(x), \tag{4.3}$$

so that $U'(\overline{x}+) \in (0, \infty]$.

Here are some common utility functions. Take $p \in (-\infty, 1) \setminus \{0\}$ and set

$$U^{(p)}(x) \overset{\Delta}{=} \begin{cases} x^p/p, & x > 0, \\ \lim_{\xi \downarrow 0} \xi^p/p, & x = 0, \\ -\infty, & x < 0. \end{cases} \tag{4.4}$$

For $p = 0$, set

$$U^{(0)}(x) \overset{\Delta}{=} \begin{cases} \log x, & x > 0, \\ -\infty, & x \le 0. \end{cases} \tag{4.5}$$

The *Arrow–Pratt index of risk aversion*, $-\frac{xU''(x)}{U'(x)}$ for $U^{(p)}$, is $1 - p$. Other utility functions are $U^{(p)}(x - \overline{x})$, where $\overline{x}$ is a positive constant.

Let U be a utility function with $\overline{x}$ given by (4.2). The strictly decreasing, continuous function $U': (\overline{x}, \infty) \overset{onto}{\longrightarrow} (0, U'(\overline{x}+))$ has a strictly decreasing, continuous inverse $I: (0, U'(\overline{x}+)) \overset{onto}{\longrightarrow} (\overline{x}, \infty)$. We set $I(y) = \overline{x}$ for $U'(\overline{x}+) \le y \le \infty$, so that I is defined, finite, and continuous on the extended half-line $(0, \infty]$, and

$$U'(I(y)) = \begin{cases} y, & 0 < y < U'(\overline{x}+), \\ U'(\overline{x}+), & U'(\overline{x}+) \le y \le \infty, \end{cases} \tag{4.6}$$

$$I(U'(x)) = x, \quad \overline{x} < x < \infty. \tag{4.7}$$

Definition 4.2: Let U be a utility function. The *convex dual* of U is the function

$$\widetilde{U}(y) \overset{\Delta}{=} \sup_{x \in \mathbb{R}}\{U(x) - xy\}, \quad y \in \mathbb{R}. \tag{4.8}$$

Except for the presence of some minus signs, $\widetilde{U}$ is the Legendre–Fenchel transform of U (Rockafellar (1970), Ekeland and Temam (1976)). Indeed, if we define the convex function

$$f(x) \overset{\Delta}{=} -U(x), \quad x \in \mathbb{R}, \tag{4.9}$$

then the Legendre–Fenchel transform of f is

$$f^*(y) \overset{\Delta}{=} \sup_{x \in \mathbb{R}}\{xy - f(x)\} = \widetilde{U}(-y), \quad y \in \mathbb{R}. \tag{4.10}$$

Lemma 4.3: *Let U and $\overline{x}$ be as in Definition 4.1, I as in (4.6), (4.7), and let $\widetilde{U}$ be the convex dual of U. Then $\widetilde{U} : \mathbb{R} \to (-\infty, \infty]$ is convex, nonincreasing, lower semicontinuous, and satisfies*

(i)

$$\widetilde{U}(y) = \begin{cases} U\left(I(y)\right) - yI(y), & y > 0, \\ U(\infty) \overset{\triangle}{=} \lim_{x \to \infty} U(x), & y = 0, \\ \infty, & y < 0. \end{cases} \quad (4.11)$$

(ii) *The derivative $\widetilde{U}'$ is defined, continuous, and nondecreasing on $(0, \infty)$, and*

$$\widetilde{U}'(y) = -I(y), \quad 0 < y < \infty. \quad (4.12)$$

(iii) *For all $x \in \mathbb{R}$,*

$$U(x) = \inf_{y \in \mathbb{R}} \{\widetilde{U}(y) + xy\}. \quad (4.13)$$

(iv) *For fixed $x \in (\overline{x}, \infty)$, the function $y \mapsto \widetilde{U}(y) + xy$ is uniquely minimized over $\mathbb{R}$ by $y = U'(x)$; i.e.,*

$$U(x) = \widetilde{U}\left(U'(x)\right) + xU'(x). \quad (4.14)$$

PROOF. According to Rockafellar (1970), Theorem 12.2, the function $\widetilde{U}$ is lower semicontinuous, convex, takes values in $(-\infty, \infty]$, and is related to U via (4.13). Equation (4.11) is easily verified directly from the definition of $\widetilde{U}$.

According to Rockafellar (1970), Theorem 23.5, the function $\xi \mapsto U(\xi) - y\xi$ is maximized at $\xi = x$ if and only if $-x \in \partial\widetilde{U}(y)$. But for $0 < y < \infty$, this function is uniquely maximized by $\xi = I(y)$, whence (4.12) holds. From (4.12) we see that $\widetilde{U}'$ is continuous and nonincreasing on $(0, \infty)$.

Finally, for fixed $x \in (\overline{x}, \infty)$, the convex function $y \mapsto \widetilde{U}(y) + xy$ has derivative $-I(y) + x$ for $y \in (0, \infty)$. This derivative is zero at $y = U'(x)$ (see (4.7)), which gives us (iv). $\square$

Remark 4.4: We shall usually consider utility functions U for which $\overline{x}$ given by (4.2) is zero. For such a function, we shall impose sometimes the following additional conditions:

(a) $x \mapsto xU'(x)$ is nondecreasing on $(0, \infty)$;   (4.15)
(b) For some $\beta \in (0, 1), \gamma \in (1, \infty)$, we have

$$\beta U'(x) \geq U'(\gamma x) \quad \forall x \in (0, \infty). \quad (4.16)$$

The first of these conditions is equivalent to

(c) $y \mapsto y\widetilde{U}'(y)$ is nondecreasing on $(0, U'(0+))$
       (set $y = U'(x)$ and use (4.7), (4.12)).   (4.15')

Condition (c) implies that

$$z \mapsto \widetilde{U}(e^z) \quad \text{is convex on } \mathbb{R}; \quad (4.15'')$$

Furthermore, for a utility function U of class $C^2(0, \infty)$, condition (4.4) implies that the so-called *Arrow–Pratt index of relative risk aversion*

$$J(x) \triangleq -\frac{xU''(x)}{U'(x)} \qquad (4.17)$$

does not exceed 1.

Condition (4.16) is equivalent to having

$$\widetilde{U}'(\beta y) \geq \gamma \widetilde{U}'(y) \quad \forall y \in (0, U'(0+)) \text{ for some } \beta \in (0, 1), \gamma \in (1, \infty) \tag{4.16'}$$

(just replace x by $-\widetilde{U}'(y)$ in (4.16), and then apply $\widetilde{U}'$ to both sides of the resulting inequality). By iterating (4.16'), one obtains the apparently stronger statement

$$\forall \beta \in (0, 1), \exists \gamma \in (1, \infty) \text{ such that } \widetilde{U}'(\beta y) \geq \gamma \widetilde{U}'(y) \ \forall y \in (0, U'(0+)). \tag{4.16''}$$

It should also be noted, in the notation of (4.4), (4.5), that for $\gamma \geq 0$ and $0 < p < 1$, condition (4.4) is satisfied by the function

$$U(x) \triangleq \begin{cases} U^{(p)}(x + \gamma), & x \geq 0, \\ -\infty, & x < 0, \end{cases}$$

whereas (4.16) is satisfied by $U^{(p)}$ for all $p \in (-\infty, 1)$.

3.5 The Optimization Problems

We formulate three optimization problems for an agent. This agent is sometimes called a *small investor* because his actions do not affect the prices of financial assets.

Definition 5.1: A (time-separable, von Neumann–Morgenstern) *preference structure* is a pair of functions $U_1: [0, T] \times \mathbb{R} \to [-\infty, \infty)$ and $U_2: \mathbb{R} \to [-\infty, \infty)$ as described below:

(i) For each $t \in [0, T]$, $U_1(t, \cdot)$ is a utility function (Definition 4.1), and the *subsistence consumption*

$$\bar{c}(t) \triangleq \inf \{c \in \mathbb{R}; U_1(t, c) > -\infty\}, \quad 0 \leq t \leq T, \tag{5.1}$$

is a continuous function of t, with values in $[0, \infty)$;

(ii) U_1 and U_1' (where the prime denotes differentiation with respect to the second argument) are continuous on the set

$$D_1 \triangleq \{(t, c) \in [0, T] \times (0, \infty); c > \bar{c}(t)\}; \tag{5.2}$$

(iii) U_2 is a utility function, with *subsistence terminal wealth* defined by

$$\bar{x} \stackrel{\Delta}{=} \inf \left\{ x \in \mathbb{R}; U_2(x) > -\infty \right\}. \tag{5.3}$$

Let an agent have an initial endowment $x \in \mathbb{R}$ and a preference structure (U_1, U_2). The agent can consider three problems whose elements of control are the admissible consumption and portfolio processes in $\mathcal{A}(x)$ of Definition 3.2.

Problem 5.2: Find an optimal pair $(c_1, \pi_1) \in \mathcal{A}_1(x)$ for the problem

$$V_1(x) \stackrel{\Delta}{=} \sup_{(c,\pi)\in\mathcal{A}_1(x)} E \int_0^T U_1\left(t, c(t)\right) dt \tag{5.4}$$

of maximizing expected total utility from consumption over $[0, T]$, where

$$\mathcal{A}_1(x) \stackrel{\Delta}{=} \left\{ (c, \pi) \in \mathcal{A}(x); E \int_0^T \min\left[0, U_1\left(t, c(t)\right)\right] dt > -\infty \right\}. \tag{5.5}$$

Problem 5.3: Find an optimal pair $(c_2, \pi_2) \in \mathcal{A}_2(x)$ for the problem

$$V_2(x) \stackrel{\Delta}{=} \sup_{(c,\pi)\in\mathcal{A}_2(x)} E U_2\left(X^{x,c,\pi}(T)\right) \tag{5.6}$$

of maximizing expected utility from terminal wealth, where

$$\mathcal{A}_2(x) \stackrel{\Delta}{=} \left\{ (c, \pi) \in \mathcal{A}(x); E \min\left[0, U_2\left(X^{x,c,\pi}(T)\right)\right] > -\infty \right\}. \tag{5.7}$$

Problem 5.4: Find an optimal pair $(c_3, \pi_3) \in \mathcal{A}_3(x)$ for the problem

$$V_3(x) \stackrel{\Delta}{=} \sup_{(c,\pi)\in\mathcal{A}_3(x)} E\left[\int_0^T U_1\left(t, c(t)\right) dt + U_2\left(X^{x,c,\pi}(T)\right)\right] \tag{5.8}$$

of maximizing expected total utility from both consumption and terminal wealth, where

$$\mathcal{A}_3(x) \stackrel{\Delta}{=} \mathcal{A}_1(x) \cap \mathcal{A}_2(x). \tag{5.9}$$

Of course, since $\mathcal{A}(x) = \emptyset$ for $x < 0$, we have $\mathcal{A}_i(x) = \emptyset$ for $x < 0$ and $i = 1, 2, 3$. We adopt the convention that the supremum over the empty set is $-\infty$. In the next sections we shall strive to compute the *value functions* V_1, V_2, and V_3 of these problems and to characterize (or even compute) optimal pairs $(c_i, \pi_i), i = 1, 2, 3$, that attain the suprema in (5.4), (5.6), and (5.8), respectively.

Remark 5.5: In Problems 5.2–5.4, we could allow U_1 (respectively, U_2) also to depend on $\omega \in \Omega$ in an $\{\mathcal{F}(t)\}$-progressively measurable (respectively, $\mathcal{F}(T)$-measurable) manner. The analysis of the subsequent sections would be unaffected.

In the remainder of this section, we develop for future use technical results concerning some of the functions introduced here. The reader may wish to postpone this material until these results are actually used in subsequent sections.

Remark 5.6: Let (U_1, U_2) be a preference structure.

(i) Because $\bar{c}(\cdot)$ is continuous, there exists a finite number $\hat{c}$ such that $\hat{c} > \bar{x} \vee \max_{0 \leq t \leq T} \bar{c}(t)$. From the continuity of U_1 on $D_1 \supset [0, T] \times [\hat{c}, \infty)$, we have

$$\int_0^T |U_1(t, \hat{c})| \, dt + |U_2(\hat{c})| < \infty. \tag{5.10}$$

Furthermore, under the respective Assumptions 2.1–2.3, the respective quantities

$$\mathcal{X}_1(\infty) \triangleq E \int_0^T H_0(t) \bar{c}(t) \, dt, \quad \mathcal{X}_2(\infty) \triangleq E\left[H_0(T)\bar{x}\right], \tag{5.11}$$

$$\mathcal{X}_3(\infty) \triangleq E\left[\int_0^T H_0(t)\bar{c}(t) \, dt + H_0(T)\bar{x}\right] \tag{5.12}$$

are finite.

(ii) From (4.1), we have $\lim_{c \to \infty} U_1'(t, c) = 0$ for all $t \in [0, T]$. In fact, the following stronger statement holds:

$$\lim_{c \to \infty} \max_{0 \leq t \leq T} U_1'(t, c) = 0, \tag{5.13}$$

as one can see by considering, for fixed $\epsilon > 0$, the nested sequence of compact sets $K_n(\epsilon) \triangleq \{t \in [0, T]; U_1'(t, \hat{c} + n) \geq \epsilon\}$, $n = 1, 2, \ldots$. These sets have empty intersection, so for some integer n, we have $K_n(\epsilon) = \emptyset$.

Remark 5.7: For Problem 5.4, the agent must have initial wealth at least $\mathcal{X}_3(\infty)$ in order to avoid expected utility of $-\infty$. Indeed, for this problem, the preference structure forces the constraints

$$c(t) \geq \bar{c}(t), \quad \text{a.e.}\, t \in [0, T], \tag{5.14}$$

$$X^{x,c,\pi}(T) \geq \bar{x} \tag{5.15}$$

almost surely, for otherwise $E[\int_0^T U_1(t, c(t)) \, dt + U_2(X^{x,c,\pi}(T))]$ would be $-\infty$. But (5.14), (5.15), and (5.12) imply

$$E\left[\int_0^T H_0(t)c(t) \, dt + H_0(T)X^{x,c,\pi}(T)\right] \geq \mathcal{X}_3(\infty). \tag{5.16}$$

Recalling the budget constraint (3.4), we see that $V_3 = -\infty$ on the half-line $(-\infty, \mathcal{X}_3(\infty))$. For $x = \mathcal{X}_3(\infty)$, any $(c, \pi) \in \mathcal{A}_3(x)$ satisfying (5.14), (5.15) must actually satisfy $c(t) = \bar{c}(t)$, $X^{x,c,\pi}(T) = \bar{x}$. According to Theorem 3.5, there is in fact a portfolio process $\bar{\pi}$ for which $X^{\mathcal{X}_3(\infty),\bar{c},\bar{\pi}}(T) = \bar{x}$, and we conclude that

$$V_3(x) = \begin{cases} \int_0^T U_1\left(t, \bar{c}(t)\right) dt + U_2(\bar{x}), & x = \mathcal{X}_3(\infty), \\ -\infty, & x < \mathcal{X}_3(\infty). \end{cases} \qquad (5.17)$$

From (5.10), this last expression is well-defined, although it may be $-\infty$, in which case $\mathcal{A}_3\left(\mathcal{X}_3(\infty)\right) = \emptyset$ and $V_3\left(\mathcal{X}_3(\infty)\right) = -\infty$, in accordance with (5.17). Similar arguments show that

$$V_1(x) = \begin{cases} \int_0^T U_1\left(t, \bar{c}(t)\right) dt, & x = \mathcal{X}_1(\infty), \\ -\infty & x < \mathcal{X}_1(\infty), \end{cases} \qquad (5.18)$$

$$V_2(x) = \begin{cases} U_2(\bar{x}), & x = \mathcal{X}_2(\infty), \\ -\infty, & x < \mathcal{X}_2(\infty). \end{cases} \qquad (5.19)$$

Let (U_1, U_2) be a preference structure. For fixed $t \in [0, T]$, the function $I_1(t, \cdot) : (0, \infty] \xrightarrow{\text{onto}} [\bar{c}(t), \infty)$, satisfying the analogue

$$U_1'\left(t, I_1(t, y)\right) = \begin{cases} y, & 0 < y < U_1'\left(t, \bar{c}(t)+\right), \\ U_1'\left(t, \bar{c}(t)+\right), & U_1'\left(t, \bar{c}(t)+\right) \le y \le \infty \end{cases} \qquad (5.20)$$

of (4.6), is strictly decreasing on $\left(0, U_1'\left(t, \bar{c}(t)+\right)\right]$, is identically equal to $\bar{c}(t)$ on $\left[U_1'\left(t, \bar{c}(t)+\right), \infty\right]$, and is continuous on all of $(0, \infty]$. Similar remarks apply to the function $I_2 : (0, \infty] \longrightarrow [\bar{x}, \infty)$ that satisfies

$$U_2'\left(I_2(y)\right) = \begin{cases} y, & 0 < y < U_2'(\bar{x}+), \\ U_2'(\bar{x}+), & U_2'(\bar{x}+) \le y \le \infty. \end{cases} \qquad (5.21)$$

Lemma 5.8: *The function I_1 of (5.20) is jointly continuous on $[0, T] \times (0, \infty]$.*

PROOF. Let $(t_0, y_0) \in [0, T] \times (0, \infty]$ be given, and let $\{(t_n, y_n)\}_{n=1}^\infty$ be a sequence in the same set with limit (t_0, y_0). Define $c_n = I_1(t_n, y_n), n = 0, 1, \ldots$. We need to show that $c_0 = \lim_{n \to \infty} c_n$.

Let us consider separately the two cases:

Case 1: $y_0 < U_1'\left(t_0, \bar{c}(t_0)+\right)$.

For some $\gamma > 0$, we have $y_0 < U_1'\left(t_0, \bar{c}(t_0) + \gamma\right)$. Continuity of $\bar{c}$ and of U_1' on D_1 implies $y_n < U_1'\left(t_n, \bar{c}(t_n) + \gamma\right)$ for n sufficiently large, and thus $c_n \ge \bar{c}(t_n) + \gamma$ and

$$y_n = U_1'(t_n, c_n). \qquad (5.22)$$

The sequence $\{c_n\}_{n=1}^\infty$ is bounded, for otherwise (5.13) would imply $0 = \underline{\lim}_{n \to \infty} U_1'(t_n, c_n) = \underline{\lim}_{n \to \infty} y_n$, which violates the assumption $\lim_{n \to \infty} y_n = y_0 > 0$. Any accumulation point c_0^* of $\{c_n\}_{n=1}^\infty$ must satisfy $c_0^* \ge \bar{c}(t_0) + \gamma$; thus, the continuity of U_1' on D_1 and (5.22) imply $y_0 = U_1'(t_0, c_0^*)$ and $c_0^* = I_1(t_0, y_0) = c_0$. Consequently, $\lim_{n \to \infty} c_n = c_0$.

Case 2: $y_0 \geq U_1'(t_0, \bar{c}(t_0)+)$.

Now we have $c_0 = \bar{c}(t_0)$. We divide the index set into $N_1 \overset{\Delta}{=} \{n \geq 1; y_n \geq U_1'(t_n, \bar{c}(t_n)+)\}$ and $N_2 \overset{\Delta}{=} \{n \geq 1; y_n < U_1'(t_n, \bar{c}(t_n)+)\}$. For $n \in N_1$, we have $c_n = \bar{c}(t_n)$, so $\lim_{n \in N_1} c_n = \bar{c}(t_0) = c_0$, as desired. For $n \in N_2$, we have $c_n > \bar{c}(t_n)$, and thus (5.22) holds; we can argue as in Case 1 that $\{c_n\}_{n \in N_2}$ is bounded. If $\{c_n\}_{n \in N_2}$ were to have an accumulation point $c_0^* > \bar{c}(t_0)$, then the argument of Case 1 would imply $c_0^* = c_0$, which violates the Case 2 assumption. Hence, every accumulation point c_0^* of $\{c_n\}_{n \in N_2}$ satisfies $c_0^* = \bar{c}(t_0) = c_0$. $\square$

3.6 Utility from Consumption and Terminal Wealth

In Theorem 6.3 and Corollary 6.5 below we provide a complete solution to Problem 5.4 of maximizing expected utility from consumption plus expected utility from terminal wealth. We follow this with a number of examples of this solution. Theorem 6.11 begins the study of the dual value function for this problem.

Let a preference structure (U_1, U_2) be given. We define the function

$$\mathcal{X}_3(y) \overset{\Delta}{=} E\left[\int_0^T H_0(t)I_1\left(t, yH_0(t)\right) dt + H_0(T)I_2\left(yH_0(T)\right)\right], \quad 0 < y < \infty.$$
(6.1)

Assumption 6.1: $\mathcal{X}_3(y) < \infty, \quad \forall y \in (0, \infty)$.

Remarks 6.8, 6.9 below give conditions that imply the validity of this assumption.

Lemma 6.2: *Under Assumption 6.1, the function $\mathcal{X}_3$ is nonincreasing and continuous on $(0, \infty)$, and strictly decreasing on $(0, r_3)$, where $\mathcal{X}_3(0+) \overset{\Delta}{=} \lim_{y \downarrow 0} \mathcal{X}_3(y) = \infty$, $\mathcal{X}_3(\infty) \overset{\Delta}{=} \lim_{y \to \infty} \mathcal{X}_3(y)$ is given by (5.12), and*

$$r_3 \overset{\Delta}{=} \sup\{y > 0; \mathcal{X}_3(y) > \mathcal{X}_3(\infty)\} > 0.$$
(6.2)

In particular, the function $\mathcal{X}_3$ restricted to $(0, r_3)$ has a strictly decreasing inverse function $\mathcal{Y}_3: (\mathcal{X}_3(\infty), \infty) \xrightarrow{\text{onto}} (0, r_3)$, so that

$$\mathcal{X}_3(\mathcal{Y}_3(x)) = x, \quad \forall x \in (\mathcal{X}_3(\infty), \infty).$$
(6.3)

PROOF. Since $I_1(t, \cdot)$ and $I_2(\cdot)$ are nonincreasing, so is $\mathcal{X}_3(\cdot)$. The right continuity of $\mathcal{X}_3(\cdot)$ and the equation $\mathcal{X}_3(0+) = \infty$ are consequences of the monotone convergence theorem and, in the latter case, the equalities $I_1(t, 0+) = I_2(0+) = \infty$. The left continuity of $\mathcal{X}_3(\cdot)$ follows from Assumption 6.1 and the dominated convergence theorem. Likewise, agreement with

(5.12) follows from the dominated convergence theorem and the equalities $\lim_{c\to\infty} I_1(t,c) = \overline{c}(t)$, $\lim_{x\to\infty} I_2(x) = \overline{x}$.

It remains to show that $\mathcal{X}_3$ is strictly decreasing on $(0, r_3)$. For $y \in (0, r_3)$, we have $\mathcal{X}_3(y) > \mathcal{X}_3(\infty)$, and thus, either $yH_0(t,\omega) < U_1'(t, \overline{c}(t)+)$ for all (t, ω) in a set of positive product measure, or else $yH_0(T, \omega) < U_2'(\overline{x}+)$ for all ω in an event of positive P measure. But $I_1(t, \cdot)$ (respectively, $I_2(\cdot)$) is *strictly* decreasing on $(0, U_1'(t, \overline{c}(t)+))$ (respectively, on $(0, U_2'(\overline{x}+))$). Thus, either one of the above inequalities is enough to imply $\mathcal{X}_3(y - \delta) > \mathcal{X}_3(y)$ for all $\delta \in (0, y)$. □

We are now prepared to solve Problem 5.4. In light of Remark 5.7, specifically (5.17), we need only consider initial wealth x in the domain $(\mathcal{X}_3(\infty), \infty)$ of $\mathcal{Y}_3(\cdot)$. For such an x, we know from (3.4) and Theorem 3.5 that Problem 5.4 amounts to maximizing $E[\int_0^T U_1(t, c(t))\,dt + U_2(\xi)]$ over pairs (c, ξ), consisting of a consumption process $c(\cdot)$ and a nonnegative $\mathcal{F}(T)$-measurable random variable ξ, that satisfy the budget constraint (3.4), namely, $E[\int_0^T H_0(t)c(t)\,dt + H_0(T)\xi] \leq x$. Now, if $y > 0$ is a "Lagrange multiplier" that enforces this constraint, the problem reduces to the unconstrained maximization of

$$E\left[\int_0^T U_1(t, c(t))\,dt + U_2(\xi)\right] + y\left(x - E\left[\int_0^T H_0(t)c(t)\,dt + H_0(T)\xi\right]\right).$$

But this expression is

$$xy + E\int_0^T [U_1(t, c(t)) - yH_0(t)c(t)]\,dt + E[U_2(\xi) - yH_0(T)\xi]$$

$$\leq xy + E\left[\int_0^T \widetilde{U}_1(t, yH_0(t))\,dt + \widetilde{U}_2(yH_0(T))\right],$$

with equality if and only if

$$c(t) = I_1(t, yH_0(t)),\ 0 \leq t \leq T\ \text{ and }\ \xi = I_2(yH_0(T))$$

(recall (4.8) and Lemma 4.3(i)). Quite clearly, $y = \mathcal{Y}_3(x)$ is the only value of $y > 0$ for which the above pair (c, ξ) satisfies the budget constraint with equality. Thus, for every $x \in (\mathcal{X}_3(\infty), \infty)$, we are led to the *candidate optimal terminal wealth*

$$\xi_3 \overset{\Delta}{=} I_2(\mathcal{Y}_3(x)H_0(T)) \tag{6.4}$$

and the *candidate optimal consumption process*

$$c_3(t) \overset{\Delta}{=} I_1(t, \mathcal{Y}_3(x)H_0(t)),\quad 0 \leq t \leq T. \tag{6.5}$$

From (6.1), (6.3), we have

$$E\left[\int_0^T H_0(u)c_3(u)\,du + H_0(T)\xi_3\right] = \mathcal{X}_3(\mathcal{Y}_3(x)) = x, \tag{6.6}$$

and Theorem 3.5 guarantees the existence of a candidate optimal portfolio process $\pi_3(\cdot)$ such that $(c_3, \pi_3) \in \mathcal{A}(x)$ and $\xi_3 = X^{x,c_3,\pi_3}(T)$.

Theorem 6.3: *Suppose that both Assumptions 2.3 and 6.1 hold, let $x \in (\mathcal{X}_3(\infty), \infty)$ be given, let ξ_3 and $c_3(\cdot)$ be given by (6.4), (6.5), and let $\pi_3(\cdot)$ be such that $(c_3, \pi_3) \in \mathcal{A}(x)$, $\xi_3 = X^{x,c_3,\pi_3}(T)$. Then $(c_3, \pi_3) \in \mathcal{A}_3(x)$, and (c_3, π_3) is optimal for Problem 5.4:*

$$V_3(x) = E\left[\int_0^T U_1(t, c_3(t))\, dt + U_2\left(X^{x,c_3,\pi_3}(T)\right)\right]. \tag{6.7}$$

PROOF. We first show that $(c_3, \pi_3) \in \mathcal{A}_3(x)$. With $\hat{c}$ as in Remark 5.6, Lemma 4.3(i) and Definition 4.2 imply

$$U_1(t, c_3(t)) - \mathcal{Y}_3(x)H_0(t)c_3(t) = \tilde{U}_1(t, \mathcal{Y}_3(x)H_0(t))$$
$$\geq U_1(t, \hat{c}) - \mathcal{Y}_3(x)H_0(t)\hat{c},$$

$$U_2(\xi_3) - \mathcal{Y}_3(x)H_0(T)\xi_3 = \tilde{U}_2(\mathcal{Y}_3(x)H_0(T))$$
$$\geq U_2(\hat{c}) - \mathcal{Y}_3(x)H_0(T)\hat{c},$$

and consequently,

$$E\left\{\int_0^T \min\left[0, U_1(t, c_3(t))\right]\, dt + \min\left[0, U_2(\xi_3)\right]\right\}$$
$$\geq \int_0^T \min\left[0, U_1(t, \hat{c})\right]\, dt + \min\left[0, U_2(\hat{c})\right]$$
$$-\mathcal{Y}_3(x)\hat{c}E\left\{\int_0^T H_0(t)\, dt + H_0(T)\right\}$$
$$> -\infty.$$

Next, we show that (c_3, π_3) attains the supremum in (5.8). Let (c, π) be another pair in $\mathcal{A}_3(x)$. Using Lemma 4.3(i) again, we have

$$U_1(t, c_3(t)) - \mathcal{Y}_3(x)H_0(t)c_3(t) \geq U_1(t, c(t)) - \mathcal{Y}_3(x)H_0(t)c(t), \tag{6.8}$$
$$U_2(t, \xi_3) - \mathcal{Y}_3(x)H_0(T)\xi_3 \geq U_2\left(X^{x,c,\pi}(T)\right)$$
$$-\mathcal{Y}_3(x)H_0(T)X^{x,c,\pi}(T), \tag{6.9}$$

and thus

$$E\left[\int_0^T U_1(t, c_3(t))\, dt + U_2(t, \xi_3)\right]$$
$$\geq E\left[\int_0^T U_1(t, c(t))\, dt + U_2\left(X^{x,c,\pi}(T)\right)\right]$$
$$+\mathcal{Y}_3(x)E\left[\int_0^T H_0(t)c_3(t)\, dt + H_0(T)\xi_3\right]$$

$$-\mathcal{Y}_3(x)E\left[\int_0^T H_0(t)c(t)\,dt + H_0(T)X^{x,c,\pi}(T)\right]$$

$$\geq E\left[\int_0^T U_1((t,c(t))\,dt + U_2\left(X^{x,c,\pi}(T)\right)\right],$$

because of (6.6) and the budget constraint (3.4) satisfied by (c,π).   □

Remark 6.4: Assume that $V_3(x) < \infty$. Inequality (6.8) is strict unless $c(t) = c_3(t)$, and likewise (6.9) is strict unless $X^{x,c,\pi}(T) = \xi_3$. It follows that up to almost-everywhere equivalence under the product of Lebesgue measure and P, $c_3(\cdot)$ is the *unique* optimal consumption process and ξ_3 is the *unique* optimal terminal wealth. This implies also that $\pi_3(\cdot)$ is the *unique* optimal portfolio process, again up to almost-everywhere equivalence.

Corollary 6.5: *Under the assumptions of Theorem 6.3, the optimal wealth process* $X_3(t) = X^{x,c_3,\pi_3}(t)$ *is*

$$X_3(t) = \frac{1}{H_0(t)}E\left[\int_t^T H_0(u)c_3(u)\,du + H_0(T)\xi_3\,\middle|\,\mathcal{F}(t)\right],\quad 0\leq t\leq T.$$
$$(6.10)$$

Furthermore, the optimal portfolio π_3 *is given by*

$$\sigma'(t)\pi_3(t) = \frac{\psi_3(t)}{H_0(t)} + X_3(t)\theta(t),\qquad (6.11)$$

in terms of the integrand $\psi_3(\cdot)$ *in the stochastic integral representation* $M_3(t) = x + \int_0^t \psi_3'(u)\,dW(u)$ *of the martingale*

$$M_3(t) \triangleq E\left[\int_0^T H_0(u)c_3(u)\,du + H_0(T)\xi_3\,\middle|\,\mathcal{F}(t)\right],\quad 0\leq t\leq T.\quad (6.12)$$

The value function V_3 *is then given as*

$$V_3(x) = G_3\left(\mathcal{Y}_3(x)\right),\quad \mathcal{X}_3(\infty) < x < \infty,\qquad (6.13)$$

where

$$G_3(y) \triangleq E\left[\int_0^T U_1\left(t, I_1\left(t, yH_0(t)\right)\right)\,dt + U_2\left(I_2\left(yH_0(T)\right)\right)\right],\ 0 < y < \infty.$$
$$(6.14)$$

PROOF.   The formula for $X_3(\cdot)$ comes directly from (3.6), which also provides the formula for $M_3(\cdot)$ in terms of $H_0(\cdot)$ and converts (3.7) to (6.11). Equations (6.13), (6.14) are just restatements of (6.4), (6.5), and (6.7).   □

Example 6.6: $U_1(t,x) = U_2(x) = \log x,\ \forall(t,x) \in [0,T] \times (0,\infty)$.
   In this case, $I_1(t,y) = I_2(y) = 1/y$ for $0 < y < \infty$, $\mathcal{X}_3(y) = (T+1)/y$ for $0 < y < \infty$, and $\mathcal{Y}_3(x) = (T+1)/x$ for $0 < x < \infty$. The optimal terminal

wealth, consumption, and wealth processes are given respectively by

$$\xi_3 = \frac{x}{(T+1)H_0(T)},$$

$$c_3(t) = \frac{x}{(T+1)H_0(t)}, \quad X_3(t) = \frac{(T+1-t)x}{(T+1)H_0(t)}; \quad 0 \le t \le T.$$

In particular, the martingale $M_3(\cdot)$ of (6.12) is identically equal to x, so $\psi(\cdot) \equiv 0$, and the optimal portfolio, given by (6.11), is

$$\pi_3(t) = (\sigma(t)\sigma'(t))^{-1} [b(t) + \delta(t) - r(t)\underline{1}] X_3(t), \quad 0 \le t \le T. \qquad (6.15)$$

Furthermore,

$$G_3(y) = -(T+1)\log y - E \int_0^T \log H_0(t)\, dt - E \log H_0(T), \quad 0 < y < \infty,$$

$$V_3(x) = (T+1)\log\left(\frac{x}{T+1}\right) - E \int_0^T \log H_0(t)\, dt - E \log H_0(T),$$

$$0 < x < \infty.$$

This example is extended in Example 7.11.

Example 6.7: $U_1(t,x) = U_2(x) = \frac{1}{p}x^p, \quad \forall (t,x) \in [0,T] \times (0,\infty),$ with $p < 1$, $p \ne 0$.

We have $I_1(t,y) = I_2(y) = y^{1/(p-1)}$ for $0 < y < \infty$, and

$$\mathcal{X}_3(y) = y^{\frac{1}{p-1}} E\left[\int_0^T (H_0(t))^{p/(p-1)}\, dt + (H_0(T))^{p/(p-1)}\right]$$

$$= \mathcal{X}_3(1)y^{1/(p-1)}, \quad 0 < y < \infty,$$

$$\mathcal{Y}_3(x) = \left(\frac{x}{\mathcal{X}_3(1)}\right)^{p-1}, \quad 0 < x < \infty.$$

The optimal terminal wealth and the optimal consumption process are given as

$$\xi_3 = \frac{x}{\mathcal{X}_3(1)} (H_0(T))^{1/(p-1)}, \qquad c_3(t) = \frac{x}{\mathcal{X}_3(1)} (H_0(t))^{1/(p-1)},$$

and

$$X_3(t) = \frac{x}{\mathcal{X}_3(1)H_0(t)} E\left[\int_t^T (H_0(u))^{p/(p-1)}\, du + (H_0(T))^{p/p-1}\,\bigg|\, \mathcal{F}(t)\right].$$

Finally,

$$G_3(y) = \frac{1}{p}\mathcal{X}_3(1)y^{p/(p-1)}, \qquad 0 < y < \infty,$$

$$V_3(x) = \frac{1}{p}(\mathcal{X}_3(1))^{1-p} x^p, \qquad 0 < x < \infty.$$

We may write $(H_0(t))^{p/(p-1)} = m(t)\Lambda(t)$, where

$$m(t) \triangleq \exp\left\{ \frac{p}{1-p}\left[A(t) + \int_0^t r(u)\, du \right] + \frac{p}{2(1-p)^2} \int_0^t \|\theta(u)\|^2\, du \right\},$$

$$\Lambda(t) \triangleq \exp\left\{ \frac{p}{1-p} \int_0^t \theta'(u)\, dW(u) - \frac{p^2}{2(1-p)^2} \int_0^t \|\theta(u)\|^2\, du \right\}.$$

If $r(\cdot)$, $A(\cdot)$, and $\theta(\cdot)$ are deterministic, then $m(\cdot)$ is deterministic and $\Lambda(\cdot)$ is a martingale, so

$$E\left[(H_0(u))^{p/(p-1)} \Big| \mathcal{F}(t) \right] = m(u)\Lambda(t), \quad 0 \leq t \leq u \leq T.$$

With $N(t) \triangleq \int_0^t m(s)\, ds + m(t)$, we have $\mathcal{X}_3(1) = N(T)$ and

$$X_3(t) = \frac{x\Lambda(t)}{N(T)H_0(t)} \left[\int_t^T m(u)\, du + m(T) \right],$$

$$c_3(t) = \frac{m(t)X_3(t)}{\int_t^T m(u)\, du + m(T)},$$

$$M_3(t) = \frac{x}{N(T)} \left[\Lambda(t) \left(\int_t^T m(u)\, du + m(T) \right) \right.$$
$$\left. + \int_0^t m(u)\Lambda(u)\, du \right],$$

$$dM_3(t) = \frac{x}{N(T)} \left(\int_t^T m(u)\, du + m(T) \right) d\Lambda(t)$$

$$= H_0(t)X_3(t)\frac{p}{1-p}\theta'(t)\, dW(t).$$

This last expression yields the optimal portfolio $\pi_3(\cdot)$ of (6.11) as

$$\pi_3(t) = \frac{1}{1-p} (\sigma(t)\sigma'(t))^{-1} [b(t) + \delta(t) - r(t)\mathbb{1}] X_3(t) \qquad (6.16)$$

for the case of deterministic $r(\cdot)$, $A(\cdot)$, and $\theta(\cdot)$.

Remark 6.8: If $r(\cdot)$, $A(\cdot)$, and $\theta(\cdot)$ are bounded uniformly in (t,ω), but *not necessarily deterministic*, then in Example 6.7, the function $m(\cdot)$ can be bounded independently of (t,ω), and $\Lambda(\cdot)$ is a martingale. This implies that

$$\mathcal{X}_3(y) = y^{1/(p-1)} E\left[\int_0^T m(t)\Lambda(t)\, dt + m(T)\Lambda(T) \right] < \infty, \quad \forall\, y > 0,$$

so Assumption 6.1 is satisfied; in particular, $V_3(x) < \infty$ for all $x > 0$. It is then clear that for *any* utility functions U_1 and U_2 satisfying a growth condition of the form

$$U_1(t, x) + U_2(x) \leq \kappa(1 + x^p), \quad \forall (t, x) \in [0, T] \times (0, \infty),$$

where $0 < \kappa < \infty$ and $0 < p < 1$, the boundedness of $r(\cdot)$, $A(\cdot)$, and $\theta(\cdot)$ implies $V_3(x) < \infty$ for all $x > 0$.

Remark 6.9:

(i) Suppose that

$$\left\{ \begin{array}{l} \text{both } U_1(t, \cdot) \text{ and } U_2(\cdot) \text{ satisfy condition (4.16) with} \\ \text{the same constants } \beta \text{ and } \gamma, \text{ for all } t \in [0, T]. \end{array} \right\} \quad (6.17)$$

It follows then, using (4.12) and (4.16''), that if $\mathcal{X}_3(y) < \infty$ for some $y > 0$, then Assumption 6.1 holds.

(ii) Assumption 6.1 is also implied by the condition

$$\sup_{0 \leq t \leq T} I_1(t, y) + I_2(y) \leq \kappa y^{-\rho}, \quad \forall y \in (0, \infty), \quad (6.18)$$

for some $\kappa > 0$, $\rho > 0$, provided that *at least one* of the following conditions also holds:

$$0 < \rho \leq 1 \text{ and Assumption 2.3 holds,} \quad (6.19)$$

or

$$A(\cdot), r(\cdot), \text{ and } \theta(\cdot) \text{ are uniformly bounded in } (t, \omega) \in [0, T] \times \Omega. \quad (6.20)$$

Indeed, under (6.18), (6.19), we have

$$\mathcal{X}_3(y) \leq \kappa y^{-\rho} E\left[\int_0^T (1 \vee H_0(t)) \, dt + (1 \vee H_0(T)) \right] < \infty, \quad y > 0.$$

On the other hand, under (6.18) and (6.20), let us write $(H_0(t))^{1-\rho} = m(t)\Lambda(t)$, where

$$m(t) \triangleq \exp\left\{ (\rho - 1)\left[A(t) + \int_0^t r(u) \, du \right] + \frac{1}{2}\rho(\rho - 1) \int_0^t \|\theta(u)\|^2 \, du \right\},$$

and

$$\Lambda(t) \triangleq \exp\left\{ (\rho - 1) \int_0^t \theta'(u) \, dW(u) - \frac{1}{2}(\rho - 1)^2 \int_0^t \|\theta(u)\|^2 \, du \right\}$$

is a martingale; because of (6.20), $m(\cdot)$ is bounded by some constant K, and thus

$$\mathcal{X}_3(y) \leq y^{-\rho} K E\left[\int_0^T \Lambda(t) \, dt + \Lambda(T) \right] \leq K y^{-\rho}, \quad y > 0.$$

Remark 6.10: The case $\bar{x} = 0$ and $U_2'(0+) = \infty$.

In this case, $I_2(y) > 0$ for all $y > 0$, and the random variable ξ_3 of (6.4) is strictly positive almost surely, as is the optimal wealth process $X_3(t)$ of (6.10) for $0 \le t \le T$. We can define the *portfolio proportion*

$$p_3(t) \overset{\Delta}{=} \frac{\pi_3(t)}{X_3(t)}, \quad 0 \le t \le T,$$

a process which is obviously $\{\mathcal{F}(t)\}$-progressively measurable and satisfies $\int_0^T \|p_3(t)\|^2 \, dt < \infty$ almost surely. The components of $p_3(t)$ represent the proportions of wealth $X_3(t)$ invested in the respective assets at time $t \in [0, T]$, and equation (3.1) for $X_3(\cdot)$ becomes

$$\frac{X_3(t)}{S_0(t)} = x - \int_0^t \frac{c_3(u) \, du}{S_0(u)} + \int_0^t \frac{X_3(u)}{S_0(u)} p_3'(u) \sigma(u) \, dW_0(u), \quad 0 \le t \le T.$$

In (6.15) and (6.16) of Examples 6.6 and 6.7, $p_3(t)$ depends on the market processes and the utility functions, but not on the wealth of the agent.

We close this section with the observation that the value function V_3 is a utility function in the sense of Definition 4.1, and we find its derivative and convex dual. Recall from Remark 5.7 and Corollary 6.5 that

$$V_3(x) = \begin{cases} G_3\left(\mathcal{Y}_3(x)\right), & x > \mathcal{X}_3(\infty), \\ \int_0^T U_1\left(t, \bar{c}(t)\right) dt + U_2(\bar{x}), & x = \mathcal{X}_3(\infty), \\ -\infty, & x < \mathcal{X}_3(\infty). \end{cases} \tag{6.21}$$

Theorem 6.11: *Let Assumptions 2.3 and 6.1 hold, and assume $V_3(x) < \infty$ for all $x \in \mathbb{R}$. Then V_3 satisfies all the conditions of Definition 4.1, and*

$$\mathcal{X}_3(\infty) = \inf\{x \in \mathbb{R}; V_3(x) > -\infty\}, \tag{6.22}$$

$$V_3'(x) = \mathcal{Y}_3(x), \quad \forall x \in (\mathcal{X}_3(\infty), \infty), \tag{6.23}$$

$$\tilde{V}_3(y) = G_3(y) - y\mathcal{X}_3(y) \tag{6.24}$$

$$= E\left[\int_0^T \tilde{U}_1\left(t, yH_0(t)\right) dt + \tilde{U}_2\left(yH_0(T)\right)\right], \quad \forall y \in (0, \infty),$$

$$\tilde{V}_3'(y) = -\mathcal{X}_3(y), \quad \forall y \in (0, \infty), \tag{6.25}$$

where

$$\tilde{V}_3(y) \overset{\Delta}{=} \sup_{x \in \mathbb{R}}\{V_3(x) - xy\}, \quad y \in \mathbb{R}. \tag{6.26}$$

PROOF. We first prove the concavity of V_3. Let $x_1, x_2 \in [\mathcal{X}_3(\infty), \infty)$ be given, and let $(c_1, \pi_1) \in \mathcal{A}_3(x_1)$, $(c_2, \pi_2) \in \mathcal{A}_3(x_2)$ also be given. It is easily verified that for $\lambda_1, \lambda_2 \in (0, 1)$ with $\lambda_1 + \lambda_2 = 1$, the consumption/portfolio pair $(c, \pi) \overset{\Delta}{=} (\lambda_1 c_1 + \lambda_2 c_2, \lambda_1 \pi_1 + \lambda_2 \pi_2)$ is in $\mathcal{A}_3(x)$ with $x \overset{\Delta}{=} \lambda_1 x_1 + \lambda_2 x_2$ and

$$X^{x,c,\pi}(\cdot) = \lambda_1 X^{x_1, c_1, \pi_1}(\cdot) + \lambda_2 X^{x_2, c_2, \pi_2}(\cdot).$$

Consequently,

$$\lambda_1 E\left[\int_0^T U_1\left(t, c_1(t)\right) dt + U_2\left(X^{x_1, c_1, \pi_1}(T)\right)\right]$$

$$+\lambda_2 E\left[\int_0^T U_1\left(t, c_2(t)\right) dt + U_2\left(X^{x_2, c_2, \pi_2}(T)\right)\right]$$

$$\leq E\left[\int_0^T U_1\left(t, c(t)\right) dt + U_2\left(X^{x, c, \pi}(T)\right)\right]$$

$$\leq V_3(x)$$

$$= V_3(\lambda_1 x_1 + \lambda_2 x_2).$$

Maximizing over $(c_1, \pi_1) \in \mathcal{A}_3(x_1)$ and $(c_2, \pi_2) \in \mathcal{A}_3(x_2)$, we obtain

$$\lambda_1 V_3(x_1) + \lambda_2 V_3(x_2) \leq V_3(\lambda_1 x_1 + \lambda_2 x_2).$$

It is easily seen that V_3 is nondecreasing. Furthermore, for each $x \in (\mathcal{X}_3(\infty), \infty)$, we constructed in Theorem 6.3 a policy $(c_3, \pi_3) \in \mathcal{A}_3(x)$, which shows that $V_3 > -\infty$ on $(\mathcal{X}_3(\infty), \infty)$; (6.22) follows.

A concave function is continuous on the interior of the set where it is finite. Therefore, to establish the upper semicontinuity of V_3, we need only show that

$$\lim_{x \downarrow \mathcal{X}_3(\infty)} V_3(x) = \int_0^T U_1\left(t, \bar{c}(t)\right) dt + U_2(\bar{x}). \tag{6.27}$$

But $\lim_{x \downarrow \mathcal{X}_3(\infty)} V_3(x) = \lim_{y \uparrow r_3} G_3(y)$, where r_3, defined by (6.2), possesses the property that $\mathcal{X}_3$ is constant on $[r_3, \infty)$. We consider separately the cases $r_3 = \infty$ and $r_3 < \infty$. If $r_3 = \infty$, then

$$\lim_{y \uparrow r_3} I_1\left(t, yH_0(t)\right) = \bar{c}(t), \quad \lim_{y \uparrow r_3} I_2\left(yH_0(T)\right) = \bar{x}, \tag{6.28}$$

so that the monotone convergence theorem and the finiteness of $G_3(y) = V_3(\mathcal{X}_3(y))$ for $0 < y < r_3$ imply

$$\lim_{y \uparrow r_3} G_3(y) = \int_0^T U_1\left(t, \bar{c}(t)\right) dt + U_2(\bar{x}),$$

as desired. If $r_3 < \infty$, then the constancy of $\mathcal{X}_3$ on $[r_3, \infty)$ implies $r_3 H_0(t) \geq U_1'\left(t, \bar{c}(t)+\right)$ for Lebesgue-almost-every $t \in [0, T]$ and $r_3 H_0(T) \geq U_2'(\bar{x}+)$ almost surely. This implies (6.28), and the rest follows.

We turn to (6.24). The second equation in (6.24) follows directly from (4.11), (6.1), and (6.14). For the first, let $Q(y) = G_3(y) - y\mathcal{X}_3(y)$ for $0 < y < \infty$, and observe from (4.8) that

$$U_1\left(t, c(t)\right) \leq \tilde{U}_1\left(t, yH_0(t)\right) + yH_0(t)c(t), \quad 0 \leq t \leq T,$$

$$U_2\left(X^{x, c, \pi}(T)\right) \leq \tilde{U}_2\left(yH_0(T)\right) + yH_0(T)X^{x, c, \pi}(T)$$

hold almost surely for any $y > 0$, $x \geq \mathcal{X}_3(\infty)$, and $(c, \pi) \in \mathcal{A}_3(x)$. Consequently, from the budget constraint (3.4), we have

$$E\left[\int_0^T U_1\left(t, c(t)\right)\, dt + U_2\left(X^{x,c,\pi}(T)\right)\right]$$

$$\leq Q(y) + yE\left[\int_0^T H_0(t)c(t)\, dt + H_0(T)X^{x,c,\pi}(T)\right]$$

$$\leq Q(y) + xy, \tag{6.29}$$

with equality if and only if

$$c(t) = I_1\left(t, yH_0(t)\right), \qquad X^{x,c,\pi}(T) = I_2\left(yH_0(T)\right), \tag{6.30}$$

and

$$E\left[\int_0^T H_0(t)c(t)\, dt + H_0(T)X^{x,c,\pi}(T)\right] = x.$$

Taking the supremum in (6.29) over $(c, \pi) \in \mathcal{A}(x)$, we obtain $V_3(x) \leq Q(y) + xy$ for all $x \in \mathbb{R}$, and thus $\tilde{V}_3(y) \leq Q(y)$ for all $y > 0$. For the reverse inequality, observe that equality holds in (6.29) if (6.30) is satisfied and $x = \mathcal{X}_3(y)$. This gives $Q(y) = V_3\left(\mathcal{X}_3(y)\right) - y\mathcal{X}_3(y) \leq \tilde{V}_3(y)$. This completes the proof of (6.24) and shows that for $y > 0$, the maximum in (6.26) is attained by $x = \mathcal{X}_3(y)$.

To prove (6.25), we use (4.11) and (4.12) to write for any utility function U and for $0 < z < y < \infty$,

$$yI(y) - zI(z) - \int_z^y I(\xi)\, d\xi = yI(y) - zI(z) + \tilde{U}(y) - \tilde{U}(z)$$

$$= U\left(I(y)\right) - U\left(I(z)\right). \tag{6.31}$$

Therefore,

$$y\mathcal{X}_3(y) - z\mathcal{X}_3(z) - \int_z^y \mathcal{X}_3(\lambda)\, d\lambda$$

$$= E\int_0^T\left[yH_0(t)I_1(t, yH_0(t)) - zH_0(t)I_1(t, zH_0(t)) - \int_{zH_0(t)}^{yH_0(t)} I_1(t, \xi)\, d\xi\right] dt$$

$$+ E\left[yH_0(T)I_2(yH_0(T)) - zH_0(T)I_2(zH_0(T)) - \int_{zH_0(T)}^{yH_0(T)} I_2(\xi)\, d\xi\right]$$

$$= E\int_0^T \left[U_1(t, I_1(t, yH_0(t))) - U_1(t, I_1(t, zH_0(t)))\right] dt$$

$$+ E[U_2(I_2(yH_0(T))) - U_2(I_2(zH_0(T)))]$$

$$= G_3(y) - G_3(z), \tag{6.32}$$

or equivalently

$$\tilde{V}_3(y) - \tilde{V}_3(z) = -\int_z^y \mathcal{X}_3(\lambda)\, d\lambda, \quad 0 < z < y < \infty, \tag{6.33}$$

and (6.25) follows.

According to Rockafellar (1970), Theorem 23.5, for $x^* > \mathcal{X}_3(\infty)$ and $y > 0$ we have $y \in \partial V_3(x^*)$ if and only if x^* attains the maximum in (6.26). We have already seen that this maximum is attained by $\mathcal{X}_3(y)$, so the sole element in $\partial V_3(x^*)$ is $\mathcal{Y}_3(x^*)$. Equation (6.23) follows, and implies that V_3' is continuous, positive, and strictly decreasing on $(\mathcal{X}_3(\infty), \infty)$, and $\lim_{x \to \infty} V_3'(x) = \lim_{x \to \infty} \mathcal{Y}_3(x) = 0$. Thus V_3 satisfies property (ii) of Definition 4.1 and is a utility function. □

Remark 6.12: From (6.13) we have $G_3(y) = V_3(\mathcal{X}_3(y))$ for all $y \in (0, \mathcal{X}_3(\infty))$. If $\mathcal{X}_3'(y)$ exists, then $G_3'(y)$ also exists and is given by the formula

$$G_3'(y) = V_3'(\mathcal{X}_3(y))\, \mathcal{X}_3'(y) = y\mathcal{X}_3'(y), \quad 0 < y < \mathcal{X}_3(\infty), \tag{6.34}$$

where we have used (6.23).

3.7 Utility from Consumption or Terminal Wealth

Theorem 7.3 below provides a complete solution to Problem 5.2 of maximization of expected utility from consumption alone, and Theorem 7.6 does the same for Problem 5.3 of maximization of expected utility from terminal wealth alone. This section also contains examples of these solutions and examines the dual value functions for Problems 5.2 and 5.3. Theorem 7.10 shows how to combine the solutions of these two problems to obtain the solution of Problem 5.4 of maximization of expected utility from consumption plus expected utility from terminal wealth. In particular, the dual value function for Problem 5.4 is the sum of the dual value functions for Problems 5.2 and 5.3.

Let a preference structure (U_1, U_2) be given. We define the functions

$$\mathcal{X}_1(y) \triangleq E\left[\int_0^T H_0(t)I_1(t, yH_0(t))\, dt\right], \quad 0 < y < \infty, \tag{7.1}$$

$$\mathcal{X}_2(y) \triangleq E\left[H_0(T)I_2(yH_0(T))\right], \quad 0 < y < \infty. \tag{7.2}$$

Assumption 7.1: $\mathcal{X}_1(y) < \infty, \quad \forall y \in (0, \infty)$.

Assumption 7.2: $\mathcal{X}_2(y) < \infty, \quad \forall y \in (0, \infty)$.

Remarks 6.8 and 6.9 can be trivially modified to provide sufficient conditions for Assumptions 7.1 and 7.2 to hold. Similarly, Remark 6.8 can be modified to obtain conditions that guarantee $V_1(x) < \infty$ and $V_2(x) < \infty$.

Just as we proved Lemma 6.2, we can show that for each $i = 1, 2$, under Assumption 7.i, the function $\mathcal{X}_i$ is nonincreasing and continuous on $(0, \infty)$ with $\mathcal{X}_i(0+) = \infty$ and with $\mathcal{X}_i(\infty) = \lim_{y \to \infty} \mathcal{X}_i(y)$ given by (5.11). With

$$r_i \triangleq \sup\{y > 0; \mathcal{X}_i(y) > \mathcal{X}_i(\infty)\} > 0, \quad i = 1, 2, \tag{7.3}$$

$\mathcal{X}_i$ is strictly decreasing on $(0, r_i)$ and, when restricted to $(0, r_i)$, has a strictly decreasing inverse function $\mathcal{Y}_i \colon (\mathcal{X}_i(\infty), \infty) \xrightarrow{\text{onto}} (0, r_i)$.

The proof of the following theorem parallels the proof of Theorem 6.3 and Corollary 6.5.

Theorem 7.3 (Maximization of the expected utility from consumption): *Let Assumptions 2.1 and 7.1 hold, let $x \in (\mathcal{X}_1(\infty), \infty)$ be given, and define*

$$c_1(t) \triangleq I_1(t, \mathcal{Y}_1(x) H_0(t)), \quad 0 \le t \le T. \tag{7.4}$$

(i) *There exists a portfolio $\pi_1(\cdot)$ such that $(c_1, \pi_1) \in \mathcal{A}_1(x)$, $X^{x, c_1, \pi_1}(T) = 0$, and the pair (c_1, π_1) is optimal for Problem 5.2, i.e.,*

$$V_1(x) = E \int_0^T U_1(t, c_1(t)) \, dt.$$

(ii) *The optimal wealth process $X_1(t) = X^{x, c_1, \pi_1}(t)$ is*

$$X_1(t) = \frac{1}{H_0(t)} E\left[\int_t^T H_0(u) c_1(u) \, du \,\middle|\, \mathcal{F}(t) \right], \quad 0 \le t \le T. \tag{7.5}$$

(iii) *The optimal portfolio $\pi_1(\cdot)$ is given by*

$$\sigma'(t) \pi_1(t) = \frac{\psi_1(t)}{H_0(t)} + X_1(t) \theta(t), \tag{7.6}$$

where $\psi_1(\cdot)$ is the integrand in the stochastic integral representation $M_1(t) = x + \int_0^t \psi_1'(u) \, dW(u)$ of the martingale

$$M_1(t) \triangleq E\left[\int_0^T H_0(u) c_1(u) \, du \,\middle|\, \mathcal{F}(t) \right].$$

(iv) *The value function V_1 is given by $V_1(x) = G_1(\mathcal{Y}_1(x))$ for all $x > \mathcal{X}_1(\infty)$, where*

$$G_1(y) \triangleq E \int_0^T U_1(t, I_1(t, y H_0(t))) \, dt, \quad 0 < y < \infty. \tag{7.7}$$

It is now not difficult to see that the value function V_1 is given by (cf. (6.21))

$$V_1(x) = \begin{cases} G_1(\mathcal{Y}_1(x)), & x > \mathcal{X}_1(\infty), \\ E \int_0^T U_1(t, \bar{c}(t)) \, dt, & x = \mathcal{X}_1(\infty), \\ -\infty, & x < \mathcal{X}_1(\infty). \end{cases} \tag{7.8}$$

Imitating the proof of Theorem 6.11, we can show that V_1 has the following properties.

Theorem 7.4: *Let Assumptions 2.1 and 7.1 hold, and assume $V_1(x) < \infty$ for all $x \in \mathbb{R}$. Then V_1 satisfies all the conditions of Definition 4.1, and*

$$\mathcal{X}_1(\infty) = \inf\{x \in \mathbb{R}; V_1(x) > -\infty\},$$
$$V_1'(x) = \mathcal{Y}_1(x), \quad \forall x \in (\mathcal{X}_1(\infty), \infty),$$
$$\tilde{V}_1(y) = G_1(y) - y\mathcal{X}_1(y)$$
$$= E\left[\int_0^T \tilde{U}_1(t, yH_0(t))\, dt\right], \quad \forall y \in (0, \infty),$$
$$\tilde{V}_1'(y) = -\mathcal{X}_1(y), \quad \forall y \in (0, \infty), \tag{7.9}$$

where

$$\tilde{V}_1(y) \stackrel{\Delta}{=} \sup_{x \in \mathbb{R}}\{V_1(x) - xy\}, \quad y \in \mathbb{R}.$$

Example 7.5 *(Subsistence consumption):* Suppose

$$U_1(c) = \begin{cases} \log(c - \bar{c}), & \bar{c} < c < \infty, \\ -\infty, & -\infty < c \leq \bar{c}, \end{cases}$$

where $\bar{c}$ is a positive constant that consumption must exceed at all times. Then $I_1(y) = \bar{c} + (1/y)$ and $\mathcal{X}_1(y) = \bar{c}h_1 + (T/y)$ for $0 < y < \infty$, where $h_1 \stackrel{\Delta}{=} E\int_0^T H_0(t)\, dt$. In order to ensure that consumption can exceed $\bar{c}$ at all times, the initial endowment x must exceed $\mathcal{X}_1(\infty) = \bar{c}h_1$. We have $\mathcal{Y}_1(x) = T/(x - \bar{c}h_1)$ for $x > \bar{c}h_1$, and the optimal consumption and wealth processes from (7.4), (7.5) are

$$c_1(t) = \frac{x - \bar{c}h_1}{TH_0(t)} + \bar{c},$$

$$X_1(t) = \frac{1}{H_0(t)}\left\{\frac{T - t}{T}(x - \bar{c}h_1) + \bar{c}E\left[\int_t^T H_0(u)\, du\,\middle|\, \mathcal{F}(t)\right]\right\}.$$

Finally, $G_1(y) = -T\log y - E\int_0^T \log H_0(t)\, dt$, so

$$V_1(x) = T\log(x - \bar{c}h_1) - T\log T - E\int_0^T \log H_0(t)\, dt, \quad x > \bar{c}h_1.$$

If $S_0(\cdot)$ is deterministic, we can derive the optimal portfolio explicitly. Under this condition,

$$X_1(t) = \frac{T - t}{TH_0(t)}(x - \bar{c}h_1) + \bar{c}S_0(t)\left(\int_t^T \frac{du}{S_0(u)}\right),$$

$$M_1(t) = \bar{c}E\left[\left.\int_0^T \frac{Z_0(u)}{S_0(u)}\,du\right|\mathcal{F}(t)\right] + x - \bar{c}h_1$$

$$= \bar{c}\int_0^t H_0(u)\,du + \bar{c}Z_0(t)\int_t^T \frac{du}{S_0(u)} + x - \bar{c}h,$$

and

$$dM_1(t) = -\bar{c}\left(\int_t^T \frac{du}{S_0(u)}\right)Z_0(t)\theta'(t)\,dW(t).$$

It follows from (7.6) that

$$\pi_1(t) = (\sigma'(t))^{-1}\left[X_1(t) - \bar{c}S_0(t)\int_t^T \frac{du}{S_0(u)}\right]\theta(t)$$

$$= \frac{T-t}{TH_0(t)}(x - \bar{c}h_1)(\sigma(t)\sigma'(t))^{-1}\left(b(t) + \delta(t) - r(t)\underline{1}\right).$$

The analogues of Theorems 7.3, 7.4 for Problem 5.3 are the following.

Theorem 7.6 (Maximization of utility from terminal wealth): *Let Assumptions 2.2 and 7.2 hold, let $x \in (\mathcal{X}_2(\infty), \infty)$ be given, and let*

$$\xi_2 = I_2\left(\mathcal{Y}_2(x)H_0(T)\right).$$

(i) *With $c_2 \equiv 0$, there exists a portfolio $\pi_2(\cdot)$ such that $(c_2, \pi_2) \in \mathcal{A}_2(x)$, $X^{x,c_2,\pi_2}(T) = \xi_2$, and the pair (c_2, π_2) is optimal for Problem 5.3, i.e.,*

$$V_2(x) = EU_2(X^{x,c_2,\pi_2}(T)).$$

(ii) *The optimal wealth process $X_2(t) = X^{x,c_2,\pi_2}(t)$ is*

$$X_2(t) = \frac{1}{H_0(t)}E[H_0(T)\xi_2|\mathcal{F}(t)], \quad 0 \le t \le T.$$

(iii) *The optimal portfolio $\pi_2(\cdot)$ is given by*

$$\sigma'(t)\pi_2(t) = \frac{\psi_2(t)}{H_0(t)} + X_2(t)\theta(t),$$

where ψ_2 is the integrand in the stochastic integral representation $M_2(t) = x + \int_0^t \psi_2'(u)\,dW(u)$ of the martingale

$$M_2(t) \overset{\Delta}{=} H_0(t)X_2(t).$$

(iv) *The value function V_2 is given by $V_2(x) = G_2(\mathcal{Y}_2(x))$ for all $x > \mathcal{X}_2(\infty)$, where*

$$G_2(y) \overset{\Delta}{=} EU_2\left(I_2(yH_0(T))\right), \quad 0 < y < \infty.$$

Summarizing information about V_2, we have the formula

$$V_2(x) = \begin{cases} G_2(\mathcal{Y}_2(x)), & x > \mathcal{X}_2(\infty), \\ U_2(\overline{x}), & x = \mathcal{X}_2(\infty), \\ -\infty, & x < \mathcal{X}_2(\infty). \end{cases} \qquad (7.10)$$

Theorem 7.7: *Let Assumptions 2.2 and 7.2 hold, and assume $V_2(x) < \infty$ for all $x \in \mathbb{R}$. Then V_2 satisfies all the conditions of Definition 4.1, and*

$$\mathcal{X}_2(\infty) = \inf \{ x \in \mathbb{R}; V_2(x) > -\infty \},$$
$$V_2'(x) = \mathcal{Y}_2(x), \quad \forall x \in (\mathcal{X}_2(\infty), \infty),$$
$$\widetilde{V}_2(y) = G_2(y) - y\mathcal{X}_2(y)$$
$$\qquad = E\widetilde{U}_2(yH_0(T)), \quad \forall y \in (0, \infty),$$
$$\widetilde{V}_2'(y) = -\mathcal{X}_2(y) \quad \forall y \in (0, \infty), \qquad (7.11)$$

where

$$\widetilde{V}_2(y) \overset{\Delta}{=} \sup_{x \in \mathbb{R}} \{ V_2(x) - xy \}, \quad y \in \mathbb{R}.$$

Remark 7.8: If $\mathcal{X}_1'(y)$ and $\mathcal{X}_2'(y)$ exist, then just as in Remark 6.12, we have for $i = 1, 2, 3$,

$$G_i'(y) = y\mathcal{X}_i'(y), \quad 0 < y < \mathcal{X}_i(\infty). \qquad (7.12)$$

Example 7.9 *(Portfolio insurance):* Suppose

$$U_2(x) = \begin{cases} \log(x - \overline{x}), & \overline{x} < x < \infty, \\ -\infty, & -\infty < x \leq \overline{x}, \end{cases}$$

where $\overline{x}$ is a positive constant below which terminal wealth is not permitted to fall. We have $\mathcal{X}_2(y) = \overline{x}h_2 + (1/y)$ for $0 < y < \infty$, where $h_2 \overset{\Delta}{=} EH_0(T)$. In order to ensure that terminal wealth can exceed $\overline{x}$, the initial wealth must exceed $\mathcal{X}_2(\infty) = \overline{x}h_2$. We have $\mathcal{Y}_2(x) = 1/(x - \overline{x}h_2)$ for $x > \overline{x}h_2$, and the optimal consumption and wealth processes are $c_2(t) \equiv 0$ and

$$X_2(t) = \frac{1}{H_0(t)} \{ x - \overline{x}h_2 + \overline{x}E[H_0(T)|\mathcal{F}(t)] \}.$$

Finally, $G_2(y) = -\log y - E \log H_0(T)$, so

$$V_2(x) = \log(x - \overline{x}h_2) - E \log H_0(T), \quad x > \overline{x}h_2.$$

As in Example 7.5, we can derive the optimal portfolio explicitly *when $S_0(\cdot)$ is deterministic.* Under this condition,

$$M_2(t) \overset{\Delta}{=} H_0(t)X_2(t) = x - \overline{x}h_2 + \frac{\overline{x}Z_0(t)}{S_0(T)}$$

and

$$dM_2(t) = d(H_0(t)X_2(t)) = -\frac{\overline{x}}{S_0(T)} Z_0(t)\theta'(t)\, dW(t).$$

It follows that

$$\pi_2(t) = (\sigma'(t))^{-1} \left[X_2(t) - \frac{\overline{x}S_0(t)}{S_0(T)} \right] \theta(t)$$

$$= \frac{x - \overline{x}h_2}{H_0(t)} (\sigma(t)\sigma'(t))^{-1} (b(t) + \delta(t) - r(t)\underline{1}).$$

In the remainder of this section we examine the relationship among the value functions and the optimal policies for Problems 5.2–5.4. Consider an agent with initial endowment $x > \mathcal{X}_3(\infty)$ who divides this wealth into two pieces, $x_1 > \mathcal{X}_1(\infty)$ and $x_2 > \mathcal{X}_2(\infty)$, so that $x_1 + x_2 = x$. For the piece x_1, he constructs the optimal policy $(c_1, \pi_1) \in \mathcal{A}_1(x_1)$ of Theorem 7.3 for the problem of maximization of utility from consumption only. With the piece x_2, he constructs the optimal policy $(c_2, \pi_2) \in \mathcal{A}_2(x_2)$ of Theorem 7.6 for the problem of maximization of utility from terminal wealth only. Note that $X^{x_1,c_1,\pi_1}(T) = 0$ and $c_2(\cdot) \equiv 0$, so the superposition $(c, \pi) = (c_1 + c_2, \pi_1 + \pi_2)$ of the policies (c_1, π_1), (c_2, π_2) is in $\mathcal{A}_3(x)$, results in the wealth process $X^{x,c,\pi}(t) = X^{x_1,c_1,\pi_1}(t) + X^{x_2,c_2,\pi_2}(t)$, and satisfies

$$V_1(x_1) + V_2(x_2) = E\left[\int_0^T U_1(t, c(t))\, dt + U_2(X^{x,c,\pi}(T)) \right] \leq V_3(x).$$

Therefore,

$$\sup\{V_1(x_1) + V_2(x_2); x_1 \in \mathbb{R}, x_2 \in \mathbb{R}, x_1 + x_2 = x\} \leq V_3(x), \quad \forall x \in \mathbb{R},$$
$$(7.13)$$

where we have used (6.21), (7.8), and (7.10) to extend these considerations to all x, x_1, and x_2 in $\mathbb{R}$.

Moreover, the reverse of inequality (7.13) holds. Again, we consider only $x > \mathcal{X}_3(\infty)$, relying on (6.21), (7.8), and (7.10) for the rest. For $x > \mathcal{X}_3(\infty)$, let $(c_3, \pi_3) \in \mathcal{A}_3(x)$ be the optimal policy of Theorem 6.3 for the problem of maximization of utility from consumption *and* terminal wealth. Define

$$x_1 \overset{\Delta}{=} E \int_0^T H_0(t)c_3(t)\, dt, \quad x_2 \overset{\Delta}{=} E\left[H_0(T)X^{x,c_3,\pi_3}(T)\right],$$

so that $x_1 + x_2 = x$ (see (6.6)). Theorem 3.5 guarantees the existence of a portfolio process $\hat{\pi}_1(\cdot)$ such that $X^{x_1,c_3,\hat{\pi}_1}(T) = 0$ and $(c_3, \hat{\pi}_1) \in \mathcal{A}_1(x_1)$; therefore, $E \int_0^T U_1(t, c_3(t))\, dt \leq V_1(x_1)$. This same theorem guarantees the existence of a portfolio process $\hat{\pi}_2(\cdot)$ such that with $\hat{c}_2 \equiv 0$, we have $X^{x_2,\hat{c}_2,\hat{\pi}_2}(T) = X^{x,c_3,\pi_3}(T)$ and $(\hat{c}_2, \hat{\pi}_2) \in \mathcal{A}_2(x_2)$; therefore, $EU_2(X^{x,c_3,\pi_3}(T)) \leq V_2(x_2)$. We have then

$$V_3(x) = E \int_0^T U_1(t, c_3(t))\, dt + EU_2(X^{x,c_3,\pi_3}(T))$$
$$\leq V_1(x_1) + V_2(x_2).$$

Theorem 7.10: *Let Assumptions 2.3 and 6.1 hold. Then*

$$V_3(x) = \sup\{V_1(x_1) + V_2(x_2); x_1 + x_2 = x\} \quad \forall x \in \mathbb{R}. \tag{7.14}$$

If, in addition, $V_i(x) < \infty$ for all $x \in \mathbb{R}$ and $i = 1, 2$, then for each $x \in (\mathcal{X}_3(\infty), \infty)$ the supremum in (7.14) is attained by $x_1 = \mathcal{X}_1(\mathcal{Y}_3(x))$, $x_2 = \mathcal{X}_2(\mathcal{Y}_3(x))$. In particular,

$$V_3(x) = V_1\left(\mathcal{X}_1(\mathcal{Y}_3(x))\right) + V_2\left(\mathcal{X}_2(\mathcal{Y}_3(x))\right), \quad \forall x \in (\mathcal{X}_3(\infty), \infty) \tag{7.15}$$

and

$$\widetilde{V}_3(y) = \widetilde{V}_1(y) + \widetilde{V}_2(y), \quad \forall y \in (0, \infty). \tag{7.16}$$

PROOF. For fixed $x \in (\mathcal{X}_3(\infty), \infty)$, let us consider the concave function $f: \mathbb{R} \to [-\infty, \infty)$ defined by $f(x_1) \triangleq V_1(x_1) + V_2(x - x_1)$. Under the assumption that $V_i < \infty$ for $i = 1, 2$, f is finite on the open interval $(\mathcal{X}_1(\infty), x - \mathcal{X}_2(\infty))$. Outside the closure of this interval, f takes the value $-\infty$. Now, $f'(x_1) = \mathcal{Y}_1(x_1) - \mathcal{Y}_2(x - x_1)$ for $\mathcal{X}_1(\infty) < x_1 < x - \mathcal{X}_2(\infty)$, and f' is continuous and strictly decreasing on this interval, with

$$f'\left(\mathcal{X}_1(\infty)+\right) = r_1 - \mathcal{Y}_2(x - \mathcal{X}_1(\infty)),$$
$$f'\left((x - \mathcal{X}_2(\infty))-\right) = \mathcal{Y}_1(x - \mathcal{X}_2(\infty)) - r_2,$$

where r_1 and r_2 are given by (7.3). There are three possibilities:

(i) $f'\left(\mathcal{X}_1(\infty)+\right) \leq 0$,
(ii) $f'\left(\mathcal{X}_1(\infty)+\right) > 0$, $f'\left((x - \mathcal{X}_2(\infty))-\right) < 0$,
(iii) $f'\left((x - \mathcal{X}_2(\infty))-\right) \geq 0$.

In case (ii), the maximum in (7.14) is attained by the unique value $x_1 \in (\mathcal{X}_1(\infty), x - \mathcal{X}_2(\infty))$ where $f'(x_1) = 0$, i.e., $\mathcal{Y}_1(x_1) = \mathcal{Y}_2(x - x_1)$. We check that $x_1 = \mathcal{X}_1(\mathcal{Y}_3(x))$ solves this equation, to wit,

$$\begin{aligned}
\mathcal{Y}_1\left(\mathcal{X}_1(\mathcal{Y}_3(x))\right) &= \mathcal{Y}_3(x) \\
&= \mathcal{Y}_2\left(\mathcal{X}_2(\mathcal{Y}_3(x))\right) \\
&= \mathcal{Y}_2\left(\mathcal{X}_3(\mathcal{Y}_3(x)) - \mathcal{X}_1(\mathcal{Y}_3(x))\right) \\
&= \mathcal{Y}_2\left(x - \mathcal{X}_1(\mathcal{Y}_3(x))\right).
\end{aligned}$$

In case (i), the supremum in (7.14) is attained by $x_1 = \mathcal{X}_1(\infty)$. In this case, we have $r_1 \leq \mathcal{Y}_2(x - \mathcal{X}_1(\infty))$, or equivalently, $\mathcal{X}_2(r_1) \geq x - \mathcal{X}_1(\infty)$. This implies $\mathcal{X}_3(r_1) = \mathcal{X}_1(r_1) + \mathcal{X}_2(r_1) = \mathcal{X}_1(\infty) + \mathcal{X}_2(r_1) \geq x$, so that $r_1 \leq \mathcal{Y}_3(x)$ and $\mathcal{X}_1(\infty) = \mathcal{X}_1(r_1) \geq \mathcal{X}_1(\mathcal{Y}_3(x)) \geq \mathcal{X}_1(\infty)$ because $\mathcal{X}_1$ is nonincreasing. Thus $\mathcal{X}_1(\infty) = \mathcal{X}_1(\mathcal{Y}_3(x))$, as claimed.

Case (iii) is dispatched by interchanging the subscripts 1 and 2 in the argument for case (i).

We have shown that the supremum in (7.14) is attained by $x_1 = \mathcal{X}_1(\mathcal{Y}_3(x))$, $x_2 = x - \mathcal{X}_1(\mathcal{Y}_3(x)) = \mathcal{X}_3(\mathcal{Y}_3(x)) - \mathcal{X}_1(\mathcal{Y}_3(x)) = \mathcal{X}_2(\mathcal{Y}_3(x))$; i.e., (7.15) holds. Equation (7.16) follows immediately from (6.24), (7.9), (7.11), and the definitions of G_i and $\mathcal{X}_i$ for $i = 1, 2, 3$. $\qquad\square$

Example 7.11: Let

$$U_1(c) = \begin{cases} \log(c - \bar{c}), & \bar{c} < c < \infty, \\ -\infty, & -\infty < c \leq \bar{c}, \end{cases}$$

as in Example 7.5 (subsistence consumption), and let

$$U_2(x) = \begin{cases} \log(x - \bar{x}), & \bar{x} < x < \infty, \\ -\infty, & -\infty < x \leq \bar{x}, \end{cases}$$

as in Example 7.9 (portfolio insurance). Then

$$\mathcal{X}_3(y) = \mathcal{X}_1(y) + \mathcal{X}_2(y) = \bar{c}h_1 + \bar{x}h_2 + (T+1)/y,$$

so $\mathcal{Y}_3(x) = (T+1)/(x - \bar{c}h_1 - \bar{x}h_2)$ and

$$\mathcal{X}_1(\mathcal{Y}_3(x)) = \frac{Tx + \bar{c}h_1 - T\bar{x}h_2}{T+1}, \quad \mathcal{X}_2(\mathcal{Y}_3(x)) = \frac{x - \bar{c}h_1 + T\bar{x}h_2}{T+1}.$$

It follows from Theorem 7.10 that

$$V_3(x) = V_1\left(\mathcal{X}_1(\mathcal{Y}_3(x))\right) + V_2\left(\mathcal{X}_2(\mathcal{Y}_3(x))\right)$$

$$= (T+1)\log\left(\frac{x - \bar{c}h_1 - \bar{x}h_2}{T+1}\right) - E\int_0^T \log H_0(t)\,dt - E\log H_0(T).$$

When $\bar{c} = \bar{x} = 0$, we recover the formula for V_3 obtained in Example 6.6.

3.8 Deterministic Coefficients

In this section we specialize the results of Section 3.6 to the case of $A(\cdot) \equiv 0$ and continuous, *deterministic* functions $r(\cdot): [0,T] \to \mathbb{R}$, $\theta(\cdot): [0,T] \to \mathbb{R}^N$ and $\sigma(\cdot): [0,T] \to L(\mathbb{R}^N; \mathbb{R}^N)$, the set of $N \times N$ matrices. In this case, stock prices and the money-market price become Markov processes. We will focus on obtaining an explicit formula for the optimal portfolio $\pi_3(\cdot)$ of Theorem 6.3, whose existence was established there but for which no useful representation apart from (6.11) was provided. We shall show that the value function for Problem 5.4 is a solution to the nonlinear, second-order parabolic Hamilton–Jacobi–Bellman partial differential equation one would expect (Theorem 8.11), and that subject to a growth condition, the dual value function is the *unique* solution of a *linear* second-order parabolic partial differential equation (Theorem 8.12). Several examples are provided.

We shall represent both the optimal portfolio $\pi_3(\cdot)$ and also the optimal consumption rate process $c_3(\cdot)$ in "feedback form" on the level of wealth $X_3(\cdot)$ of (6.10), i.e.,

$$c_3(t) = C(t, X_3(t)), \quad \pi_3(t) = \Pi(t, X_3(t)), \quad 0 \leq t \leq T, \tag{8.1}$$

for suitable functions $C: [0,T] \times (0,\infty) \to [0,\infty)$ and $\Pi: [0,T] \times (0,\infty) \to \mathbb{R}^N$ (cf. Theorem 8.8), which do not depend on the initial wealth. Such a representation shows that in the case of deterministic coefficients, the current level of wealth is a *sufficient statistic* for the utility maximization Problem 5.4: an investor who computes his optimal strategy at time t on the basis of his current wealth only can do just as well as an investor who keeps track of the whole past and present information $\mathcal{F}(t)$ about the market! Similar results hold for Problems 5.2 and 5.3; the interested reader will find their derivation to be straightforward.

Throughout this section, the following two assumptions will be in force. These are not the weakest assumptions that support the subsequent analysis, but they will permit us to proceed with a minimum of technical fuss. For more general results, the reader is referred to Ocone and Karatzas (1991), Section 6.

Assumption 8.1:   *We have $A(\cdot) \equiv 0$, and the processes $r(\cdot)$, $\theta(\cdot)$, and $\sigma(\cdot)$ are nonrandom, continuous (and hence bounded) functions on $[0, T]$, and $r(\cdot)$ and $\|\theta(\cdot)\|$ are in fact Hölder continuous, i.e., for some $K > 0$ and $\rho \in (0,1)$ we have*

$$|r(t_1) - r(t_2)| \leq K|t_1 - t_2|^\rho, \quad \left|\|\theta(t_1)\| - \|\theta(t_2)\|\right| \leq K|t_1 - t_2|^\rho$$

for all $t_1, t_2 \in [0, T]$. Furthermore, $\|\theta(\cdot)\|$ is bounded away from zero. In particular, there are positive constants κ_1, κ_2 such that

$$0 < \kappa_1 \leq \|\theta(t)\| \leq \kappa_2 < \infty, \quad \forall t \in [0, T]$$

almost surely.

Because of Novikov's condition (e.g., Karatzas and Shreve (1991) Section 3.5D), Assumption 8.1 guarantees that the local martingale $Z_0(\cdot)$ of (2.5) is in fact a martingale. This permits the construction of the martingale measure P_0 of (1.5.3) under which the process $W_0(\cdot)$ is a Brownian motion. We have not needed this probability measure in previous sections, but we shall make use of it in this section.

Assumption 8.2:   *The agent's preference structure (U_1, U_2) satisfies*

(i) *(polynomial growth of I_1 and I_2) there is a constant $\gamma > 0$ such that*

$$I_1(t, y) \leq \gamma + y^{-\gamma} \quad \forall (t, y) \in [0, T] \times (0, \infty),$$
$$I_2(y) \leq \gamma + y^{-\gamma} \quad \forall y \in (0, \infty);$$

(ii) *(polynomial growth of $U_1 \circ I_1$ and $U_2 \circ I_2$) there is a constant $\gamma > 0$ such that*

$$U_1(t, I_1(t, y)) \geq -\gamma - y^\gamma \quad \forall (t, y) \in [0, T] \times (0, \infty),$$
$$U_2(I_2(y)) \geq -\gamma - y^\gamma \quad \forall y \in (0, \infty);$$

(iii) *(Hölder continuity of I_1) for each $y_0 \in (0, \infty)$, there exist constants $\epsilon(y_0) > 0$, $K(y_0) > 0$, and $\rho(y_0) \in (0, 1)$ such that*

$$|I_1(t,y) - I_1(t,y_0)| \le K(y_0)|y - y_0|^{\rho(y_0)} \quad \forall t \in [0,T],$$
$$\forall y \in (0,\infty) \cap (y_0 - \epsilon(y_0), y_0 + \epsilon(y_0));$$

(iv) *either for each $t \in [0,T]$, $I_1'(t,y) \triangleq \frac{\partial}{\partial y} I_1(t,y)$ is defined and strictly negative for all y in a set of positive Lebesgue measure, or else $I_2'(y)$ is defined and strictly negative for all y in a set of positive Lebesgue measure.*

Remark 8.3:   Because $I_1(t,\cdot)$ and I_2 are nonincreasing, Assumption 8.2(i), (ii) and Remark 5.6(ii) imply the existence of a constant $\gamma > 0$ such that

$$|U_1(t, I_1(t,y))| \le \gamma + y^\gamma + y^{-\gamma} \quad \forall (t,y) \in [0,T] \times (0,\infty),$$
$$|U_2(I_2(y))| \le \gamma + y^\gamma + y^{-\gamma} \quad \forall y \in (0,\infty).$$

Furthermore, for each $y_0 \in (0,\infty)$ and $\epsilon(y_0)$, $K(y_0)$, and $\rho(y_0)$ as in Assumption 8.2(iii), the mean value theorem implies for all $y \in (0,\infty) \cap (y_0 - \epsilon(y_0), y_0 + \epsilon(y_0))$ that

$$|U_1(t, I_1(t,y)) - U_1(t, I_1(t,y_0))| \le U_1'(t, \iota(t))|I_1(t,y) - I_1(t,y_0)|$$
$$\le MK(y_0)|y - y_0|^{\rho(y_0)},$$

where $\iota(t)$ takes values between $I_1(t,y)$ and $I_1(t,y_0)$ and M is a bound on the continuous function $U_1'(t, I_1(t,\eta))$ as (t,η) ranges over the set $[0,T] \times [(0,\infty) \cap (y_0 - \epsilon(y_0), y_0 + \epsilon(y_0))]$. In other words, $U_1 \circ I_1$ enjoys the same kind of Hölder continuity posited in Assumption 8.2(iii) for I_1.

We introduce the process

$$Y^{(t,y)}(s) \triangleq y \exp\left\{ -\int_t^s r(u)\, du - \int_t^s \theta'(u)\, dW(u) - \frac{1}{2} \int_t^s \|\theta(u)\|^2\, du \right\}$$

$$= y \exp\left\{ -\int_t^s r(u)\, du - \int_t^s \theta'(u)\, dW_0(u) \right.$$

$$\left. + \frac{1}{2} \int_t^s \|\theta(u)\|^2\, du \right\}, \quad t \le s \le T, \tag{8.2}$$

for any given $(t,y) \in [0,T] \times (0,\infty)$. The process $Y^{(t,y)}(\cdot)$ is a diffusion with linear dynamics:

$$dY^{(t,y)}(s) = Y^{(t,y)}(s)[-r(s)\, ds - \theta'(s)\, dW(s)]$$
$$= Y^{(t,y)}(s)[-r(s)\, ds + \|\theta(s)\|^2\, ds - \theta'(s)\, dW_0(s)],$$

$Y^{(t,y)}(t) = y$, and $Y^{(t,y)}(s) = yY^{(t,1)}(s) = yH_0(s)/H_0(t)$, where $H_0(s)$ is given by (2.6). With these properties in mind, and using the Markov property for $Y^{(t,y)}(\cdot)$ under the martingale measure P_0, as well as "Bayes's rule" of Lemma 3.5.3 in Karatzas and Shreve (1991), we may rewrite the expression (6.10) for the optimal wealth as follows:

$$X_3(t) = \frac{1}{Z_0(t)} E\left[\int_t^T Z_0(s) e^{-\int_t^s r(u)du} I_1(s, zY^{(0,1)}(s))\, ds\right.$$

$$+ Z_0(T) e^{-\int_t^T r(u)du} I_2(zY^{(0,1)}(T)) \left.\bigg|\, \mathcal{F}(t)\right]$$

$$= E_0\left[\int_t^T e^{-\int_t^s r(u)du} I_1(s, Y^{(0,z)}(s))\, ds\right.$$

$$+ e^{-\int_t^T r(u)du} I_2(Y^{(0,z)}(T)) \left.\bigg|\, \mathcal{F}(t)\right]$$

$$= \mathcal{X}(t, Y^{(0,z)}(t)), \quad 0 \le t \le T, \tag{8.3}$$

where $z = \mathcal{Y}_3(x)$, E_0 denotes expectation with respect to the martingale measure P_0, and $\mathcal{X}: [0,T] \times (0,\infty) \to (0,\infty)$ is given by

$$\mathcal{X}(t,y) \triangleq E_0\left[\int_t^T e^{-\int_t^s r(u)du} I_1(s, yY^{(t,1)}(s))\, ds\right.$$

$$+ e^{-\int_t^T r(u)du} I_2(yY^{(t,1)}(T)) \left.\vphantom{\int_t^T}\right]. \tag{8.4}$$

The Markov property for $Y^{(t,y)}(\cdot)$ under P implies that

$$E\left[\int_t^T Y^{(t,y)}(s) I_1(s, Y^{(t,y)}(s))\, ds + Y^{(t,y)}(T) I_2(Y^{(t,y)}(T)) \bigg|\, \mathcal{F}(t)\right]$$

is a function of $Y^{(t,y)}(t) = y$, i.e., is deterministic. Therefore,

$$\mathcal{X}(t,y) = E_0\left\{E_0\left[\int_t^T e^{-\int_t^s r(u)du} I_1(s, Y^{(t,y)}(s))\, ds\right.\right.$$

$$+ e^{-\int_t^T r(u)du} I_2(Y^{(t,y)}(T)) \left.\left.\bigg|\, \mathcal{F}(t)\right]\right\}$$

$$= E\left\{\frac{Z_0(T)}{Z_0(t)} E\left[\int_t^T e^{-\int_t^s r(u)du} Z_0(s) I_1(s, Y^{(t,y)}(s))\, ds\right.\right.$$

$$+ e^{-\int_t^T r(u)du} Z_0(T) I_2(Y^{(t,y)}(T)) \left.\left.\bigg|\, \mathcal{F}(t)\right]\right\}$$

$$= \frac{1}{y} E\left\{Z_0(T) E\left[\int_t^T Y^{(t,y)}(s) I_1(s, Y^{(t,y)}(s))\, ds\right.\right.$$

$$+ Y^{(t,y)}(T) I_2(Y^{(t,y)}(T)) \Big| \mathcal{F}(t) \Big] \bigg\}$$

$$= \frac{1}{y} E \left[\int_t^T Y^{(t,y)}(s) I_1(s, Y^{(t,y)}(s)) \, ds \right.$$

$$\left. + Y^{(t,y)}(T) I_2(Y^{(t,y)}(T)) \right], \tag{8.5}$$

and $\mathcal{X}(\cdot, \cdot)$ is an extension of the function $\mathcal{X}_3(\cdot) = \mathcal{X}(0, \cdot)$ defined by (6.1). We should properly write $\mathcal{X}_3(t, y)$ rather than $\mathcal{X}(t, y)$, to indicate that this function is associated with Problem 5.4. However, we do not carry out an analysis for Problems 5.2 and 5.3 under the assumption of deterministic coefficients, and hence permit ourselves the convenience of suppressing the subscript.

Lemma 8.4: *Under Assumptions 8.1, 8.2, the function $\mathcal{X}$ defined by (8.4) is of class $C([0, T] \times (0, \infty)) \cap C^{1,2}([0, T) \times (0, \infty))$ and solves the Cauchy problem*

$$\mathcal{X}_t(t, y) + \frac{1}{2} \|\theta(t)\|^2 y^2 \mathcal{X}_{yy}(t, y) + \left(\|\theta(t)\|^2 - r(t) \right) y \mathcal{X}_y(t, y) - r(t) \mathcal{X}(t, y)$$

$$= -I_1(t, y) \quad \text{on } [0, T) \times (0, \infty), \tag{8.6}$$

$$\mathcal{X}(T, y) = I_2(y) \quad \text{on } (0, \infty). \tag{8.7}$$

Furthermore, for each $t \in [0, T)$, $\mathcal{X}(t, \cdot)$ is strictly decreasing with $\mathcal{X}(t, 0+) = \infty$ and

$$\mathcal{X}(t, \infty) \triangleq \lim_{y \to \infty} \mathcal{X}(t, y) = \int_t^T \exp \left(- \int_t^s r(u) du \right) \bar{c}(s) \, ds$$

$$+ \exp \left(- \int_t^T r(u) du \right) \bar{x}. \tag{8.8}$$

Consequently, for $t \in [0, T)$, $\mathcal{X}(t, \cdot)$ has a strictly decreasing inverse function $\mathcal{Y}(t, \cdot) \colon (\mathcal{X}(t, \infty), \infty) \xrightarrow{\text{onto}} (0, \infty)$, i.e.,

$$\mathcal{X}(t, \mathcal{Y}(t, x)) = x, \quad \forall x \in (\mathcal{X}(t, \infty), \infty), \tag{8.9}$$

and $\mathcal{Y}$ is of class $C^{1,2}$ on the set

$$D \triangleq \{ (t, x) \in [0, T) \times \mathbb{R}; \quad x > \mathcal{X}(t, \infty) \}. \tag{8.10}$$

For $t = T$, we have $\mathcal{X}(T, \cdot) = I_2(\cdot)$, which is strictly decreasing on the interval $(0, U_2'(\bar{x}+))$, and we have $\mathcal{X}(T, \infty) = \bar{x}$. The inverse of $\mathcal{X}(T, \cdot)$ is $\mathcal{Y}(T, \cdot) \triangleq U_2'(\cdot)$, which also satisfies (8.9). The function $\mathcal{Y}$ is continuous on the set $\{ (t, x) \in [0, T] \times \mathbb{R}; x > \mathcal{X}(t, \infty) \}$.

PROOF.   Consider the Cauchy problem

$$u_t(t, \eta) + \frac{1}{2}\|\theta(t)\|^2 u_{\eta\eta}(t, \eta) + \left(\frac{1}{2}\|\theta(t)\|^2 - r(t)\right) u_\eta(t, \eta) - r(t)u(t, \eta)$$
$$= -I_1(t, e^\eta), \quad 0 \le t < T, \quad \eta \in \mathbb{R}, \tag{8.11}$$
$$u(T, \eta) = I_2(e^\eta), \quad \eta \in \mathbb{R}. \tag{8.12}$$

The classical theory of partial differential equations (e.g., Friedman (1964), Section 1.7) implies that there is a function u of class $C([0, T] \times \mathbb{R}) \cap C^{1,2}([0, T) \times \mathbb{R})$ satisfying (8.11), (8.12). Furthermore, for each $\epsilon > 0$, there is a constant $C(\epsilon)$ such that

$$|u(t, \eta)| \le C(\epsilon)e^{\epsilon\eta^2}, \quad \forall \eta \in \mathbb{R}. \tag{8.13}$$

We fix $(t, y) \in [0, T) \times (0, \infty)$ and use Itô's rule in conjunction with (8.2) and (8.11) to compute

$$d\left[e^{-\int_t^s r(u)du} u(s, \log Y^{(t,y)}(s))\right]$$
$$= -e^{-\int_t^s r(u)du} I_1(s, Y^{(t,y)}(s)) \, ds \tag{8.14}$$
$$- e^{-\int_t^s r(u)du} u_\eta(s, \log Y^{(t,y)}(s))\theta'(s) \, dW_0(s).$$

For each positive integer n, we define

$$\tau_n \triangleq \left(T - \frac{1}{n}\right) \wedge \inf\left\{s \in [t, T]; \left|\log Y^{(t,y)}(s)\right| \ge n\right\},$$

so that $\max_{0 \le s \le \tau_n} |u_\eta(s, \log Y^{(t,y)}(s))|$ is bounded, uniformly in $\omega \in \Omega$. Integrating (8.14) and taking expectations, we obtain

$$u(t, \log y) = E_0 \int_t^{\tau_n} \exp\left(-\int_t^s r(u)du\right) I_1(s, Y^{(t,y)}(s)) \, ds$$
$$+ E_0 \exp\left(-\int_t^{\tau_n} r(u)du\right) u(\tau_n, \log Y^{(t,y)}(\tau_n)).$$

The monotone convergence theorem implies

$$u(t, \log y) = E_0 \int_t^T e^{-\int_t^s r(u)du} I_1(s, Y^{(t,y)}(s)) \, ds$$
$$+ \lim_{n \to \infty} E_0 e^{-\int_t^{\tau_n} r(u)du} u(\tau_n, \log Y^{(t,y)}(\tau_n)). \tag{8.15}$$

Now,

$$\lim_{n \to \infty} e^{-\int_t^{\tau_n} r(u)du} u(\tau_n, \log Y^{(t,y)}(\tau_n)) = e^{-\int_t^T r(u)du} I_2(Y^{(t,y)}(T)) \tag{8.16}$$

almost surely, and we wish to prove that

$$\lim_{n \to \infty} E_0 e^{-\int_t^{\tau_n} r(u)du} u(\tau_n, \log Y^{(t,y)}(\tau_n)) = E_0 e^{-\int_t^T r(u)du} I_2(Y^{(t,y)}(T)). \tag{8.17}$$

To obtain (8.17) from (8.16), we need a dominating function. We can use (8.2) and (8.13) to write

$$\left| e^{-\int_t^{\tau_n} r(u)\,du} u\left(\tau_n, \log Y^{(t,y)}(\tau_n)\right) \right|$$

$$\leq C(\epsilon) e^{\int_t^T |r(u)|\,du} e^{\epsilon(\log Y^{(t,y)}(\tau_n))^2}$$

$$\leq C(\epsilon) e^{\int_t^T |r(u)|\,du} \exp\left\{ \epsilon\left[|\log y| + \int_t^T \left| -r(u) + \frac{1}{2}\|\theta(u)\|^2 \right| du \right. \right.$$

$$\left. \left. + \sup_{t\leq s\leq T} \left| \int_t^s \theta'(u)\,dW_0(u) \right| \right]^2 \right\}.$$

Equation (8.17) will follow from the dominated convergence theorem, once we show that

$$E_0\left[\exp\left\{ \epsilon \sup_{t\leq s\leq T} \left| \int_t^s \theta'(u)\,dW_0(u) \right|^2 \right\} \right] < \infty. \qquad (8.18)$$

We may extend $\theta(\cdot)$ beyond $[0,T]$ by setting $\theta(t) \triangleq \theta(T)$ for $t \geq T$, and we can set

$$M(s) \triangleq \int_t^{t+s} \theta'(u)\,dW_0(u), \quad s \geq 0.$$

We have

$$\langle M\rangle(s) = \int_t^{t+s} \|\theta(u)\|^2 \,du \leq \kappa_2^2 s, \quad \forall s \in [0,\infty),$$

and $\langle M\rangle(\cdot)$ is strictly increasing. Under P_0, $B(\tau) \triangleq M(\langle M\rangle^{-1}(\tau))$, $0 \leq \tau < \infty$, is a standard Brownian motion (e.g., Karatzas and Shreve (1991), Theorem 3.4.6). Moreover, with $\bar{\tau} \triangleq \kappa_2^2(T-t)$, we have

$$B_* \triangleq \sup_{0\leq\tau\leq\bar{\tau}} |B(\tau)| \geq \sup_{0\leq s\leq T-t} |M(s)| = \sup_{t\leq s\leq T} \left| \int_t^s \theta'(u)\,dW_0(u) \right|.$$

We show that for $\epsilon > 0$ sufficiently small,

$$E_0 e^{\epsilon B_*^2} < \infty.$$

Let $B_+ = \sup_{0\leq\tau\leq\bar{\tau}} B(\tau)$ and $B_- = \sup_{0\leq\tau\leq\bar{\tau}}(-B(\tau))$, so that $B_* = \max\{B_+, B_-\}$. The density for both B_+ and B_- is (e.g., Karatzas and Shreve (1991), Remark 2.8.3)

$$f(b)\,db = P_0\{B_\pm \in db\} = \frac{2}{\sqrt{2\pi\bar{\tau}}} e^{-b^2/(2\bar{\tau})}\,db, \quad b > 0.$$

We define $F(b) = \int_b^\infty f(x)\,dx$, and write

$$E_0 e^{\epsilon B_*^2} = -\int_0^\infty e^{\epsilon b^2}\,dP_0\{B_* > b\}$$

$$\leq 1 + \int_0^\infty P_0\{B_* > b\} \, d(e^{\epsilon b^2})$$

$$\leq 1 + 2 \int_0^\infty F(b) \, d(e^{\epsilon b^2})$$

$$\leq 1 + 2 \lim_{b \to \infty} F(b) e^{\epsilon b^2} + \int_0^\infty f(b) e^{\epsilon b} \, db,$$

which is finite for $0 < \epsilon < 1/(2\bar{T})$.

Choosing $\epsilon \in (0, 1/(2\bar{T}))$ so that (8.18) holds, we obtain (8.17), and (8.15) yields

$$\mathcal{X}(t, y) = u(t, \log y), \quad \forall (t, y) \in [0, T] \times (0, \infty). \tag{8.19}$$

It follows immediately that $\mathcal{X}$ is of class $C([0, T] \times (0, \infty)) \cap C^{1,2}([0, T) \times (0, \infty))$ and satisfies (8.6) and (8.7).

We next use Assumption 8.2(iv) to show that $\mathcal{X}_y(t, y) < 0$. For specificity, let us assume that $I_2'(y)$ is defined and strictly negative for all y in a set $N \subset (0, \infty)$ having positive Lebesgue measure. Because $I_1(t, \cdot)$ and I_2 are nonincreasing, we have for $t \in [0, T)$, $y > 0$, and $h > 0$ that

$$\frac{1}{h}[\mathcal{X}(t, y) - \mathcal{X}(t, y + h)]$$

$$\geq e^{-\int_t^T r(u)\,du} E_0 \frac{1}{h}[I_2(yY^{(t,1)}(T)) - I_2((y+h)Y^{(t,1)}(T))].$$

Under P_0, the random variable $\int_t^T \theta'(u) \, dW_0(u) = B(\langle M \rangle(T - t))$ is normally distributed, with mean zero and standard deviation $\rho \stackrel{\Delta}{=} \sqrt{\langle M \rangle(T-t)}$. Setting $m \stackrel{\Delta}{=} \int_t^T \left(-r(u) + \|\theta(u)\|^2/2\right) \, du$, we have

$$E_0 \frac{1}{h}[I_2(yY^{(t,1)}(T)) - I_2((y+h)Y^{(t,1)}(T))]$$

$$\geq \frac{1}{\sqrt{2\pi}} \int_N \frac{1}{h} \left[I_2\left(ye^{m-\rho w}\right) - I_2\left((y+h)e^{m-\rho w}\right)\right] e^{-w^2/2} \, dw.$$

Letting $h \downarrow 0$ and using Fatou's lemma, we obtain

$$-\mathcal{X}_y(t, y) \geq -\frac{1}{\sqrt{2\pi}} \int_N e^{m-\rho w} I_2'\left(ye^{m-\rho w}\right) e^{-w^2/2} \, dw > 0.$$

From the implicit function theorem we have the existence of the function $\mathcal{Y}$ that satisfies (8.9) for all $t \in [0, T]$, is of class $C^{1,2}$ on D, and is continuous on $\{(t, x) \in [0, T] \times \mathbb{R}; x > \mathcal{X}(t, \infty)\}$. Relation (8.8) follows from the definitions of $I_1(t, \cdot)$ and I_2 in Section 3.5 and the dominated convergence theorem. $\square$

Remark 8.5: From (8.14) and (8.19), we have

$$d\left(e^{-\int_0^s r(u)\,du} \mathcal{X}(s, Y^{(0,y)}(s))\right) = e^{-\int_0^s r(u)\,du}[-I_1(s, Y^{(0,y)}(s))$$

$$- Y^{(0,y)}(s)\mathcal{X}_y(s, Y^{(0,y)}(s))\theta'(s) \, dW_0(s)],$$

which leads to the following useful integral formula for $0 \le t \le T$, $y > 0$:

$$e^{-\int_0^t r(u)du} \mathcal{X}(t, Y^{(0,y)}(t)) + \int_0^t e^{-\int_0^s r(u)du} I_1(s, Y^{(0,y)}(s))\, ds$$

$$= \mathcal{X}(0,y) - \int_0^t e^{-\int_0^s r(u)du} Y^{(0,y)}(s) \mathcal{X}_y(s, Y^{(0,y)}(s)) \theta'(s)\, dW_0(s). \quad (8.20)$$

Remark 8.6: The proof of Lemma 8.4 also shows that $\mathcal{X}$ is the *unique* $C([0,T] \times (0,\infty)) \cap C^{1,2}([0,T) \times (0,\infty))$ solution to the Cauchy problem (8.6), (8.7) among those functions f satisfying the growth condition

$$\forall \epsilon > 0, \; \exists C(\epsilon) \text{ such that } |f(t,y)| \le C(\epsilon) e^{\epsilon(\log y)^2}, \quad \forall(t,y) \in [0,T] \times (0,\infty). \quad (8.21)$$

Indeed, if f is a solution to (8.6), (8.7) satisfying (8.21), then $u(t,\eta) \triangleq f(t,e^\eta)$ is a solution to (8.11), (8.12) satisfying the growth conditon (8.13) for every $\epsilon > 0$. From (8.19) we see that f agrees with $\mathcal{X}$.

Remark 8.7: We noted in Remark 5.7 that if an agent's initial wealth $X_3(0)$ lies below $\mathcal{X}(0,\infty)$, then every consumption/portfolio process pair results in an expected utility of $-\infty$. If $X_3(0) = \mathcal{X}(0,\infty)$, then one should take $c_3(t) = \bar{c}(t)$, $0 \le t \le T$, choose $\pi_3(\cdot)$ such that $X_3(T) = \bar{x}$, and this results in expected utility $\int_0^T U_1(t, \bar{c}(t))\, dt + U_2(\bar{x})$, which is either finite or $-\infty$. Under Assumption 8.1, the portfolio $\pi_3(\cdot)$ that produces this result is $\pi_3 \equiv 0$. Indeed, with this choice of $c_3(\cdot)$ and $\pi_3(\cdot)$, the wealth equation (3.1) becomes

$$X^{\mathcal{X}(0,\infty),c_3,\pi_3}(t) = e^{\int_0^t r(u)du} \mathcal{X}(0,\infty) - \int_0^t e^{-\int_t^s r(u)du} \bar{c}(s)\, ds$$

$$= \mathcal{X}(t,\infty),$$

where we have used (8.8). We have then the feedback form (8.1) for optimal consumption and investment when wealth at time t is $\mathcal{X}(t,\infty)$:

$$C(t, \mathcal{X}(t,\infty)) = \bar{c}(t), \quad \Pi(t, \mathcal{X}(t,\infty)) = 0, \quad 0 \le t \le T. \quad (8.22)$$

We now derive the feedback form for optimal consumption and investment when wealth at time t exceeds $\mathcal{X}(t,\infty)$.

Theorem 8.8: *Under the Assumptions 8.1 and 8.2, the feedback form (8.1) for the optimal consumption/portfolio process pair (c_3, π_3) for Problem 5.4 is given by*

$$C(t,x) \triangleq I_1(t, \mathcal{Y}(t,x)), \quad (8.23)$$

$$\Pi(t,x) \triangleq -(\sigma'(t))^{-1}\theta(t)\frac{\mathcal{Y}(t,x)}{\mathcal{Y}_x(t,x)}, \quad (8.24)$$

for $0 \le t \le T$ and $x \in (\mathcal{X}(t,\infty),\infty)$.

PROOF. We have from (8.3) that when the initial wealth x at time 0 exceeds $\mathcal{X}(0,\infty)$, then the optimal wealth $X_3(t)$ at time $t \in [0,T]$ is $\mathcal{X}(t, Y^{(0,\mathcal{Y}_3(x))}(t))$. In other words,

$$Y^{(0,\mathcal{Y}(0,x))}(t) = \mathcal{Y}(t, X_3(t)),$$

and (6.5) becomes

$$c_3(t) = I_1(t, \mathcal{Y}(0,x)Y^{(0,1)}(t)) = I_1\left(t, \mathcal{Y}(t, X_3(t))\right),$$

which establishes (8.23). With $y = \mathcal{Y}(0,x)$ and using (8.3), we may write (8.20) as

$$\frac{X_3(t)}{S_0(t)} + \int_0^t \frac{c_3(s)}{S_0(s)}\, ds = x - \int_0^t \frac{\mathcal{Y}(s, X_3(s))}{S_0(s)} \mathcal{X}_y\left(s, \mathcal{Y}(s, X_3(s))\right) \theta'(s)\, dW_0(s).$$

$$(8.25)$$

But from (8.9), we have $\mathcal{X}_y(t, \mathcal{Y}(t,x)) = 1/\mathcal{Y}_x(t,x)$ for all $x > \mathcal{X}(t,\infty)$, and comparison of (8.25) with the wealth equation (3.1) shows that the optimal portfolio satisfies

$$\pi_3'(t)\sigma(t) = -\theta'(t)\frac{\mathcal{Y}(t, X_3(t))}{\mathcal{Y}_x(t, X_3(t))},$$

justifying (8.24). □

Remark 8.9 *(Merton's mutual fund theorem):* Formula (8.24) for the optimal portfolio shows that under the assumptions of Theorem 8.8, the agent should always invest in stocks according to the proportions

$$(\sigma'(t))^{-1}\theta(t) = (\sigma(t)\sigma'(t))^{-1}[b(t) + \delta(t) - r(t)\mathbf{1}],$$

independently of the utility functions U_1, U_2. This permits the formation of a *mutual fund* so that independently of his wealth and preference structure, the agent is indifferent whether he invests in the assets individually or invests only in the mutual fund and the money market. For example, we may form a mutual fund by imagining an agent who begins with initial wealth 1 and seeks to maximize $E \log X_2(T)$. The behavior of this agent is described in Example 7.9 with $\bar{x} = 0$; the optimal wealth is $X_2(t) = 1/H_0(t)$, and the optimal portfolio is $\pi_2(t) = (\sigma'(t))^{-1}\theta(t)/H_0(t)$. (We show in Section 10 that as $T \to \infty$, this agent is maximizing the growth rate of wealth.) We think of $X_2(t)$ as the price per share of a mutual fund that holds a portfolio $\pi_2(t)$ in the N stocks and $(1/H_0(t)) - \pi_2'(t)\mathbf{1}$ in the money market. In particular, each dollar invested in the mutual fund results in the vector $(\sigma'(t))^{-1}\theta(t)$ of dollar investments in the stocks. The essence of Theorem 8.8 is that any other agent solving Problem 5.4 is satisfied to have the only investment opportunities be the money market and this mutual fund. At time t and with wealth level x, this other agent invests $-\mathcal{Y}(t,x)/\mathcal{Y}_x(t,x)$ dollars in the mutual fund. This amount depends on the agent's wealth x and preference structure (U_1, U_2), but *the mutual fund itself does not.*

Finally, we develop the *Hamilton–Jacobi–Bellman (HJB) equation* associated with Problem 5.4. To do that, we must extend the value function V_3 of (5.8) to include the time variable. Given $(t, x) \in [0, T] \times \mathbb{R}$, and given a consumption/portfolio process pair $(c(\cdot), \pi(\cdot))$, the wealth process $X^{t,x,c,\pi}(\cdot)$ corresponding to (c, π) with initial condition (t, x) is given by (cf. (3.1))

$$
e^{-\int_t^s r(u)du} X^{t,x,c,\pi}(s) = x - \int_t^s e^{-\int_t^u r(v)dv} c(u)du
$$
$$
+ \int_t^s e^{-\int_t^u r(v)dv} \pi'(u)\sigma(u)\, dW_0(u),
$$
$$
t \leq s \leq T. \tag{8.26}
$$

We say that (c, π) is *admissible at* (t, x) and write $(c, \pi) \in \mathcal{A}(t, x)$ if $X^{t,x,c,\pi}(s) \geq 0$ almost surely for all $s \in [t, T]$. We set

$$
\mathcal{A}_3(t, x) \triangleq \left\{ (c, \pi) \in \mathcal{A}(t, x); \quad E \int_t^T \min[0, U_1(s, c(s))]\, ds \right.
$$
$$
\left. + E\left(\min\left[0, U_2\left(X^{t,x,c,\pi}(T)\right)\right]\right) > -\infty \right\}
$$

and define

$$
V(t, x) \triangleq \sup_{(c,\pi) \in \mathcal{A}_3(t,x)} E\left[\int_t^T U_1(s, c(s))\, ds + U_2\left(X^{t,x,c,\pi}(T)\right)\right]. \tag{8.27}
$$

Because we do not consider the time-dependent variations of Problems 5.2 and 5.3, we allow ourselves the convenience of writing $V(t, x)$ rather than $V_3(t, x)$ in (8.27).

By analogy with (6.14), we introduce the function

$$
G(t, y) \triangleq E\left[\int_t^T U_1\left(s, I_1(s, yY^{(t,1)}(s))\right) ds + U_2\left(I_2(yY^{(t,1)}(T))\right)\right],
$$
$$
(t, y) \in [0, T] \times (0, \infty), \tag{8.28}
$$

so that $G(0, \cdot) = G_3(\cdot)$ of (6.14). Under Assumptions 8.1 and 8.2, we have (cf. (6.21))

$$
V(t, x) = \begin{cases} G(t, \mathcal{Y}(t, x)), & \text{if } x > \mathcal{X}(t, \infty), \\ \int_t^T U_1(s, \bar{c}(s))\, ds + U_2(\bar{x}), & \text{if } x = \mathcal{X}(t, \infty), \\ -\infty, & \text{if } x < \mathcal{X}(t, \infty). \end{cases} \tag{8.29}
$$

Of course,

$$
V(T, x) = U_2(x), \quad \forall x \in \mathbb{R}. \tag{8.30}
$$

In particular, $V(t,x) < \infty$ for all $(t,x) \in [0,T] \times \mathbb{R}$. Moreover (cf. (6.27)),

$$\lim_{x \downarrow \mathcal{X}(t,\infty)} V(t,x) = \int_t^T U_1(s,\overline{c}(s))\,ds + U_2(\overline{x}), \quad \forall t \in [0,T]. \tag{8.31}$$

Lemma 8.10: *Under Assumptions 8.1 and 8.2, the function G defined by (8.28) is of class $C([0,T] \times (0,\infty)) \cap C^{1,2}([0,T) \times (0,\infty))$, and among such functions that also satisfy the growth condition (8.21), G is the unique solution to the Cauchy problem*

$$G_t(t,y) + \frac{1}{2}\|\theta(t)\|^2 y^2 G_{yy}(t,y) - r(t)yG_y(t,y) \tag{8.32}$$

$$= -U_1(t,I_1(t,y)) \quad on\ [0,T) \times (0,\infty),$$
$$G(T,y) = U_2(I_2(y)) \quad on\ (0,\infty). \tag{8.33}$$

Furthermore,

$$G(t,y) - G(t,z) = y\mathcal{X}(t,y) - z\mathcal{X}(t,z)$$
$$- \int_z^y \mathcal{X}(t,\lambda)\,d\lambda,\ 0 < z < y < \infty, \tag{8.34}$$

$$G_y(t,y) = y\mathcal{X}_y(t,y),$$
$$G_{yy}(t,y) = \mathcal{X}_y(t,y) + y\mathcal{X}_{yy}(t,y),\ 0 \leq t < T, \quad y > 0. \tag{8.35}$$

PROOF.   The proof of (8.32) and (8.33) is like the proof of (8.6) and (8.7), except that now we use Remark 8.3 and take $u: [0,T] \times \mathbb{R} \to \mathbb{R}$ to be the $C([0,T] \times (0,\infty)) \cap C^{1,2}([0,T) \times (0,\infty))$ solution of the Cauchy problem

$$u_t(t,\eta) + \frac{1}{2}\|\theta(t)\|^2 u_{\eta\eta}(t,\eta) - \left(r(t) + \frac{1}{2}\|\theta(t)\|^2\right)u_\eta(t,\eta) \tag{8.36}$$

$$= -U_1(t,I_1(t,e^\eta)), \quad 0 \leq t < T, \quad \eta \in \mathbb{R},$$
$$u(T,\eta) = U_2(t,I_2(t,e^\eta)), \quad \eta \in \mathbb{R}. \tag{8.37}$$

Itô's rule, (8.2), and (8.36) imply that (cf. (8.14))

$$du(s,\log Y^{(t,y)}(s)) = -U_1(s,I_1(s,Y^{(t,y)}(s)))\,ds$$
$$- u_\eta(s,\log Y^{(t,y)}(s))\theta'(s)\,dW(s),$$

and so (cf. (8.15), (8.17))

$$u(t,\log y) = E\int_t^T U_1(s,I_1(s,Y^{(t,y)}(s)))\,ds + EU_2(I_2(Y^{(t,y)}(T)))$$
$$= G(t,y).$$

Consequently, G solves the Cauchy problem (8.32), (8.33).

Equation (8.34) is just (6.32) with initial time t rather than initial time zero. Equation (8.35) follows from differentiation of (8.34). Uniqueness follows as in Remark 8.6.   □

Theorem 8.11 (Hamilton–Jacobi–Bellman equation):   *Under Assumptions 8.1 and 8.2, the value function $V(t,x)$ of (8.29), (8.30) is of class $C^{1,2}$ on the set D of (8.10), continuous on the set $\{(t,x) \in [0,T] \times (0,\infty); x > \mathcal{X}(t,\infty)\}$, and satisfies the boundary conditions (8.30), (8.31) (where $V(t,\mathcal{X}(t,\infty)+)$ may be $-\infty$). Furthermore, V satisfies the Hamilton–Jacobi–Bellman equation of dynamic programming:*

$$V_t(t,x) + \max_{\substack{0 \le c < \infty \\ \pi \in \mathbb{R}^N}} \left[\frac{1}{2} \|\sigma'(t)\pi\|^2 V_{xx}(t,x) \right.$$

$$\left. + (r(t)x - c + \pi'\sigma(t)\theta(t))V_x(t,x) + U_1(t,c) \right] = 0 \quad on \ D. \qquad (8.38)$$

In particular, the value function $V_3(\cdot)$ of (5.8) is $V(0,\cdot)$, and the maximization in (8.38) is achieved by the pair $(C(t,x), \Pi(t,x))$ of (8.23), (8.24).

PROOF.   Differentiating (8.9) and (8.29) and using the formula (8.35), we obtain for $(t,x) \in D$,

$$\mathcal{X}_t(t, \mathcal{Y}(t,x)) + \mathcal{X}_y(t, \mathcal{Y}(t,x))\mathcal{Y}_t(t,x) = 0,$$
$$\mathcal{X}_y(t, \mathcal{Y}(t,x))\mathcal{Y}_x(t,x) = 1,$$
$$V_t(t,x) = G_t(t, \mathcal{Y}(t,x))$$
$$+ G_y(t, \mathcal{Y}(t,x))\mathcal{Y}_t(t,x),$$
$$V_x(t,x) = \mathcal{Y}(t,x),$$
$$V_{xx}(t,x) = \mathcal{Y}_x(t,x).$$

Using these formulas, we can rewrite the left-hand side of (8.38) as

$$G_t(t, \mathcal{Y}(t,x)) + G_y(t, \mathcal{Y}(t,x))\mathcal{Y}_t(t,x) + r(t)x\mathcal{Y}(t,x)$$
$$+ \max_{0 \le c < \infty} [U_1(t,c) - c\mathcal{Y}(t,x)]$$
$$+ \max_{\pi \in \mathbb{R}^N} \left[\frac{1}{2} \|\sigma'(t)\pi\|^2 \mathcal{Y}_x(t,x) + \pi'\sigma(t)\theta(t)\mathcal{Y}(t,x) \right]. (8.39)$$

Both expressions to be maximized are strictly concave. Setting their derivatives equal to zero, we verify that (8.23) and (8.24) provide the maximizing values of c and π, respectively. Substitution of these values converts the expression of (8.39) into

$$G_t(t, \mathcal{Y}(t,x)) + G_y(t, \mathcal{Y}(t,x))\mathcal{Y}_t(t,x) + r(t)x\mathcal{Y}(t,x)$$
$$+ U_1(t, I_1(t, \mathcal{Y}(t,x))) - \mathcal{Y}(t,x)I_1(t, \mathcal{Y}(t,x))$$
$$- \frac{1}{2} \|\theta(t)\|^2 \frac{\mathcal{Y}^2(t,x)}{\mathcal{Y}_x(t,x)}.$$

Setting $y = \mathcal{Y}(t,x)$, so that $x = \mathcal{X}(t,y)$, we can use (8.35) and (8.32) to write this in the simpler form

$$G_t(t,y) - y\mathcal{X}_t(t,y) + r(t)y\mathcal{X}(t,y) + U_1(t, I_1(t,y))$$

$$- yI_1(t,y) - \frac{1}{2}\|\theta(t)\|^2 y^2 \mathcal{X}_y(t,y)$$

$$= -\frac{1}{2}\|\theta(t)\|^2 y^2 G_{yy}(t,y) + r(t)yG_y(t,y) - y\mathcal{X}_t(t,y)$$

$$+ r(t)y\mathcal{X}(t,y) - yI_1(t,y) - \frac{1}{2}\|\theta(t)\|^2 y^2 \mathcal{X}_y(t,y)$$

$$= -y\left[\mathcal{X}_t(t,y) + \frac{1}{2}\|\theta(t)\|^2 y^2 \mathcal{X}_{yy}(t,y) + (\|\theta(t)\|^2 - r(t))y\mathcal{X}_y(t,y)\right.$$

$$\left. - r\mathcal{X}(t,y) + I_1(t,y)\right].$$

According to Lemma 8.4, this last expression is zero. $\square$

Theorem 8.11 provides only a *necessary condition* for the value function V; it is not claimed that V is the only function that is of class $C^{1,2}$ on D and satisfies (8.38) with boundary conditions (8.30), (8.31). In order to make such a uniqueness assertion, one would have also to impose some growth condition as x approaches ∞. Instead of pursuing this approach, it is easier to derive a necessary and sufficient condition for the *convex dual* of V, defined by the formula

$$\widetilde{V}(t,y) \triangleq \sup_{x \in \mathbb{R}}\{V(t,x) - xy\}, \quad y \in \mathbb{R}.$$

In contrast to the *nonlinear* partial differential equation (8.38), which governs the value function V, the dual value function $\widetilde{V}$ satisfies the *linear* partial differential equation (8.44) below. The function V can be recovered from $\widetilde{V}$ by the Legendre transform inversion formula (cf. (4.13))

$$V(t,x) = \inf_{y \in \mathbb{R}}\{\widetilde{V}(t,y) + xy\}, \quad x \in \mathbb{R}.$$

Theorem 8.12 (Convex dual of $V(t,\cdot)$): *Let Assumptions 8.1 and 8.2 hold. Then, for each $t \in [0,T]$, the function $V(t,\cdot)$ satisfies all the conditions of Definition 4.1, and*

$$\mathcal{X}(t,\infty) = \inf\{x \in \mathbb{R}; V(t,x) > -\infty\}, \tag{8.40}$$

$$V_x(t,x) = \mathcal{Y}(t,x), \quad \forall x \in (\mathcal{X}(t,\infty), \infty), \tag{8.41}$$

$$\widetilde{V}(t,y) = G(t,y) - y\mathcal{X}(t,y) \tag{8.42}$$

$$= E\left[\int_t^T \widetilde{U}_1(s, yY^{(t,1)}(s))ds + \widetilde{U}_2(yY^{(t,1)}(T))\right], \quad \forall y \in (0,\infty),$$

$$\widetilde{V}_y(t,y) = -\mathcal{X}(t,y), \quad \forall y \in (0,\infty). \tag{8.43}$$

Moreover, $\widetilde{V}$ is of class $C([0,T] \times (0,\infty)) \cap C^{1,2}([0,T) \times (0,\infty))$ and solves the Cauchy problem

$$\widetilde{V}_t(t,y) + \frac{1}{2}\|\theta(t)\|^2 y^2 \widetilde{V}_{yy}(t,y) - r(t)y\widetilde{V}_y(t,y) = -\widetilde{U}_1(t,y) \qquad (8.44)$$

$$\text{on } [0,T) \times (0,\infty),$$

$$\widetilde{V}(T,y) = \widetilde{U}_2(y), \qquad (8.45)$$

$$y \in (0,\infty).$$

If $\tilde{v}$ is a function satisfying (8.44), (8.45), $\tilde{v}_y$ is of class $C([0,T] \times (0,\infty)) \cap C^{1,2}([0,T) \times (0,\infty))$, and $\tilde{v}$ and $\tilde{v}_y$ satisfy the growth condition (8.21), then

$$\tilde{v} = \widetilde{V}, \quad -\tilde{v}_y = \mathcal{X}, \quad G = \tilde{v} - y\tilde{v}_y.$$

PROOF.   All the claims (8.40)–(8.43) made here for fixed $t \in [0,T)$ are contained in Theorem 6.11, taking T in that theorem to be $T - t$ here. When $t = T$, (8.40)–(8.43) and (8.45) follow directly from the definitions.

Equation (8.42), Lemma 8.4, and Lemma 8.10 show that $\widetilde{V}$ has the claimed degree of smoothness. Equations (8.42), (8.32), (8.6), and (4.11) yield (8.44).

If $\tilde{v}$ has the properties stated, then differentiation of (8.44) and (8.45), using (4.12), shows that $-\tilde{v}_y$ satisfies (8.6), (8.7). According to Remark 8.6, $-\tilde{v}_y = \mathcal{X}$. Furthermore, $\tilde{v} - y\tilde{v}_y$ solves (8.32), (8.33), and from Lemma 8.10 we see that $G = \tilde{v} - y\tilde{v}_y$. From (8.42) we now have $\widetilde{V} = \tilde{v}$.   □

The following examples illustrate the use of Theorem 8.12 to compute the value function and the optimal consumption and portfolio processes in feedback form.

Example 8.13:   Fix $p \in (-\infty, 1) \setminus \{0\}$ and set

$$U_1(t,c) = U^{(p)}(c - \bar{c}(t)), \quad U_2(x) = U^{(p)}(x - \bar{x}),$$

where $U^{(p)}$ is defined by (4.4) and $\bar{c}: [0,T] \to [0,\infty)$ is continuous. Then

$$I_1(t,y) = \bar{c}(t) + y^{1/(p-1)},$$

$$\widetilde{U}_1(t,y) = \frac{1-p}{p} y^{p/(p-1)} - \bar{c}(t)y, \quad 0 \le t \le T, \quad y > 0,$$

$$I_2(t,y) = \bar{x} + y^{1/(p-1)},$$

$$\widetilde{U}_2(y) = \frac{1-p}{p} y^{p/(p-1)} - \bar{x}y, \quad y > 0.$$

We seek a solution $\tilde{v}$ of (8.44), (8.45) of the form

$$\tilde{v}(t,y) = \frac{1-p}{p} k(t)y^{p/(p-1)} - \ell(t)y. \qquad (8.46)$$

This function solves (8.44), (8.45) if and only if

$$k'(t) + \alpha(t)k(t) = -1, \quad \ell'(t) - r(t)\ell(t) = -\bar{c}(t), \quad 0 \le t \le T,$$

where

$$\alpha(t) \triangleq \frac{p}{(1-p)^2} \left[\frac{1}{2} \|\theta(t)\|^2 + r(t)(1-p) \right] \tag{8.47}$$

and

$$k(T) = 1, \ \ell(T) = \overline{x}.$$

From these conditions, we see that

$$k(t) = e^{\int_t^T \alpha(s)\,ds} \left[1 + \int_t^T e^{-\int_s^T \alpha(u)\,du}\, ds \right], \tag{8.48}$$

$$\ell(t) = e^{-\int_t^T r(s)\,ds} \left[\overline{x} + \int_t^T e^{\int_s^T r(u)\,du}\, \overline{c}(s)\, ds \right]. \tag{8.49}$$

The function $\tilde{v}: [0, T] \times (0, \infty) \to \mathbb{R}$ defined by (8.46)–(8.49) satisfies (8.44), (8.45), and

$$\tilde{v}_y(t, y) = -k(t)y^{1/(p-1)} - \ell(t)$$

is of class $C([0, T] \times (0, \infty)) \cap C^{1,2}([0, T) \times (0, \infty))$. Furthermore, both $\tilde{v}$ and $\tilde{v}_y$ satisfy the growth condition (8.21). According to Theorem 8.12, $\tilde{v}$ agrees with $\widetilde{V}$,

$$\mathcal{X}(t, y) = k(t)y^{1/(p-1)} + \ell(t), \ G(t, y) = \frac{1}{p}k(t)y^{p/(p-1)}, \ 0 \le t \le T, y > 0,$$

and consequently, for $0 \le t \le T$,

$$\mathcal{X}(t, \infty) = \ell(t),$$

$$\mathcal{Y}(t, x) = \left(\frac{x - \ell(t)}{k(t)} \right)^{p-1}, \ \ \forall x > \ell(t),$$

$$V(t, x) = \frac{1}{p}k(t) \left(\frac{x - \ell(t)}{k(t)} \right)^p, \ \ \forall x > \ell(t),$$

and the optimal consumption and portfolio in feedback form (8.23), (8.24) are

$$C(t, x) = \overline{c}(t) + \frac{x - \ell(t)}{k(t)}, \ \ \forall x \ge \ell(t),$$

$$\Pi(t, x) = (\sigma'(t))^{-1}\theta(t)\frac{x - \ell(t)}{1 - p}, \ \ \forall x \ge \ell(t).$$

Example 8.14: Set

$$U_1(t, c) = U^{(0)}(c - \overline{c}(t)), \ \ U_2(x) = U^{(0)}(x - \overline{x}),$$

where $U^{(0)}$ is defined by (4.5) and $\bar{c}\colon [0,T] \to [0,\infty)$ is continuous. Then

$$I_1(t,y) = \bar{c}(t) + \frac{1}{y},$$

$$\tilde{U}_1(t,y) = -\log y - 1 - \bar{c}(t)y, \quad 0 \le t \le T, y > 0,$$

$$I_2(y) = \bar{x} + \frac{1}{y}, \quad \tilde{U}_2(y) = -\log y - 1 - \bar{x}y, \quad y > 0.$$

We seek a solution $\tilde{v}$ of (8.44), (8.45) of the form

$$\tilde{v}(t,y) = -k(t)\log y - m(t) - \ell(t)y.$$

This function solves (8.44), (8.45) if and only if

$$k(t) = T - t + 1,$$

$$m(t) = 1 + \int_t^T \left[1 - (T - s + 1)(r(s) + \|\theta(s)\|^2/2)\right] ds,$$

and $\ell(\cdot)$ is given by (8.49). Again, we see that $\tilde{v}$ and

$$\tilde{v}_y(t,y) = -\frac{k(t)}{y} - \ell(t)$$

satisfy the growth condition (8.21), so Theorem 8.12 implies that $\tilde{v}$ agrees with $\tilde{V}$ and

$$\mathcal{X}(t,y) = \frac{k(t)}{y} + \ell(t), \quad G(t,y) = k(t)(1 - \log y) - m(t).$$

Consequently, for $0 \le t \le T$,

$$\mathcal{X}(t,\infty) = \ell(t),$$

$$\mathcal{Y}(t,x) = \frac{k(t)}{x - \ell(t)} \quad \forall x > \ell(t),$$

$$V(t,x) = k(t)\log\left(\frac{x - \ell(t)}{k(t)}\right) + k(t) - m(t) \quad \forall x > \ell(t),$$

$$C(t,x) = \bar{c}(t) + \frac{x - \ell(t)}{k(t)} \quad \forall x \ge \ell(t),$$

$$\Pi(t,x) = (\sigma'(t))^{-1}\theta(t)(x - \ell(t)) \quad \forall x \ge \ell(t).$$

Remark 8.15: If we take r, θ, and σ to be constant in Examples 8.13, 8.14, we have

$$\frac{\partial}{\partial t}\Pi(t,x) = -(\sigma')^{-1}\theta\ell'(t) = (\sigma')^{-1}\theta(-r\ell(t) + \bar{c}(t)).$$

It can easily happen that some or all of the components of $\frac{\partial}{\partial t}\Pi(t,x)$ are positive, a somewhat counterintuitive situation. In particular, suppose that there is only one stock and σ and θ have the same sign (as they will whenever the mean rate of return on the stock exceeds the interest rate). Then

$\frac{\partial}{\partial t}\Pi(t,x)$ has the same sign as $-r\ell(t) + \bar{c}(t)$. If $\bar{c}(T) > r\bar{x} = r\ell(T)$, then $\frac{\partial}{\partial t}\Pi(t,x) > 0$ for t near T. In this case, as the terminal time approaches, for a fixed level of wealth the optimal portfolio invests more heavily in the stock.

Example 8.16 *(Constant coefficients):* Consider the case that $r(\cdot) = r > 0$, $\theta(\cdot) = \theta \neq 0$, and $\sigma(\cdot) = \sigma$ are constants, and $A(\cdot) \equiv 0$. Set $\gamma = \frac{1}{2}\|\theta\|^2 > 0$. Assume that

$$U_1(t,x) = e^{-\alpha t}u_1(x), \quad U_2(x) = e^{-\alpha T}u_2(x), \quad 0 \leq t \leq T, \quad x > 0, \quad (8.50)$$

where $\alpha > 0$ and $u_1\colon (0,\infty) \to \mathbb{R}$, $u_2\colon (0,\infty) \to \mathbb{R}$ are thrice continuously differentiable utility functions

$$\lim_{x\to\infty} \frac{(u_k'(x))^a}{u_k''(x)} = 0 \quad \text{for some } a > 2, \quad \lim_{x\downarrow 0} \frac{(u_k'(x))^2}{u_k''(x)} \text{ exists}, \quad (8.51)$$

$$u_k(0) > -\infty. \quad (8.52)$$

for $k = 1,2$. Let i_k denote the inverse of u_k', $k = 1,2$.

The functions of Theorem 8.12 can be computed explicitly, following Karatzas, Lehoczky, and Shreve (1987), as follows. Denote by λ_+ and λ_- the respective positive and negative roots of the quadratic equation $\gamma\lambda^2 - (r - \alpha - \gamma)\lambda - r = 0$, and set

$$J_\pm(y) \triangleq \int_0^{i_1(y)} (u_1'(\eta))^{-\lambda_\pm} d\eta,$$

$$g(y) \triangleq \frac{1}{\alpha}u_1(i_1(y)) - \frac{1}{\gamma(\lambda_+ - \lambda_-)}\left[\frac{y^{1+\lambda_+}}{1+\lambda_+}J_+(y) - \frac{y^{1+\lambda_-}}{1+\lambda_-}J_-(y)\right],$$

$$s(y) \triangleq \frac{y}{r}i_1(y) - \frac{1}{\gamma(\lambda_+ - \lambda_-)}\left[\frac{y^{1+\lambda_+}}{\lambda_+}J_+(y) - \frac{y^{1+\lambda_-}}{\lambda_-}J_-(y)\right],$$

$$h(y) \triangleq g(y) - s(y)$$
$$= \frac{1}{\alpha}u_1(i_1(y)) - \frac{y}{r}i_1(y) + \frac{1}{\gamma(\lambda_+ - \lambda_-)}\left[\frac{y^{1+\lambda_+}}{\lambda_+(1+\lambda_+)}J_+(y)\right.$$
$$\left. - \frac{y^{1+\lambda_-}}{\lambda_-(1+\lambda_-)}J_-(y)\right],$$

as well as

$$\rho_\pm(t,x;q) \triangleq \frac{1}{\sqrt{2\gamma t}}\left[\log\frac{x}{q} + (r - \beta \pm \gamma)t\right], \quad t,x,q > 0,$$

$$v(t,x;q) = \begin{cases} xe^{-\alpha t}\Phi(\rho_+(t,x;q)) - qe^{-rt}\Phi(\rho_-(t,x;q)), & 0 < t \leq T, \\ (x-q)^+, & t = 0, \end{cases}$$

by analogy with Example 2.4.1. Then

$$y\mathcal{X}(t,y) = s(y) + \int_0^\infty (\eta i_2(\eta) - s(\eta))''v(T - t, \eta; y)d\eta, \qquad (8.53)$$

$$G(t,y) = g(y) + \left(u_2(0) - \frac{u_1(0)}{\alpha}\right)^{-\alpha(T-t)}$$

$$+ \int_0^\infty (u_2(i_2(\eta)) - g(\eta))''v(T - t, \eta; y)d\eta, \qquad (8.54)$$

and thus

$$\widetilde{V}(t,y) = G(t,y) - y\mathcal{X}(t,y)$$

$$= h(y) + \left(u_2(0) - \frac{u_1(0)}{\alpha}\right)^{-\alpha(T-t)}$$

$$+ \int_0^\infty (\tilde{u}_2(\eta) - h(\eta))''v(T - t, \eta; y)d\eta, \qquad (8.55)$$

where $\tilde{u}_2$ is the convex dual (Definition 4.2) of u_2. See Karatzas, Lehoczky, and Shreve (1987) for details.

Conditions (8.51), (8.52) and Assumption 8.2 are satisfied by

$$u_k(x) = \frac{1}{p}x^p, \quad x > 0,$$

for any $p \in (0, 1)$ (p may depend on k). Although condition (8.52) is not satisfied by $u_k(x) = \log x$, the formulas obtained above still hold and simplify considerably for this case (see Remarks 4.7, 5.5 in Karatzas, Lehoczky, and Shreve (1987)).

3.9   Consumption and Investment on an Infinite Horizon

In this section and the next we consider a complete, standard financial market on an infinite horizon, as set forth in Section 1.7. In particular, on an underlying probability space $(\Omega, \mathcal{F}, P)$, there is an N-dimensional Brownian motion $W = \{W(t); 0 \le t < \infty\}$, and we shall use the notion of *restricted progressive measurability* (Definition 1.7.1) relative to this Brownian motion.

For this market we shall be interested in Problem 9.5 below of maximizing expected utility from consumption over the infinite planning horizon. The solution of this problem is similar to that obtained from Problem 5.2 of maximizing expected utility from consumption over a finite planning horizon. After developing the expected results for the market with general coefficient processes, we turn our attention to the market with constant coefficients and utility function of the form (9.21). In this case, explicit computations become possible (e.g., Theorems 9.14, 9.18, 9.21). The Hamilton–Jacobi–Bellman equation takes the form of a nonlinear, second-order ordinary differential equation (Theorem 9.20), and the dual value function satisfies a *linear*, second-order ordinary differential equation (Theorem 9.21).

Definition 9.1: A *consumption process on the infinite planning horizon* is a nonnegative, restrictedly progressively measurable process satisfying $\int_0^T c(t)\, dt < \infty$ almost surely for every $T \in [0, \infty)$.

An agent with initial $x \geq 0$ who chooses a consumption process $c(\cdot)$ will have a cumulative income process $\Gamma(t) \triangleq x - \int_0^t c(u)\, du, 0 \leq t < \infty$. If this investor chooses a $\Gamma(\cdot)$-financed portfolio process $\pi(\cdot)$, then his corresponding wealth process $X^{x,c,\pi}(\cdot)$ will be governed by equation (1.7.6):

$$\frac{X^{x,c,\pi}(t)}{S_0(t)} = x - \int_0^t \frac{c(u)\, du}{S_0(u)} + \int_0^t \frac{1}{S_0(u)} \pi'(u)\sigma(u)\, dW_0(u),$$

$$0 \leq t < \infty. \quad (9.1)$$

Definition 9.2: Given $x \geq 0$, we say that a consumption and portfolio process pair (c, π) on the infinite planning horizon is *admissible at x*, and write $(c, \pi) \in \mathcal{A}(x)$, if the wealth process $X^{x,c,\pi}(\cdot)$ corresponding to x, c, π satisfies

$$X^{x,c,\pi}(t) \geq 0, \quad 0 \leq t < \infty$$

almost surely. For $x < 0$, we set $\mathcal{A}(x) = \emptyset$.

Remark 9.3: Just as in Remark 3.3, we have for any $(c, \pi) \in \mathcal{A}(x)$ that $E \int_0^T H_0(u)c(u)\, du \leq x$ for every $T \in [0, \infty)$. Letting $T \to \infty$ and using the monotone convergence theorem, we obtain the *infinite horizon budget constraint*

$$E \int_0^\infty H_0(u)c(u)\, du \leq x. \quad (9.2)$$

Theorem 9.4: *Let $x \geq 0$ be given and let $c(\cdot)$ be a consumption process such that*

$$E \int_0^\infty H_0(u)c(u)\, du = x. \quad (9.3)$$

Then there exists a portfolio process $\pi(\cdot)$ such that (c, π) is admissible at x. The corresponding wealth process is

$$X^{x,c,\pi}(t) = \frac{1}{H_0(t)} E\left[\int_t^\infty H_0(u)c(u)\, du \,\Big|\, \mathcal{F}(t) \right], \quad 0 \leq t \leq \infty. \quad (9.4)$$

PROOF.   Given $T \in [0, \infty)$, define

$$\xi_T = \frac{1}{H_0(T)} E\left[\int_T^\infty H_0(u)c(u)\,du \,\Big|\, \mathcal{F}(T)\right].$$

Then

$$E\left[\int_0^T H_0(u)c(u)\,du + H_0(T)\xi_T\right] = x,$$

and Theorem 3.5 implies the existence of a portfolio process $\pi_T = \{\pi_T(t); 0 \le t \le T\}$ such that the corresponding wealth process

$$X^{x,c,\pi_T}(t) = S_0(t)\left[x - \int_0^t \frac{c(u)\,du}{S_0(u)} + \int_0^t \frac{1}{S_0(u)}\pi_T'(u)\sigma(u)\,dW_0(u)\right],$$
$$0 \le t \le T$$

satisfies $X^{x,c,\pi_T}(t) \ge 0$ almost surely for $0 \le t \le T$ and $X^{x,c,\pi_T}(T) = \xi_T$. According to Remark 3.3,

$$M_T(t) \triangleq H_0(t)X^{x,c,\pi_T}(t) + \int_0^t H_0(u)c(u)\,du, \quad 0 \le t \le T$$

is a nonnegative supermartingale under P, but since

$$EM_T(T) = E\left\{E\left[\int_T^\infty H_0(u)c(u)\,du \,\Big|\, \mathcal{F}(T)\right] + \int_0^T H_0(u)c(u)\,du\right\}$$
$$= x$$
$$= M_T(0),$$

the process $\{M_T(t); 0 \le t \le T\}$ is in fact a martingale. Consequently, for $0 \le t \le T' < T < \infty$, we have the almost sure equalities

$$H_0(t)X^{x,c,\pi_T}(t) = M_T(t) - \int_0^t H_0(u)c(u)\,du$$
$$= E\left[M_T(T) - \int_0^t H_0(u)c(u)\,du \,\Big|\, \mathcal{F}(t)\right]$$
$$= E\left[H_0(T)\xi_T + \int_t^T H_0(u)c(u)\,du \,\Big|\, \mathcal{F}(t)\right]$$
$$= E\left[\int_t^\infty H_0(u)c(u)\,du \,\Big|\, \mathcal{F}(t)\right]$$
$$= E\left[H_0(T')\xi_{T'} + \int_t^{T'} H_0(u)c(u)\,du \,\Big|\, \mathcal{F}(t)\right]$$
$$= E\left[M_{T'}(T') - \int_0^t H_0(u)c(u)\,du \,\Big|\, \mathcal{F}(t)\right]$$

$$= M_{T'}(t) - \int_0^t H_0(u)c(u) \, du$$

$$= H_0(t)X^{x,c,\pi_{T'}}(t). \tag{9.5}$$

Since both $X^{x,c,\pi_T}(\cdot)$ and $X^{x,c,\pi_{T'}}(\cdot)$ are continuous, we have $X^{x,c,\pi_T}(t) = X^{x,c,\pi_{T'}}(t)$ for all $t \in [0, T']$ almost surely. This implies

$$\int_0^t \frac{1}{S_0(u)}(\pi_T(u) - \pi_{T'}(u))'\sigma(u) \, dW_0(u) = 0, \quad \forall t \in [0, T'],$$

and thus $\pi_T(t) = \pi_{T'}(t)$ for Lebesgue-almost-every $t \in [0, T']$ almost surely. We now define

$$\pi(t) \overset{\Delta}{=} \sum_{n=1}^{\infty} 1_{[n-1,n)}(t)\pi_n(t), \quad 0 \le t < \infty,$$

and we have

$$X^{x,c,\pi}(t) = \sum_{n=1}^{\infty} 1_{[n-1,n)}(t)X^{x,c,\pi_n}(t) \ge 0, \quad \forall t \in [0, \infty)$$

almost surely. For fixed $t \ge 0$, choose the integer n such that $n - 1 \le t < n$. Then (9.5) implies

$$X^{x,c,\pi}(t) = X^{x,c,\pi_n}(t) = \frac{1}{H_0(t)}E\left[\int_t^{\infty} H_0(u)c(u)du \,\middle|\, \mathcal{F}(t)\right]$$

almost surely, which establishes (9.4). □

Let a function $U_1 \colon [0, \infty) \times \mathbb{R} \to [-\infty, \infty)$ be given such that:

(i) For each $t \in [0, \infty)$, $U_1(t, \cdot)$ is a utility function in the sense of Definition 4.1, and the *subsistence consumption*

$$\bar{c}(t) \overset{\Delta}{=} \inf\{c \in \mathbb{R}; U_1(t, c) > -\infty\}, \quad 0 \le t < \infty,$$

is a continuous function of t, with values in $[0, \infty)$;

(ii) U_1 and U_1' (where prime denotes differentiation with respect to the second argument) are continuous on the set

$$D_{\infty} \overset{\Delta}{=} \{(t, c) \in [0, \infty) \times (0, \infty); c > \bar{c}(t)\}.$$

For each $t \in [0, \infty)$, we construct $I_1(t, \cdot) \colon (0, \infty] \overset{\text{onto}}{\longrightarrow} [\bar{c}(t), \infty)$ satisfying (5.20). Lemma 5.8 extends to show that I_1 is jointly continuous on $[0, \infty) \times (0, \infty]$.

Let an agent have an initial endowment $x \in \mathbb{R}$. The problem of this section is the following.

Problem 9.5: Find an optimal pair $(c_{\infty}, \pi_{\infty}) \in \mathcal{A}_{\infty}(x)$ for the problem

$$V_{\infty}(x) \overset{\Delta}{=} \sup_{(c,\pi) \in \mathcal{A}_{\infty}(x)} E \int_0^{\infty} U_1(t, c(t)) \, dt$$

of maximizing expected total utility from consumption over $[0, \infty)$, where

$$\mathcal{A}_\infty(x) \triangleq \left\{ (c, \pi) \in \mathcal{A}(x); \quad E \int_0^\infty \min[0, U_1(t, c(t))] \, dt > -\infty \right\}.$$

We recall that $\mathcal{A}(x) = \emptyset$ for $x < 0$ and that the supremum over the empty set is $-\infty$. To avoid trivialities, we impose the following condition throughout.

Assumption 9.6: *There is at least one $\hat{x} \geq 0$ such that $V_\infty(\hat{x})$ is finite.*

Remark 9.7: With $\hat{x}$ as in Assumption 9.6, there must exist some $(\hat{c}, \hat{\pi}) \in \mathcal{A}_\infty(\hat{x})$ such that $E \int_0^\infty \min[0, U_1(t, \hat{c}(t))] \, dt > -\infty$, and, in addition, $E \int_0^\infty U_1(t, \hat{c}(t)) \, dt \leq V(\hat{x}) < \infty$. Simply stated, $\hat{c}(\cdot)$ satisfies

$$E \int_0^\infty |U_1(t, \hat{c}(t))| \, dt < \infty. \tag{9.6}$$

Remark 9.8: It is not difficult to compute $V_\infty(x)$ in the case that $x \leq E \int_0^\infty H_0(t) \bar{c}(t) \, dt$. Indeed, we have

$$V_\infty(x) = \begin{cases} \displaystyle\int_0^\infty U_1(t, \bar{c}(t)) \, dt, & x = E \int_0^\infty H_0(t) \bar{c}(t) \, dt, \\[3mm] -\infty, & x < E \int_0^\infty H_0(t) \bar{c}(t) \, dt. \end{cases} \tag{9.7}$$

To verify this, note that when $x < E \int_0^\infty H_0(t) \bar{c}(t) \, dt$, the budget constraint (9.2) shows that $\mathcal{A}_\infty(x) = \emptyset$, and so $V_\infty(x) = -\infty$. If on the other hand $x = E \int_0^\infty H_0(t) \bar{c}(t) \, dt$, then (9.2) shows that the only possible admissible consumption process is $\bar{c}(\cdot)$ itself. Theorem 9.4 guarantees the existence of a portfolio $\bar{\pi}(\cdot)$ such that $(\bar{c}, \bar{\pi}) \in \mathcal{A}(x)$. If $\int_0^\infty \min[0, U_1(t, \bar{c}(t))] \, dt > -\infty$, then $(\bar{c}, \bar{\pi})$ is the sole member of $\mathcal{A}_\infty(x)$, and hence is optimal. If $\int_0^\infty \min[0, U_1(t, \bar{c}(t))] \, dt = -\infty$, then $\int_0^\infty U_1(t, \bar{c}(t)) \, dt$ is defined and equal to $-\infty$ because $\bar{c}(\cdot)$ is dominated by $\hat{c}(\cdot)$ satisfying (9.6). Moreover, in this case $\mathcal{A}_\infty(x) = \emptyset$, so $V_\infty(x) = -\infty$. In either case, we have (9.7).

We now define

$$\mathcal{X}_\infty(y) \triangleq E \int_0^\infty H_0(t) I_1(t, y H_0(t)) \, dt, \quad 0 < y < \infty. \tag{9.8}$$

A sufficient condition for the following assumption is given in Propostion 9.14.

Assumption 9.9: $\mathcal{X}_\infty(y) < \infty, \quad \forall y \in (0, \infty)$.

The following lemma is proved in the same way as Lemma 6.2.

Lemma 9.10: *Under Assumption 9.9, the function $\mathcal{X}_\infty$ is nonincreasing and continuous on $(0, \infty)$ and strictly decreasing on $(0, r_\infty)$, where $\mathcal{X}_\infty(0+) \triangleq \lim_{y \downarrow 0} \mathcal{X}_\infty(y) = \infty$, $\mathcal{X}_\infty(\infty) \triangleq \lim_{y \to \infty} \mathcal{X}_\infty(y) =$*

$\int_0^\infty H_0(t)\bar{c}(t)\,dt$, and

$$r_\infty \overset{\Delta}{=} \sup\{y > 0; \mathcal{X}_\infty(y) > \mathcal{X}_\infty(\infty)\} > 0. \tag{9.9}$$

In particular, the function $\mathcal{X}_\infty$ restricted to $(0, r_\infty)$ has a strictly decreasing inverse function $\mathcal{Y}_\infty \colon (\mathcal{X}_\infty(\infty), \infty) \overset{\text{onto}}{\longrightarrow} (0, r_\infty)$, so that

$$\mathcal{X}_\infty(\mathcal{Y}_\infty(x)) = x, \quad \forall x \in (\mathcal{X}_\infty(\infty), \infty). \tag{9.10}$$

For $x \in (\mathcal{X}_\infty(\infty), \infty)$, we define the candidate optimal consumption process

$$c_\infty(t) \overset{\Delta}{=} I_1(t, \mathcal{Y}_\infty(x)H_0(t)), \quad 0 \le t < \infty. \tag{9.11}$$

From (9.8), (9.10), we have

$$E \int_0^\infty H_0(u)c_\infty(u)\,du = x,$$

and Theorem 9.4 guarantees the existence of a candidate optimal portfolio policy $\pi_\infty(\cdot)$ such that $(c_\infty, \pi_\infty) \in \mathcal{A}(x)$.

Theorem 9.11: *Suppose that both Assumptions 9.6 and 9.9 hold, let $x \in (\mathcal{X}_\infty(\infty), \infty)$ be given, let $c_\infty(\cdot)$ be given by (9.11), and let π_∞ be such that $(c_\infty, \pi_\infty) \in \mathcal{A}(x)$. Then $(c_\infty, \pi_\infty) \in \mathcal{A}_\infty(x)$ and (c_∞, π_∞) is optimal for Problem 9.5; i.e.*

$$V_\infty(x) = E \int_0^\infty U_1(t, c_\infty(t))\,dt. \tag{9.12}$$

The optimal wealth process $X_\infty(\cdot) = X^{x, c_\infty, \pi_\infty}(\cdot)$ is

$$X_\infty(t) = \frac{1}{H_0(t)} E\left[\int_t^\infty H_0(u)c_\infty(u)\,du \,\bigg|\, \mathcal{F}(t)\right], \quad 0 \le t < \infty, \tag{9.13}$$

and the value function V_∞ is given as

$$V_\infty(x) = G_\infty(\mathcal{Y}_\infty(x)), \quad \mathcal{X}_\infty(\infty) < x < \infty, \tag{9.14}$$

where

$$G_\infty(y) \overset{\Delta}{=} E \int_0^\infty U_1(t, I_1(t, yH_0(t)))\,dt, \quad 0 < y < \infty. \tag{9.15}$$

PROOF. The proof of the optimality of (c_∞, π_∞) is the same as the proof of Theorem 6.3, except that now we use the process $\hat{c}(\cdot)$ of Remark 9.7 in place of the constant $\hat{c}$ of the proof of Theorem 6.3. Equation (9.13) comes from (9.4). Equation (9.14) is just a restatement of (9.12). □

As in Section 3.6, we examine the convex dual of V_∞, defined by

$$\tilde{V}_\infty(y) \overset{\Delta}{=} \sup_{x \in \mathbb{R}}\{V_\infty(x) - xy\}, \quad y \in \mathbb{R}.$$

Recall from Remark 9.8 and Theorem 9.11 that

$$V_\infty(x) = \begin{cases} G_\infty(\mathcal{Y}_\infty(x)), & x > \mathcal{X}_\infty(\infty), \\ \int_0^\infty U_1(t, c(t)) \, dt, & x = \mathcal{X}_\infty(\infty), \\ -\infty, & x < \mathcal{X}_\infty(\infty). \end{cases} \tag{9.16}$$

The proof of Theorem 6.11 is easily adapted to prove the following result.

Theorem 9.12: *Let Assumptions 9.6 and 9.9 hold, and assume $V_\infty(x) < \infty$ for all $x \in \mathbb{R}$. Then V_∞ satisfies all the conditions of Definition 4.1, and*

$$\mathcal{X}_\infty(\infty) = \inf\{x \in \mathbb{R}; V_\infty(x) > -\infty\}, \tag{9.17}$$

$$V_\infty'(x) = \mathcal{Y}_\infty(x), \quad \forall x \in (\mathcal{X}_\infty(\infty), \infty), \tag{9.18}$$

$$\widetilde{V}_\infty(y) = G_\infty(y) - y\mathcal{X}_\infty(y) \tag{9.19}$$

$$= E \int_0^\infty \widetilde{U}_1(t, yH_0(t)) \, dt, \quad \forall y \in (0, \infty),$$

$$\widetilde{V}_\infty'(y) = -\mathcal{X}_\infty(y), \quad \forall y \in (0, \infty). \tag{9.20}$$

For the remainder of this section, we impose the following condition.

Assumption 9.13: *The processes $r(\cdot) \equiv r$ and $\theta(\cdot) \equiv \theta$ are constants,*

$$r > 0, \quad \gamma \triangleq \frac{1}{2}\|\theta\|^2 > 0,$$

the process $A(\cdot)$ is identically zero, and the function U_1 is of the form

$$U_1(t, c) = e^{-\beta t} U(c), \quad t \in [0, \infty), \ c \in \mathbb{R}, \tag{9.21}$$

where U is a utility function (Definition 4.1) and β is a positive discount factor.

Under Assumption 9.13, we have

$$I_1(t, y) = I\left(e^{\beta t} y\right), \quad t \in [0, \infty), \ 0 < y < \infty, \tag{9.22}$$

where I is related to U by (4.6), (4.7). We set (cf. (4.2))

$$\bar{c} \triangleq \inf\{c \in \mathbb{R}; U(c) = -\infty\}.$$

Because $r > 0$, $\beta > 0$, and $\gamma > 0$, the quadratic equation

$$\gamma \rho^2 - (r - \beta + \gamma)\rho - \beta = 0 \tag{9.23}$$

has two roots, one negative and the other greater than 1. We denote the negative root of (9.23) by ρ_1 and the positive root by ρ_2. More specifically,

$$\rho_i \triangleq \frac{1}{2\gamma}\left[(r - \beta + \gamma) + (-1)^i \sqrt{(r - \beta + \gamma)^2 + 4\gamma\beta}\right], \quad i = 1, 2. \tag{9.24}$$

Theorem 9.14: *Let Assumption 9.13 hold. Then the condition*

$$\int_0^1 \eta^{-\rho_1} I(\eta) \, d\eta < \infty \tag{9.25}$$

is equivalent to Assumption 9.9. Under this condition,

$$\mathcal{X}_\infty(y) = \frac{1}{\gamma(\rho_2 - \rho_1)} \left[y^{\rho_1 - 1} \int_0^y \eta^{-\rho_1} I(\eta)\, d\eta + y^{\rho_2 - 1} \int_y^\infty \eta^{-\rho_2} I(\eta)\, d\eta \right],$$

$$0 < y < \infty,$$
$$(9.26)$$

is finite, twice continuously differentiable, satisfies $\mathcal{X}_\infty'(y) < 0$ for all $y \in (0, \infty)$, and

$$\frac{1}{2} \|\theta\|^2 y^2 \mathcal{X}_\infty''(y) + \left(\|\theta\|^2 - r + \beta \right) y \mathcal{X}_\infty'(y) - r \mathcal{X}_\infty(y) = -I(y),$$

$$0 < y < \infty. \ (9.27)$$

PROOF.   Under Assumption 9.13, equation (9.8) becomes

$$\mathcal{X}_\infty(y) = E \int_0^\infty \exp\{-(r + \gamma)t - \theta' W(t)\}$$
$$\cdot I\left(y \exp\{-(r - \beta + \gamma)t - \theta' W(t)\} \right) dt$$
$$= \int_{-\infty}^\infty \int_0^\infty \frac{1}{\sqrt{2\pi}} \exp\left\{ -(r + \gamma)t - w\sqrt{2\gamma t} - w^2/2 \right\}$$
$$\cdot I\left(y \exp\left\{ -(r - \beta + \gamma)t - w\sqrt{2\gamma t} \right\} \right) dt\, dw.$$

Holding $t > 0$ fixed, we can make the change of variable $z = (r - \beta + \gamma)t + w\sqrt{2\gamma t}$ in the outer integral and then use the Laplace transform formula

$$\int_0^\infty \frac{1}{\sqrt{\pi t}} e^{-a/(4t)} e^{-pt}\, dt = \frac{e^{-\sqrt{ap}}}{\sqrt{p}}, \qquad a > 0,\ p > 0, \qquad (9.28)$$

to obtain

$$\mathcal{X}_\infty(y) = \int_{-\infty}^\infty \frac{1}{2\sqrt{\gamma}} \exp\left\{ \frac{z(r - \beta - \gamma)}{2\gamma} \right\} I\left(y e^{-z} \right)$$
$$\cdot \int_0^\infty \frac{1}{\sqrt{\pi t}} \exp\left\{ -\frac{z^2}{4\gamma t} - \frac{(r - \beta + \gamma)^2 + 4\gamma\beta}{4\gamma} t \right\} dt\, dz$$
$$= \int_{-\infty}^\infty \frac{1}{\sqrt{(r - \beta + \gamma)^2 + 4\gamma\beta}} I\left(y e^{-z} \right)$$
$$\cdot \exp\left\{ \frac{1}{2\gamma} \left[z(r - \beta - \gamma) - |z|\sqrt{(r - \beta + \gamma)^2 + 4\gamma\beta} \right] \right\} dz$$
$$= \frac{1}{\gamma(\rho_2 - \rho_1)} \left[\int_0^\infty e^{(\rho_1 - 1)z} I\left(y e^{-z} \right) dz \right.$$
$$\left. + \int_{-\infty}^0 e^{(\rho_2 - 1)z} I\left(y e^{-z} \right) dz \right].$$

The change of variable $\eta = y e^{-z}$ leads to equation (9.26). The integral $\int_y^\infty \eta^{-\rho_2} I(\eta)\, d\eta$ converges because I is nonnegative and nonincreasing and $\rho_2 > 1$. The equivalence of (9.25) and Assumption 9.9 is now apparent.

From (9.26) it is clear that $\mathcal{X}_\infty$ is twice continuously differentiable. Direct computation verifies (9.27). To show that $\mathcal{X}'_\infty(y) < 0$, we use integration by parts for Riemann–Stieltjes integrals to compute

$$\mathcal{X}'_\infty(y) = \frac{1}{\gamma(\rho_2 - \rho_1)} \left[(\rho_1 - 1) y^{\rho_1 - 2} \int_0^y \eta^{-\rho_1} I(\eta) \, d\eta \right.$$

$$\left. + (\rho_2 - 1) y^{\rho_2 - 2} \int_y^\infty \eta^{-\rho_2} I(\eta) \, d\eta \right]$$

$$= \frac{1}{\gamma(\rho_2 - \rho_1)} \left[y^{\rho_1 - 2} \liminf_{\eta \downarrow 0} \left(\eta^{-\rho_1 + 1} I(\eta) \right) + y^{\rho_1 - 2} \int_0^y \eta^{-\rho_1 + 1} \, dI(\eta) \right.$$

$$\left. + y^{\rho_2 - 2} \int_y^\infty \eta^{-\rho_2 + 1} \, dI(\eta) \right].$$

But (9.25) implies $\liminf_{\eta \downarrow 0} \left(\eta^{-\rho_1 + 1} I(\eta) \right) = 0$, and both the above Riemann–Stieltjes integrals are strictly negative because I is strictly decreasing. $\qquad\square$

Under Assumptions 9.9 and 9.13, $\mathcal{X}_\infty$ is a bijection from $(0, \infty)$ to $(0, \mathcal{X}_\infty(\infty))$, and so its inverse $\mathcal{Y}_\infty$ maps $(0, \mathcal{X}_\infty(\infty))$ onto $(0, \infty)$. The implicit function theorem implies that $\mathcal{Y}_\infty$ is continuously differentiable, and in fact,

$$\mathcal{X}'_\infty(\mathcal{Y}_\infty(x)) \mathcal{Y}'_\infty(x) = 1, \quad \forall x \in (\mathcal{X}_\infty(\infty), \infty). \tag{9.29}$$

Corollary 9.15: *Under Assumptions 9.9 and 9.13, the feedback form for the optimal consumption/portfolio process pair (c_∞, π_∞) for Problem 9.5 is given by*

$$C(x) = I(\mathcal{Y}_\infty(x)), \tag{9.30}$$

$$\Pi(x) = -(\sigma')^{-1} \theta \frac{\mathcal{Y}_\infty(x)}{\mathcal{Y}'_\infty(x)} \tag{9.31}$$

for $x \in (\mathcal{X}_\infty(\infty), \infty)$.

PROOF. We introduce the process

$$Y^{(y)}(s) \triangleq y \exp\{-(r + \gamma)s - \theta' W(s)\}$$

$$= y \exp\{-(r - \gamma)s - \theta' W_0(s)\},$$

which satisfies

$$dY^{(y)}(s) = Y^{(y)}(s)[(2\gamma - r) \, ds - \theta' \, dW_0(s)],$$

$$Y^{(y)}(s) = y Y^{(1)}(s) = y H_0(s).$$

In terms of this process, we have the representation

$$y \mathcal{X}_\infty(y) = E \left[\int_0^\infty Y^{(y)}(s) I(e^{\beta s} Y^{(y)}(s)) \, ds \right],$$

and the Markov property implies that

$$E\left[\int_t^\infty Y^{(y)}(s)I(e^{\beta(s-t)}Y^{(y)}(s))\,ds\,\Big|\,\mathcal{F}(t)\right] = Y^{(y)}(t)\mathcal{X}_\infty(Y^{(y)}(t)).$$

Let $x > \mathcal{X}_\infty(\infty)$ be given and set $y = \mathcal{Y}_\infty(x)$. The optimal wealth process with initial condition x, given by (9.13), is (see (9.11), (9.22))

$$
\begin{aligned}
X_\infty(t) &= \frac{1}{H_0(t)}E\left[\int_t^\infty H_0(s)I\left(ye^{\beta s}H_0(s)\right)ds\,\Big|\,\mathcal{F}(t)\right]\\
&= \frac{1}{Y^{(ye^{\beta t})}(t)}E\left[\int_t^\infty Y^{(ye^{\beta t})}(s)I\left(e^{\beta(s-t)}Y^{(ye^{\beta t})}(s)\right)\,\Big|\,\mathcal{F}(t)\right]\\
&= \mathcal{X}_\infty(Y^{(ye^{\beta t})}(t))\\
&= \mathcal{X}_\infty(e^{\beta t}Y^{(y)}(t))\\
&= \mathcal{X}_\infty\left(ye^{\beta t}H_0(t)\right).
\end{aligned}
\tag{9.32}
$$

In particular, the optimal consumption process is

$$c_\infty(t) = I\left(ye^{\beta t}H_0(t)\right) = I\left(\mathcal{Y}_\infty(X_\infty(t))\right),\tag{9.33}$$

which verifies (9.30).

From Itô's rule in conjunction with (9.27) and (9.33) applied to (9.32), we have

$$
\begin{aligned}
d\left(e^{-rt}X_\infty(t)\right) &= d(e^{-rt}\mathcal{X}_\infty(e^{\beta t}Y^{(y)}(t)))\\
&= -e^{-rt}c_\infty(t)\,dt - e^{(\beta-r)t}Y^{(y)}(t)\mathcal{X}_\infty'(e^{\beta t}Y^{(y)}(t))\theta'\,dW_0(t),
\end{aligned}
$$

and comparison with (9.1) shows that the optimal portfolio process is

$$\pi_\infty(t) = -(\sigma)^{-1}\theta e^{\beta t}Y^{(y)}(t)\mathcal{X}_\infty'(e^{\beta t}Y^{(y)}(t)).$$

But $e^{\beta t}Y^{(y)}(t) = \mathcal{Y}_\infty(X_\infty(t))$, and because of (9.29), $\mathcal{X}_\infty'(e^{\beta t}Y^{(y)}(t)) = 1/\mathcal{Y}_\infty'(X_\infty(t))$. This proves (9.31). $\quad\square$

Remark 9.16: Just as in the finite-horizon model with deterministic coefficients, discussed in Section 3.8, Merton's mutual fund theorem (Remark 8.9) holds for the infinite-horizon, constant-coefficient model of Corollary 9.15.

We next compute the function G_∞ of (9.15). The following lemma enables us to establish the finiteness of this function.

Lemma 9.17: *Under Assumption 9.13, we have*

$$\int_y^\infty \eta^{-\rho_2-1}|U(I(\eta))|\,d\eta < \infty \quad \forall y \in (0,\infty).\tag{9.34}$$

If, in addition, Assumption 9.9 (or equivalently, relation (9.25), holds), then

$$\int_0^y \eta^{-\rho_1-1}|U(I(\eta))|\,d\eta < \infty, \quad \forall y \in (0,\infty). \tag{9.35}$$

PROOF.   Fix $y \in (0,\infty)$. From (6.31) we have for any $\eta \in (0,\infty)$ that

$$U(I(\eta)) = U(I(y)) - yI(y) + \eta I(\eta) + \int_\eta^y I(\xi)\,d\xi. \tag{9.36}$$

Therefore,

$$\int_y^\infty \eta^{-\rho_2-1}|U(I(\eta))|\,d\eta \le \frac{1}{\rho_2} y^{-\rho_2}|U(I(y)) - yI(y)|$$

$$+ \int_y^\infty \eta^{-\rho_2} I(\eta)\,d\eta + \int_y^\infty \int_y^\eta \eta^{-\rho_2-1} I(\xi)\,d\xi\,d\eta.$$

Fubini's theorem implies

$$\int_y^\infty \int_y^\eta \eta^{-\rho_2-1} I(\xi)\,d\xi\,d\eta = \frac{1}{\rho_2} \int_y^\infty \xi^{-\rho_2} I(\xi)\,d\xi$$

$$\le \frac{1}{\rho_2} I(y) \int_y^\infty \xi^{-\rho_2}\,d\xi$$

$$< \infty,$$

because $\rho_2 > 1$. Relation (9.34) follows.

From (9.36), we also have

$$\int_0^y \eta^{-\rho_1-1}|U(I(\eta))|\,d\eta \le -\frac{1}{\rho_1} y^{-\rho_1}|U(I(y)) - yI(y)|$$

$$+ \int_0^y \eta^{-\rho_1} I(\eta)\,d\eta + \int_0^y \int_\eta^y \eta^{-\rho_1-1} I(\xi)\,d\xi\,d\eta,$$

and Fubini's theorem implies

$$\int_0^y \int_\eta^y \eta^{-\rho_1-1} I(\xi)\,d\xi\,d\eta = -\frac{1}{\rho_1} \int_0^y \xi^{-\rho_1} I(\xi)\,d\xi.$$

From Assumption 9.9 in the form (9.25), we have (9.35).   □

Theorem 9.18: *Let Assumptions 9.9 and 9.13 hold. Then the function G_∞ of (9.15) is given by*

$$G_\infty(y) = \frac{1}{\gamma(\rho_2 - \rho_1)} \left[y^{\rho_1} \int_0^y \eta^{-\rho_1-1} U(I(\eta))\,d\eta \right.$$

$$\left. + y^{\rho_2} \int_y^\infty \eta^{-\rho_2-1} U(I(\eta))\,d\eta \right], \quad 0 < y < \infty, \tag{9.37}$$

is finite, twice continuously differentiable, and satisfies

$$\frac{1}{2}\|\theta\|^2 y^2 G_\infty''(y) + (\beta - r)yG_\infty'(y) - \beta G_\infty(y) = -U(I(y)), \quad 0 < y < \infty.$$
$$(9.38)$$

Furthermore,

$$G_\infty(y) - G_\infty(z) = y\mathcal{X}_\infty(y) - z\mathcal{X}_\infty(z)$$
$$- \int_z^y \mathcal{X}_\infty(\lambda)\, d\lambda, \quad 0 < z < y < \infty, \qquad (9.39)$$
$$G_\infty'(y) = y\mathcal{X}_\infty'(y), \quad G_\infty''(y) = \mathcal{X}_\infty'(y) + y\mathcal{X}_\infty''(y), \quad y > 0. \qquad (9.40)$$

PROOF.   By computations similar to those in the proof of Theorem 9.14, using again the Laplace transform formula (9.28), we have for $0 < y < \infty$,

$$G_\infty(y) = E \int_0^\infty e^{-\beta t} U\left(I\left(ye^{\beta t} H_0(t)\right)\right) dt$$

$$= \int_{-\infty}^\infty \int_0^\infty \frac{1}{\sqrt{2\pi}} \exp\left\{-\beta t - \frac{w^2}{2}\right\}$$
$$\cdot U\left(I(y\exp\{-(r - \beta + \gamma)t - w\sqrt{2\gamma t}\}\right) dt\, dw$$

$$= \int_{-\infty}^\infty \frac{1}{2\sqrt{\gamma}} \exp\left\{\frac{z(r - \beta + \gamma)}{2\gamma}\right\} U\left(I(ye^{-z})\right)$$
$$\cdot \int_0^\infty \frac{1}{\sqrt{\pi t}} \exp\left\{-\frac{z^2}{4\gamma t} - \frac{(r - \beta + \gamma)^2 + 4\gamma\beta}{4\gamma}t\right\} dt\, dz$$

$$= \int_{-\infty}^\infty \frac{1}{\sqrt{(r - \beta + \gamma)^2 + 4\gamma\beta}} U\left(I(ye^{-z})\right)$$
$$\cdot \exp\left\{\frac{1}{2\gamma}\left[z(r - \beta + \gamma) - |z|\sqrt{(r - \beta + \gamma)^2 + 4\gamma\beta}\right]\right\} dz$$

$$= \frac{1}{\gamma(\rho_2 - \rho_1)} \left[\int_0^\infty e^{\rho_1 z} U\left(I(ye^{-z})\right) dz\right.$$
$$\left. + \int_{-\infty}^0 e^{\rho_2 z} U\left(I(ye^{-z})\right) dz\right],$$

and the change of variable $\eta = ye^{-z}$ gives us (9.37). Finiteness of G_∞ follows from Lemma 9.17, and (9.38) can be verified by direct computation. Equation (9.39) is proved in the same manner as (6.32), and from (9.39) we obtain (9.40). $\qquad \square$

Corollary 9.19:   *Under Assumptions 9.9 and 9.13, the value function $V_\infty(x)$ given by (9.16) is finite for all $x \in (\mathcal{X}_\infty(\infty), \infty)$.*

Theorem 9.20 (Hamilton–Jacobi–Bellman equation):   *Let Assumptions 9.9 and 9.13 hold. Then the value function V_∞ is twice continuously* *differentiable on* $(\mathcal{X}_\infty(\infty), \infty)$ *and satisfies the* Hamilton–Jacobi–Bellman equation of dynamic programming,

$$-\beta V_\infty(x) + \max_{\substack{0 \le c < \infty \\ \pi \in \mathbb{R}^N}} \left[\frac{1}{2} \|\sigma'\pi\|^2 V_\infty''(x) + (rx - c + \pi'\sigma\theta)V_\infty'(x) + U(c) \right] = 0,$$

$$x > \mathcal{X}_\infty(\infty). \tag{9.41}$$

PROOF. Equations (9.16), (9.29), and (9.40) imply

$$V_\infty'(x) = \mathcal{Y}_\infty(x), \quad V_\infty''(x) = \mathcal{Y}_\infty'(x), \quad x > \mathcal{X}_\infty(\infty).$$

We may thus rewrite the left-hand side of (9.41) as

$$- \beta G_\infty(\mathcal{Y}_\infty(x)) + rx\mathcal{Y}_\infty(x) + \max_{0 \le c < \infty} [U(c) - c\mathcal{Y}_\infty(x)]$$

$$+ \max_{\pi \in \mathbb{R}^N} \left[\frac{1}{2} \|\sigma'\pi\|^2 \mathcal{Y}_\infty'(x) + \pi'\sigma\theta\mathcal{Y}_\infty(x) \right].$$

The maximizing values of c and π are given by (9.30) and (9.31), respectively. Substituting these values into (9.41) and setting $y = \mathcal{Y}_\infty(x)$, so $x = \mathcal{X}_\infty(y)$, we obtain

$$-\beta G_\infty(y) + ry\mathcal{X}_\infty(y) + U(I(y)) - yI(y) - \frac{1}{2}\|\theta\|^2 y^2 \mathcal{X}_\infty'(y).$$

Using (9.38) and (9.40), we reduce this expression to

$$- \frac{1}{2}\|\theta\|^2 y^2 G_\infty''(y) - (\beta - r)yG_\infty'(y) + ry\mathcal{X}_\infty(y) - yI(y) - \frac{1}{2}\|\theta\|^2 y^2 \mathcal{X}_\infty'(y)$$

$$= -y \left[\frac{1}{2}\|\theta\|^2 y^2 \mathcal{X}_\infty''(y) + (\|\theta\|^2 - r + \beta) y\mathcal{X}_\infty'(y) - r\mathcal{X}_\infty(y) + I(y) \right],$$

which is zero because of (9.27). $\square$

Theorem 9.21: *Let Assumptions 9.9 and 9.13 hold. Then*

$$\widetilde{V}_\infty(y) = \frac{1}{\gamma(\rho_2 - \rho_1)} \left[y^{\rho_1} \int_0^y \eta^{-\rho_1 - 1} \widetilde{U}(\eta)\, d\eta + y^{\rho_2} \int_y^\infty \eta^{-\rho_2 - 1} \widetilde{U}(\eta)\, d\eta \right],$$

$$0 < y < \infty, \tag{9.42}$$

is finite, twice continuously differentiable, and

$$\frac{1}{2}\|\theta\|^2 y^2 \widetilde{V}_\infty''(y) + (\beta - r)y\widetilde{V}_\infty'(y) - \beta\widetilde{V}_\infty(y) = -\widetilde{U}(y), \quad 0 < y < \infty. \tag{9.43}$$

PROOF. According to Theorems 9.12, 9.14, and 9.18,

$$\widetilde{V}_\infty(y) = G_\infty(y) - y\mathcal{X}_\infty(y)$$

$$= \frac{1}{\gamma(\rho_2 - \rho_1)} \left[y^{\rho_1} \int_0^y \eta^{-\rho_1 - 1} \left(U(I(\eta)) - \eta I(\eta) \right) d\eta \right.$$
$$\left. + y^{\rho_2} \int_y^\infty \eta^{-\rho_2 - 1} \left(U(I(\eta)) - \eta I(\eta) \right) d\eta \right], \quad 0 < y < \infty.$$

Equation (9.42) follows from (4.11). Equation (9.43) can be proved by direct computation. □

Example 9.22: For $p < 1$, $p \neq 0$, take $U(c) = U^{(p)}(c - \bar{c})$, given by (4.4), where $\bar{c} \geq 0$ is constant. Then

$$I(y) = y^{1/(p-1)} + \bar{c}, \quad U(I(y)) = \frac{1}{p} y^{p/(p-1)}, \quad \tilde{U}(y) = \frac{1-p}{p} y^{p/(p-1)} - \bar{c}y,$$

for all $y > 0$. Let

$$\delta \triangleq \beta - rp - \frac{\gamma p}{1 - p}.$$

Then (9.25) is equivalent to $\delta > 0$, which is certainly the case if $p < 0$, but can fail to be if $0 < p < 1$. Under the assumption $\delta > 0$, we have

$$\mathcal{X}_\infty(y) = \frac{1-p}{\delta} y^{1/(p-1)} + \frac{\bar{c}}{r}, \quad y > 0,$$

$$\mathcal{Y}_\infty(x) = \left[\frac{\delta}{1-p} \left(x - \frac{\bar{c}}{r} \right) \right]^{p-1}, \quad x > \frac{\bar{c}}{r},$$

$$G_\infty(y) = \frac{1-p}{\delta p} y^{p/(p-1)}, \quad y > 0,$$

$$V_\infty(x) = \frac{1}{p} \left(\frac{1-p}{\delta} \right)^{1-p} \left(x - \frac{\bar{c}}{r} \right)^p, \quad x > \frac{\bar{c}}{r},$$

$$\tilde{V}_\infty(y) = \frac{(1-p)^2}{\delta p} y^{p/(p-1)} - \frac{\bar{c}y}{r}, \quad y > 0.$$

The optimal consumption and portfolio processes in feedback form are

$$C(x) = \frac{\delta}{1-p} \left(x - \frac{\bar{c}}{r} \right) + \bar{c}, \quad \Pi(x) = \frac{1}{1-p} (\sigma')^{-1} \theta \left(x - \frac{\bar{c}}{r} \right), \quad x \geq \frac{\bar{c}}{r}.$$

The optimal wealth process, given by (9.32), is

$$X_\infty(t) = \left(x - \frac{\bar{c}}{r} \right) \exp \left\{ \frac{1}{1-p} (r - \beta + \gamma)t + \frac{1}{1-p} \theta' W(t) \right\} + \frac{\bar{c}}{r}, \quad t \geq 0.$$

Remark 9.23: For $0 < p < 1$ and $U(c) = U^{(p)}(c - \bar{c})$, where $\bar{c} \geq 0$ is constant, condition (9.25), or equivalently, the inequality

$$\delta(\beta) \triangleq \beta - rp - \frac{\gamma p}{1 - p} > 0,$$

is not only sufficient but also necessary for the value function $V_\infty(x)$ to be finite for $x > \bar{c}/r$. We see this by varying the discount factor $\beta > 0$. Let us write V_∞^β rather than V_∞ to indicate explicitly the dependence of the value function on β. Because $e^{-\beta t}U(c)$ is decreasing in β, the value function V_∞^β must also be decreasing in β. For $\beta > \beta_0 \overset{\Delta}{=} rp + (\gamma p/(1-p))$, Example 9.22 shows that

$$V_\infty^\beta(x) = \frac{1}{p}\left(\frac{1-p}{\delta(\beta)}\right)^{1-p}\left(x - \frac{\bar{c}}{r}\right)^p, \qquad x > \frac{\bar{c}}{r}.$$

Consequently,

$$\lim_{\beta\downarrow\beta_0} V_\infty^\beta(x) = \infty, \qquad x > \frac{\bar{c}}{r},$$

which implies $V_\infty^\beta(x) = \infty$ for all $x > \bar{c}/r$ and $\beta \le \beta_0$.

Example 9.24: Take $U(c) = U^{(0)}(c - \bar{c})$, given by (4.5), where $\bar{c} \ge 0$ is constant. Then

$$I(y) = \frac{1}{y} + \bar{c}, \quad U(I(y)) = -\log y, \quad \widetilde{U}(y) = -\log y - \bar{c}y - 1, \quad y > 0.$$

Condition (9.25) holds, and we have

$$\mathcal{X}_\infty(y) = \frac{1}{\beta y} + \frac{\bar{c}}{r}, \qquad y > 0,$$

$$\mathcal{Y}_\infty(x) = \frac{1}{\beta}\left(x - \frac{\bar{c}}{r}\right)^{-1}, \qquad x > \frac{\bar{c}}{r},$$

$$G_\infty(y) = -\frac{1}{\beta}\log y + \frac{r - \beta + \gamma}{\beta^2}, \qquad y > 0,$$

$$V_\infty(x) = \frac{1}{\beta}\log\beta\left(x - \frac{\bar{c}}{r}\right) + \frac{r - \beta + \gamma}{\beta^2}, \qquad x > \frac{\bar{c}}{r},$$

$$\widetilde{V}_\infty(y) = -\frac{1}{\beta}\log y - \frac{\bar{c}y}{r} + \frac{r - 2\beta + \gamma}{\beta^2}, \qquad y > 0.$$

The optimal consumption and portfolio processes in feedback form are

$$C(x) = \beta\left(x - \frac{\bar{c}}{r}\right) + \bar{c}, \quad \Pi(x) = (\sigma')^{-1}\theta\left(x - \frac{\bar{c}}{r}\right), \qquad x \ge \frac{\bar{c}}{r}.$$

The optimal wealth process, given by (9.32), is

$$X_\infty(t) = \left(x - \frac{\bar{c}}{r}\right)\exp\{(r - \beta + \gamma)t + \theta'W(t)\} + \frac{\bar{c}}{r}, \qquad t \ge 0.$$

3.10   Maximization of the Growth Rate of Wealth

Let us consider as a special case of Problem 5.3 the maximization of $E\log\left(X^{x,c,\pi}(T)\right)$ over consumption/portfolio process pairs (c,π). Since there is no utility from consumption, we may, without loss of generality, restrict attention to pairs (c, π) for which $c(\cdot) \equiv 0$ (see Theorem 7.6(i)). Example 7.9 with $\bar{x}$ set equal to zero shows that for any initial endowment $x > 0$, the optimal wealth process $X_2(\cdot)$ and the optimal portfolio process $\pi_2(\cdot)$ are given by

$$X_2(t) = x/H_0(t), \tag{10.1}$$

$$\pi_2(t) = (\sigma(t)\sigma'(t))^{-1} \left(b(t) + \delta(t) - r(t)\underline{1} \right) X_2(t). \tag{10.2}$$

Note that *the expressions (10.1), (10.2) do not depend on the terminal time T.* This is a very special property of the logarithmic utility function; we shall exploit it below to show that the portfolio process $\pi_2(\cdot)$ of (10.2) solves the problem of maximization of the growth rate of wealth over the infinite horizon $[0, \infty)$.

We use the complete, standard financial market on an infinite horizon, as set forth in Section 1.7. Let $x > 0$ be given. Recall from Definition 9.2 that a consumption/portfolio process pair (c, π) is said to be *admissible at x* if the corresponding wealth process is almost surely nonnegative at all times; the set of all such processes is denoted by $\mathcal{A}(x)$. Since we shall only be considering $c(\cdot) \equiv 0$, we shall simplify the notation by writing $\pi \in \mathcal{A}(x)$ rather than $(0, \pi) \in \mathcal{A}(x)$ and $X^{x,\pi}(\cdot)$ rather than $X^{x,0,\pi}(\cdot)$. Within the class $\mathcal{A}(x)$ of portfolio processes, there is the smaller class of portfolios $\pi(\cdot)$ for which $E \min\{0, \log X^{x,\pi}(T)\} > -\infty$ for all $T \in [0, \infty)$, and we shall denote by $\mathcal{A}_{2,\infty}(x)$ this collection. For $\pi \in \mathcal{A}_{2,\infty}(x)$, the expectation $E \log X^{x,\pi}(T)$ is well-defined for all $T \in [0, \infty)$. Since

$$\log X^{x,\pi_2}(T) = \log x + A(T) + \int_0^T \left[r(s) + \frac{1}{2}\|\theta(s)\|^2 \right] ds$$

$$+ \int_0^T \theta'(s) \, dW(s), \tag{10.3}$$

a sufficient condition for $\pi_2(\cdot)$ to be in $\mathcal{A}_{2,\infty}(x)$ is that both

$$\left\{ \begin{array}{l} \forall T \in [0, \infty) \text{ there exists a nonrandom constant } \kappa_T > -\infty \\ \text{such that } A(T) + \int_0^T \left[r(s) + \frac{1}{2}\|\theta(s)\|^2 \right] ds > -\kappa_T \text{ a.s.} \end{array} \right\}, \tag{10.4}$$

$$E \int_0^T \|\theta(s)\|^2 \, ds < \infty, \quad \forall \ T \in [0, \infty) \tag{10.5}$$

hold. Condition (10.4) implies for each fixed T that $S_0(T)$ is bounded away from zero, which implies in turn that Assumption 2.2 holds.

When (10.4), (10.5) hold, it is straightforward to verify that $\pi_2(\cdot)$ of (10.2) maximizes the *expected* growth rate of wealth

$$\limsup_{T \to \infty} \frac{1}{T} E \log X^{x,\pi}(T)$$

over all $\pi \in \mathcal{A}_{2,\infty}(x)$. Indeed, from the first paragraph of this section we know that $E \log X^{x,\pi}(T) \le E \log X^{x,\pi_2}(T)$ for all $\pi \in \mathcal{A}_{2,\infty}(x)$ for each fixed $T \in [0, \infty)$, whence

$$\limsup_{T \to \infty} \frac{1}{T} E \log X^{x,\pi}(T) \le \limsup_{T \to \infty} \frac{1}{T} E \log X^{x,\pi_2}(T)$$

$$= \limsup_{T \to \infty} \frac{1}{T} E \left\{ A(T) \right.$$

$$\left. + \int_0^T \left[r(s) + \frac{1}{2} \|\theta(s)\|^2 \right] ds \right\}.$$

The following theorem shows that $\pi_2(\cdot)$ is also optimal in the *almost sure sense* for this long-term growth problem. Because this is not a statement about expectations, it is not necessary to assume (10.4), (10.5), and we can show that $\pi_2(\cdot)$ is optimal in the class $\mathcal{A}(x)$ rather than the smaller class $\mathcal{A}_{2,\infty}(x)$.

Theorem 10.1: *Let $x > 0$ be given. For any portfolio process $\pi \in \mathcal{A}(x)$, we have almost surely*

$$\limsup_{T \to \infty} \frac{1}{T} \log X^{x,\pi}(T) \le \limsup_{T \to \infty} \frac{1}{T} \log X^{x,\pi_2}(T). \tag{10.6}$$

In other words, for P-almost every $\omega \in \Omega$, the portfolio process $\pi_2(\cdot)$ maximizes the actual rate of growth $\limsup_{T \to \infty} \frac{1}{T} \log X^{x,\pi}(T, \omega)$ of wealth from investment over all admissible portfolios $\pi(\cdot)$.

PROOF.   For any $\pi \in \mathcal{A}(x)$, the ratio

$$R(t) \overset{\Delta}{=} X^{x,\pi}(t)/X^{x,\pi_2}(t) = \frac{1}{x} H_0(t) X^{x,\pi}(t)$$

satisfies (cf. Remark 3.3)

$$dR(t) = \frac{1}{x} H_0(t)[\sigma'(t)\pi(t) - X^{x,\pi}(t)\theta(t)]' \, dW(t)$$

and is, therefore, a nonnegative local martingale and supermartingale. As such, $R(\cdot)$ satisfies the inequality

$$e^{\delta n} P \left\{ \sup_{n \le t < \infty} R(t) > e^{\delta n} \right\} \le ER(n) \le R(0) = 1, \quad \forall n \in \mathbb{N}, \ 0 < \delta < 1$$

(cf. Karatzas and Shreve (1991), Problem 1.3.16 and Theorem 1.3.8(ii)). Fix $\delta \in (0, 1)$. Then

$$\sum_{n=1}^{\infty} P \left\{ \sup_{n \le t < \infty} \log R(t) > \delta n \right\} \le \sum_{n=1}^{\infty} e^{-\delta n} < \infty,$$

and the Borel–Cantelli lemma implies the existence of an integer-valued random variable N_δ such that

$$\log R(t,\omega) \leq \delta n \leq \delta t \quad \forall n \geq N_\delta(\omega) \quad \forall t \geq n$$

for P-almost every $\omega \in \Omega$. In particular, for all such ω, we have

$$\sup_{t \geq n} \frac{1}{t} \log R(t,\omega) \leq \delta \quad \forall n \geq N_\delta(\omega),$$

hence

$$\limsup_{t \to \infty} \frac{1}{t} \log X^{x,\pi}(t,\omega) \leq \limsup_{t \to \infty} \frac{1}{t} \log X^{x,\pi_2}(t,\omega) + \delta,$$

and inequality (10.6) follows from the arbitrariness of $\delta \in (0,1)$.   □

Corollary 10.2: *Assume that the processes $r(\cdot)$ and $\theta(\cdot)$ are constants r and θ, and that the process $A(\cdot)$ is identically zero. Then the optimal rate of growth for the wealth process, given by the right-hand side of (10.6), is $r + \frac{1}{2}\|\theta\|^2$.*

PROOF. Use (10.3) and the observation that $\lim_{T \to \infty} \frac{1}{T} \theta' W(T) = 0$ almost surely.   □

3.11   Notes

The modern mathematical theory of finance begins with Markowitz (1952, 1959), who conceived the idea of trading off the mean return of a portfolio against its variance. In a one-step, discrete-time model, one can buy an initial portfolio of stocks, and the value of this portfolio after one step is a random variable. Dividing the difference between this random variable and the initial value of the portfolio by the initial value of the portfolio, one obtains the (random) *return* associated with the portfolio. A given portfolio is said to be *efficient* if every portfolio that has mean return greater than that of the given portfolio also has a greater variance of return. Markowitz argues that one should hold only efficient portfolios. Tobin (1958) extends the portfolios of risky assets considered by Markowitz, to include linear combinations of these portfolios with a risk-free asset. There is then a distinguished portfolio of risky assets, called the *market portfolio*, such that any other portfolio can be dominated in the mean–variance sense by a linear combination of the market portfolio and the risk-free asset. This result is often called a *separation theorem*, because the problem of optimal investment separates into the two problems of (1) finding the market portfolio and (2) determining the optimal allocation between the market portfolio and the risk-free asset.

There may be several ways to construct the market portfolio from the risky assets, but regardless of the particular assets that are used to build the market portfolio, the mean and variance of its return are uniquely determined. Furthermore, the covariance of the return between any other risky asset and the market portfolio (the so-called β of that asset, when normalized by the variance of the market portfolio) is also uniquely determined. This leads to the *capital asset pricing model* of Sharpe (1964), Lintner (1965), Mossin (1966). According to this model, every risky asset must have a mean return that exceeds the risk-free return by a certain *risk premium*, which can be computed from that asset's β. If an asset had a risk premium higher than this computed value, the definition of market portfolio would be contradicted. If an asset had risk premium lower than predicted, it could not be part of any market portfolio and would therefore not be demanded; this would cause its current price to fall, which would increase its return, bringing its risk premium into line with the capital asset pricing formula.

The Sharpe–Lintner–Mossin capital asset pricing model is *static*: investments are made once, and then a return is realized. The assumption underlying the model is that the vector of returns on risky assets has a multivariate normal distribution, or else all agents have quadratic utility functions. (However, Ross (1976) provides a derivation of the capital asset pricing model from arbitrage rather than utility theory.) It is not surprising, therefore, that the risk premia computed from the model have been found not to conform well to real data; see, e.g., Jensen (1972).

In an attempt to consider more realistic, dynamic models, Mossin (1968) and Samuelson (1969) apply dynamic programming to solve multiperiod problems of portfolio management. Hakansson (1970) obtains a separation theorem in this context. However, the set of utility functions and asset price models for which the discrete-time backward recursion of dynamic programming can be executed analytically is rather limited. An early work on optimal consumption in continuous time is Mirrlees (1974), which actually dates from 1965. This paper presents a heuristic argument that the marginal utility of consumption should equal the marginal value of wealth along an optimal trajectory (cf. (8.23), (8.41)).

In two landmark papers, Merton (1969, 1971) introduces Itô calculus and the methods of continuous-time stochastic optimal control to the problem of capital asset pricing. (We refer the reader to Merton (1990) for a compilation of Merton's papers and for essays that place them in context.) By assuming a model with constant coefficients and solving the relevant Hamilton–Jacobi–Bellman equation, Merton (1969) produces solutions to both the finite- and infinite-horizon models when the utility function is a power function (Examples 8.13, 9.22), the logarithm (Examples 8.14, 9.24), or of the form $1 - e^{-\eta x}$ for some positive constant η. The mutual fund theorem (Remarks 8.9, 9.16), a separation theorem described above, appears in Merton (1971). It is often called the *two-fund theorem*, because the investor is content to have his only investment opportunities be Merton's mutual fund and the money market. Merton (1973b) introduces a Markov stochastic interest rate and a *three-fund theorem*, according to which the investor requires a second mutual fund to hedge against fluctuations in the interest rate. Richard (1979) generalizes this result to a market with an underlying N-dimensional Markov *state* and obtains an $(N + 2)$-*fund theorem*. Sethi and Taksar (1988) resolve some boundary issues in Merton's model, and Merton (1989) returns to this topic. Richard (1975) introduces a random time horizon to Merton's model. Khanna and Kulldorff (1998) obtain the two-fund theorem under very weak assumptions.

The original analysis of Merton's model was wedded to the Hamilton–Jacobi–Bellman equation and its requirement of an underlying Markov state process. In a non-Markov model of optimal consumption without portfolio control, Foldes (1978a,b) proves the existence of an optimal consumption process and shows that the marginal utility of consumption is, up to a discount factor, a martingale (our $Z_0(\cdot)$ of (2.5)). In the continuous-time model for both consumption and portfolio selection, Bismut (1975) obtains the key formula (6.5) for optimal consumption using his stochastic duality theory (Bismut (1973)) rather than relying on the Hamilton–Jacobi–Bellman equation. With the appearance of the papers by Harrison and Kreps (1979) and Harrison and Pliska (1981, 1983), which provide a martingale characterization of the set of terminal wealths that can be attained by investment in a complete market, it became possible to solve the optimal control problem of maximizing the expected utility of terminal wealth (by appropriate choice of portfolio) without the assumption of Markov asset prices. This was accomplished by Pliska (1986). The application of the Harrison–Kreps–Pliska martingale methodology to reproduce the Bismut (1973) formula for optimal consumption was worked out independently in Cox and Huang (1989, 1991) and Karatzas, Lehoczky, and Shreve (1987); both these papers show how to decompose the nonlinear Hamilton–Jacobi–Bellman equation into linear partial differential equations. The presentation in this chapter follows the latter reference, with the addition of Bismut's stochastic duality theory. The fact that the dual value function satisfies a *linear* partial differential equation was discovered by Xu (1990). Connections with the stochastic maximum principle appear in Back and Pliska (1987), Cadenillas (1992), Cadenillas and Karatzas (1995). See Brock and Magill (1979) for another application of Bismut's stochastic duality theory to economics. Kramkov and Schachermayer (1998) provide a necessary and sufficient condition on the utility function U_2 for V_2 to be a utility function and for an optimal portfolio process to exist; this would replace our Assumption 7.2.

The further extension of the martingale methodology to the infinite-horizon problem (Section 3.9) appears in Huang and Pagès (1992). The constant-coefficient computations for the infinite-horizon model are due to Karatzas, Lehoczky, Sethi, and Shreve (1986). Foldes (1990, 1991a,b, 1992), Jacka (1984), and Lakner and Slud (1991) treat the infinite-horizon problem with discontinuous asset price processes.

Example 7.9 (portfolio insurance) is inspired by the work of Başak (1993, 1995, 1996a). Some other papers addressing portfolio insurance are Black and Perold (1992), Brennan and Schwartz (1988, 1989), Cvitanić and Karatzas (1995), Grossman and Zhou (1993, 1996), Korn and Trautmann (1995).

Maximization of the growth rate of a portfolio (Section 3.10) goes back to Breiman (1961). Hakansson (1970) contains a discrete-time version of the results we present here; see also Thorp (1971). We take the results of Section 3.10 from Karatzas (1989). Aase and Øksendal (1988) extend these results to allow stock prices to jump. Taksar, Klass, and Assaf (1988) and Pliska and Selby (1994) address this problem in the presence of transaction costs. Algoet and Cover (1988) and Cover (1984, 1991) provide algorithms for maximizing of the growth rate of a portfolio in a very general discrete-time model. Jamshidian (1991) examines the behavior of this algorithm in a continuous-time model. Because it leads to maximization of the growth rate, the logarithmic utility would seem a natural choice for money managers. To temper the enthusiasm for this criterion, Merton and Samuelson (1974) point out that maximization of the growth rate does not necessarily maximize even approximately the expectation of a nonlogarithmic utility at any finite time, and Samuelson (1979) argues in words of literally one syllable that maximization of nonlogarithmic utility at a finite time is the more desirable goal. Kulldorff (1993) and Heath (1993) solve the related problem of maximizing the probability of reaching a goal by a fixed time.

In a continuous-time capital asset pricing model with an underlying N-dimensional Markov state process, the risk premia of assets can be computed theoretically from their covariances with a set of $N+1$ mutual funds. Breeden (1979) shows that rather than using the set of all covariances, one can in principle compute risk premia from the covariance of assets with the consumption process of an optimally behaving investor. Like the simple mean–variance capital asset pricing model, this *consumption-based capital asset pricing model* does not conform well to actual data. In particular, individuals' consumption patterns are smoother than predicted by the model; see Cornell (1981), Hansen and Singleton (1982, 1983), Mehra and Prescott (1985), Dunn and Singleton (1986), Singleton (1987). To address this so-called *equity premium puzzle*, several generalizations of the basic time-additive utility function maximization (considered in this chapter) have been proposed. One of these, which models *habit formation* of consumers, constructs a utility function that at each time depends on the current level of consumption and on an average of previous levels of consumption; see, e.g., Alvarez (1994), Constantinides (1990), Detemple and Zapatero (1991, 1992), Dybvig (1993), Hindy and Huang (1989, 1992, 1993), Hindy, Huang, and Kreps (1992), Hindy, Huang, and Zhu (1993), Sundaresan (1989), Uzawa (1968). A more radical approach is the construction of *recursive utility*, to disentangle agents' aversion to risk from their feelings about smoothness of consumption over time, an idea due to Kreps and

Porteus (1978). Some papers related to recursive utility in the context of dynamic optimal consumption and investment are Bergman (1985), Chew and Epstein (1991), Duffie and Epstein (1992a,b), Duffie and Lions (1990), Duffie and Skiadas (1994), Epstein and Zin (1989), Kan (1991), Ma (1991), El Karoui, Peng and Quenez (1997), Schroder and Skiadas (1997); see also the survey by Epstein (1990). Another way to account for the smoothness of observed consumption is the assumption of *transaction costs* for changes in level of consumption; see Grossman and Laroque (1990), Heston (1990), Cuoco and Liu (1997). He and Huang (1991, 1994) provide conditions on a consumption/portfolio policy that guarantee that it is optimal for some time-additive utility function; see also Lazrak (1996).

The value function in problems of optimal consumption and investment is quite sensitive to the introduction of *transaction costs*; see Dumas and Luciano (1989), Fleming, Grossman, Vila, and Zariphopoulou (1990), Shreve and Soner (1994). The notes to Chapter 2 survey the literature on utility-based models for option pricing in the presence of transaction costs. Some papers dealing with the problem of an investor who seeks to maximize expected utility of wealth and/or consumption and incurs transaction costs for changes in portfolio are Akian, Menaldi, and Sulem (1996), Cadenillas and Pliska (1997), Constantinides (1979, 1986), Davis and Norman (1990), Magill (1976), Magill and Constantinides (1976), Cvitanić and Karatzas (1996), Korn (1998), Morton and Pliska (1995), Shreve, Soner, and Xu (1991), Shreve and Soner (1994), Weerasinghe (1996), Zariphopoulou (1992). When transaction costs or other market frictions (e.g., borrowing constraints, different rates for borrowing and lending) are introduced, one can study the optimal consumption and investment problem by probabilistic techniques (e.g., Xu (1990), Shreve and Xu (1992), Cvitanić and Karatzas (1992, 1993, 1996), Jouini and Kallal (1995a,b), Karatzas and Kou (1996)) or, in a Markovian framework, by a viscosity solution analysis of the corresponding Hamilton–Jacobi–Bellman equation (see, in addition to the several papers already mentioned, Duffie, Fleming, Soner and Zariphopoulou (1997), Fitzpatrick and Fleming (1991), Fleming and Zariphopoulou (1991), Vila and Zariphopoulou (1991), Zariphopoulou (1989)). Cuoco and Cvitanić (1998) consider optimal consumption for an investor whose actions affect the market. Other work on optimal consumption and/or investment in incomplete markets is cited in the notes to Chapter 6.

Extension of the optimal consumption/investment model to allow for several consumption goods can be found in Breeden (1979), Madan (1988), Lakner (1989) and the references therein. Ocone and Karatzas (1991) use ideas from the Malliavin calculus to compute optimal portfolios. Pikovsky and Karatzas (1996) use enlargement of filtration techniques to study a version of the consumption/investment problem in which the investor has some "inside" information about the behavior of future prices; see also Elliott, Geman, and Korkie (1997), Amendinger, Imkeller, and Schweizer (1997), and Pikovsky (1998), as well as Kyle (1985), Duffie and Huang (1986), and

Back (1992, 1993) for earlier work in a similar vein. Kuwana (1993, 1995) and Lakner (1995) consider the mixed control/filtering problem, in which the investor must estimate the mean rate of return of the assets; see also Karatzas (1997), Karatzas and Zhao (1998). Richardson (1989), Duffie and Richardson (1991), and Schweizer (1992b) find minimal-variance portfolios that achieve desired expected returns. Xu (1989) constructs a simple example in which the optimal portfolio does not invest in the risky asset, even when its mean rate of return dominates the risk-free rate. The optimal consumption/investment model with an allowance for bankruptcy has been considered by Lehoczky, Sethi, and Shreve (1983, 1985), Presman and Sethi (1991, 1996), Sethi and Taksar (1992), Sethi, Taksar and Presman (1992). Several related papers are collected in Sethi (1997).

Adler and Dumas (1983) provide a survey of the application of the continuous-time capital asset pricing model to international finance. For general theory on stochastic control problems, the reader can consult the books by Fleming and Rishel (1975), Bertsekas and Shreve (1978), Elliott (1982), Chapter 12, Fleming and Soner (1993), as well as the lecture notes by El Karoui (1981).

4

Equilibrium in a Complete Market

4.1 Introduction

In the context of continuous-time financial markets, the *equilibrium problem* is to build a model in which *security prices are determined by the law of supply and demand*. The primitives in this model are the *endowment processes* and the *utility functions* of a finite number of agents. We shall assume in this chapter that all agents are endowed in units of the same *perishable commodity*, which arrives at some time-varying random *rate*. Agents may consume their endowment as it arrives, they may sell some portion of it to other agents, or they may buy extra endowment from other agents. The endowment, however, cannot be stored, and agents will wish to hedge the variability in their endowment processes by trading with one another. To facilitate the trading of endowment, there is a *financial market* consisting of a money market and of several stocks, in which agents may invest (positively or negatively).

Each agent takes the security prices as given, observed stochastic processes, and maximizes his expected utility from consumption over the finite time horizon of the model, subject to the condition that his wealth at the final time must be nonnegative almost surely. This problem differs from Problem 5.2 of Chapter 3 only because the agent receives his endowment *over time* rather than initially, but in the context of a *complete* financial market this difference is inconsequential. The goal is to choose the prices of the money market and of the stocks so that when each agent solves his optimal consumption and investment problem, at all times *the aggregate endowment is consumed as it enters the economy* and *all securities are in zero net supply.* The first condition codifies the concept of a "perishable" commodity; the second reflects the fact that for every buyer of a security, there must be a seller.

In the model of this chapter all securities will be denominated in units of the single perishable commodity. When the marginal utility functions of all the individual agents are infinite at their individual "subsistence levels" (see Section 4.2), then all agents' optimal consumption processes are always strictly above the subsistence level; in this case the equilibrium money market price can be described solely in terms of an interest rate, and the equilibrium prices for all stocks are determined. However, when we allow even one agent to have finite marginal utility at the subsistence level of consumption, then this agent may sometimes see his equilibrium optimal consumption fall to the subsistence level. In this case we still obtain equilibrium prices for the money market and stocks, but the money-market price can no longer be described in terms of an interest rate. More specifically, the money-market price is given by the formula

$$S_0(t) = \exp\left\{ \int_0^t r(u)\, du + A(t) \right\}$$

of (1.1.7), where $A(\cdot)$ is singularly continuous. Although $A(\cdot)$ is continuous and has zero derivative at Lebesgue-almost-every time, $A(\cdot)$ decreases at those times when an agent's equilibrium optimal consumption either falls to, or rises from, the subsistence level (see Remark 6.8). The stock prices of this equilibrium market are given by (1.1.9), which also includes the singularly continuous process $A(\cdot)$.

The inclusion of this singularly continuous component in the equilibrium security prices could be avoided by denominating security prices in terms of money rather than units of commodity (see Karatzas, Lehoczky, and Shreve (1990)). In effect, whenever an agent's consumption falls to, or rises from, the subsistence level, there is a burst of inflation in which securities become substantially less valuable in terms of commodity but not in terms of money. It is interesting to note that both falling to and rising from subsistence consumption leads to inflation; neither induces deflation.

The singularly continuous component in equilibrium security prices could also be avoided by denominating security prices in terms of the money-market price. This can be seen from (1.1.10); the process $A(\cdot)$ does not appear in the stock prices discounted by the money-market price.

When denominating security prices in units of commodity, however, a singularly continuous component in the security prices cannot be avoided. It was for this reason that we set up the market model in Section 1.1 to include this possibility. In this model, under the assumption (6.4) on the indices of risk aversion for the individual agents, we obtain uniqueness of the equilibrium allocations of the commodity, uniqueness of the equilibrium money-market price, and uniqueness of the equilibrium stock prices up to the formation of mutual funds (see Theorem 6.4 and the discussion following it). The equilibrium existence Theorem 6.3 and uniqueness Theorem 6.7 still hold when we permit agent endowments to arrive in a singularly continuous way, i.e., so that the cumulative endowment is continuous but is not necessarily described by a rate. For this reason we allow this added degree of generality, and the singularly continuous process $\xi(\cdot)$ appearing in the aggregate endowment equation (2.2) can contribute to the singularly continuous process $A(\cdot)$ in the security prices. However, even without the presence of the singularly continuous process $\xi(\cdot)$ in the aggregate endowment process, the singularly continuous process $A(\cdot)$ will appear in the security prices under the conditions discussed above.

The model of this chapter is a pure *exchange economy*, because there are no securities associated with production. Only financial securities are posited, which are in zero net supply. One could, however, use the equilibrium model to price the right to receive future endowments, and thereby have "productive assets" that are held in positive net supply (see Remark 6.6). The more challenging task of including production that can be enhanced by forgoing current consumption is not addressed here.

We conclude with a section-by-section summary of the chapter. Section 2 describes the exogenous processes and Section 3 describes the endogenous ones. Section 4 modifies the analysis of Chapter 3 to solve the optimal consumption and investment problem for an agent who receives an endowment process and acts as price-taker. Equilibrium is defined in Section 5, as is the concept of a single "representative agent" (really a utility function) who aggregates with appropriate weights the individual agents (really their utility functions). Equilibrium is characterized in terms of the representative agent's utility function via Corollary 5.4 and Theorem 5.6. This reduces the question of existence and uniqueness of equilibrium to the *finite-dimensional* problem of determining the appropriate weights in the construction of the representative agent's utility function. Theorem 6.1 establishes the existence of these weights and describes the extent to which they are uniquely determined. The remainder of Section 6 works out the ramifications of Theorem 6.1 for the existence and uniqueness of equilibrium. One of these is the *consumption-based capital asset pricing model (CCAPM)* (Remark 6.7). Section 7 contains examples in which the equilibrium consumption allocations and security prices can be exhibited explicitly.

4.2   Agents, Endowments, and Utility Functions

We consider an economy consisting of a finite number K of agents, each of whom is continuously endowed in units of a single perishable commodity. The exogenous *endowment processes* $\{\epsilon_k(t); 0 \le t \le T\}, k = 1, \ldots, K,$ of these agents are assumed to be nonnegative and progressively measurable with respect to the filtration $\{\mathcal{F}(t)\}_{0 \leq t \leq T}$ of Section 1.1, the augmentation by P-null sets of the filtration generated by the N-dimensional Brownian motion $W(\cdot)$ on the interval $[0, T]$. The *aggregate endowment*

$$\epsilon(t) \stackrel{\triangle}{=} \sum_{k=1}^{K} \epsilon_k(t), \quad 0 \leq t \leq T, \tag{2.1}$$

is assumed to be a continuous, positive, bounded semimartingale:

$$\epsilon(t) = \epsilon(0) + \int_0^t \epsilon(s)\nu(s)\, ds + \int_0^t \epsilon(s)\, d\xi(s) + \int_0^t \epsilon(s)\rho'(s)\, dW(s),$$

$$0 \leq t \leq T. \tag{2.2}$$

Here $\xi(\cdot)$ is an $\{\mathcal{F}(t)\}_{0 \leq t \leq T}$ - progressively measurable process with paths that are continuous and of finite variation on $[0, T]$, but that are singular with respect to Lebesgue measure (see Proposition B.1 in Appendix B). We take $\xi(0) = 0$ and assume that the total variation of $\xi(\cdot)$ on $[0, T]$ is almost surely *bounded*. The processes $\nu(\cdot)$ and $\rho(\cdot)$ are $\{\mathcal{F}(t)\}$-progressively measurable and *bounded*; they take values in $\mathbb{R}$ and $\mathbb{R}^N$, respectively.

In addition to his endowment, each agent k has a *utility function* $U_k : \mathbb{R} \to [-\infty, \infty)$ as described in Definition 3.4.1. We denote the *subsistence consumption for agent k* by

$$\bar{c}_k \stackrel{\triangle}{=} \inf\{c \in \mathbb{R}; U_k(c) > -\infty\} \tag{2.3}$$

(cf. (3.4.2)), and define *aggregate subsistence consumption* as

$$\bar{c} \stackrel{\triangle}{=} \sum_{k=1}^{K} \bar{c}_k. \tag{2.4}$$

Recall that $\bar{c}_k \geq 0$ for $k = 1, \ldots, K$.

Finally, the agents have a common *discount rate* $\beta : [0, T] \to \mathbb{R}$, which is a nonrandom Lebesgue-integrable function, bounded from below. Agent k will attempt to maximize his expected discounted utility from consumption

$$E \int_0^T e^{-\int_0^t \beta(u)\, du} U_k(c_k(t))\, dt$$

over the time horizon $[0, T]$, where $c_k(\cdot)$ is his consumption process. This maximization is very similar to Problem 3.5.2 with utility function $e^{-\int_0^t \beta(u)\, du} U_k(c)$, a function of both $t \in [0, T]$ and $c \in \mathbb{R}$.

The endowment processes $\{\epsilon_k(\cdot)\}_{k=1}^K$, the utility functions $\{U_k(\cdot)\}_{k=1}^K$, and the discount rate $\beta(\cdot)$ are the *primitives* of our equilibrium model. Starting with these primitives, we shall construct an equilibrium market. It is also possible to carry out this construction when each agent k has his own discount rate $\beta_k(\cdot)$, or, even more generally, when U_k is a function of both time and consumption. However, this more general construction involves a more complicated version of Itô's formula than the one we employ in Section 6. For this reason, we have restricted our attention to the situation presented above.

In order to construct an equilibrium market, we impose throughout the following conditions on the primitives.

Condition 2.1:

(i) For each $k = 1, \ldots, K$, the function $U_k(\cdot)$ is of class C^3 on $(\bar{c}_k, \infty)$, satisfies $U_k''(c) < 0$ for all $c > \bar{c}_k$, and the quantity

$$\lim_{c \downarrow \bar{c}_k} \frac{U_k'''(c)}{(U_k''(c))^2} \tag{2.5}$$

exists and is finite.

(ii) For each $k = 1, \ldots, K$, we have

$$\epsilon_k(t) \geq \bar{c}_k, \quad 0 \leq t \leq T, \tag{2.6}$$

almost surely.

(iii) There exist constants $0 < \gamma_1 < \gamma_2 < \infty$ such that

$$\bar{c} + \gamma_1 \leq \epsilon(t) \leq \gamma_2, \quad 0 \leq t \leq T, \tag{2.7}$$

almost surely.

Remark 2.2: We note that

$$\frac{1}{U_k''(c)} = \frac{1}{U_k''(\bar{c}_k + 1)} - \int_{\bar{c}_k + 1}^{c} \frac{U_k'''(\eta)}{(U_k''(\eta))^2} \, d\eta, \quad c > \bar{c}_k,$$

and so the existence of the limit (2.5) implies that $\lim_{c \downarrow \bar{c}_k} \frac{1}{U_k''(c)}$ also exists and is finite.

4.3 The Financial Market: Consumption and Portfolio Processes

To give structure to the search for an equilibrium market, we set out in this section a description of the object of our search. We seek to construct a *complete, standard financial market* $\mathcal{M}$ as in Definitions 1.5.1, 1.6.1, but without dividends. More specifically, we seek a money market price process $S_0(\cdot)$ with $S_0(0) = 1$ and

$$dS_0(t) = S_0(t)[r(t) \, dt + dA(t)], \tag{3.1}$$

as well as N stock price processes $S_1(\cdot), \ldots, S_N(\cdot)$, with $S_n(0)$ a positive constant for each n, and

$$dS_n(t) = S_n(t) \left[b_n(t)\, dt + dA(t) + \sum_{j=1}^{D} \sigma_{nj}(t)\, dW^{(j)}(t) \right], n = 1, \ldots, N.$$

(3.2)

In order to guarantee completeness, the volatility matrix $\sigma(t) = \{\sigma_{nj}(t)\}_{1 \le n,j \le N}$ must be nonsingular for Lebesgue-almost-every $t \in [0, T]$ almost surely (Theorem 1.6.6).

After we have constructed an equilibrium market as described above, we can define the *market price of risk*

$$\theta(t) = \sigma^{-1}(t)[b(t) - r(t)\underline{1}]$$

(3.3)

(see (1.6.16)), and then the processes $Z_0(\cdot)$ of (1.5.2), $W_0(\cdot)$ of (1.5.6), and the *standard martingale measure* P_0 of (1.5.3). Finally, we can define the *state price density process*

$$H_0(t) \triangleq \frac{Z_0(t)}{S_0(t)}, \quad 0 \le t \le T,$$

(3.4)

of (1.5.12). To aid in the subsequent analysis, we shall require that our equilibrium market satisfy the following condition. Unlike Condition 2.1, which concerns the primitives of the economy and is assumed throughout this chapter, *Condition 3.1 below must be verified after the candidate equilibrium market has been constructed.*

Condition 3.1:

(i) We have

$$E_0 \left[\max_{0 \le t \le T} \left(\frac{1}{S_0(t)} \right) \right] < \infty.$$

(ii) There exist constants $0 < \delta_0 < \Delta_0 < \infty$ such that we have almost surely

$$\delta_0 \le H_0(t) \le \Delta_0, \quad 0 \le t \le T.$$

Once an equilibrium market has been constructed, each agent k can choose a *consumption process* $c_k : [0, T] \times \Omega \to [0, \infty)$ and a *portfolio process* $\pi_k : [0, T] \times \Omega \to \mathbb{R}^N$. These are both $\{\mathcal{F}(t)\}$-progressively measurable; $\pi_k(\cdot)$ satisfies (1.2.5), (1.2.6); and $c_k(\cdot)$ satisfies $\int_0^T c_k(t)\, dt < \infty$ almost surely. The structure of U_k implies that agent k will be interested only in consumption processes $c_k(\cdot)$ satisfying the additional condition

$$c_k(t) \ge \bar{c}_k, \quad 0 \le t \le T,$$

(3.5)

almost surely. The *wealth process* $X_k(\cdot) = X_k^{\pi_k, c_k}(\cdot)$ as in (1.5.8), corresponding to the portfolio $\pi(\cdot)$ and the cumulative income process $\Gamma_k(t) \triangleq$

$\int_0^t (\epsilon_k(s) - c_k(s)) \, ds$, is given by

$$\frac{X_k(t)}{S_0(t)} = \int_0^t \frac{\epsilon_k(u) - c_k(u)}{S_0(u)} \, du + \int_0^t \frac{1}{S_0(u)} \pi_k'(u)\sigma(u) \, dW_0(u), \quad 0 \le t \le T.$$
(3.6)

We take $X_k(0) = 0$.

Remark 3.2: The wealth equation (3.6) can be written in the equivalent form

$$H_0(t)X_k(t) + \int_0^t H_0(s)(c_k(s) - \epsilon_k(s)) \, ds = \int_0^t H_0(s)[\sigma(s)\pi_k(s) - X_k(s)\theta'(s)]' \, dW(s)$$
(3.7)

by analogy with (3.3.3), as is seen if one applies Itô's formula to the product of the processes $X_k(\cdot)/S_0(\cdot)$ and $Z_0(\cdot)$ in (3.6) and (1.5.5), respectively.

Definition 3.3: A consumption/portfolio process pair (c_k, π_k) is *admissible for the kth agent* if the corresponding wealth process $X_k(\cdot)$ of (3.6) satisfies

$$\frac{X_k(t)}{S_0(t)} + E_0 \left[\int_t^T \frac{\epsilon_k(u)}{S_0(u)} \, du \,\Big|\, \mathcal{F}(t) \right] \ge 0, \quad 0 \le t \le T,$$
(3.8)

almost surely. The class of admissible pairs (c_k, π_k) is denoted by $\mathcal{A}_k$.

Remark 3.4: The admissibility condition says that at each time t, the present wealth $X_k(t)$ (which may be negative) plus the current value $S_0(t) \cdot E_0[\int_t^T \frac{\epsilon_k(u)}{S_0(u)} \, du | \mathcal{F}(t)]$ of future endowment (cf. Proposition 2.2.3) must be nonnegative. This condition is equivalent to

$$H_0(t)X_k(t) + E \left[\int_t^T H_0(u)\epsilon_k(u) \, du \,\Big|\, \mathcal{F}(t) \right] \ge 0, \quad 0 \le t \le T,$$
(3.9)

almost surely, as can be seen from "Bayes's rule" in Chapter 3, Lemma 5.3 of Karatzas and Shreve (1991).

Remark 3.5: Condition 3.1(i) guarantees that for every given $(c_k, \pi_k) \in \mathcal{A}_k$, the local P_0-martingale

$$M^{\pi_k}(t) \overset{\triangle}{=} \int_0^t \frac{1}{S_0(u)} \pi_k'(u)\sigma(u) \, dW_0(u), \quad 0 \le t \le T,$$
(3.10)

is also a P_0-supermartingale. To see this, use Condition 2.1(iii) and (3.6), (3.8) to write

$$M^{\pi_k}(t) \ge \frac{X_k(t)}{S_0(t)} - \int_0^t \frac{\epsilon_k(u)}{S_0(u)} \, du$$

$$\ge -E_0 \left[\int_0^T \frac{\epsilon_k(u)}{S_0(u)} \, du \,\Big|\, \mathcal{F}(t) \right]$$

$$\geq -\gamma_2 T \cdot E_0 \left[\max_{0 \leq u \leq T} \left(\frac{1}{S_0(u)} \right) \mid \mathcal{F}(t) \right].$$

Under the probability measure P_0, the expression $Y(t) \triangleq -\gamma_2 T \cdot E_0[\max_{0 \leq u \leq T}(\frac{1}{S_0(u)}) \mid \mathcal{F}(t)]$ is a martingale; being a nonnegative local martingale, $M^{\pi_k}(\cdot) - Y(\cdot)$ is also a supermartingale, so $M^{\pi_k}(\cdot)$ is a supermartingale.

From this, (3.6), and (3.8) with $t = T$, it develops that $c_k(\cdot)$ must satisfy the *budget constraint*

$$E_0 \int_0^T \frac{c_k(t)}{S_0(t)} \, dt \leq E_0 \int_0^T \frac{\epsilon_k(t)}{S_0(t)} \, dt, \tag{3.11}$$

or equivalently,

$$E \int_0^T H_0(t) c_k(t) \, dt \leq E \int_0^T H_0(t) \epsilon_k(t) \, dt. \tag{3.12}$$

In any market satisfying Condition 3.1(i), and for any consumption process $c_k(\cdot)$ such that $(c_k, \pi_k) \in \mathcal{A}_k$ for some portfolio process $\pi_k(\cdot)$, *the value of an agent's consumption cannot exceed the value of his endowment,* where value is determined using the state price density process $H_0(\cdot)$ for that particular market.

We have the following counterpart to Theorem 3.3.5.

Theorem 3.6: *Suppose we have constructed a complete, standard financial market satisfying Condition 3.1(i). Let $c_k(\cdot)$ be a consumption process in this market that satisfies (3.11) (or equivalently (3.12)) with equality, namely*

$$E_0 \int_0^T \frac{c_k(t)}{S_0(t)} \, dt = E_0 \int_0^T \frac{\epsilon_k(t)}{S_0(t)} \, dt. \tag{3.11'}$$

Then there exists a portfolio process $\pi_k(\cdot)$ such that $(c_k, \pi_k) \in \mathcal{A}_k$, and the corresponding wealth process is given by

$$X_k(t) = \frac{1}{H_0(t)} E \left[\int_t^T H_0(s)(c_k(s) - \epsilon_k(s)) \, ds \mid \mathcal{F}(t) \right], \quad 0 \leq t \leq T. \tag{3.13}$$

PROOF.   In Proposition 1.6.2, take $B = S_0(T) \cdot \int_0^T \frac{c_k(u) - \epsilon_k(u)}{S_0(u)} \, du$, so that

$$E_0 \left[\frac{|B|}{S_0(T)} \right] \leq E_0 \left[\int_0^T \frac{c_k(u) + \epsilon_k(u)}{S_0(u)} \, du \right] \leq 2 E_0 \left[\int_0^T \frac{\epsilon_k(u)}{S_0(u)} \, du \right]$$

$$\leq 2\gamma_2 T E_0 \left[\max_{0 \leq t \leq T} \left(\frac{1}{S_0(t)} \right) \right] < \infty.$$

From the assumption of market completeness, we have the existence of a martingale-generating portfolio process $\pi_k(\cdot)$ such that

$$\int_0^T \frac{c_k(u) - \epsilon_k(u)}{S_0(u)} \, du = M^{\pi_k}(T), \tag{3.14}$$

where $M^{\pi_k}(\cdot)$ is given by (3.10). Taking conditional expectations in (3.14), we obtain

$$E_0 \left[\int_t^T \frac{c_k(u) - \epsilon_k(u)}{S_0(u)} \, du \mid \mathcal{F}(t) \right] = \int_0^t \frac{\epsilon_k(u) - c_k(u)}{S_0(u)} \, du$$

$$+ \int_0^t \frac{1}{S_0(u)} \pi_k'(u) \sigma(u) \, dW_0(u),$$

$$0 \leq t \leq T.$$

Comparison of this equation with (3.6) reveals that

$$X_k(t) = S_0(t) \cdot E_0 \left[\int_t^T \frac{c_k(u) - \epsilon_k(u)}{S_0(u)} \, du \mid \mathcal{F}(t) \right], \quad 0 \leq t \leq T \tag{3.15}$$

(the initial condition $X_k(0) = 0$ is a consequence of (3.11$'$)), and (3.13) follows from "Bayes's rule" (cf. Karatzas and Shreve (1991), Lemma 3.5.3). From (3.15) we see that the admissibility condition (3.8) is satisfied. □

4.4   The Individual Optimization Problems

Suppose that we have constructed a complete, standard financial market satisfying Condition 3.1, as described in the previous section. In this market, each agent, say the kth agent, will be faced with the following problem.

Problem 4.1: Find an optimal pair $(\hat{c}_k, \hat{\pi}_k)$ for the problem of maximizing expected discounted utility from consumption

$$E \int_0^T e^{-\int_0^t \beta(u)} U_k(c_k(t)) \, dt$$

over consumption/portfolio process pairs in the set

$$\mathcal{A}_k' \triangleq \left\{ (c_k, \pi_k) \in \mathcal{A}_k; \ E \int_0^T e^{-\int_0^t \beta(u)\,du} \min[0, U_k(c_k(t))] \, dt > -\infty \right\}. \tag{4.1}$$

From (3.5) and the budget constraint (3.12), we see that this problem is interesting only if the *feasibility condition*

$$E \int_0^T H_0(t)\epsilon_k(t) \, dt \geq \bar{c}_k \cdot E \int_0^T H_0(t) \, dt \tag{4.2}$$

is satisfied; otherwise, $\mathcal{A}'_k = \emptyset$. If equality holds in (4.2), then the only candidate optimal consumption process is $\hat{c}_k(\cdot) \equiv \bar{c}_k$. According to Theorem 3.6, there is then a portfolio process $\hat{\pi}_k(\cdot)$ such that $(\hat{c}_k, \hat{\pi}_k) \in \mathcal{A}_k$. This consumption/portfolio process pair is in $\mathcal{A}'_k$ if and only if $U_k(\bar{c}_k) > -\infty$. Regardless of whether this is the case or not, $\int_0^T e^{-\int_0^t \beta(u)\, du} U_k(\bar{c}_k)\, dt$ is well-defined (even though it may be $-\infty$) and is the value of Problem 4.1, whenever (4.2) holds as equality. This leads us to adopt the following convention, even though $\mathcal{A}'_k$ may be empty.

Convention 4.2: If the *nonstrict feasibility condition*

$$E \int_0^T H_0(t)\epsilon_k(t)\, dt = \bar{c}_k \cdot E \int_0^T H_0(t)\, dt \qquad (4.3)$$

holds, we say that the optimal consumption process for Problem 4.1 is $\hat{c}_k(\cdot) \equiv \bar{c}_k$. There exists then a portfolio process $\hat{\pi}_k(\cdot)$ such that $(\bar{c}_k, \hat{\pi}_k) \in \mathcal{A}_k$.

We consider now the case of *strict feasibility*:

$$E \int_0^T H_0(t)\epsilon_k(t)\, dt > \bar{c}_k \cdot E \int_0^T H_0(t)\, dt. \qquad (4.4)$$

To treat this case, we define as in Section 3.4 the nonincreasing, continuous function $I_k : (0, \infty] \overset{\text{onto}}{\longrightarrow} [\bar{c}_k, \infty)$ which, when restricted to $(0, U'_k(\bar{c}_k))$, is the (strictly decreasing) inverse of $U'_k : (\bar{c}_k, \infty) \overset{\text{onto}}{\longrightarrow} (0, U'_k(\bar{c}_k))$. On the interval $[U'(\bar{c}_k), \infty]$, $I_k(\cdot)$ is identically equal to $\bar{c}_k$.

Agent k uses the time-dependent utility function $e^{-\int_0^t \beta(u)\, du} U_k(c)$ in Problem 4.1, and the inverse of $e^{-\int_0^t \beta(u)\, du} U'_k(\cdot)$ is $y \mapsto I_k(y e^{\int_0^t \beta(u)\, du})$. Following (3.7.1), we define

$$\mathcal{X}_k(y) \overset{\triangle}{=} E \int_0^T H_0(t) I_k\left(y e^{\int_0^t \beta(u)\, du} H_0(t)\right)\, dt, \quad 0 < y < \infty. \qquad (4.5)$$

Lemma 4.3: *Under Condition 3.1(ii), the function $\mathcal{X}_k(\cdot)$ is finite, nonincreasing, and continuous on $(0, \infty)$. We define $\mathcal{X}_k(0+) \overset{\triangle}{=} \lim_{y \downarrow 0} \mathcal{X}_k(y)$, $\mathcal{X}_k(\infty) \overset{\triangle}{=} \lim_{y \to \infty} \mathcal{X}_k(y)$, and*

$$r_k \overset{\triangle}{=} \sup\{y > 0; \mathcal{X}_k(y) > \mathcal{X}_k(\infty)\}. \qquad (4.6)$$

Then

$$\mathcal{X}_k(\infty) = \bar{c}_k \cdot E \int_0^T H_0(t)\, dt, \qquad (4.7)$$

$r_k > 0$, and $\mathcal{X}_k$ restricted to $(0, r_k)$ is strictly decreasing; thus, this function has a continuous and strictly decreasing inverse $\mathcal{Y}_k : (\mathcal{X}_k(\infty), \infty) \overset{\text{onto}}{\longrightarrow} (0, r_k)$, which satifies

$$\mathcal{X}_k(\mathcal{Y}_k(x)) = x, \quad \forall x \in (\mathcal{X}_k(\infty), \infty). \qquad (4.8)$$

PROOF.   Condition 3.1(ii) implies $\mathcal{X}_k(y) < \infty$ for all $y \in (0, \infty)$. The other properties of $\mathcal{X}_k(\cdot)$ now follow from the arguments used to prove Lemma 3.6.2.   □

Remark 4.4: If r_k in (4.6) is finite, then $\mathcal{X}_k(\cdot)$ is identically equal to $\bar{c}_k \cdot E \int_0^T H_0(t) \, dt$ on $[r_k, \infty)$. But $I_k(r_k e^{\int_0^t \beta(u) \, du} H_0(t)) \geq \bar{c}_k$ for $0 \leq t \leq T$ almost surely, and so we must actually have

$$I_k \left(r_k e^{\int_0^t \beta(u) \, du} H_0(t) \right) = \bar{c}_k, \quad 0 \leq t \leq T, \tag{4.9}$$

almost surely. In other words,

$$r_k e^{\int_0^t \beta(u) \, du} H_0(t) \geq U_k'(\bar{c}_k), \quad 0 \leq t \leq T, \tag{4.10}$$

almost surely. If $r_k = \infty$, (4.9) and (4.10) still hold.

The omitted proof of the following theorem uses Theorem 3.6 and is otherwise a minor modification of the proofs of Theorem 3.6.3 and Corollary 3.6.5 (see also Remark 3.6.4). The statement of the theorem is close to that of Theorem 3.7.3.

Theorem 4.5: *Suppose that we have constructed a complete, standard financial market satisfying Condition 3.1. Under the strict feasibility condition (4.4), the unique optimal consumption/portfolio pair* $(\hat{c}_k, \hat{\pi}_k) \in \mathcal{A}_k$ *for Problem 4.1 and the corresponding wealth process* $\hat{X}_k(\cdot)$ *are given for* $0 \leq t \leq T$ *by*

$$\hat{c}_k(t) = I_k \left(y_k e^{\int_0^t \beta(u) \, du} H_0(t) \right), \tag{4.11}$$

$$\hat{X}_k(t) = \frac{1}{H_0(t)} E \left[\int_t^T H_0(s)(\hat{c}_k(s) - \epsilon_k(s)) \, ds \,\bigg|\, \mathcal{F}(t) \right], \tag{4.12}$$

$$\sigma'(t)\hat{\pi}_k(t) = \frac{\psi_k(t)}{H_0(t)} + \hat{X}_k(t)\theta(t), \tag{4.13}$$

where

$$y_k = \mathcal{Y}_k \left(E \int_0^T H_0(t)\epsilon_k(t) \, dt \right) \in (0, r_k) \tag{4.14}$$

and $\psi_k(\cdot)$ *is the integrand in the representation*

$$M_k(t) = \int_0^t \psi_k'(s) \, dW(s) \tag{4.15}$$

of the zero-mean P-martingale

$$M_k(t) = E \left[\int_0^T H_0(s)(\hat{c}_k(s) - \epsilon_k(s)) \, ds \,\bigg|\, \mathcal{F}(t) \right]. \tag{4.16}$$

4.5   Equilibrium and the Representative Agent

We are now in a position to state the properties of the complete standard financial market $\mathcal{M}$ we shall be seeking.

Definition 5.1: Let the endowment processes and utility functions $\{\epsilon_k, U_k\}_{k=1}^K$ and the discount rate $\beta(\cdot)$ of Section 2 be given. We say that a financial market $\mathcal{M}$ as described in Section 3 and satisfying Condition 3.1 is an *equilibrium market* (for the economic primitives $\{\epsilon_k, U_k\}_{k=1}^K, \beta$), if the following conditions hold.

(i) *Feasibility for the agents:*

$$E \int_0^T H_0(t)\epsilon_k(t)\, dt \geq \bar{c}_k \cdot E \int_0^T H_0(t)\, dt, \quad k = 1, \ldots, K. \quad (5.1)$$

(ii) *Clearing of the commodity market:*

$$\sum_{k=1}^K \hat{c}_k(t) = \epsilon(t), \quad 0 \leq t \leq T, \quad (5.2)$$

almost surely.

(iii) *Clearing of the stock markets:*

$$\sum_{k=1}^K \hat{\pi}_k(t) = \underset{\sim}{0}, \quad 0 \leq t \leq T, \quad (5.3)$$

almost surely, where $\underset{\sim}{0}$ is the origin in $\mathbb{R}^N$.

(iv) *Clearing of the money market:*

$$\sum_{k=1}^K (\hat{X}_k(t) - \hat{\pi}'_k(t)\underset{\sim}{1}) = 0, \quad 0 \leq t \leq T, \quad (5.4)$$

almost surely, where $\underset{\sim}{1}$ is the vector $(1, \ldots, 1)'$ in $\mathbb{R}^N$.

Here $\hat{c}_k(\cdot), \hat{\pi}_k(\cdot), \hat{X}_k(\cdot)$ are the unique optimal processes for Problem 4.1; these are given by (4.11)–(4.14) if the strict feasibility condition (4.4) holds for agent k, and by Convention 4.2 in the case of the nonstrict feasibility condition (4.3).

For the remainder of this section we shall focus on characterizing such an equilibrium market. Aided by this characterization, we shall establish in the next section the existence of an equilibrium market and examine the extent to which this equilbruim market is uniquely determined by the economic primitives $\{\epsilon_k, U_k\}_{k=1}^K$ and β.

We note immediately from (3.5) and (5.2) that a necessary condition for an equilibrium market to exist is that $\epsilon(t) \geq \bar{c}$, $0 \leq t \leq T$; in other words, *aggregate endowment must always dominate aggregate sub-*

sistence consumption. We have imposed a somewhat stronger assumption in Condition 2.1(iii).

We now provide a simple characterization of an equilibrium market.

Theorem 5.2: *If $\mathcal{M}$ is an equilibrium market, then*

$$\epsilon(t) = \sum_{k=1}^{K} I_k \left(\frac{1}{\lambda_k} e^{\int_0^t \beta(u)\,du} H_0(t) \right), \quad 0 \le t \le T, \tag{5.5}$$

where $\lambda_k \in [0, \infty), k = 1, \ldots, K$, satisfy the system of equations

$$E \int_0^T H_0(t) \left[I_k \left(\frac{1}{\lambda_k} e^{\int_0^t \beta(u)\,du} H_0(t) \right) - \epsilon_k(t) \right] dt = 0, \quad k = 1, \ldots, K. \tag{5.6}$$

(If $\lambda_k = 0$, we adopt the convention $I_k(\frac{1}{\lambda_k} e^{\int_0^t \beta(u)\,du} H_0(t)) = I_k(\infty) = \bar{c}_k$.) *Conversely, if $\mathcal{M}$ is a standard, complete financial market satisfying Condition 3.1, and there exists a vector $\Lambda = (\lambda_1, \ldots, \lambda_K) \in [0, \infty)^K$ satisfying (5.5) and (5.6), then $\mathcal{M}$ is an equilibrium market. In either case, the optimal consumption processes for the individual agents are given by*

$$\hat{c}_k(t) = I_k \left(\frac{1}{\lambda_k} e^{\int_0^t \beta(u)\,du} H_0(t) \right), \quad 0 \le t \le T, \quad k = 1, \ldots, K. \tag{5.7}$$

PROOF. First, let us assume that $\mathcal{M}$ is an equilibrium market. If the *strict* feasibility condition (4.4) holds for agent k, then this agent's optimal consumption process (4.11) is given by (5.7), with $\lambda_k \in (\frac{1}{r_k}, \infty)$ and $\mathcal{X}_k(\frac{1}{\lambda_k}) = E \int_0^T H_0(t)\epsilon_k(t)\,dt$ (see (4.14)); this last equation is (5.6). If the *nonstrict* feasibility condition (4.3) holds for agent k, then (4.6) shows that (5.6) is equivalent to $\mathcal{X}_k(\frac{1}{\lambda_k}) = \mathcal{X}_k(\infty)$, and this equation is solved by any $\lambda_k \in [0, \frac{1}{r_k}]$; with such a choice of λ_k, we see from Convention 4.2 and Remark 4.4 that (5.7) holds, where now $\hat{c}_k(t) \equiv \bar{c}_k$. Summing (5.7) over k and using the commodity market clearing condition (5.2), we obtain (5.5).

For the second part of the theorem, we assume that $\mathcal{M}$ is a standard, complete financial market satisfying Condition 3.1, and that there exists a vector $\Lambda = (\lambda_1, \ldots, \lambda_k) \in [0, \infty)^K$ satisfying (5.5) and (5.6). Since $I_k(\frac{1}{\lambda_k} e^{\int_0^T \beta(u)\,du} H_0(t)) \ge \bar{c}_k, 0 \le t \le T$, (5.6) implies the "feasibility for agents" condition (5.1). Under this feasibility condition, we have just seen that the optimal consumption process for each agent k is given by (5.7), and now (5.5) implies the clearing of commodity markets (5.2). We sum (4.16) over k and use (5.2) to obtain $\sum_{k=1}^{K} M_k(t) \equiv 0$. From (4.15) we see that $\int_0^t \sum_{k=1}^{K} \psi_k'(s)\,dW(s) \equiv 0$, which implies $\sum_{k=1}^{K} \psi_k(t) \equiv 0$. Summing first (4.12) and then (4.13) over k, we conclude first that $\sum_{k=1}^{K} \hat{X}_k(t) \equiv 0$, and next that $\sigma'(t) \sum_{k=1}^{K} \hat{\pi}_k(t) \equiv 0$. Since $\sigma'(t)$ is nonsingular, we have the clearing of the stock markets (5.3) and then the clearing of the money market (5.4). $\qquad\square$

Remark 5.3: *If $\underset{\sim}{\Lambda} \in [0,\infty)^K$ satisfies either (5.5) or (5.6), then $\underset{\sim}{\Lambda}$ cannot be the zero vector.* For suppose it were; then the right-hand side of (5.5) would be $\sum_{k=1}^{K} \bar{c}_k = \bar{c}$, which is different from $\epsilon(t)$ because of Condition 2.1(iii), and (5.6) would become

$$E \int_0^T H_0(t)(\bar{c}_k - \epsilon_k(t))\, dt = 0, \quad k = 1, \ldots, K,$$

so that summing up over k we would again obtain a contradiction to Condition 2.1(iii). To simplify notation, we define

$$^*[0,\infty)^K = [0,\infty)^K \backslash \{\underset{\sim}{0}\} \tag{5.8}$$

to be the K-dimensional nonnegative orthant with the origin $\underset{\sim}{0} = (0, \ldots, 0)$ removed.

Theorem 5.2 reduces the search for an equilibrium market to *the search for a vector $\underset{\sim}{\Lambda} \in^* [0,\infty)^K$, and for a market with state price density $H_0(\cdot)$, so that (5.5) and (5.6) are satisfied.* We can further simplify the search by inverting (5.5), writing the sought $H_0(\cdot)$ as a function of the given aggregate endowment process $\epsilon(\cdot)$; cf. (5.16) below.

Let $\underset{\sim}{\Lambda} \in^* [0,\infty)^K$ be given. For $k = 1, \ldots, K$, the function $y \mapsto I_k(\frac{y}{\lambda_k})$ is identically equal to $\bar{c}_k$ if $\lambda_k = 0$; but if $\lambda_k > 0$, it is continuous on $(0,\infty)$ as well as strictly decreasing on $(0, \lambda_k U_k'(\bar{c}_k)]$, and maps $(0, \lambda_k U_k'(\bar{c}_k)]$ onto $[\bar{c}_k, \infty)$. We set

$$m(\underset{\sim}{\Lambda}) \overset{\triangle}{=} \max_{\{k; \lambda_k > 0\}} (\lambda_k U_k'(\bar{c}_k)), \tag{5.9}$$

which is strictly positive since $\underset{\sim}{\Lambda}$ is not the zero vector. The function

$$I(y; \underset{\sim}{\Lambda}) \overset{\triangle}{=} \sum_{k=1}^{K} I_k \left(\frac{y}{\lambda_k}\right), \quad 0 < y \leq \infty, \tag{5.10}$$

is continuous on $(0,\infty]$, is strictly decreasing on $(0, m(\underset{\sim}{\Lambda})]$, and maps $(0, m(\underset{\sim}{\Lambda})]$ onto $[\bar{c}, \infty)$ (recall that $\bar{c} \overset{\triangle}{=} \sum_{k=1}^{K} \bar{c}_k$). For $\underset{\sim}{\Lambda} \in^* [0,\infty)^k$, we define

$$\mathcal{H}(\cdot; \underset{\sim}{\Lambda}) : [\bar{c}, \infty) \xrightarrow{\text{onto}} (0, m(\underset{\sim}{\Lambda})] \tag{5.11}$$

to be the (continuous, strictly decreasing) inverse of

$$I(\cdot; \underset{\sim}{\Lambda}) : (0, m(\underset{\sim}{\Lambda})] \xrightarrow{\text{onto}} [\bar{c}, \infty). \tag{5.12}$$

We note that $I(y; \underset{\sim}{\Lambda})$ is also defined for $y \geq m(\underset{\sim}{\Lambda})$, and in fact,

$$I(y; \underset{\sim}{\Lambda}) = \bar{c}, \quad \forall y \in [m(\underset{\sim}{\Lambda}), \infty]. \tag{5.13}$$

We have

$$\mathcal{H}(I(y; \underset{\sim}{\Lambda}); \underset{\sim}{\Lambda}) = y, \quad \forall y \in (0, m(\underset{\sim}{\Lambda})], \tag{5.14}$$

$$I(\mathcal{H}(c; \underset{\sim}{\Lambda}); \underset{\sim}{\Lambda}) = c, \quad \forall c \in [\bar{c}, \infty). \tag{5.15}$$

Because of Condition 2.1(iii), $\epsilon(t)$ is in the domain of $\mathcal{H}(\cdot; \underset{\sim}{\Lambda})$, and we can invert (5.5) to obtain

$$H_0(t) = e^{-\int_0^t \beta(u)\,du} \mathcal{H}(\epsilon(t); \underset{\sim}{\Lambda}), \quad 0 \le t \le T. \tag{5.16}$$

This leads to the following corollary of Theorem 5.2.

Corollary 5.4: *A standard, complete financial market $\mathcal{M}$ satisfying Condition 3.1 is an equilibrium market if and only if its state price density process $H_0(\cdot)$ is given by (5.16), where $\underset{\sim}{\Lambda} = (\lambda_1, \ldots, \lambda_K) \in {}^*[0, \infty)^K$ is a solution to the system of equations*

$$E \int_0^T e^{-\int_0^t \beta(u)\,du} \mathcal{H}(\epsilon(t); \underset{\sim}{\Lambda}) \left[I_k \left(\frac{1}{\lambda_k} \mathcal{H}(\epsilon(t); \underset{\sim}{\Lambda}) \right) - \epsilon_k(t) \right] dt = 0,$$

$$k = 1, \ldots, K. \tag{5.17}$$

In this case, the optimal consumption process for the kth agent is

$$\hat{c}_k(t) = I_k \left(\frac{1}{\lambda_k} \mathcal{H}(\epsilon(t); \underset{\sim}{\Lambda}) \right), \quad 0 \le t \le T. \tag{5.18}$$

In Section 6 we shall establish the existence of a solution $\underset{\sim}{\Lambda}$ to (5.17). Although (5.17) does not have a unique solution, the optimal consumption processes $\hat{c}_1(\cdot), \ldots, \hat{c}_K(\cdot)$ given by (5.18) are uniquely determined. The following lemma examines one type of nonuniqueness that can occur in the solution of (5.17).

Lemma 5.5: *Define*

$$\mathcal{T} = \{k \in \{1, \ldots, K\}; \quad \epsilon_k(t) = \bar{c}_k, \quad 0 \le t \le T, a.s.\},$$
$$\mathcal{T}^c = \{1, \ldots, K\} \backslash \mathcal{T}. \tag{5.19}$$

Because of Condition 2.1(iii), $\mathcal{T}^c$ is nonempty.
Suppose $\underset{\sim}{\Lambda} \in {}^*[0, \infty)^K$ satisfies (5.17). Then $k \in \mathcal{T}$ if and only if

$$\lambda_k \le \frac{\mathcal{H}(\epsilon(t); \underset{\sim}{\Lambda})}{U_k'(\bar{c}_k)}, \quad 0 \le t \le T, \tag{5.20}$$

almost surely. Define $\underset{\sim}{\Lambda}^* = (\lambda_1^*, \ldots, \lambda_K^*)$ by

$$\lambda_k^* \overset{\triangle}{=} \begin{cases} P\text{-ess inf} \left(\min_{0 \le t \le T} \frac{\mathcal{H}(\epsilon(t); \underset{\sim}{\Lambda})}{U_k'(\bar{c}_k)} \right), & \text{if} \quad k \in \mathcal{T}, \\ \lambda_k, & \text{if} \quad k \in \mathcal{T}^c, \end{cases} \tag{5.21}$$

and note that $\lambda_k^* \ge \lambda_k$. Then $m(\underset{\sim}{\Lambda}) = m(\underset{\sim}{\Lambda}^*)$,

$$\mathcal{H}(\epsilon(t); \underset{\sim}{\Lambda}) = \mathcal{H}(\epsilon(t); \underset{\sim}{\Lambda}^*), \quad 0 \le t \le T, \tag{5.22}$$

$$I_k \left(\frac{1}{\lambda_k} \mathcal{H}(\epsilon(t); \underset{\sim}{\Lambda}) \right) = I_k \left(\frac{1}{\lambda_k^*} \mathcal{H}(\epsilon(t); \underset{\sim}{\Lambda}^*) \right),$$

$$0 \le t \le T, \quad k = 1, \ldots, K, \tag{5.23}$$

and $\underset{\sim}{\Lambda}^*$ also satisfies (5.17).

PROOF.   From (5.17) and the fact that $I_k(y) \geq \bar{c}_k$ for all $y \in (0, \infty]$, we see that $k \in \mathcal{T}$ if and only if $I_k(\frac{1}{\lambda_k}\mathcal{H}(\epsilon(t); \Lambda)) = \bar{c}_k, 0 \leq t \leq T$, almost surely. This equality is equivalent to (5.20). Using the convention $0 \cdot (\pm\infty) = 0$, we have

$$\lambda_k^* U_k'(\bar{c}_k) \leq \mathcal{H}(\epsilon(t); \Lambda), \quad 0 \leq t \leq T, \quad \forall k \in \mathcal{T}, \tag{5.20'}$$

almost surely, and since $(0, m(\Lambda)]$ is the range of $\mathcal{H}(\cdot; \Lambda)$, this implies

$$\lambda_k U_k'(\bar{c}_k) \leq \lambda_k^* U_k'(\bar{c}_k) \leq m(\Lambda) = \max\{\lambda_j U_j'(\bar{c}_j); \lambda_j > 0\}.$$

It is apparent that $m(\Lambda^*) = m(\Lambda)$.

We compare $I(y; \Lambda) = \sum_{k=1}^K I_k(\frac{y}{\lambda_k})$ with $I(y; \Lambda^*) = \sum_{k=\mathcal{T}} I_k(\frac{y}{\lambda_k^*}) + \sum_{k \in \mathcal{T}^c} I_k(\frac{y}{\lambda_k})$. These two expressions agree if and only if $\lambda_k^* U_k'(\bar{c}_k) \leq y, \forall k \in \mathcal{T}$. But λ_k^* is defined so that $(5.20)'$ holds, and from (5.15) we obtain

$$\epsilon(t) = I(\mathcal{H}(\epsilon(t); \Lambda); \Lambda) = I(\mathcal{H}(\epsilon(t); \Lambda); \Lambda^*).$$

Applying $\mathcal{H}(\cdot; \Lambda^*)$ to both sides and using (5.14), we obtain (5.22).

If $k \in \mathcal{T}^c$, (5.23) follows from (5.22) because $\lambda_k^* = \lambda_k$. If $k \in \mathcal{T}$, we have

$$\frac{1}{\lambda_k}\mathcal{H}(\epsilon(t); \Lambda) \geq \frac{1}{\lambda_k^*}\mathcal{H}(\epsilon(t); \Lambda) \geq U_k'(\bar{c}_k),$$

and (5.23) holds with both sides identically equal to $\bar{c}_k$. Equations (5.22) and (5.23) imply that Λ^* is a solution of the system (5.17). $\square$

The remainder of this section develops properties of the function $\mathcal{H}(\cdot; \Lambda)$ of (5.11). We shall see, in particular, that $\mathcal{H}(\cdot; \Lambda)$ is the derivative of the function

$$U(c; \Lambda) \stackrel{\triangle}{=} \max_{\substack{c_1 \geq \bar{c}_1, \ldots, c_K \geq \bar{c}_K \\ c_1 + \cdots + c_K = c}} \sum_{k=1}^K \lambda_k U_k(c_k), \quad c \in \mathbb{R}. \tag{5.24}$$

(We use here and elsewhere the convention $0 \cdot (\pm\infty) = 0$.)

The next theorem shows that $U(\cdot; \Lambda)$ is itself a utility function. It plays the role of the utility function for a *"representative agent"* who assigns "weights" $\lambda_1, \ldots, \lambda_k$ to the various agents and, with proper choice of $\Lambda = (\lambda_1, \ldots, \lambda_k)$, has optimal consumption equal to the aggregate endowment. The weights $\lambda_1, \ldots, \lambda_k$ correspond to the "relative importance" of the individual agents in the equilibrium market. The maximizing values $c_1, \ldots, c_k$ in (5.24) give the optimal consumptions of the individual agents when the aggregate endowment is c.

The reader may wish to skip on first reading the (long and technical) proof of Theorem 5.6.

Theorem 5.6: Let $\Lambda \in^* [0, \infty)^K$ be given. Then the function $U(\cdot; \Lambda)$ of (5.24) is a utility function as set forth in Definition 3.4.1, and $\bar{c} \stackrel{\triangle}{=} \sum_{k=1}^K \bar{c}_k$ satisfies

$$\bar{c} = \inf\{c \in \mathbb{R}; \quad U(c; \Lambda) > -\infty\}. \tag{5.25}$$

Moreover, $U(\cdot; \underset{\sim}{\Lambda})$ is continuously differentiable on $(\bar{c}, \infty)$ with

$$U'(c; \underset{\sim}{\Lambda}) = \mathcal{H}(c; \underset{\sim}{\Lambda}), \quad c > \bar{c}. \tag{5.26}$$

For $k = 1, \ldots, K$, define

$$\alpha_k \overset{\triangle}{=} I(\lambda_k U'_k(\bar{c}_k); \underset{\sim}{\Lambda})$$

so that

$$\min_{1 \le k \le K} \alpha_k = I(m(\underset{\sim}{\Lambda}); \underset{\sim}{\Lambda}) = \bar{c},$$

and set

$$\mathcal{D} \overset{\triangle}{=} \{\alpha_k\}_{k=1}^K \backslash \{\bar{c}\}$$

so that $\mathcal{D} \subset (\bar{c}, \infty)$. The derivatives $U''(\cdot; \underset{\sim}{\Lambda})$ and $U'''(\cdot; \underset{\sim}{\Lambda})$ exist and are continuous on $(\bar{c}, \infty) \backslash \mathcal{D}$, and their one-sided limits exist and are finite at all points in $\mathcal{D}$. Moreover,

$$U''(\alpha+; \underset{\sim}{\Lambda}) - U''(\alpha-; \underset{\sim}{\Lambda}) \ge 0, \quad \forall \alpha \in \mathcal{D}. \tag{5.27}$$

PROOF. Using the convention that the maximum over the empty set is $-\infty$, we see that

$$U(c; \underset{\sim}{\Lambda}) = -\infty \quad \forall c \in (-\infty, \bar{c}). \tag{5.28}$$

If $c > \bar{c}$, then the numbers $c_k \overset{\triangle}{=} \bar{c}_k + \frac{1}{K}(c - \bar{c})$ satisfy $c_k > \bar{c}_k$ and $c_1 + \cdots + c_K = c$. From the definition (5.24) of $U(c; \underset{\sim}{\Lambda})$, we have $U(c; \underset{\sim}{\Lambda}) \ge \sum_{k=1}^K \lambda_k U_k(c_k) > -\infty$. Relation (5.25) follows from this inequality and (5.28).

Now let $c \ge \bar{c}$ be given. For each k, set

$$\hat{c}_k \overset{\triangle}{=} I_k \left(\frac{1}{\lambda_k} \mathcal{H}(c; \underset{\sim}{\Lambda}) \right). \tag{5.29}$$

Then $\sum_{k=1}^K \hat{c}_k = I(\mathcal{H}(c; \underset{\sim}{\Lambda}); \underset{\sim}{\Lambda}) = c$, from (5.15). Furthermore, $\hat{c}_k \ge \bar{c}_k$ for each k. If $\hat{c}_k > \bar{c}_k$ for some k, then $\lambda_k > 0$ and $U'_k(\hat{c}_k) = \frac{1}{\lambda_k} \mathcal{H}(c; \underset{\sim}{\Lambda})$, whereas if $\hat{c}_k = \bar{c}_k$, we know only that $U'_k(\hat{c}_k) \le \frac{1}{\lambda_k} \mathcal{H}(c; \underset{\sim}{\Lambda})$. Suppose $c_1 \ge \bar{c}_1, \ldots, c_K \ge \bar{c}_K$ and $c_1 + \cdots + c_K = c$. We have from the concavity of each $U_k(\cdot)$ that

$$\sum_{k=1}^K \lambda_k U_k(c_k) \le \sum_{k=1}^K \lambda_k [U_k(\hat{c}_k) + (c_k - \hat{c}_k) U'_k(\hat{c}_k)]$$

$$= \sum_{k=1}^K \lambda_k U_k(\hat{c}_k) + \mathcal{H}(c; \underset{\sim}{\Lambda}) \cdot \sum_{\{k; \hat{c}_k > \bar{c}_k\}} (c_k - \hat{c}_k)$$

$$+ \sum_{\{k; \hat{c}_k = \bar{c}_k\}} \lambda_k U'_k(\hat{c}_k)(c_k - \hat{c}_k)$$

$$\leq \sum_{k=1}^{K} \lambda_k U_k(\hat{c}_k) + \mathcal{H}(c; \underset{\sim}{\Lambda}) \sum_{k=1}^{K} (c_k - \hat{c}_k) = \sum_{k=1}^{K} \lambda_k U_k(\hat{c}_k).$$

In other words, $\hat{c}_1, \ldots, \hat{c}_K$ attain the maximum in (5.24), and we obtain the representation

$$U(c; \underset{\sim}{\Lambda}) = \sum_{k=1}^{K} \lambda_k U_k \left(I_k \left(\frac{1}{\lambda_k} \mathcal{H}(c; \underset{\sim}{\Lambda}) \right) \right), \quad c \in (\bar{c}, \infty). \tag{5.30}$$

We develop the differentiability properties of $I_k(\cdot), I(\cdot; \underset{\sim}{\Lambda})$, and $\mathcal{H}(\cdot; \underset{\sim}{\Lambda})$. If $U_k'(\bar{c}_k) = \infty$, then $I_k(\cdot)$ is of class C^2 on $(0, \infty)$, with $I_k'(y) < 0$ for all $y \in (0, \infty)$. However, if $U_k'(\bar{c}_k) < \infty$, then $I_k(y) = \bar{c}_k$ for $y \geq U_k'(\bar{c}_k)$, so $I_k'(\cdot) = I_k''(\cdot) = 0$ on $(U_k'(\bar{c}_k), \infty)$. The relation $I_k(U_k'(c)) = c$ for $c > \bar{c}_k$ implies

$$I_k'(U_k'(c)) = \frac{1}{U_k''(c)}, \quad I_k''(U_k'(c)) = -\frac{U_k'''(c)}{(U_k''(c))^3} \quad \text{for } c > \bar{c}_k,$$

and Remark 2.2 and Condition 2.1(i) guarantee that when $U_k'(\bar{c}_k) < \infty$, we have

$$-\infty < I_k'(U_k'(\bar{c}_k)-) \leq 0 = I_k'(U_k'(\bar{c}_k)+), \tag{5.31}$$

$$-\infty < I_k''(U_k'(\bar{c}_k)-) < \infty, \quad I_k''(U_k'(\bar{c}_k)+) = 0. \tag{5.32}$$

In particular, each $I_k(\cdot)$ is piecewise C^2 on $(0, \infty)$. The function $I(y; \underset{\sim}{\Lambda}) = \sum_{k=1}^{K} I_k(\frac{y}{\lambda_k})$ is continuous on $(0, \infty)$, and the derivative formulas

$$I'(y; \underset{\sim}{\Lambda}) = \sum_{\{k; \lambda_k > 0\}} \frac{1}{\lambda_k} I_k'\left(\frac{y}{\lambda_k}\right), \quad I''(y; \underset{\sim}{\Lambda}) = \sum_{\{k; \lambda_k > 0\}} \frac{1}{\lambda_k^2} I_k''\left(\frac{y}{\lambda_k}\right)$$

show that $I(\cdot; \underset{\sim}{\Lambda})$ is piecewise C^2 on $(0, \infty)$. The points of possible discontinuity of $I'(\cdot; \underset{\sim}{\Lambda})$ and $I''(\cdot; \underset{\sim}{\Lambda})$ are $\lambda_1 U_1'(\bar{c}_1), \ldots, \lambda_K U_K'(\bar{c}_K)$. At any of these points that is also contained in $(0, m(\underset{\sim}{\Lambda}))$, (5.31) implies

$$-\infty < I'(\lambda_k U_k'(\bar{c}_k)-; \underset{\sim}{\Lambda}) \leq I'(\lambda_k U_k'(\bar{c}_k)+; \underset{\sim}{\Lambda}) < 0. \tag{5.33}$$

In particular, $I'(\cdot; \underset{\sim}{\Lambda})$ is bounded below and bounded away from zero on each closed, bounded subinterval of $(0, m(\underset{\sim}{\Lambda}))$.

Because $\mathcal{H}(I(y; \underset{\sim}{\Lambda}); \underset{\sim}{\Lambda}) = y$ for $0 < y < m(\underset{\sim}{\Lambda})$,

$$\mathcal{H}'(I(y; \underset{\sim}{\Lambda}); \underset{\sim}{\Lambda}) = \frac{1}{I'(y; \underset{\sim}{\Lambda})}, \quad y \in (0, m(\Lambda)) \backslash \{\lambda_k U_k'(\bar{c}_k)\}_{k=1}^{K}. \tag{5.34}$$

This shows that $\mathcal{H}$ is piecewise C^1. From (5.33), we also have the inequality

$$\mathcal{H}'(\alpha+; \underset{\sim}{\Lambda}) \geq \mathcal{H}'(\alpha-; \underset{\sim}{\Lambda}), \quad \forall \alpha \in \mathcal{D}. \tag{5.35}$$

Differentiation of (5.34) leads to

$$\mathcal{H}''(I(y; \underset{\sim}{\Lambda}); \underset{\sim}{\Lambda}) = -\frac{I''(y; \underset{\sim}{\Lambda})}{I'(y; \underset{\sim}{\Lambda})^3}, \quad y \in (0, m(\Lambda)) \backslash \{\lambda_k U_k'(\bar{c}_k)\}_{k=1}^{K}.$$

From (5.32) and (5.33), we see that $\mathcal{H}''(\alpha\pm;\underline{\Lambda})$ exists for all $\alpha \in \mathcal{D}$; i.e., $\mathcal{H}''(\cdot;\underline{\Lambda})$ is piecewise C^2 on $(0, m(\underline{\Lambda}))$.

Let $\alpha_k \in \mathcal{D}$ be given. Since $\mathcal{H}(\cdot;\underline{\Lambda})$ is strictly decreasing on $(\bar{c},\infty)$ and $\mathcal{H}(\alpha_k;\underline{\Lambda}) = \lambda_k U'_k(\bar{c}_k)$, we have

$$\frac{1}{\lambda_k}\mathcal{H}(c;\underline{\Lambda}) > U'_k(\bar{c}_k), \quad \forall c \in (\bar{c}, \alpha_k), \tag{5.36}$$

$$0 < \frac{1}{\lambda_k}\mathcal{H}(c;\underline{\Lambda}) < U'_k(\bar{c}_k), \quad \forall c \in (\alpha_k, \infty). \tag{5.37}$$

For $c \in (\alpha_k, \infty)\backslash\mathcal{D}$, we compute

$$\frac{d}{dc}\left[\lambda_k U_k\left(I_k\left(\frac{1}{\lambda_k}\mathcal{H}(c;\underline{\Lambda})\right)\right)\right] = U'_k\left(I_k\left(\frac{1}{\lambda_k}\mathcal{H}(c;\underline{\Lambda})\right)\right)$$

$$\cdot I'_k\left(\frac{1}{\lambda_k}\mathcal{H}(c;\underline{\Lambda})\right)\cdot \mathcal{H}'(c;\underline{\Lambda})$$

$$= \frac{1}{\lambda_k}\mathcal{H}(c;\underline{\Lambda})\cdot I'_k\left(\frac{1}{\lambda_k}\mathcal{H}(c;\underline{\Lambda})\right)\cdot \mathcal{H}'(c;\underline{\Lambda})$$

$$= \mathcal{H}(c;\underline{\Lambda})\cdot \frac{d}{dc}\left[I_k\left(\frac{1}{\lambda_k}\mathcal{H}(c;\underline{\Lambda})\right)\right],$$

where we have used (5.37) and the fact that $U'_k : (\bar{c}_k,\infty) \xrightarrow{\text{onto}} (0, U'_k(\bar{c}_k))$ is the inverse of $I_k : (0, U'_k(\bar{c}_k)) \xrightarrow{\text{onto}} (\bar{c}_k,\infty)$. For $c \in (\bar{c}, \alpha_k)\backslash\mathcal{D}$, we have from (5.36) that $I_k(\frac{1}{\lambda_k}\mathcal{H}(c;\underline{\Lambda})) = \bar{c}_k$ and $I'_k(\frac{1}{\lambda_k}\mathcal{H}(c;\underline{\Lambda})) = 0$, so once again

$$\frac{d}{dc}\left[\lambda_k U_k\left(I_k\left(\frac{1}{\lambda_k}\mathcal{H}(c;\underline{\Lambda})\right)\right)\right] = \mathcal{H}(c;\underline{\Lambda})\cdot \frac{d}{dc}\left[I_k\left(\frac{1}{\lambda_k}\mathcal{H}(c;\underline{\Lambda})\right)\right], \tag{5.38}$$

but now with both sides equal to zero.

We have established (5.38) for all $c \in (\bar{c},\infty)\backslash\{\alpha_k\}$, provided that $\alpha_k \neq \bar{c}$. If $\alpha_k = \bar{c}$, then

$$\lambda_k U'_k(\bar{c}_k) = \mathcal{H}(\alpha_k;\Lambda) = m(\underline{\Lambda}) > 0. \tag{5.39}$$

Therefore, $\lambda_k > 0$ and (5.39) implies (5.37), which leads to (5.38) as before.

We may now sum (5.38) over $k \in \{1, \ldots, K\}$ and use the representation (5.30) to obtain

$$U'(c;\underline{\Lambda}) = \mathcal{H}(c;\Lambda)\cdot \frac{d}{dc}I(\mathcal{H}(c;\underline{\Lambda});\underline{\Lambda}) = \mathcal{H}(c;\underline{\Lambda}), \quad \forall c \in (\bar{c},\infty)\backslash\mathcal{D}.$$

This implies

$$U(c;\underline{\Lambda}) = U(\bar{c}+1,\underline{\Lambda}) + \int_{\bar{c}+1}^{c} \mathcal{H}(\eta;\underline{\Lambda})\,d\eta, \quad \forall c \in (\bar{c},\infty),$$

and since $\mathcal{H}(\cdot;\underline{\Lambda})$ is continuous, differentiation yields 5.26.

Because $\mathcal{H}(\cdot;\underline{\Lambda})$ is continuous, positive, and strictly decreasing on $(\bar{c},\infty)$, the function $U(\cdot;\underline{\Lambda})$ is strictly concave and increasing on this set. Moreover,

$\mathcal{H}(\cdot; \Lambda)$ is continuous from the right at $\bar{c}$, and the representation (5.30), combined with the right continuity (upper semicontinuity) of each $U_k(\cdot)$ at $\bar{c}_k$, establishes the right continuity (upper semicontinuity) of $U(\cdot; \Lambda)$ at $\bar{c}$. Finally,

$$\lim_{c \to \infty} U'(c; \Lambda) = \lim_{c \to \infty} \mathcal{H}(c; \Lambda) = 0.$$

This concludes the proof that $U(\cdot; \Lambda)$ has all the properties required of utility functions by Definition 3.4.1.

The piecewise continuity of $U''(\cdot; \Lambda)$ and $U'''(\cdot; \Lambda)$ on $(\bar{c}, \infty)$ follows from the properties proved for $\mathcal{H}'(\cdot; \Lambda)$ and $\mathcal{H}''(\cdot; \Lambda)$. Inequality (5.35) is (5.27).   □

4.6   Existence and Uniqueness of Equilibrium

In light of Corollary 5.4, the key remaining step in the construction of an equilibrium market is the solution of the system of equations (5.17) for $\Lambda \in {}^*[0, \infty)^K$. In contrast to the original problem of determining equilibrium price processes, the problem at hand is *finite-dimensional*.

Lemma 5.5 shows that *we should not expect the system of equations (5.17) to have a unique solution* $\Lambda \in {}^*[0, \infty)^K$, since $\mathcal{H}(\epsilon(t); \Lambda)$ and $\mathcal{H}(\epsilon(t); \Lambda^*)$ can agree even though $\Lambda \neq \Lambda^*$. However, the equality (5.23) guarantees that both Λ and Λ^* result in the same optimal consumption processes for the individual agents, given by (5.18).

There is, however, an additional kind of nonuniqueness in (5.17). The representation (5.26) of the function $\mathcal{H}$, along with the definition (5.24) of the "representative agent" utility function, allows us to deduce the *positive homogeneity* properties for $c \geq \bar{c}, \eta > 0$, and $\Lambda \in {}^*[0, \infty)^K$:

$$U(c; \eta\Lambda) = \eta U(c; \Lambda), \tag{6.1}$$

$$\mathcal{H}(c; \eta\Lambda) = \eta \mathcal{H}(c; \Lambda). \tag{6.2}$$

It follows from (6.2) that if a vector $\Lambda \in {}^*[0, \infty)^K$ satisfies the equations (5.17), then so does every other vector $\eta\Lambda$ with $\eta \in (0, \infty)$, on the same ray through the origin; for all such vectors, the optimal consumption processes are the same:

$$\hat{c}_k(t) = I_k \left(\frac{1}{\lambda_k} \mathcal{H}(\epsilon(t); \Lambda) \right)$$

$$= I_k \left(\frac{1}{\eta\lambda_k} \mathcal{H}(\epsilon(t); \eta\Lambda) \right), \quad 0 \leq t \leq T, \quad k = 1, \ldots, K. \tag{6.3}$$

The first result of this section guarantees the existence of a solution to (5.17) and provides a condition under which the equilibrium optimal consumption processes of the individual agents are uniquely determined.

Theorem 6.1:   *There exists a vector $\underline{\Lambda} = (\lambda_1, \ldots, \lambda_k) \in {}^*[0, \infty)^K$ satisfying the system of equations (5.17). Suppose, moreover, that for each agent k, the Arrow–Pratt "index of risk-aversion" is less than or equal to one, namely*

$$-\frac{cU_k''(c)}{U_k'(c)} \leq 1 \quad \text{for all} \quad c \in (\bar{c}_k, \infty), \quad k = 1, \ldots, K, \qquad (6.4)$$

(see Remark 3.4.4), and let $\underline{M} = (\mu_1, \ldots, \mu_K) \in {}^*[0, \infty)^K$ be any other solution of (5.17); then for some positive constant η, we have almost surely

$$\eta \mathcal{H}(\epsilon(t); \underline{\Lambda}) = \mathcal{H}(\epsilon(t); \underline{M}), \quad 0 \leq t \leq T, \qquad (6.5)$$

and

$$\hat{c}_k(t) = I_k \left(\frac{1}{\lambda_k} \mathcal{H}(\epsilon(t); \underline{\Lambda}) \right)$$

$$= I_k \left(\frac{1}{\mu_k} \mathcal{H}(\epsilon(t); \underline{M}) \right), \quad 0 \leq t \leq T, \quad k = 1, \ldots, K. \qquad (6.6)$$

PROOF.   We first establish *existence*. Let $\{\underline{e}_1, \ldots, \underline{e}_K\}$ be the standard basis of unit vectors in $\mathbb{R}^K$, and let $\mathbb{K} = \{1, \ldots, K\}$. For any nonempty set $\mathbb{B} \subset \mathbb{K}$, denote by

$$\mathcal{S}_{\mathbb{B}} \triangleq \left\{ \sum_{k \in \mathbb{B}} \lambda_k \underline{e}_k; \quad \lambda_k \geq 0 \quad \forall k \in \mathbb{B} \text{ and } \sum_{k \in \mathbb{B}} \lambda_k = 1 \right\} \subset {}^*[0, \infty)^K$$

the convex hull of $\{\underline{e}_k\}_{k \in \mathbb{B}}$. For every $k \in \mathbb{K}$, define $R_k : \mathcal{S}_{\mathbb{K}} \to \mathbb{R}$ by

$$R_k(\underline{\Lambda}) \triangleq E \int_0^T e^{-\int_0^t \beta(u)\,du} \mathcal{H}(\epsilon(t); \underline{\Lambda}) \left[I_k \left(\frac{1}{\lambda_k} \mathcal{H}(\epsilon(t); \underline{\Lambda}) \right) - \epsilon_k(t) \right] dt, \qquad (6.7)$$

where as usual, $I_k(\frac{1}{\lambda_k} \mathcal{H}(\epsilon(t); \underline{\Lambda})) \triangleq \bar{c}_k$ if $\lambda_k = 0$. The function R_k is continuous, and hence the set $F_k \triangleq \{\underline{\Lambda} \in \mathcal{S}_{\mathbb{K}}; R_k(\underline{\Lambda}) \geq 0\}$ is closed. Note from (5.15) that

$$\sum_{k \in \mathbb{K}} R_k(\underline{\Lambda}) = E \int_0^T e^{-\int_0^t \beta(u)\,du} \mathcal{H}(\epsilon(t); \underline{\Lambda})[I(\mathcal{H}(\epsilon(t); \underline{\Lambda}); \underline{\Lambda}) - \epsilon(t)]\,dt = 0$$

$$(6.8)$$

holds for every $\underline{\Lambda} \in {}^*[0, \infty)$. We claim that

$$\mathcal{S}_{\mathbb{B}} \subset \bigcup_{k \in \mathbb{B}} F_k, \quad \forall \mathbb{B} \subset \mathbb{K}, \quad \mathbb{B} \neq \emptyset. \qquad (6.9)$$

Indeed, suppose $\underline{\Lambda} \in \mathcal{S}_{\mathbb{B}}$ but $\underline{\Lambda} \notin \cup_{k \in \mathbb{B}} F_k$; then $R_k(\underline{\Lambda}) < 0$ for all $k \in \mathbb{B}$. Condition (2.1(ii)) implies $R_k(\underline{\Lambda}) \leq 0$ for all $k \in \mathbb{K} \backslash \mathbb{B}$, so that (6.8) is contradicted. From (6.9) and the lemma of Knaster–Kuratowski–Mazurkiewicz (1929) (e.g., Border (1985), p. 26), we conclude that $\cap_{k \in \mathbb{K}} F_k \neq \emptyset$. Let $\underline{\Lambda}$ be in this set. The definition of F_k, combined with (6.8), implies $R_k(\underline{\Lambda}) = 0$ for all $k \in \mathbb{K}$; hence $\underline{\Lambda}$ satisfies the system (5.17).

We next *assume the risk-aversion condition (6.4)* and characterize the set of all solutions of (5.17) in terms of a particular solution Λ. Let $\mathcal{T}$ and Λ^* be as in Lemma 5.5. Suppose that $M = (\mu_1, \ldots, \mu_K)$ is another solution of (5.17), and let $M^* = (\mu_1^*, \ldots, \mu_K^*)$ be defined by the analogue of (5.21):

$$\mu_k^* \triangleq \begin{cases} P\text{-ess inf} \left(\min_{0 \le t \le T} T \dfrac{\mathcal{H}(\epsilon(t); M)}{U_k'(\bar{c}_k)} \right), & \text{if } k \in \mathcal{T}, \\ \mu_k, & \text{if } k \in \mathcal{T}^c, \end{cases} \tag{6.10}$$

and note that $\mu_k^* \ge \mu_k$.

If $\lambda_k^* = 0$ for some k, then Lemma 5.5 implies $k \in \mathcal{T}$, and thus

$$P\text{-ess inf} \left(\min_{0 \le t \le T} \frac{\mathcal{H}(\epsilon(t); \Lambda)}{U_k'(\bar{c}_k)} \right) = 0.$$

But $\mathcal{H}(\epsilon(t); \Lambda)$ is bounded from below by $\mathcal{H}(\gamma_2; \Lambda) > 0$ (Condition 2.1(iii)), so we must have $U_k'(\bar{c}_k) = \infty$ and thus $\mu_k^* = 0$ from (4.6). Now define

$$\eta \triangleq \max \left\{ \frac{\mu_k^*}{\lambda_k^*}; k \in \mathbb{K}, \lambda_k^* \neq 0 \right\}. \tag{6.11}$$

Because $\lambda_k^* = 0$ implies $\mu_k^* = 0$, we have

$$\mu_k^* \le \eta \lambda_k^*, \quad \forall k \in \mathbb{K}. \tag{6.12}$$

Furthermore, there is an index $\bar{k} \in \mathbb{K}$ such that $\mu_{\bar{k}}^* = \eta \lambda_{\bar{k}}^* > 0$. From (6.12) we have

$$I(y; \eta \Lambda^*) = \sum_{k=1}^{K} I_k \left(\frac{y}{\eta \lambda_k^*} \right) \ge \sum_{k=1}^{K} I_k \left(\frac{y}{\mu_k^*} \right) = I(y; M^*), \quad y > 0,$$

and thus, almost surely,

$$\eta \mathcal{H}(\epsilon(t); \Lambda^*) \ge \mathcal{H}(\epsilon(t); M^*), \quad 0 \le t \le T,$$

or equivalently,

$$\frac{1}{\lambda_{\bar{k}}^*} \mathcal{H}(\epsilon(t); \Lambda^*) \ge \frac{1}{\mu_{\bar{k}}^*} \mathcal{H}(\epsilon(t); M^*), \quad 0 \le t \le T. \tag{6.13}$$

Consider the function $\varphi_k(y) = y I_k(y), 0 < y < U_k'(\bar{c}_k)$. With $c = I_k(y)$, we have the derivative formula

$$\varphi_k'(y) = I_k(y) + y I_k'(y) = c + \frac{U_k'(c)}{U_k''(c)}, \quad 0 < y < U_k'(\bar{c}_k),$$

and (6.4) implies that φ_k is nonincreasing. Because both Λ^* and M^* satisfy (5.17), we have, in the notation of (6.7),

$$0 = \frac{1}{\lambda_{\bar{k}}^*} R_{\bar{k}}(\Lambda^*)$$

$$= E \int_0^T e^{-\int_0^t \beta(u)\, du} \left[\varphi_{\bar{k}} \left(\frac{1}{\lambda_{\bar{k}}^*} \mathcal{H}(\epsilon(t); \Lambda^*) \right) - \frac{1}{\lambda_{\bar{k}}^*} \mathcal{H}(\epsilon(t); \Lambda^*) \epsilon_{\bar{k}}(t) \right] dt$$

$$\leq E \int_0^T e^{-\int_0^t \beta(u)\,du} \left[\varphi_{\bar{k}} \left(\frac{1}{\mu_{\bar{k}}^*} \mathcal{H}(\epsilon(t); \underline{M}^*) \right) - \frac{1}{\mu_{\bar{k}}^*} \mathcal{H}(\epsilon(t); \underline{M}^*) \epsilon_{\bar{k}}(t) \right] dt$$

$$= \frac{1}{\mu_{\bar{k}}^*} R_{\bar{k}}(\underline{M}^*) = 0.$$

This shows that equality must hold in (6.13) almost surely, which yields (6.5).

For any $k \in \mathbb{K}$, (6.5) and (6.12) imply

$$I_k \left(\frac{1}{\lambda_k^*} \mathcal{H}(\epsilon(t); \underline{\Lambda}) \right) = I_k \left(\frac{1}{\eta \lambda_k^*} \mathcal{H}(\epsilon(t); \underline{M}) \right)$$

$$\geq I_k \left(\frac{1}{\mu_k^*} \mathcal{H}(\epsilon(t); \underline{M}) \right), \quad 0 \leq t \leq T. \quad (6.14)$$

But (5.17) for both $\underline{\Lambda}^*$ and $\underline{M}^*$ and (6.5) gives

$$E \int_0^T e^{-\int_0^t \beta(u)\,du} \mathcal{H}(\epsilon(t); \underline{\Lambda}^*) I_k \left(\frac{1}{\lambda_k^*} \mathcal{H}(\epsilon(t); \underline{\Lambda}^*) \right) dt$$

$$= E \int_0^T e^{-\int_0^t \beta(u)\,du} \mathcal{H}(\epsilon(t); \underline{\Lambda}^*) \epsilon_k(t)\,dt$$

$$= \frac{1}{\eta} E \int_0^T e^{-\int_0^t \beta(u)\,du} \mathcal{H}(\epsilon(t); \underline{M}^*) \epsilon_k(t)\,dt$$

$$= \frac{1}{\eta} E \int_0^T e^{-\int_0^t \beta(u)\,du} \mathcal{H}(\epsilon(t); \underline{M}^*) I_k \left(\frac{1}{\mu_k^*} \mathcal{H}(\epsilon(t); \underline{M}^*) \right) dt$$

$$= E \int_0^T e^{-\int_0^t \beta(u)\,du} \mathcal{H}(\epsilon(t); \underline{\Lambda}^*) I_k \left(\frac{1}{\mu_k^*} \mathcal{H}(\epsilon(t); \underline{M}^*) \right) dt.$$

Since $e^{-\int_0^t \beta(u)\,du} \mathcal{H}(\epsilon(t); \underline{\Lambda}^*)$ is always positive, equality must hold in (6.14) almost surely; this and Lemma 5.5 imply (6.6). $\square$

Remark 6.2: Although we do not use this observation, it is interesting to note that $\underline{\Lambda}^*$ and $\underline{M}^*$ in the proof of Theorem 6.1 are related via $\underline{M}^* = \eta \underline{\Lambda}^*$. In light of (6.12), we need only rule out the possibilty $\eta \lambda_k^* > \mu_k^*$. From (6.5) and the definitions (5.21), (6.10) of λ_k^*, μ_k^* for $k \in \mathcal{T}$, such a k must be in $\mathcal{T}^c$. Lemma 5.5 applied to $\underline{M}$ shows that $k \in \mathcal{T}^c$ if and only if the inequality

$$\mu_k^* \leq \frac{\mathcal{H}(\epsilon(t); \underline{M}^*)}{U_k'(\bar{c}_k)}, \quad 0 \leq t \leq T,$$

does not hold almost surely, i.e., $P \left[U_k'(\bar{c}_k) > \min_{0 \leq t \leq T} \frac{1}{\mu_k^*} \mathcal{H}(\epsilon(t); \underline{M}^*) \right] > 0$. If $\eta \lambda_k^* > \mu_k^*$, then the above inequality and (6.5), (5.22) imply

$$P \left[U_k'(\bar{c}_k) > \min_{0 \leq t \leq T} \frac{1}{\lambda_k^*} \mathcal{H}(\epsilon(t); \underline{\Lambda}^*) \right] > 0;$$

but for t and ω chosen such that $U_k'(\bar{c}_k) > \frac{1}{\lambda_k^*}\mathcal{H}(\epsilon(t,\omega);\underline{\Lambda}^*)$, we have the strict inequality

$$I_k\left(\frac{1}{\lambda_k^*}\mathcal{H}(\epsilon(t,\omega);\underline{\Lambda}^*)\right) > I_k\left(\frac{\eta}{\mu_k^*}\mathcal{H}(\epsilon(t,\omega);\underline{\Lambda}^*)\right) = I_k\left(\frac{1}{\mu_k^*}\mathcal{H}(\epsilon(t,\omega);\underline{M}^*)\right),$$

and (6.6) fails.

To complete the construction of an equilibrium market, we appeal to Theorem 6.1 and choose a vector $\underline{\Lambda} \in {}^*[0,\infty)^K$ satisfying the system of equations (5.17). The positive homogeneity properties of (6.1), (6.2) permit us to assume without loss of generality that

$$\mathcal{H}(\epsilon(0);\underline{\Lambda}) = 1. \tag{6.15}$$

Let us consider the process

$$\eta(t) \overset{\triangle}{=} \mathcal{H}(\epsilon(t);\underline{\Lambda}) = U'(\epsilon(t);\underline{\Lambda}), \quad 0 \le t \le T. \tag{6.16}$$

An application of Itô's rule for differences of convex functions of martingales (e.g., Karatzas and Shreve (1991), Theorems 3.6.22, 3.7.1 and Problem 3.6.24) yields

$$\eta(t) = 1 + \int_0^t \left[U''(\epsilon(s);\underline{\Lambda})\epsilon(s)\nu(s) + \frac{1}{2}U'''(\epsilon(s);\underline{\Lambda})\|\rho(s)\|^2\epsilon^2(s)\right]ds$$

$$+ \int_0^t U''(\epsilon(s);\underline{\Lambda})\epsilon(s)\,d\xi(s)$$

$$+ \sum_{k=1}^K [U''(\alpha_k+;\underline{\Lambda}) - U''(\alpha_k-;\underline{\Lambda})]L_t(\alpha_k)$$

$$+ \int_0^t U''(\epsilon(s);\underline{\Lambda})\epsilon(s)\rho'(s)\,dW(s), \quad 0 \le t \le T, \tag{6.17}$$

in conjunction with Theorem 5.6 and equation (2.2). Here $L_t(\alpha_k)$ is the local time at α_k of the semimartingale $\epsilon(\cdot)$, accumulated during $[0,t]$. On the other hand, if $H_0(\cdot)$ is the state price density in a standard, complete financial market $\mathcal{M}$, then the process

$$\zeta(t) \overset{\triangle}{=} H_0(t)e^{\int_0^t \beta(u)\,du}$$

$$= Z_0(t)\exp\left\{-A(t) + \int_0^t (\beta(u) - r(u))\,du\right\}, \quad 0 \le t \le T \tag{6.18}$$

satisfies the integral equation

$$\zeta(t) = 1 - \int_0^t \zeta(s)\,dA(s) + \int_0^t \zeta(s)(\beta(s) - r(s))\,ds$$

$$- \int_0^t \zeta(s)\theta'(s)\,dW(s), \quad 0 \le t \le T. \tag{6.19}$$

Corollary 5.4 asserts that $\mathcal{M}$ is an equilibrium market if and only if $\eta(\cdot) \equiv \zeta(\cdot)$, or equivalently, in light of the decompositions (6.17) and (6.19), if and only if

$$r(t) = \beta(t) - \frac{1}{U'(\epsilon(t); \Lambda)} \left[U''(\epsilon(t); \Lambda)\epsilon(t)\nu(t) \right.$$

$$\left. + \frac{1}{2} U'''(\epsilon(t); \Lambda)\|\rho(t)\|^2\epsilon^2(t) \right], \tag{6.20}$$

$$\theta(t) = -\frac{U''(\epsilon(t); \Lambda)}{U'(\epsilon(t); \Lambda)}\epsilon(t)\rho(t), \tag{6.21}$$

$$A(t) = -\int_0^t \frac{U''(\epsilon(s); \Lambda)}{U'(\epsilon(s); \Lambda)}\epsilon(s)\,d\xi(s)$$

$$- \sum_{k=1}^K \frac{U''(\alpha_k+; \Lambda) - U''(\alpha_k-; \Lambda)}{U'(\alpha_k; \Lambda)}L_t(\alpha_k) \tag{6.22}$$

for $0 \le t \le T$.

Theorem 6.3 (Existence of an equilibrium market): *Choose $\Lambda^* \in$ *$[0, \infty)^K$ to satisfy (5.17) and (6.15). Define $r(\cdot), \theta(\cdot)$, and $A(\cdot)$ by (6.20)–(6.22). Let $\sigma(t) = \{\sigma_{nj}(t)\}_{1 \le n,j \le N}$ be an arbitrary, nonsingular, matrix-valued process satisfying the integrability condition (vii) of Definition 1.1.3, and define*

$$b(t) \stackrel{\triangle}{=} r(t)1_N + \sigma(t)\theta(t). \tag{6.23}$$

Let the initial stock prices be any vector $S(0) = (S_1(0), \ldots, S_N(0))$ of positive constants. Then the market $\mathcal{M} = (r(\cdot), b(\cdot), \sigma(\cdot), S(0), A(\cdot))$ is an equilibrium market.

PROOF. Because of Corollary 5.4, we need only verify that $\mathcal{M}$ is a standard, complete financial market satisfying Condition 3.1. (Recall that we are omitting dividends from the markets in this chapter.) Condition 2.1(iii), the integrability of $\beta(\cdot)$, and the boundedness of $\nu(\cdot)$ and $\rho(\cdot)$ ensure the integrability of $r(\cdot)$ and the boundedness of $\theta(\cdot)$. Together with the integrability condition on $\sigma(\cdot)$, this guarantees that $\int_0^T \|b(t)\|dt < \infty$ a.s. Therefore, $\mathcal{M}$ is a standard, complete financial market (Definitions 1.1.3, 1.5.1 and Theorem 1.4.2, 1.6.6).

Because $U(\cdot; \Lambda)$ is piecewise C^3 on $(\bar{c}, \infty)$ (Theorem 5.6) and $U'(\cdot; \Lambda) = \mathcal{H}(\cdot; \Lambda)$ is strictly positive and continuous on $(\bar{c}, \infty)$, the bounds in Condition 2.1(iii) imply that $H_0(t) = e^{-\int_0^t \beta(u)\,du}\mathcal{H}(\epsilon(t); \Lambda)$ satisfies Condition 3.1(ii). To verify Condition 3.1(i), we consider

$$\frac{1}{S_0(t)} = e^{-\int_0^t r(u)\,du - A(t)}, \quad 0 \le t \le T.$$

Now, $\beta(\cdot)$ was assumed in Section 2 to be bounded from below, and all the other terms appearing in (6.20) are bounded, so $\max_{0 \le t \le T} e^{-\int_0^t r(u)\,du}$ is bounded from above. It remains to establish

$$E\left[\max_{0\leq t\leq T}\left(e^{-A(t)}\right)\right] < \infty. \tag{6.24}$$

From (6.17) we see that for some real constant C_1, we have

$$\sum_{k=1}^{K}[U''(\alpha_k+;\underline{\Lambda}) - U''(\alpha_k-;\underline{\Lambda})]L_T(\alpha_k)$$

$$\leq C_1 - \int_0^T U''(\epsilon(s);\underline{\Lambda})\epsilon(s)\rho'(s)\, dW(s)$$

almost surely. For $\alpha_k \notin (\bar{c},\infty)$ we have $L_t(\alpha_k) \equiv 0$, and inequality (5.27) for $\alpha_k \in (\bar{c},\infty)$ shows that

$$\max_{0\leq t\leq T}\sum_{k=1}^{K}[U''(\alpha_k+;\underline{\Lambda}) - U''(\alpha_k-;\underline{\Lambda})]L_t(\alpha_k)$$

$$\leq \sum_{k=1}^{K}[U''(\alpha_k+;\underline{\Lambda}) - U''(\alpha_k-;\underline{\Lambda})]L_T(\alpha_k).$$

These inequalities and equation (6.22) imply

$$\max_{0\leq t\leq T}(-A(t)) \leq C_2 + C_3\sum_{k=1}^{K}[U''(\alpha_k+;\underline{\Lambda}) - U''(\alpha_k-;\underline{\Lambda})]L_T(\alpha_k)$$

$$\leq C_4 - C_5\int_0^T U''(\epsilon(s);\Lambda)\epsilon(s)\rho'(s)\, dW(s),$$

for appropriate constants C_2, C_3, C_4, C_5. Condition 6.24 follows.   □

Theorem 6.4 (Uniqueness of the equilibrium market): *Assume that (6.4) holds. Then the equilibrium money market process $S_0(\cdot)$, the state price density process $H_0(\cdot)$, and the market price of risk process $\theta(\cdot)$, are uniquely determined, as are the optimal consumption processes $\hat{c}_1(\cdot),\ldots, \hat{c}_K(\cdot)$ of the individual agents.*

PROOF.   The uniqueness of $H_0(\cdot)$ follows from Corollary 5.4, Theorem 6.1, and the initial condition $H_0(0) = 1$. The uniqueness of $\hat{c}_1(\cdot),\ldots,\hat{c}_K(\cdot)$ also follows from Theorem 6.1. The semimartingale $\log H_0(t)$ can be decomposed uniquely as the sum of a finite-variation process $F(\cdot)$ plus a local martingale $M(\cdot)$. But

$$\log H_0(t) = -\int_0^t r(u)\, du - A(t) - \frac{1}{2}\int_0^t \|\theta(u)\|^2 du - \int_0^t \theta(u)\, dW(u),$$

and the equation $M(t) = -\int_0^t \theta(u)\, dW(u), 0 \leq t \leq T$ determines the process $\theta(\cdot)$. Knowing $\theta(\cdot)$ and using the equation

$$F(t) = - \int_0^t r(u)\, du - A(t) - \frac{1}{2}\int_0^t \|\theta(u)\|^2\, du,$$

we determine $-\int_0^t r(u)\, du - A(t)$, and hence $S_0(\cdot)$.   □

We should not expect the stock mean rate of return vector $b(\cdot)$ and the volatility matrix $\sigma(\cdot)$ to be determined by equilibrium conditions, *because of the possibility of replacing stocks by mutual funds.* Given a market $\mathcal{M} = (r(\cdot), b(\cdot), \sigma(\cdot), S(0), A(\cdot))$, we can form a mutual fund by specifying an initial value $\tilde{S}(0) > 0$ and the proportion $p_j(t)$ of the fund that is to be invested in each stock j at time t. The proportion $1 - \sum_{j=1}^N p_j(t)$ (which may be negative, or may exceed 1) is invested in the money market. The value of the mutual fund will then evolve according to the equation

$$d\tilde{S}(t) = \tilde{S}(t)[(1 - p'(t)1_N)(r(t)\, dt + dA(t))$$
$$+ p'(t)(b(t)\, dt + 1_N dA(t)) + p'(t)\sigma(t)\, dW(t)]$$
$$= \tilde{S}(t)[r(t)\, dt + dA(t) + p'(t)(b(t) - r(t)1_N)\, dt + p'(t)\sigma(t)\, dW(t)].$$

This is just (1.3.3) with $\Gamma(\cdot) = \delta(\cdot) \equiv 0$, $X(\cdot) = \tilde{S}(\cdot)$, and $\pi(\cdot) = \tilde{S}(\cdot)p(\cdot)$.

Now let us choose a set of N mutual funds $p_1(\cdot) = (p_{11}(\cdot), \ldots, p_{1N}(\cdot))', \ldots, p_N(\cdot) = (p_{N1}(\cdot), \ldots, p_{NN}(\cdot))'$ such that the matrix $P(t) = (p_{ij}(t))_{1 \le i,j \le N}$ is nonsingular for all $t \in [0, T]$ almost surely. Then the values $\tilde{S}(t) = (\tilde{S}_1(t), \ldots, \tilde{S}_N(t))'$ for these funds evolve according to the stochastic differential equation

$$d\tilde{S}(t) = diag(\tilde{S}(t)) \cdot [(r(t)\, dt + dA(t))1_N$$
$$+ P(t)(b(t) - r(t)1)\, dt + P(t)\sigma(t)\, dW(t)],$$

where $diag(\tilde{S}(t))$ denotes the $N \times N$ diagonal matrix with $\tilde{S}_1(t), \ldots, \tilde{S}_N(t)$ in the diagonal positions. We may regard $\tilde{S}_1(\cdot), \ldots, \tilde{S}_N(\cdot)$ as a complete set of stocks with mean rate of return vector

$$\tilde{b}(t) \overset{\triangle}{=} r(t)1_N + P(t)[b(t) - r(t)1_N] \tag{6.25}$$

and volatility matrix

$$\tilde{\sigma}(t) \overset{\triangle}{=} P(t)\sigma(t). \tag{6.26}$$

The associated market price of risk is

$$\hat{\theta}(t) \overset{\triangle}{=} \tilde{\sigma}^{-1}(t)[\tilde{b}(t) - r(t)1_N] = \sigma^{-1}(t)[b(t) - r(t)1_N] = \theta(t),$$

the same as the market price of risk associated with the original set of stocks $S_1(\cdot), \ldots, S_N(\cdot)$.

If $\mathcal{M} = (r(\cdot), b(\cdot), \sigma(\cdot), S(\cdot), A(\cdot))$ is an equilibrium market for the primitives introduced in Section 2, then so is $\tilde{\mathcal{M}} = (r(\cdot), \tilde{b}(\cdot), \tilde{\sigma}(\cdot), \tilde{S}(\cdot), A(\cdot))$. Indeed, since $\hat{\theta}(\cdot) = \theta(\cdot)$, the markets have a common state price density process $H_0(\cdot)$, and this is all that matters (Theorem 5.2). Thus, equilibrium considerations cannot determine the processes $b(\cdot)$ and $\sigma(\cdot)$.

It follows from Theorem 6.4, however, that under the risk-aversion condition (6.4), *the equilibrium market is unique up to the formation of mutual funds.* If $\mathcal{M} = (r(\cdot), b(\cdot), \sigma(\cdot), S(0), A(\cdot))$ and $\tilde{\mathcal{M}} = (r(\cdot), \tilde{b}(\cdot), \tilde{\sigma}(\cdot), \tilde{S}(0), A(\cdot))$ are two equilibrium markets, the uniqueness of the market price of risk implies

$$\sigma^{-1}(t)[b(t) - r(t)1_N] = \tilde{\sigma}^{-1}(t)[\tilde{b}(t) - r(t)1_N].$$

Setting $P(t) = \tilde{\sigma}(t)\sigma^{-1}(t)$, we have (6.25), (6.26).

Remark 6.5: Under condition (6.4), the representative agent utility function (5.24) that results in equilibrium is determined (up to an irrelevant multiplicative constant) *purely endogenously*, by the individual agents' utility functions $U_1, \ldots, U_k$, the discount rate $\beta(\cdot)$, and the *distribution* of the vector of endowment processes $\mathcal{E}(\cdot) = (\epsilon_K(\cdot), \ldots, \epsilon_k(\cdot))$.

The *paths* of the equilibrium market processes $r(\cdot), A(\cdot)$, and $\theta(\cdot)$, as well as the individual agents' optimal consumption processes $\hat{c}_k(\cdot)$, depend on the representative agent's utility function, the discount rate function $\beta(\cdot)$, the paths of the aggregate endowment process $\epsilon(\cdot)$, and the paths of the processes $\nu(\cdot), \rho(\cdot)$, and $\xi(\cdot)$ used in the model (2.2) of $\epsilon(\cdot)$. More generally, $r(\cdot), A(\cdot), \theta(\cdot)$ and $\hat{c}_k(\cdot)$ are adapted to the filtration $\{\mathcal{F}(t)\}_{0 \le t \le T}$, the augmentation by null sets of the filtration generated by the N-dimensional Brownian motion $W(\cdot)$.

A more satisfactory result would be for $r(\cdot), A(\cdot), \theta(\cdot)$, and $\hat{c}_k(\cdot)$ to be adapted to the filtration $\{\mathcal{F}^{\mathcal{E}}(t)\}_{0 \le t \le T}$ generated by the vector $\mathcal{E}(\cdot) = (\epsilon_1(\cdot), \ldots, \epsilon_K(\cdot))$ of endowment processes. This is indeed the case, under the following conditions.

Assume that instead of (2.2), the individual agents' endowment processes are given by the system of *functional stochastic differential equations*

$$d\epsilon_k(t) = \epsilon_k(t)\left[\nu_k(t, \mathcal{E}(\cdot))\,dt + \sum_{j=1}^{N}\rho_{kj}(t, \mathcal{E}(\cdot))\,dW^{(j)}(t)\right], \quad k = 1, \ldots, K,$$

where $\nu_k : [0, T] \times C([0, T])^K \to \mathbb{R}$ and $p_{kd} : [0, T] \times C([0, T])^K \to \mathbb{R}$ are *progressively measurable functionals* as in Definition 3.5.15 of Karatzas and Shreve (1991). If these functionals $\nu(t, y) = \{\nu_k(t, y)\}_{k=1,\ldots,K}$ and $\rho(t, y) = \{\rho_{kj}(t, y)\}_{\substack{1 \le k \le K \\ 1 \le j \le N}}$ are bounded and satisfy the Lipschitz condition

$$\|\nu(t, y) - \nu(t, z)\| + \|\rho(t, y) - \rho(t, z)\| \le L(1 + \sup_{0 \le u \le t}\|y(u) - z(u)\|)$$

for every $t \in [0, T]$ and y, z in $C([0, T])^K$, then the system (6.27) has a pathwise unique, strong solution $\mathcal{E}(t), 0 \le t \le T$. The proof is a straightforward modification of the standard iterative construction (e.g., Karatzas and Shreve (1991), Theorem 5.2.9).

The solution to (6.27) satisfies

$$\epsilon_k(t) = \epsilon_k(0) \exp \left\{ \int_0^t \left[\nu_k(t, \mathcal{E}(\cdot)) - \frac{1}{2} \sum_{d=1}^N \rho_{kd}^2(t, \mathcal{E}(\cdot)) \right] dt \right.$$

$$\left. + \sum_{j=1}^N \rho_{kj}(t, \mathcal{E}(\cdot)) \, dW^{(j)}(t) \right\}$$

and hence is nonnegative, and positive if $\epsilon_k(0)$ is positive. Provided that at least one $\epsilon_k(0)$ is positive, we may write the differential of $\epsilon(t) = \sum_{k=1}^K \epsilon_k(t)$ as

$$d\epsilon(t) = \epsilon(t)\nu(t) \, dt + \epsilon(t) \sum_{j=1}^N \rho_j(t) \, dW^{(j)}(t),$$

where

$$\nu(t) = \sum_{k=1}^K \frac{\epsilon_k(t)}{\epsilon(t)} \nu_k(t, \mathcal{E}(\cdot)),$$

$$\rho_j(t) = \sum_{k=1}^K \frac{\epsilon_k(t)}{\epsilon(t)} \rho_{kj}(t, \mathcal{E}(\cdot)), \quad j = 1, \ldots, N.$$

We are now in the setting of (2.2) with $\xi(\cdot) \equiv 0$, *except that now all processes are adapted to the filtration* $\{\mathcal{F}^{\mathcal{E}}(t)\}_{0 \le t \le T}$.

Remark 6.6: The equilibrium market in this chapter is constructed so that the money market and all stocks are in zero net supply (cf. (5.3), (5.4)). Within the framework of this chapter, other assets can be defined and priced, and these can be in positive net supply. For example, the right to receive agent k's endowment process is in positive net supply. The value of this right at time t is

$$\frac{1}{H_0(t)} \cdot E \left[\int_t^T H_0(u)\epsilon_k(u) \, du \;\middle|\; \mathcal{F}(t) \right],$$

where $H_0(\cdot)$ is determined by equilibrium. Any other value would result in an arbitrage opportunity.

Remark 6.7 *(The Consumption-based Capital Asset Pricing Model):* Suppose there exists a unique equilibrium market. From (6.16)–(6.22) we have

$$\frac{dU'(\epsilon(t); \Lambda)}{U'(\epsilon(t); \Lambda)} = \beta(t) \, dt - r(t) \, dt - dA(t) - \frac{U''(\epsilon(t); \Lambda)}{U'(\epsilon(t); \Lambda)} \epsilon(t)\rho'(t) \, dW(t)$$

$$= \beta(t) \, dt - \frac{dS_0(t)}{S_0(t)} - \theta'(t) \, dW(t). \tag{6.27}$$

We see that if there is no discounting ($\beta(\cdot) \equiv 0$), then *the rate of growth of the instantaneously risk-free asset is the negative of the growth rate of the representative agent's marginal utility from consumption.* Note also that with

$$J(x; \underset{\sim}{\Lambda}) \overset{\triangle}{=} -\frac{xU''(x; \underset{\sim}{\Lambda})}{U'(x; \underset{\sim}{\Lambda})}, \tag{6.28}$$

we have

$$b_n(t) - r(t) = \sum_{j=1}^{N} \sigma_{nj}(t)\theta_j(t)$$

$$= J(\epsilon(t); \Lambda) \sum_{j=1}^{N} \sigma_{nj}(t)\rho_j(t)$$

$$= J(\epsilon(t); \Lambda) \frac{d\langle S_n, \epsilon\rangle(t)}{S_n(t)\epsilon(t)\,dt}. \tag{6.29}$$

In other words, *the risk premium associated with each risky asset is proportional to the relative covariance between the price of that asset and the aggregate consumption; the proportionality constant is independent of the particular asset and equals the "index of relative risk-aversion" for the representative agent.*

The above two observations are referred to as the *consumption-based capital asset-pricing model (CCAPM)* for an economy.

Remark 6.8: Formula (6.22) suggests that even if the singularly continuous component $\xi(\cdot)$ of the aggregate endowment process is identically zero, the singularly continuous component $A(\cdot)$ of the money market price can be nonzero. In this case, *movements in the equilibrium money market price cannot be captured by the interest rate process $r(\cdot)$ alone.* If $\rho(\cdot)$ is nonzero, then the local-time process $t \mapsto L_t(\alpha_k)$ strictly increases each time $\epsilon(t) = \alpha_k$. If in addition, $\xi(\cdot) \equiv 0$, then

$$A(t) = -\sum_{k=1}^{K} \frac{U''(\alpha_k+; \underset{\sim}{\Lambda}) - U''(\alpha_k-; \underset{\sim}{\Lambda})}{U'(\alpha_k; \underset{\sim}{\Lambda})} L_t(\alpha_k) \tag{6.30}$$

strictly *decreases* (recall (5.27)) each time $\epsilon(t) = \alpha_k$ for some k, and is constant on each open interval in the complement of the set $\{t \in [0, T]; \epsilon(t) = \alpha_k$ for some $k\}$. The set $\{t \in [0, T]; \epsilon(t) = \alpha_k\}$ is empty if $U'_k(\bar{c}_k) = \infty$; but if $U'_k(\bar{c}_k) < \infty$, then $\{t \in [0, T]; \epsilon(t) = \alpha_k\}$ is precisely the set of time points at which the optimal consumption process for the kth agent "falls to" or "rises from" the subsistence level $\bar{c}_k$. (Of course, "falling to" or "rising from" subsistence consumption $\bar{c}_k$ is not a simple concept here, since every point of the set $\{t \in [0, T]; \epsilon(t) = \alpha_k\}$ is a cluster point of this set.)

Example 7.7 in the next section demonstrates that the preceding phenomenon does occur.

4.7 Examples

This section comprises several examples in which the various processes of the equilibrium market can be computed more or less explicitly. In Example 7.6 there are two agents with completely different utility functions. In Example 7.7 there are two agents with related utility functions, except that the optimal equilibrium consumption of one agent sometimes falls to zero, whereas this quantity for the other agent is always positive. When optimal equilibrium consumption of an agent falls to zero, or rises from zero, the money market price *decreases* in a singularly continuous manner, i.e., the money market price is continuous but cannot be represented by an interest rate. Example 7.8 considers an ergodic aggregate endowment process.

Example 7.1 *(Logarithmic utility with subsistence consumption):* Let $U_k(c) = \log(c - \bar{c}_k)$, for $c > c_k, k = 1, \ldots, K$, where each $\bar{c}_k$ is a nonnegative constant. Then

$$U'(c; \underset{\sim}{\Lambda}) = \mathcal{H}(c; \underset{\sim}{\Lambda}) = \frac{1}{c - \bar{c}} \sum_{k=1}^{K} \lambda_k, \quad c > \bar{c}.$$

We normalize $\underset{\sim}{\Lambda}$ by setting $\sum_{k=1}^{K} \lambda_k = \epsilon(0) - \bar{c}$, a strictly positive quantity because of Condition 2.1(iii); then $\mathcal{H}(\epsilon(0); \Lambda) = 1$. Equation (5.17) becomes

$$\lambda_k = \frac{(\epsilon(0) - \bar{c}) E \int_0^T e^{-\int_0^t \beta(u)\, du} \left(\frac{\epsilon_k(t) - \bar{c}_k}{\epsilon(t) - \bar{c}} \right) dt}{\int_0^T e^{-\int_0^t \beta(u)\, du}\, du}. \tag{7.1}$$

With λ_k defined by (7.1), we have

$$H_0(t) = \mathcal{H}(\epsilon(t); \underset{\sim}{\Lambda}) = \frac{\epsilon(0) - \bar{c}}{\epsilon(t) - \bar{c}}, \tag{7.2}$$

and the optimal consumption process for agent k is

$$\hat{c}_k(t) = I_k \left(\frac{\epsilon(0) - \bar{c}}{\lambda_k(\epsilon(t) - \bar{c})} \right) = \frac{\lambda_k(\epsilon(t) - \bar{c})}{\epsilon(0) - \bar{c}} + \bar{c}_k, \quad k = 1, \ldots, K.$$

For each agent, $\hat{c}_k(t) > \bar{c}_k$ for all t, almost surely. The equilibrium market coefficients of (6.20)–(6.22) become

$$r(t) = \beta(t) + \frac{\epsilon(t)\nu(t)}{\epsilon(t) - \bar{c}} - \frac{\epsilon^2(t)\|\rho(t)\|^2}{(\epsilon(t) - \bar{c})^2},$$

$$\theta(t) = \frac{\epsilon(t)\rho(t)}{\epsilon(t) - \bar{c}}, \quad A(t) = \int_0^t \frac{\epsilon(s)}{\epsilon(s) - \bar{c}}\, d\xi(s).$$

Example 7.2 *(Power utility with subsistence consumption):* Let $U_k(c) = \frac{1}{p}(c - \bar{c}_k)^p$ for $c > \bar{c}_k, k = 1, \ldots, K$, where $p < 1, p \neq 0$, and each $\bar{c}_k$ is a nonnegative constant. Then

$$U'(c; \underset{\sim}{\Lambda}) = \mathcal{H}(c; \underset{\sim}{\Lambda}) = \left[\frac{\sum_{k=1}^{K} \lambda_k^{\frac{1}{1-p}}}{c - \bar{c}} \right]^{1-p}, \quad c > \bar{c}.$$

We normalize $\underset{\sim}{\Lambda}$ by setting $\sum_{k=1}^{K} \lambda_k^{\frac{1}{1-p}} = \epsilon(0) - \bar{c}$, so that $\mathcal{H}(\epsilon(0); \underset{\sim}{\Lambda}) = 1$. Equation (5.17) becomes

$$\lambda_k^{\frac{1}{1-p}} = \frac{(\epsilon(0) - \bar{c}) E \int_0^T e^{-\int_0^t \beta(u)\, du} \frac{\epsilon_k(t) - \bar{c}_k}{(\epsilon(t) - \bar{c})^{1-p}} \, dt}{E \int_0^T e^{-\int_0^t \beta(u)\, du} (\epsilon(t) - \bar{c})^p \, dt}. \tag{7.3}$$

With λ_k defined by (7.3), we have

$$H_0(t) = \mathcal{H}(\epsilon(t); \underset{\sim}{\Lambda}) = \left(\frac{\epsilon(0) - \bar{c}}{\epsilon(t) - \bar{c}} \right)^{1-p},$$

and the optimal consumption process for agent k is

$$\hat{c}_k(t) = I_k \left(\frac{1}{\lambda_k} \left(\frac{\epsilon(0) - \bar{c}}{\epsilon(t) - \bar{c}} \right)^{1-p} \right) = \lambda_k^{\frac{1}{1-p}} \left(\frac{\epsilon(t) - \bar{c}}{\epsilon(0) - \bar{c}} \right) + \bar{c}_k, \ k = 1, \dots, K.$$

For each agent, $\hat{c}_k(t) > \bar{c}_k$ for all t, almost surely. The equilibrium market coefficients of (6.20)–(6.22) become

$$r(t) = \beta(t) + \frac{(1-p)\epsilon(t)\nu(t)}{\epsilon(t) - \bar{c}} - \frac{(1-p)(2-p)\epsilon^2(t)\|\rho(t)\|^2}{2(\epsilon(t) - \bar{c})},$$

$$\theta(t) = \frac{(1-p)\epsilon(t)\rho(t)}{\epsilon(t) - \bar{c}}, \quad A(t) = (1-p) \int_0^t \frac{\epsilon(s)}{\epsilon(s) - \bar{c}} \, d\xi(s).$$

The logarithmic formulas of Example 7.1 are recovered by setting $p = 0$ in this example.

Remark 7.3: In Example 7.2 and with Λ given by (7.3), the ray of vectors $\{\eta\Lambda\}_{0 < \eta < \infty}$ is the locus of solutions to the system of equations (7.3), even for negative powers p. This shows that *condition (6.4) is not necessary for uniqueness in Theorem 7.1*, since in this example

$$-\frac{cU_k''(c)}{U_k'(c)} = \frac{(1-p)c}{c - \bar{c}_k} > 1 \quad \text{for} \quad c > \bar{c}_k, \quad p < 0.$$

Remark 7.4: Condition 2.1(ii) is not necessary for the construction the equilibria in Examples (7.1) and (7.2). All that is required is that the expressions appearing on the right-hand sides of (7.1), (7.3) be non-negative for all k. They will be positive for at least one k because of Condition 2.1(iii).

Example 7.5 *(Constant aggregate endowment):* If the aggregate endowment $\epsilon > \bar{c}$ is constant, then the unique vector Λ satisfying the normalization $\mathcal{H}(\epsilon; \Lambda) = 1$ is

$$\Lambda = \left(\frac{1}{U_1'(\hat{c}_1)}, \ldots, \frac{1}{U_K'(\hat{c}_k)} \right),$$

where the constants $\hat{c}_k \geq \bar{c}_k$ are the optimal consumption rates

$$\hat{c}_k = \frac{E \int_0^T e^{-\int_0^t \beta(u)\, du} \epsilon_k(t)\, dt}{\int_0^T e^{-\int_0^t \beta(u)\, du}\, dt}, \quad k = 1, \ldots, K.$$

Constant aggregate endowment implies $\nu(\cdot) \equiv 0, \xi(\cdot) \equiv 0, \rho(\cdot) \equiv 0$ in (2.2), and the local time of $\epsilon(\cdot)$ at every point is zero. Therefore, the equilibrium market coefficients (6.20)–(6.22) are

$$r(t) = \beta(t), \quad \theta(t) = 0, \quad A(t) = 0, \quad 0 \leq t \leq T.$$

It should be noted that the individual agents' endowments can be random and time-varying, which means that agents may still have to trade with one another in order to finance their constant rates of consumption.

Example 7.6 $(K = 2, U_1(c) = \log c, U_2(c) = \sqrt{c}.)$: In this case, we have

$$U'(c; \Lambda) = \mathcal{H}(c; \Lambda) = \frac{\lambda_1}{2c} \left[1 + \sqrt{1 + c \left(\frac{\lambda_2}{\lambda_1} \right)^2} \right], \quad c > 0,$$

and the optimal consumption rates become

$$\hat{c}_1(t) = \frac{2\epsilon(t)}{1 + \sqrt{1 + \epsilon(t)(\frac{\lambda_2}{\lambda_1})^2}}, \quad \hat{c}_2(t) = \left(\frac{\frac{\lambda_2}{\lambda_1}\epsilon(t)}{1 + \sqrt{1 + \epsilon(t)(\frac{\lambda_2}{\lambda_1})^2}} \right)^2.$$

The positive constants λ_1 and λ_2 are uniquely determined by (5.17) with $k = 1$:

$$2 \int_0^T e^{-\int_0^t \beta(u)\, du}\, dt$$

$$= E \int_0^T e^{-\int_0^t \beta(u)\, du} \left[1 + \sqrt{1 + \epsilon(t) \left(\frac{\lambda_2}{\lambda_1} \right)^2} \right] \frac{\epsilon_1(t)}{\epsilon(t)}\, dt, \quad (7.4)$$

and the normalization condition (6.15) gives

$$\lambda_1 = \frac{2\epsilon(0)}{1 + \sqrt{1 + \epsilon(0)(\frac{\lambda_2}{\lambda_1})^2}}. \quad (7.5)$$

Indeed, (7.4) determines $\frac{\lambda_2}{\lambda_1}$, and then λ_1 is found from (7.5).

With the vector $\Lambda = (\lambda_1, \lambda_2) \in (0, \infty)^2$ thus determined, the formulae

$$r(t) = \beta(t) + J(\epsilon(t); \Lambda)\nu(t) + \frac{1}{2}\|\rho(t)\|^2 K(\epsilon(t); \Lambda),$$

$$\theta(t) = J(\epsilon(t); \Lambda)\rho(t), \quad A(t) = \int_0^t J(\epsilon(s); \Lambda)\, d\xi(s)$$

of (6.20)–(6.22), with

$$J(c; \Lambda) = -cU''(c; \Lambda)/U'(c; \Lambda), \quad K(c; \Lambda) = -cU'''(c; \Lambda)/U'(c; \Lambda),$$

provide the coefficients of the equilibrium market model.

Example 7.7 *(Money market not represented by an interest rate):* This example shows that the equilibrium money-market price can have a non-trivial singularly continuous component $A(\cdot)$, even though the singularly continuous component $\xi(\cdot)$ of the aggregate endowment process is identically zero. There is a discussion of this phenomenon in Remark 6.8. Here we set up a particular model exhibiting the behavior of interest.

We consider two agents $(K = 2)$ with utility functions

$$U_1(c) = \begin{cases} \log c, & c > 0, \\ -\infty, & c \leq 0, \end{cases}$$

$$U_2(c) = \begin{cases} \log(c+1), & c \geq 0, \\ -\infty, & c < 0, \end{cases}$$

so that $\bar{c}_1 = \bar{c}_2 = \bar{c} = 0$. Then

$$I_1(y) = \frac{1}{y}, \quad I_2(y) = \begin{cases} (1/y) - 1, & 0 < y \leq 1, \\ 0, & y \geq 1, \end{cases}$$

$$I(y; \Lambda) = \begin{cases} \frac{\lambda_1 + \lambda_2}{y} - 1, & 0 < y \leq \lambda_2, \\ \frac{\lambda_1}{y}, & y \geq \lambda_2, \end{cases}$$

$$U'(c; \Lambda) = \mathcal{H}(c; \Lambda) = \begin{cases} \frac{\lambda_1}{c}, & 0 < c < \frac{\lambda_1}{\lambda_2}, \\ \frac{\lambda_1 + \lambda_2}{1+c}, & c \geq \frac{\lambda_1}{\lambda_2}. \end{cases}$$

In the notation of Theorem 5.6, $\alpha_1 = 0$, $\alpha_2 = \frac{\lambda_1}{\lambda_2}$, and $\mathcal{D} = \{\frac{\lambda_1}{\lambda_2}\}$. In the notation of (5.29), (5.30),

$$\hat{c}_1 = \begin{cases} c, & 0 < c < \frac{\lambda_1}{\lambda_2}, \\ \frac{\lambda_1(1+c)}{\lambda_1 + \lambda_2}, & c \geq \frac{\lambda_1}{\lambda_2}, \end{cases} \tag{7.6}$$

$$\hat{c}_2 = \begin{cases} c, & 0 < c < \frac{\lambda_1}{\lambda_2}, \\ \frac{\lambda_2(1+c)}{\lambda_1 + \lambda_2} - 1, & c \geq \frac{\lambda_1}{\lambda_2}, \end{cases} \tag{7.7}$$

$$U(c; \Lambda) = \begin{cases} \lambda_1 \log c, & 0 < c < \frac{\lambda_1}{\lambda_2}, \\ (\lambda_1 + \lambda_2)\log(1+c) + \lambda_1 \log\left(\frac{\lambda_1}{\lambda_2 + \lambda_2}\right) \\ \quad + \lambda_2 \log\left(\frac{\lambda_2}{\lambda_1 + \lambda_2}\right), & c \geq \frac{\lambda_1}{\lambda_2}, \end{cases}$$

and we observe that (cf. (5.27))

$$U'' \left(\frac{\lambda_1}{\lambda_2} +; \Lambda \right) - U'' \left(\frac{\lambda_1}{\lambda_2} -; \Lambda \right) = \frac{\lambda_3^2}{\lambda_1(\lambda_1 + \lambda_2)}.$$

We set $\beta(\cdot) \equiv 0$. For the aggregate endowment, we take the process

$$\epsilon(t) = 1 + \exp \left\{ W(t \wedge \tau) - \frac{1}{2}(t \wedge \tau)^2 \right\}, \quad 0 \leq t \leq T,$$

where $\tau = \inf\{t \in [0,T); W(t) = 1\} \wedge T$. Then $\epsilon(\cdot)$ is a continuous martingale bounded strictly between 1 and $1 + e$, and

$$d\epsilon(t) = (\epsilon(t) - 1)1_{\{t \leq \tau\}} \, dW(t), \quad \epsilon(0) = 2.$$

This is of the form (2.2) with $\nu(\cdot) \equiv 0, \xi(\cdot) \equiv 0$, and $\rho(t) = \frac{\epsilon(t)-1}{\epsilon(t)}1_{\{t \leq \tau\}}$. Condition 2.1(iii) is satisfied.

Because $\epsilon(t) > 1$ for $0 \leq t \leq T$, almost surely, we have $E \int_0^T \frac{2\epsilon(t)}{1+\epsilon(t)} \, dt > T$. Choose $\kappa \in (0,1)$ to satisfy

$$\kappa \cdot E \int_0^T \frac{2\epsilon(t)}{1 + \epsilon(t)} \, dt > T \tag{7.8}$$

and set $\epsilon_1(t) = \kappa\epsilon(t), \epsilon_2(t) = (1 - \kappa)\epsilon(t)$, so that Condition 2.1(ii) is also satisfied. Equation (5.17) with $k = 1$ or 2 reduces to

$$\frac{T}{\kappa} = f \left(\frac{\lambda_1}{\lambda_2} \right), \tag{7.9}$$

where

$$f(\alpha) \overset{\triangle}{=} E \int_0^T 1_{\{\epsilon(t)<\alpha\}} \, dt + \left(1 + \frac{1}{\alpha} \right) \cdot E \int_0^T \frac{\epsilon(t)}{1 + \epsilon(t)}1_{\{\epsilon(t) \geq \alpha\}} \, dt. \tag{7.10}$$

Note that

$$f(\alpha) = E \int_0^T 1_{\{\epsilon(t)<\alpha\}} \, dt + E \int_0^T \left(1 - \frac{1}{1 + \epsilon(t)} \right) 1_{\{\epsilon(t) \geq \alpha\}} \, dt$$

$$+ \frac{1}{\alpha} E \int_0^T \frac{\epsilon(t)}{1 + \epsilon(t)}1_{\{\epsilon(t) \geq \alpha\}} \, dt$$

$$= T + \frac{1}{\alpha} E \int_0^T \frac{(\epsilon(t) - \alpha)^+}{1 + \epsilon(t)} \, dt. \tag{7.11}$$

From (7.10) we have $f(1) = 2E \int_0^T \frac{\epsilon(t)}{1+\epsilon(t)} \, dt > \frac{T}{\kappa}$ and $f(1 + e) = T < \frac{T}{\kappa}$. Because f is continuous, there must exist $\alpha \in (1, 1 + e)$ such that

$$f(\alpha) = \frac{T}{\kappa}. \tag{7.12}$$

Let α^* be the smallest such α. Because α^* solves (7.12), (7.11) implies $E \int_0^T 1_{\{\epsilon(t)>\alpha^*\}} \, dt > 0$, and $f(\alpha) < f(\alpha^*)$ for all $\alpha > \alpha^*$. For $0 < \alpha < 1$, it is apparent from (7.8) that $f(\alpha) \geq f(1) > \frac{T}{\kappa}$. Therefore, α^* is the unique solution in $(0, \infty)$ to (7.12).

From (7.9) we see that the equilibrium vector $\Lambda = (\lambda_1, \lambda_2)$ must satisfy $\frac{\lambda_1}{\lambda_2} = \alpha^*$. We determine λ_1 and λ_2 individually from the normalization condition

$$1 = \mathcal{H}(\epsilon(0); \Lambda) = \begin{cases} \frac{\lambda_1}{\epsilon(0)}, & \text{if } 0 < \epsilon(0) < \alpha^*, \\ \frac{\lambda_1 + \lambda_2}{1 + \epsilon(0)}, & \text{if } \epsilon(0) \geq \alpha^*. \end{cases}$$

We have already seen that $E \int_0^T 1_{\{\epsilon(t) > \alpha^*\}} \, dt > 0$. We must also have $E \int_0^T 1_{\{\epsilon(t) \leq \alpha^*\}} \, dt > 0$, because $P[\inf_{0 \leq t \leq T} \epsilon(t) = 1] > 0$ by construction. It follows that *the process $\epsilon(\cdot)$ crosses the level α^* during the interval $[0, T]$ with positive probability.* Being a continuous martingale, $\epsilon(\cdot)$ is a time-changed Brownian motion (Karatzas and Shreve (1991), Section 3.4B), and hence $L_t(\alpha^*)$ increases at each t satisfying $\epsilon(t) = \alpha^*$ (*ibid*, Problem 6.13(iv)). The equilibrium market coefficient processes (6.20)–(6.22) are

$$r(t) = -\left[\frac{1}{\epsilon^2(t)} 1_{\{\epsilon(t) < \alpha^*\}} + \frac{1}{(1 + \epsilon(t))^2} 1_{\{\epsilon(t) \geq \alpha^*\}} \right] (\epsilon(t) - 1)^2 1_{\{t \leq \tau\}},$$

$$\theta(t) = \left[\frac{1}{\epsilon(t)} 1_{\{\epsilon(t) < \alpha^*\}} + \frac{1}{1 + \epsilon(t)} 1_{\{\epsilon(t) \geq \alpha^*\}} \right] (\epsilon(t) - 1) 1_{\{t \leq \tau\}},$$

$$A(t) = -\frac{L_t(\alpha^*)}{\alpha^*(1 + \alpha^*)},$$

and $A(\cdot)$ is nontrivial. According to (6.6) and (7.6), (7.7), the optimal consumption processes are

$$\hat{c}_1(t) = \epsilon(t) 1_{\{\epsilon(t) < \alpha^*\}} + \frac{\alpha^*(1 + \epsilon(t))}{1 + \alpha^*} 1_{\{\epsilon(t) \geq \alpha^*\}},$$

$$\hat{c}_2(t) = \frac{\epsilon(t) - \alpha^*}{1 + \alpha^*} 1_{\{\epsilon(t) \geq \alpha^*\}}.$$

Example 7.8 *(Ergodic aggregate endowment):* Let us suppose that each agent k has utility function U_k with $\bar{c}_k = 0$ and $U_k'(0) = \infty$, so $\bar{c} = 0$. Let us further suppose that the aggregate endowment process $\epsilon(\cdot)$ is a time-homogeneous diffusion on an interval $\mathcal{I} = (\gamma_1, \gamma_2)$ with $0 < \gamma_1 < \gamma_2 < \infty$; i.e.,

$$d\epsilon(t) = \epsilon(t) \nu(\epsilon(t)) \, dt + \epsilon(t) \rho(\epsilon(t)) \, dW(t),$$

where the functions $\nu : \mathcal{I} \to \mathbb{R}$ and $\rho : \mathcal{I} \to \mathbb{R}^N$ are bounded on compact subintervals of $\mathcal{I}$. We assume also that $\|\rho(\cdot)\|$ is bounded away from zero on compact subintervals of $\mathcal{I}$.

We introduce the *scale function*

$$p(c) = \int_\gamma^c \exp\left\{ -2 \int_\gamma^y \frac{\nu(z) \, dz}{z \|\rho(z)\|^2} \right\} dy, \quad c \in \mathcal{I},$$

and the *speed measure*

$$m(dc) = \frac{2dc}{c^2\|\rho(c)\|^2 p'(c)}, \quad c \in \mathcal{I},$$

where γ is a fixed point in $\mathcal{I}$, and assume that

$$p(\gamma_1) = -\infty, \quad p(\gamma_2) = \infty, \quad m(\mathcal{I}) < \infty. \tag{7.13}$$

Then the diffusion process $\epsilon(\cdot)$ is *ergodic* with invariant measure $m(dc)/m(\mathcal{I})$ (cf. Proposition 5.5.22 and Exercise 5.5.40 in Karatzas and Shreve (1991)).

Finally, suppose that $\beta(\cdot) = \beta$ is constant. Then (6.20)–(6.22) give $A(\cdot) \equiv 0$ and

$$r(t) = \beta - \frac{1}{U'(\epsilon(t))}\left[U''(\epsilon(t))\epsilon(t)\nu(\epsilon(t)) + \frac{1}{2}U'''(\epsilon(t))\|\rho(\epsilon(t))\|^2\epsilon^2(t)\right],$$

$$\theta(t) = \epsilon(t)\rho(\epsilon(t))\left(-\frac{U''(\epsilon(t))}{U'(\epsilon(t))}\right),$$

where $U(\cdot) = U(\cdot; \Lambda)$ is the representative agent utility function of (5.24). In particular,

$$r(t) + \frac{1}{2}\|\theta(t)\|^2 = \beta - \frac{1}{2}\epsilon^2(t)\|\rho(\epsilon(t))\|^2\left[\frac{U'(\epsilon(t))U'''(\epsilon(t)) - (U''(\epsilon(t)))^2}{(U'(\epsilon(t)))^2}\right.$$

$$\left. + \frac{U''(\epsilon(t))}{U'(\epsilon(t))} \cdot \frac{2\nu(\epsilon(t))}{\epsilon(t)\|\rho(\epsilon(t))\|^2}\right], \tag{7.14}$$

and the *maximal growth rate from investment* in this market (corresponding to the "optimal logarithmic portfolio" $\hat{\pi}(\cdot) = (\sigma'(\cdot))^{-1}\theta(\cdot)\hat{X}(\cdot)$ of (3.10.2)) *is equal to the discount rate* β, as we show below.

We note first from (3.10.3) that the wealth process $\hat{X}(\cdot)$ corresponding to the optimal logarithmic portfolio, regardless of the positive initial condition $\hat{X}(0)$, satisfies

$$\lim_{T\to\infty}\frac{1}{T}\log\hat{X}(T) = \lim_{T\to\infty}\frac{1}{T}\int_0^T\left[r(s) + \frac{1}{2}\|\theta(s)\|^2\right]ds,$$

because $\theta(\cdot)$ is bounded, which implies $\lim_{T\to\infty}\frac{1}{T}\int_0^T\theta'(s)\,dW(s) = 0$. The ergodic property for $\epsilon(\cdot)$, in conjunction with (7.13), implies

$$\lim_{T\to\infty}\frac{1}{T}\int_0^T\left[r(s) + \frac{1}{2}\|\theta(s)\|^2\right]ds$$

$$= \beta - \frac{1}{2m(\mathcal{I})}\int_{\gamma_1}^{\gamma_2}c^2\|\rho(c)\|^2\left[\frac{U'(c)U'''(c) - (U''(c))^2}{(U'(c))^2} + \frac{U''(c)}{U'(c)}\right.$$

$$\left. \cdot \frac{2\nu(c)}{c\|\rho(c)\|^2}\right]m(dc)$$

$$= \beta - \frac{1}{m(\mathcal{I})}\int_{\gamma_1}^{\gamma_2}\left[\frac{U'(c)U'''(c) - (U''(c))^2}{(U'(c))^2} - \frac{U''(c)}{U'(c)} \cdot \frac{p''(c)}{p'(c)}\right]\frac{dc}{p'(c)}$$

$$= \beta - \frac{1}{m(\mathcal{I})} \int_{\gamma_1}^{\gamma_2} \left[\left(\frac{U''(c)}{U'(c)} \right)' \cdot \frac{1}{p'(c)} - \frac{U''(c)}{U'(c)} \cdot \frac{p''(c)}{(p'(c))^2} \right] dc$$

$$= \beta - \frac{1}{m(\mathcal{I})} \int_{\gamma_1}^{\gamma_2} \frac{d}{dc} \left(\frac{U''(c)}{U'(c)} \cdot \frac{1}{p'(c)} \right) dc$$

$$= \beta - \frac{1}{m(\mathcal{I})} \left[\frac{U''(\gamma_2)}{U'(\gamma_2)p'(\gamma_2-)} - \frac{U''(\gamma_1)}{U'(\gamma_1)p'(\gamma_1+)} \right] = \beta,$$

because $p'(\gamma_2-) = p'(\gamma_1+) = \infty$ from (7.13).

4.8 Notes

Models of *competitive equilibrium*, i.e., of the way in which demand for goods determines prices, have occupied economists for more than a century. One of the oldest works on this subject is Walras (1874/77). A mathematical treatment of the existence and uniqueness of solutions to Walras's equations was given by Wald (1936), in whose work can be found an early version of the risk-aversion index condition 6.4 (see also Rothschild and Stiglitz (1971) for another use of this condition). The first complete proof for the existence of equilibrium in an economy with several agents and finitely many commodities was given by Arrow and Debreu (1954). A perspective on this and related work can be obtained from Arrow (1970, 1983), Debreu (1982), and the surveys by Debreu (1959, 1983).

The classical reference on competitive equilibrium with an infinite-dimensional commodity space is Bewley (1972). The commodity space in this work is L_∞, a space whose positive orthant has nonempty interior, and this fact is necessary for the separating hyperplane argument at the heart of the paper. To remove this interiority condition, Mas-Colell (1986) introduced the concept of "uniform properness" for the preferences of agents. In the context of the model of this chapter, and with $\bar{c}_k = 0$ for all k, uniform properness requires that $U_k'(0)$ be finite for every k. A survey of equilibrium existence theory in infinite-dimensional spaces is Mas-Colell and Zame (1991).

The models in the above papers are not explicitly either dynamic or stochastic. Models that are *both stochastic and dynamic* have the interesting feature that individuals can achieve equilibrium allocations by trading in securities. This role of securities in *spanning uncertainty in a complete market* was already recognized by Arrow (1952). Radner (1972) established existence of equilibrium in a discrete-time dynamical market with several agents who trade with one another. Lucas (1978) set up a discrete-time Markov model in which the optimal consumption/production of a representative agent leads to equilibrium. This work was presaged by LeRoy (1973). Prescott and Mehra (1980) extended the work of Lucas (1978) to a setup with several identical agents. The analysis of a representative agent or several identical agents leads to the *consumption-based capital asset pricing model (CCAPM)* of Merton (1973b), Breeden (1979), and Cox, Ingersoll, and Ross (1985a); see Remark 6.7. Empirical tests of this model performed by Breeden, Gibbons, and Litzenberger (1989) found partial agreement with data.

The issue of existence and uniqueness of equilibrium in a continuous-time stochastic model with nonidentical (heterogeneous) agents is inherently infinite-dimensional, because consumption is indexed by both time and by the "state of nature" $\omega \in \Omega$. Duffie (1986) gave conditions sufficient for the existence of such an equilibrium, but CCAPM was not obtained because the analysis required the uniform properness condition of Mas-Colell (1986); see also Duffie and Huang (1987). Duffie and Huang (1985) showed that *if* a continuous-time stochastic model with heterogeneous agents has an equilibrium, *then* this equilibrium can be implemented by trading in securities. Huang (1987) provided conditions under which such an equilibrium leads to prices that are functions of a diffusion state-process. The missing piece in this puzzle was supplied by Duffie and Zame (1989), and independently by Araujo and Monteiro (1989a,b), who provided functional-analytic proofs of the existence of equilibrium without the uniform properness condition of Mas-Colell (1986).

The approach to the questions of existence and uniqueness of equilibrium followed in this chapter is taken from the papers Karatzas, Lehoczky, and Shreve (1990, 1991) and Karatzas, Lakner, Lehoczky, and Shreve (1991). The fundamental idea of assigning weights to the different agents, and thereby reducing the problem to one of finding the proper weights, was apparently first used by Negishi (1960). Other authors, including Magill (1981) and Constantinides (1982), have used this method. In the model of this chapter the Negishi method turns the infinite-dimensional problem of finding equilibrium consumption processes into the *finite-dimensional problem* of finding the proper weights. Extensions of this approach permit a study of *equilibrium in the presence of several commodities* (e.g., Lakner (1989)), of an agent who takes into account the *effect of his actions on prices* (Cuoco and Cvitanić (1996), Başak (1996b)), of *heterogeneous beliefs or information* for agents (cf. Detemple (1986a,b), Detemple and Murthy (1993), Başak (1996b,c), DeMarzo and Skiadas (1996, 1997), Pikovsky (1998)), of *restrictions on stock-market participation* (Detemple and Murthy (1996)), Başak (1996c), Başak and Cuoco (1998), Cuoco (1997)), of *stochastic differential utility* (Duffie, Geoffard and Skiadas (1994)), and of the effect that *portfolio insurers* have on the market (Başak (1993, 1995, 1996a), Grossman and Zhou (1996)). Dana and Pontier (1992) and Dana (1993a,b) have extended some of the results in Karatzas, Lehoczky, and Shreve (1990) and have cast the arguments of this chapter into more traditional economic terms. Dumas (1989) is similar to, but more detailed than, our Example 7.6. Example 7.8 grew out of conversations with Dean Foster.

When markets are *incomplete*, equilibrium analysis becomes more difficult. Hart (1975) provided an example of a discrete-time model in which equilibrium does not exist. Following this, Kreps (1982), McManus (1984), Magill and Shafer (1984), and Duffie and Shafer (1985, 1986) demonstrated that for *nearly all* discrete-time models, equilibrium does exist. Again in a discrete-time model, Duffie (1987) established the existence of equilibrium under the assumption that all securities are purely financial, i.e., represent claims to monetary dividends rather than claims to goods. Rubinstein (1974) provided conditions on agents' utility functions that lead to existence of equilibrium as if the market were complete. Lucas (1978), Bewley (1986), Duffie, Geanakoplos, Mas-Colell, and McLennan (1994), and Karatzas, Shubik, and Sudderth (1994, 1997) study discrete-time stationary Markovian equilibria in infinite-horizon models. For more information about discrete-time equilibrium results for incomplete market models, the reader can consult the recent monograph by Magill and Quinzii (1996) and its references.

Cuoco and He (1994) have extended the methodology of this chapter to incomplete continuous-time markets; see also Cuoco and He (1993), He and Pagès (1993), Cuoco (1997) and the references therein. Equilibrium analysis in this setting depends on the ability to solve the single-agent optimal consumption/investment problem of Chapter 6, but in the presence of a random endowment stream rather than an initial capital at $t = 0$; see the notes at the end of that chapter.

5

Contingent Claims in Incomplete Markets

5.1 Introduction

The subject of this chapter is the arbitrage pricing and almost sure hedging of contingent claims in markets which are incomplete due to portfolio constraints. It often occurs in such markets that a given contingent claim cannot be hedged perfectly, no matter how large the initial wealth of the would-be hedging agent. However, it can be the case that with sufficient initial wealth, a hedging agent can construct a portfolio which respects the constraints and still leads to a final wealth that dominates almost surely the payoff of the contingent claim. This chapter distinguishes these two cases and shows how, when possible, to construct the *superreplicating* portfolio of the second case.

In Section 2 of this chapter we set up the financial market model $\mathcal{M}(K)$ with constraints. In particular, there is a nonempty, closed convex set $K \subseteq \mathbb{R}^N$, and the investing agent's vector of wealth proportions in the N stocks is required to lie in this set. The model includes such common situations as prohibition or restriction on short-selling, prohibition or restriction on borrowing from the money market, and incompleteness in the sense that some stocks (or other sources of uncertainty) are unavailable for investment. Given a contingent claim, which in this chapter pays off a nonnegative amount at the final time only, we define $h_{up}(K)$, the *upper hedging price* of the claim, to be the least initial wealth that permits construction of a superreplicating portfolio in the constrained market $\mathcal{M}(K)$. To compute $h_{up}(K)$, we introduce a family of *dual processes* $\mathcal{D}$, and with each dual process $\nu \in \mathcal{D}$ we construct an auxiliary market $\mathcal{M}_\nu$ related to the original one. There are no portfolio constraints in the auxiliary markets, and so the contingent claim in these markets can be priced and hedged using the theory of Chapter 2. Let u_ν be the price of the contingent claim in the (unconstrained) market $\mathcal{M}_\nu$. Theorem 6.2, the principal result of this chapter, is that

$$h_{up}(K) = \sup_{\nu \in \mathcal{D}} u_\nu. \tag{6.3}$$

This supremum is infinite if and only if superreplication in the constrained market $\mathcal{M}(K)$ is not possible.

In the special case of a market with constant coefficients and of a contingent claim whose payoff is a function of the stock prices at the final time, the supremum in (6.3) turns out to be the price of a related contingent claim in the original market without portfolio constraints. This result is obtained in Section 7.

Section 8 is a study of conditions under which the supremum in (6.3) is attained by a so-called *optimal dual process* $\hat{\nu}(\cdot)$. When an optimal dual process $\hat{\nu}(\cdot)$ exists, the hedging portfolio process in the unconstrained market $\mathcal{M}_{\hat{\nu}}$ satisfies the portfolio constraints in the constrained market $\mathcal{M}(K)$ and replicates exactly the contingent claim almost surely.

The discussion so far has concerned the *seller* of a contingent claim, who receives an initial sum of money and wishes to invest in the constrained market $\mathcal{M}(K)$ so as to superreplicate the contingent claim. In Section 9 we take up the problem of the *buyer* of a contingent claim, who initially either borrows from the money market or sells stock short in order to raise capital to buy the contingent claim. The buyer wishes to manage his debt so that the payoff of the contingent claim at the final time is sufficient to cover this debt. The buyer also invests in a constrained market, although the modeling of his constraint is complicated by the fact that his wealth prior to the final time is nonpositive. We require in Section 9 that the vector of the buyer's wealth proportions invested in the N stocks lie in a nonempty, closed, convex set K_- related to K. We define $h_{low}(K_-)$ to be the largest sum the buyer can borrow and still have the payoff from the contingent claim cover his debt almost surely at the final time. For the buyer's problem, our principal result is Theorem 9.10:

$$h_{low}(K_-) = \inf_{\nu \in \mathcal{D}^{(b)}} u_\nu, \tag{9.29}$$

where $\mathcal{D}^{(b)}$ is the set of *bounded* processes in $\mathcal{D}$. Actually, the supremum in (6.3) could have been restricted to bounded processes $\nu \in \mathcal{D}^{(b)}$ as well.

It is clear from (6.3) and (9.29) that $h_{low}(K_-) \le h_{up}(K)$. Arbitrage arguments show that the price of the contingent claim cannot lie outside the interval $[h_{low}(K_-), h_{up}(K)]$, but are incapable of determining a single price inside the interval, unless this interval contains only one point.

Section 10 develops the formula for the lower hedging price for contingent claims that are functions of the final stock prices in a market with constant coefficients. This formula is analogous to the one derived in Section 7 for the upper hedging price.

5.2   The Model

In this chapter and the next, we shall work in the context of a *financial market* $\mathcal{M} = (r(\cdot), b(\cdot), \delta(\cdot), \sigma(\cdot), S(0), A(\cdot))$ as in Definition 1.1.3, governed by the stochastic differential equations

$$dS_0(t) = S_0(t)[r(t)\, dt + dA(t)], \tag{2.1}$$

$$dS_n(t) = S_n(t)\left[b_n(t)\, dt + dA(t) + \sum_{d=1}^{D} \sigma_{nd}(t)\, dW^{(d)}(t) \right], \tag{2.2}$$

$$n = 1, \ldots, N,$$

for the money market and stock price processes, respectively. For this market we shall assume throughout this chapter, without further mention, that

$$N = D, \tag{2.3}$$

$$\left\{ \begin{array}{l} \text{the volatility matrix } \sigma(t,\omega) = (\sigma_{nd}(t,\omega))_{1 \le n,d \le N} \\ \text{is bounded and nonsingular, with } \sigma^{-1}(t,\omega) \\ \text{bounded, uniformly in } (t,\omega) \in [0,T] \times \Omega \end{array} \right\}, \tag{2.4}$$

$$S_0(T) \ge s_0 \text{ a.s. for some constant } s_0 > 0, \tag{2.5}$$

$$\int_0^T \|\theta(t)\|^2\, dt < \infty, \text{ a.s.,} \tag{2.6}$$

where the *market price of risk process* $\theta(\cdot)$ of (1.4.9) is

$$\theta(t) \overset{\Delta}{=} \sigma^{-1}(t)[b(t) + \delta(t) - r(t)\mathbf{1}], \quad 0 \le t \le T, \tag{2.7}$$

and $\delta(\cdot)$ is the vector of dividend rate processes. We will have a complete financial market (Theorem 1.6.6) if the positive local martingale

$$Z_0(t) \overset{\Delta}{=} \exp\left\{ -\int_0^t \theta'(s)\, dW(s) - \frac{1}{2}\int_0^t \|\theta(s)\|^2\, ds \right\}, \quad 0 \le t \le T, \tag{2.8}$$

of (1.5.2) is a martingale. We do not always assume this property, so the market may not be *standard* in the sense of Definition 1.5.1. Here and in the sequel, we retain the notation (1.5.6) and (1.5.12) for the *Brownian* *motion with drift (under P)*

$$W_0(t) \overset{\Delta}{=} W(t) + \int_0^t \theta(s)\, ds, \quad 0 \le t \le T, \tag{2.9}$$

and the *state price density process*

$$H_0(t) \overset{\Delta}{=} \frac{Z_0(t)}{S_0(t)}, \quad 0 \le t \le T. \tag{2.10}$$

Consider now an agent with cumulative income process $\Gamma(t) = x - C(t)$, $0 \le t \le T$, where $x \ge 0$ is his *initial wealth* and $C(\cdot)$ is his *cumulative consumption*, an $\{\mathcal{F}(t)\}$-adapted process with nondecreasing, right-continuous paths and $C(0) = 0$, $C(T) < \infty$ almost surely. A *portfolio process* $\pi(\cdot)$ is an $\mathbb{R}^N$-valued, $\{\mathcal{F}(t)\}$-progressively measurable process satisfying (1.2.5) and (1.2.6) (see Remark 1.2.2); since

$$\int_0^T |\pi'(t)(b(t) - \delta(t) - r(t)\underline{1})|\, dt$$

$$= \int_0^T |\pi'(t)\sigma(t)\theta(t)|\, dt$$

$$\le \left(\int_0^T \|\sigma'(t)\pi(t)\|^2\, dt \right)^{1/2} \left(\int_0^T \|\theta(t)\|^2\, dt \right)^{1/2},$$

the condition (1.2.5) follows from (1.2.6) and (2.6). Thus, under the assumptions of this chapter, a *portfolio process* is an $\mathbb{R}^N$-valued, and $\{\mathcal{F}(t)\}$-progressively measurable process almost surely satisfying (1.2.6):

$$\int_0^T \|\sigma'(t)\pi(t)\|^2\, dt < \infty. \tag{2.11}$$

The *wealth process* $X^{x,C,\pi}(\cdot)$ corresponding to the triple (x, C, π) is determined by (1.3.4) (cf. (3.3.1)):

$$\frac{X^{x,C,\pi}(t)}{S_0(t)} + \int_{(0,t]} \frac{dC(v)}{S_0(v)} = x + \int_0^t \frac{1}{S_0(v)}\pi'(v)\sigma(v)\, dW_0(v), \quad 0 \le t \le T, \tag{2.12}$$

which can also be written as

$$M_0(t) \overset{\Delta}{=} H_0(t)X^{x,C,\pi}(t) + \int_{(0,t]} H_0(v)\, dC(v)$$

$$= x + \int_0^t H_0(v)[\sigma'(v)\pi(v) - X^{x,C,\pi}(v)\theta(v)]'\, dW(v), \quad 0 \le t \le T \tag{2.13}$$

(cf. (3.3.3)). By analogy with Remark 3.6.10, the corresponding *portfolio-proportion process* is defined by

$$p(t) \overset{\Delta}{=} \begin{cases} \dfrac{\pi(t)}{X^{x,C,\pi}(t)}, & \text{if } X^{x,C,\pi}(t) \ne 0, \\ p_*, & \text{if } X^{x,C,\pi}(t) = 0, \end{cases} \tag{2.14}$$

where p_* is an arbitrary but fixed vector in K. The N-dimensional vector $p(t) = (p_1(t), \ldots, p_N(t))'$ represents the *proportions of wealth* $X^{x,C,\pi}(t)$ invested in the corresponding stocks at time t, whereas $\pi(t) = (\pi_1(t), \ldots, \pi_N(t))'$ are the *actual amounts* invested.

Let K be a nonempty, closed, convex subset of $\mathbb{R}^N$. Interpretations of various choices of K are provided in Examples 4.1.

Definition 2.1: We say that a pair (C, π) consisting of a cumulative consumption process and a portfolio process is *admissible for the initial wealth $x \geq 0$ and the constraint set K*, and we write $(C, \pi) \in \mathcal{A}(x; K)$, if the process $X^{x,C,\pi}(\cdot)$ given by (2.12) satisfies

$$X^{x,C,\pi}(t) \geq 0, \quad \forall t \in [0, T] \tag{2.15}$$

almost surely, and the portfolio-proportion process $p(\cdot)$ defined by (2.14) satisfies

$$p(t) \in K \quad \text{for Lebesgue-a.e. } t \in [0, T] \tag{2.16}$$

almost surely. We denote by $\mathcal{M}(K)$ the financial market $\mathcal{M}$ of (2.1)–(2.6) when agents are constrained to choose (C, π) so that (2.15) and (2.16) are satisfied.

When $x \geq 0$ and $(C, \pi) \in \mathcal{A}(x; K)$, the process $M_0(\cdot)$ of (2.13) is a nonnegative local martingale, and hence a supermartingale. Taking expectations in (2.13), we obtain the *budget constraint*

$$E\left[H_0(T)X^{x,C,\pi}(T) + \int_{(0,T]} H_0(v)\, dC(v)\right] \leq x. \tag{2.17}$$

The following result is an extension of Theorem 3.3.5; this latter is more restrictive in that it requires the cumulative consumption process $C(\cdot)$ to be of the form $C(t) = \int_0^t c(s)\, ds$ for some nonnegative, $\{\mathcal{F}(t)\}$-progressively measurable consumption process $c(\cdot)$. It is straightforward to check that the proof of Theorem 3.3.5 goes through in the more general setting of Theorem 2.2 below.

Theorem 2.2: *Let $x \geq 0$ be given and suppose that $K = \mathbb{R}^N$. Let $C(\cdot)$ be a cumulative consumption process and B a nonnegative, $\mathcal{F}(T)$-measurable random variable such that*

$$E\left[H_0(T)B + \int_{(0,T]} H_0(t)\, dC(t)\right] = x. \tag{2.18}$$

Then there exists a portfolio process $\pi(\cdot)$ such that $(C, \pi) \in \mathcal{A}(x; \mathbb{R}^N)$ and the corresponding wealth process is given by

$$X^{x,\pi,C}(t) = \frac{1}{H_0(t)} E\left[H_0(T)B + \int_{(t,T]} H_0(s)\, dC(s) \,\middle|\, \mathcal{F}(t)\right],$$

$$0 \leq t \leq T. \tag{2.19}$$

In particular, the process $H_0(t)X^{x,C,\pi}(t) + \int_{(0,t]} H_0(s)\,dC(s)$, $0 \le t \le T$, is a martingale, and $X^{x,C,\pi}(T) = B$ almost surely.

5.3   Upper Hedging Price

Let us suppose that at time $t = 0$, the agent of Section 2 agrees to make a payment of a random amount $B \ge 0$ at the future time T. The randomness in the size of his payment may come from several factors, still unresolved at time $t = 0$ and beyond the agent's control. For instance, $B = (S_1(T) - q)^+$ describes the case of selling a European call option on the first stock, with strike price $q \ge 0$; $B = (q - S_1(T))^+$ is the case of a European put option. Additional examples are presented in Chapter 2.

What is the value at time $t = 0$ of this promise to pay B at time T? To answer this question, let us argue as in the beginning of Section 1.6. Suppose that at time $t = 0$ the agent sets aside an amount $x \ge 0$ to invest in the market $\mathcal{M}(K)$. He has to obey the constraint $p(\cdot) \in K$ of this market, but wants to be certain that at time T his wealth $X(T)$ will have grown to match or exceed the size of the payment he has to make, i.e., he wants to achieve $X(T) \ge B$ almost surely. We call the smallest amount of initial capital that enables him to do this the *upper hedging price* of the contingent claim B.

We formalize this discussion with the following definition. To simplify the presentation of this chapter, we define below a *contingent claim* as a nonnegative payment at the final time only. This is a special case of Definition 2.2.1, under which a *European contingent claim* was a cash flow over an entire time interval, at any point of which the flow could be making either positive or negative payments.

Definition 3.1:

(i) In this chapter, a *contingent claim* B is defined as a nonnegative $\mathcal{F}(T)$-measurable random variable. We call

$$u_0 \stackrel{\Delta}{=} E[H_0(T)B] \tag{3.1}$$

the *unconstrained hedging price* of B in the market $\mathcal{M}$.

(ii) The *upper hedging price in* $\mathcal{M}(K)$ of the contingent claim B is defined to be

$$h_{up}(K) \tag{3.2}$$
$$\stackrel{\Delta}{=} \inf\{x \ge 0; \exists (C, \pi) \in \mathcal{A}(x; K) \text{ with } X^{x,C,\pi}(T) \ge B, \text{ a.s.}\}.$$

(iii) Finally, we say that B is *K-attainable* if $h_{up}(K) < \infty$ and if there exists a portfolio process $\pi(\cdot)$ such that

$$(0, \pi) \in \mathcal{A}(h_{up}(K); K) \text{ and } X^{h_{up}(K),0,\pi}(T) = B \text{ a.s.} \tag{3.3}$$

Here we mean 0 to be the cumulative consumption process that is identically zero.

Suppose that there are no constraints on portfolio choice, i.e., $K = \mathbb{R}^N$. Then we know from the theory of Chapter 2 (in particular, Section 2.2) that the hedging price $h_{up}(\mathbb{R}^N)$ of the contingent claim is given by the expectation of its discounted value $B/S_0(T)$ under the standard equivalent martingale measure,

$$h_{up}(\mathbb{R}^N) = E\left[Z_0(T)\frac{B}{S_0(T)}\right] = E[H_0(T)B] = u_0, \qquad (3.4)$$

at least when the process that $Z_0(\cdot)$ of (2.8) is a martingale. This martingale condition is actually superfluous; it was not used in Theorem 3.3.5, and is thus not required for its extension, Theorem 2.2. This leads to the following result.

Proposition 3.2: *If $K = \mathbb{R}^N$, the upper hedging price $h_{up}(\mathbb{R}^N)$ is given by u_0 of (3.1). The infimum of (3.2) is then attained by some $(0, \pi_0) \in \mathcal{A}(u_0; \mathbb{R}^N)$ with*

$$X^{u_0,0,\pi_0}(t) = X_0(t) \triangleq \frac{1}{H_0(t)}E[H_0(T)B|\mathcal{F}(t)], \quad 0 \le t \le T, \qquad (3.5)$$

and in particular,

$$X^{u_0,0,\pi_0}(T) = B \qquad (3.6)$$

holds almost surely.

PROOF. For any $x \in [0, \infty)$ and $(C, \pi) \in \mathcal{A}(x; \mathbb{R}^N)$ with $X^{x,C,\pi} \ge B$ almost surely, the budget constraint (2.17) gives $x \ge E[H_0(T)B] = u_0$, whence $h_{up}(\mathbb{R}^N) \ge u_0$ from (3.2). On the other hand, Theorem 2.2 with $C(\cdot) \equiv 0$ and $x = u_0$ provides the existence of a portfolio $\pi_0(\cdot)$ such that $X^{u_0,0,\pi_0}(\cdot) = X_0(\cdot)$. This implies $h_{up}(\mathbb{R}^N) \le u_0$ and the other assertions of the proposition. $\qquad \square$

Remark 3.3: We call $\pi_0(\cdot)$ in Proposition 3.2 an *unconstrained hedging portfolio*. Because of the uniqueness of the integrand $\psi_0(\cdot)$ in the stochastic integral representation of the martingale $E[H_0(T)B|\mathcal{F}(t)]$, $\pi_0(\cdot)$ is uniquely determined, provided that we do not distinguish between processes that agree for Lebesgue-almost-every $t \in [0, T]$ almost surely.

5.4 Convex Sets and Support Functions

We introduced in Section 2 the nonempty, closed, convex set K in which the random vector $p(\cdot)$ of portfolio proportions is constrained to take values. We shall review now some basic notions from convex analysis that will be useful later on, and discuss several examples of such constraint sets K.

For a given closed, convex subset $K \neq \emptyset$ of $\mathbb{R}^N$, let us define $\zeta: \mathbb{R}^N \to \mathbb{R} \cup \{+\infty\}$ by

$$\zeta(\nu) \triangleq \sup_{p \in K} (-p'\nu), \quad \nu \in \mathbb{R}^N. \tag{4.1}$$

This is the *support function* of the convex set $-K$. It is a closed (i.e., lower semicontinuous), proper (i.e., not identically $+\infty$) convex function, which is finite on its *effective domain*

$$\widetilde{K} \triangleq \{\nu \in \mathbb{R}^N; \zeta(\nu) < \infty\}, \tag{4.2}$$

a convex cone, called the *barrier cone* of $-K$ (Rockafellar (1970), p. 114). In particular, $0 \in \widetilde{K}$ and $\zeta(0) = 0$. The function ζ is *positively homogeneous*,

$$\zeta(\alpha\nu) = \alpha\zeta(\nu), \quad \forall \nu \in \mathbb{R}^N, \quad \alpha \geq 0, \tag{4.3}$$

and *subadditive*,

$$\zeta(\nu + \mu) \leq \zeta(\nu) + \zeta(\mu), \quad \forall \nu, \mu \in \mathbb{R}^N. \tag{4.4}$$

According to Rockafellar (1970), Theorem 13.1, p. 112,

$$p \in K \iff \zeta(\nu) + p'\nu \geq 0, \quad \forall \nu \in \widetilde{K}. \tag{4.5}$$

It will be assumed in this chapter and the next that ζ is bounded from below on $\mathbb{R}^N$:

$$\zeta(\nu) \geq \zeta_0, \quad \forall \nu \in \mathbb{R}^N \text{ for some } \zeta_0 \in \mathbb{R}. \tag{4.6}$$

Condition (4.6) is satisfied with $\zeta_0 = 0$ if K contains the origin of $\mathbb{R}^N$.

Examples 4.1: Let us consider the following possible constraint sets K on portfolio proportions, all of which satisfy condition (4.6).

   (i) *Unconstrained case:* $K = \mathbb{R}^N$. Then $\widetilde{K} = \{0\}$, $\zeta \equiv 0$ on $\widetilde{K}$.

  (ii) *Prohibition of short-selling:* $K = [0, \infty)^N$. Then $\widetilde{K} = K$ and $\zeta \equiv 0$ on $\widetilde{K}$.

 (iii) *Incomplete market:* $K = \{p \in \mathbb{R}^N; p_{M+1} = \cdots = p_N = 0\}$, for some $M \in \{1, \ldots, N-1\}$. Then $\widetilde{K} = \{\nu \in \mathbb{R}^N; \nu_1 = \cdots = \nu_M = 0\}$, and $\zeta \equiv 0$ on $\widetilde{K}$.

 (iv) *Incomplete market with prohibition of short-selling:* $K = \{p \in \mathbb{R}^N; p_1 \geq 0, \ldots, p_M \geq 0, p_{M+1} = \cdots = p_N = 0\}$. Then $\widetilde{K} = \{\nu \in \mathbb{R}^N; \nu_1 \geq 0, \ldots, \nu_M \geq 0\}$ and $\zeta \equiv 0$ on $\widetilde{K}$.

  (v) K *is a nonempty, closed, convex cone in* $\mathbb{R}^N$. Then $\widetilde{K} = \{\nu \in \mathbb{R}^N; \ p'\nu \geq 0, \ \forall p \in K\}$ is the polar cone of $-K$ and $\zeta \equiv 0$ on $\widetilde{K}$. This generalizes (i)–(iv).

 (vi) *Prohibition of borrowing:* $K = \{p \in \mathbb{R}^N; \sum_{n=1}^N p_n \leq 1\}$. Then $\widetilde{K} = \{\nu \in \mathbb{R}^N; \nu_1 = \cdots = \nu_N \leq 0\}$ and $\zeta(\nu) = -\nu_1$ on $\widetilde{K}$.

(vii) *Constraints on short-selling:* $K = [-\kappa, \infty)^N$ for some $\kappa > 0$. Then $\widetilde{K} = [0, \infty)^N$ and $\zeta(\nu) = \kappa \sum_{n=1}^N \nu_n$ on $\widetilde{K}$.

(viii) *Constraints on borrowing:* $K = \{p \in \mathbb{R}^N; \sum_{n=1}^{N} p_n \le \kappa\}$ for some $\kappa > 1$. Then $\widetilde{K} = \{\nu \in \mathbb{R}^N; \nu_1 = \cdots = \nu_N \le 0\}$ and $\zeta(\nu) = -\kappa\nu_1$ on $\widetilde{K}$.

(ix) *Rectangular constraints:* $K = I_1 \times \cdots \times I_N$ with $I_n = [\alpha_n, \beta_n]$, $-\infty \le \alpha_n \le 0 \le \beta_n \le \infty$ and with the understanding that I_n is open on the right (respectively, left) if $\beta_n = \infty$ (respectively, $\alpha_n = -\infty$). Then

$$\widetilde{K} = \mathbb{R}^N, \quad \zeta(\nu) = -\sum_{n=1}^{N}(\alpha_n\nu_n^+ - \beta_n\nu_n^-)$$

if all the α_n and β_n are finite. More generally,

$$\widetilde{K} = \{\nu \in \mathbb{R}^N;\ \nu_n \ge 0, \quad \forall n \in \mathcal{S}_+ \text{ and } \nu_m \le 0, \quad \forall m \in \mathcal{S}_-\},$$

where $\mathcal{S}_+ = \{n = 1, \ldots, N;\ \beta_n = \infty\}$, $\mathcal{S}_- = \{m = 1, \ldots, N; \alpha_m = -\infty\}$, and the above formula for ζ remains valid.

We shall need the following lemma in Section 6.

Lemma 4.2: *For any given $\{\mathcal{F}(t)\}$-progressively measurable process $p \colon [0,T] \times \Omega \rightarrow \mathbb{R}^N$, there exists an $\mathbb{R}^N$-valued, $\{\mathcal{F}(t)\}$-progressively measurable process $\nu(\cdot)$ such that*

$$\|\nu(t)\| \le 1, \ |\zeta(\nu(t))| \le 1, \quad 0 \le t \le T, \tag{4.7}$$

almost surely, and for all $t \in [0,T]$ we have

$$\begin{aligned} p(t) \in K &\iff \nu(t) = 0, \\ p(t) \notin K &\iff \zeta(\nu(t)) + p'(t)\nu(t) < 0 \end{aligned} \tag{4.8}$$

almost surely.

PROOF. For $n = 1, 2, \ldots$, define $\widetilde{K}_n \triangleq \{\nu \in \widetilde{K};\ \|\nu\| \le n\}$, and $f_n \colon \mathbb{R}^N \times \widetilde{K}_n \rightarrow \mathbb{R}$ by $f_n(p, \nu) = \zeta(\nu) + p'\nu$. According to the Dubins–Savage measurable selection theorem (Dubins and Savage (1965) or Bertsekas and Shreve (1978), Proposition 7.33), there is a Borel-measurable function $\varphi_n \colon \mathbb{R}^N \rightarrow \widetilde{K}_n$ such that

$$\zeta(\varphi_n(p)) + p'\varphi_n(p) = \min_{\nu \in \widetilde{K}_n}\{\zeta(\nu) + p'\nu\}, \quad \forall p \in \mathbb{R}^N.$$

For $p \in \mathbb{R}^N$, define $\varphi(p)$ to be $\varphi_n(p)$ for the smallest positive integer n satisfying $\zeta(\varphi_n(p)) + p'\varphi_n(p) < 0$; if no such n exists, define $\varphi(p) = 0$. Then $\varphi \colon \mathbb{R}^N \rightarrow \widetilde{K}$ is Borel measurable. According to the equivalence (4.5), $\varphi(p) = 0$ for every $p \in K$ and $\varphi(p) < 0$ for every $p \notin K$.

Finally, we set

$$\nu(t) \triangleq \frac{\varphi(p(t))}{1 + \|\varphi(p(t))\| + |\zeta(\varphi(p(t)))|}, \quad 0 \le t \le T.$$

Conditions (4.7) and (4.8) follow from the positive homogeneity of ζ. $\quad\square$

5.5   A Family of Auxiliary Markets

The principal problem of this chapter can be formulated as follows: Given a contingent claim $B \geq 0$, find a minimal initial wealth $x \geq 0$, a cumulative consumption process $C(\cdot)$, and a portfolio process $\pi(\cdot)$ such the corresponding portfolio-proportion process $p(\cdot)$ satisfies the constraint $p(t) \in K$ for Lebesgue-almost-every $t \in [0, T]$ almost surely, and the corresponding terminal wealth satisfies $X^{x,C,\pi}(T) \geq B$ almost surely. In order to handle the constraint $p(t) \in K$, we introduce *dual processes*, which play a role similar to Lagrange multipliers. Corresponding to each dual process there is an auxiliary market as described below, in which we construct unconstrained portfolio-proportion processes.

Definition 5.1:   Let $\mathcal{H}$ denote the Hilbert space of $\{\mathcal{F}(t)\}$-progressively measurable processes $\nu\colon [0, T] \times \Omega \to \mathbb{R}^N$ with norm $[\![\nu]\!]$ given by

$$[\![\nu]\!]^2 \triangleq E \int_0^T \|\nu(t)\|^2 \, dt < \infty. \tag{5.1}$$

We define the inner product

$$\langle \nu_1, \nu_2 \rangle = E \int_0^T \nu_1'(t) \nu_2(t) \, dt$$

on this space, and denote by $\mathcal{D}$ the subset of $\mathcal{H}$ consisting of processes $\nu\colon [0, T] \times \Omega \to \widetilde{K}$ with

$$E \int_0^T \zeta(\nu(t)) \, dt < \infty. \tag{5.2}$$

We further define $\mathcal{D}^{(b)}$ to be the set of bounded processes in $\mathcal{D}$, and $\mathcal{D}^{(m)}$ to be the set of all processes in $\mathcal{D}$ for which $Z_\nu(\cdot)$ defined by (5.10) below is a martingale.

We began in Section 2 with a market

$$\mathcal{M} = (r(\cdot), b(\cdot), \delta(\cdot), \sigma(\cdot), S(0), A(\cdot)).$$

For every process $\nu(\cdot) \in \mathcal{D}$, consider a *new interest rate* process

$$r_\nu(t) \triangleq r(t) + \zeta(\nu(t)), \quad 0 \leq t \leq T \tag{5.3}$$

as well as a *new mean rate of return* vector process

$$b_\nu(t) \triangleq b(t) + \nu(t) + \zeta(\nu(t))1_N, \quad 0 \leq t \leq T, \tag{5.4}$$

and construct the *new market* $\mathcal{M}_\nu = (r_\nu(\cdot), b_\nu(\cdot), \delta(\cdot), \sigma(\cdot), S(0), A(\cdot))$. In this new market, the money market price $S_0^{(\nu)}(\cdot)$ and the stock prices $\{S_n^{(\nu)}(\cdot)\}_{n=1}^N$ obey the equations

$$dS_0^{(\nu)}(t) = S_0^{(\nu)}(t)[(r(t) + \zeta(\nu(t))) \, dt + dA(t)], \tag{5.5}$$

$$dS_n^{(\nu)}(t) = S_n^{(\nu)}(t) \left[\left(b_n(t) + \nu_n(t) + \zeta(\nu(t)) \right) dt + dA(t) \right.$$

$$\left. + \sum_{d=1}^{N} \sigma_{nd}(t) \, dW^{(d)}(t) \right], \quad n = 1, \ldots, N, \tag{5.6}$$

with the initial conditions

$$S_0^{(\nu)}(0) = 1 \quad \text{and} \quad S_n^{(\nu)}(0) = S_n(0), \quad n = 1, \ldots, N.$$

In other words,

$$S_0^{(\nu)}(t) = S_0(t) \exp \left[\int_0^t \zeta(\nu(s)) \, ds \right], \tag{5.7}$$

$$S_n^{(\nu)}(t) = S_n(t) \exp \left[\int_0^t \left(\zeta(\nu(s)) + \nu_n(s) \right) ds \right], \quad n = 1, \ldots, N. \tag{5.8}$$

The analogues of $\theta(\cdot)$, $Z_0(\cdot)$, $W_0(\cdot)$, and $H_0(\cdot)$ in (2.7)–(2.10) are

$$\theta_\nu(t) \triangleq \sigma^{-1}(t)[b_\nu(t) + \delta(t) - r_\nu(t)\mathbb{1}] \tag{5.9}$$
$$= \theta(t) + \sigma^{-1}(t)\nu(t),$$

$$Z_\nu(t) \triangleq \exp \left[-\int_0^t \theta_\nu'(s) \, dW(s) - \frac{1}{2} \int_0^t \|\theta_\nu(s)\|^2 \, ds \right]$$
$$= Z_0(t) \exp \left[-\int_0^t (\sigma^{-1}(s)\nu(s))' \, dW_0(s) \right.$$
$$\left. - \frac{1}{2} \int_0^t \|\sigma^{-1}(s)\nu(s)\|^2 \, ds \right], \tag{5.10}$$

$$W_\nu(t) \triangleq W(t) + \int_0^t \theta_\nu(s) \, ds \tag{5.11}$$
$$= W_0(t) + \int_0^t \sigma^{-1}(s)\nu(s) \, ds,$$

$$H_\nu(t) \triangleq \frac{Z_\nu(t)}{S_0^{(\nu)}(t)}. \tag{5.12}$$

Note that with $\nu(\cdot) \equiv 0$, we recover the unconstrained model of Section 2 (i.e., $\mathcal{M} = \mathcal{M}_0$). Note also from (4.6) that

$$S_0^{(\nu)}(T) \geq s_0 e^{\zeta_0 T} \tag{5.13}$$

almost surely; i.e., (2.5) holds in the market $\mathcal{M}_\nu$, and because of Definition 1.1.3(iv), (2.4), (2.6), (5.1), and (5.2), we have $\int_0^T |r_\nu(t)| \, dt < \infty$ and $\int_0^T \|\theta_\nu(t)\|^2 \, dt < \infty$ almost surely. Consequently, the conditions of Definition 1.1.3 and the conditions (2.3)–(2.6) are satisfied by $\mathcal{M}_\nu$ for every $\nu(\cdot) \in \mathcal{D}$.

Remark 5.2: If the exponential local martingale $Z_0(\cdot)$ of (2.8), (1.5.2) is a martingale (i.e., if the process $\nu(\cdot) \equiv 0$ belongs to $\mathcal{D}^{(m)}$), then we may define the standard martingale measure P_0 by

$$P_0(A) \triangleq \int_A Z_0(T)\,dP, \quad A \in \mathcal{F}(T), \tag{5.14}$$

as in (1.5.3). If $\theta(\cdot)$ is bounded and $\nu \in \mathcal{D}^{(b)}$, then $\theta_\nu(\cdot)$ is bounded, $Z_\nu(\cdot)$ is a martingale, and we may define the standard equivalent martingale measure P_ν for the market $\mathcal{M}_\nu$ by

$$P_\nu(A) \triangleq \int_A Z_\nu(T)\,dP, \quad A \in \mathcal{F}(T). \tag{5.15}$$

Thus, if $\theta(\cdot)$ is bounded, we have $\mathcal{D}^{(b)} \subseteq \mathcal{D}^{(m)}$.

In the market $\mathcal{M}_\nu$, the wealth process $X_\nu^{x,C,\pi}(\cdot)$ corresponding to initial capital $x \geq 0$, cumulative consumption process $C(\cdot)$, and (unconstrained) portfolio process $\pi(\cdot)$ satisfies the equation

$$\frac{X_\nu^{x,C,\pi}(t)}{S_0^{(\nu)}(t)} + \int_{(0,t]} \frac{dC(s)}{S_0^{(\nu)}(s)} = x + \int_0^t \frac{1}{S_0^{(\nu)}(s)}\pi(s)\sigma(s)\,dW_\nu(s), \quad 0 \leq t \leq T \tag{5.16}$$

(cf. (2.12)), or equivalently,

$$\frac{X_\nu^{x,C,\pi}(t)}{S_0(t)} + \int_{(0,t]} \frac{dC(s)}{S_0(s)}$$

$$= x + \int_0^t \frac{1}{S_0(s)}\left[(X_\nu^{x,C,\pi}(s)\zeta(\nu(s)) + \pi'(s)\nu(s))\,ds + \pi'(s)\sigma(s)\,dW_0(s)\right]$$

$$= x + \int_0^t \frac{X_\nu^{x,C,\pi}(s)}{S_0(s)}\left[(\zeta(\nu(s)) + p'(s)\nu(s))\,ds + p'(s)\sigma(s)\,dW_0(s)\right]. \tag{5.17}$$

By analogy with (2.13), we have

$$M_\nu(t) \triangleq H_\nu(t)X_\nu^{x,C,\pi}(t) + \int_{(0,t]} H_\nu(s)\,dC(s)$$

$$= x + \int_0^t H_\nu(s)[\sigma'(s)\pi(s) - X_\nu^{x,C,\pi}(s)\theta_\nu(s)]'\,dW(s)$$

$$= x + \int_0^t H_\nu(s)X^{x,C,\pi}(s)[\sigma'(s)p(s) - \theta_\nu(s)]'\,dW(s), \quad 0 \leq t \leq T. \tag{5.18}$$

Definition 5.3: Let $\nu(\cdot) \in \mathcal{D}$ be given. We say that a pair (C, π) consisting of a cumulative consumption process and a portfolio process is *admissible* in $\mathcal{M}_\nu$ for the *initial wealth* $x \geq 0$, and we write $(C, \pi) \in \mathcal{A}_\nu(x)$, if the process $X_\nu^{x,C,\pi}(\cdot)$ determined by (5.16) satisfies almost surely

$$X_\nu^{x,C,\pi}(t) \geq 0, \quad \forall t \in [0,T]. \tag{5.19}$$

Remark 5.4: For $x \geq 0$ and $(C,\pi) \in \mathcal{A}_\nu(x)$, the local martingale $M_\nu(\cdot)$ of (5.18) is nonnegative, and hence a supermartingale. Fatou's lemma implies then the *budget constraint* (cf. (2.17))

$$E\left[H_\nu(T)X_\nu^{x,C,\pi}(T) + \int_{(0,T]} H_\nu(s)\,dC(s)\right] \leq x. \tag{5.20}$$

We have the following analogue to Theorem 2.2 concerning the existence of portfolio processes.

Theorem 5.5: *Let $\nu(\cdot) \in \mathcal{D}$ and $x \geq 0$ be given. Let $C(\cdot)$ be a cumulative consumption process, and B a nonnegative, $\mathcal{F}(T)$-measurable random variable such that*

$$E\left[H_\nu(T)B + \int_{(0,T]} H_\nu(s)\,dC(s)\right] = x. \tag{5.21}$$

Then there exists a portfolio process $\pi(\cdot)$ such that $(C,\pi) \in \mathcal{A}_\nu(x)$, and the corresponding wealth process is given by

$$X_\nu^{x,C,\pi}(t) = \frac{1}{H_\nu(t)} E\left[H_\nu(T)B + \int_{(t,T]} H_\nu(s)\,dC(s)\,\bigg|\,\mathcal{F}(t)\right], \quad 0 \leq t \leq T. \tag{5.22}$$

In particular, the process $H_\nu(t)X_\nu^{x,C,\pi}(t) + \int_{(0,t]} H_\nu(s)\,dC(s)$, $0 \leq t \leq T$, is a martingale, and $X_\nu^{x,C,\pi}(T) = B$ almost surely.

5.6 The Main Hedging Result

Definition 6.1: Consider a contingent claim B as in Definition 3.1. The *unconstrained hedging price* of B in the market $\mathcal{M}_\nu$ is

$$u_\nu \triangleq E\left[Z_\nu(T)\frac{B}{S_0^{(\nu)}(T)}\right] = E[H_\nu(T)B], \tag{6.1}$$

a nonnegative and possibly infinite quantity. If $u_\nu < \infty$, then an *unconstrained hedging portfolio* $\pi_\nu(\cdot)$ is any portfolio process satisfying

$$X_\nu^{u_\nu,0,\pi_\nu}(t) = X_\nu(t) \triangleq \frac{1}{H_\nu(t)} E[H_\nu(T)B|\mathcal{F}(t)], \quad 0 \leq t \leq T. \tag{6.2}$$

The existence of an unconstrained hedging portfolio in Definition 6.1 follows from Theorem 5.5 with $C(\cdot) \equiv 0$. As in Remark 3.3, $\pi_\nu(\cdot)$ is uniquely determined.

Theorem 6.2: *For any contingent claim B, we have the representation*

$$h_{up}(K) = \sup_{\nu \in \mathcal{D}} u_\nu \tag{6.3}$$

of the upper hedging price of Definition 3.1(ii). Furthermore, if

$$\hat{u} \overset{\Delta}{=} \sup_{\nu \in \mathcal{D}} u_\nu \tag{6.4}$$

is finite, then there exists a pair $(\widehat{C}, \hat{\pi}) \in \mathcal{A}(\hat{u}; K)$ with corresponding wealth process

$$X^{\hat{u}, \widehat{C}, \hat{\pi}}(t) = \mathrm{ess\ sup}_{\nu \in \mathcal{D}} \frac{E[H_\nu(T) B | \mathcal{F}(t)]}{H_\nu(t)}, \quad 0 \leq t \leq T, \tag{6.5}$$

and in particular,

$$X^{\hat{u}, \widehat{C}, \hat{\pi}}(T) = B \tag{6.6}$$

holds almost surely.

Remark 6.3: We call $\hat{\pi}(\cdot)$ a *superreplicating portfolio process* because it allows an agent to begin with initial wealth $h_{up}(K)$, possibly consume along the way, and end up with terminal wealth B almost surely. If $\widehat{C}(\cdot) \equiv 0$, then $\hat{\pi}(\cdot)$ is a *replicating portfolio process*, and the contingent claim B is *K-attainable* (Definition 3.1(iii)).

Definition 6.4: We call the nonnegative process

$$\widehat{X}(t) \overset{\Delta}{=} \mathrm{ess\ sup}_{\nu \in \mathcal{D}} \frac{E[H_\nu(T) B | \mathcal{F}(t)]}{H_\nu(t)}, \quad 0 \leq t \leq T \tag{6.7}$$

on the right-hand side of (6.5) the *upper hedging value process* for the contingent claim B. We shall always take a right-continuous, left-limited modification (RCLL; see Proposition 6.5 below) of this process.

We devote the remainder of this section to the proof of Theorem 6.2. Let us start with the inequality

$$h_{up}(K) \geq \hat{u}, \tag{6.8}$$

which is obvious if $h_{up}(K) = \infty$. Now assume $h_{up}(K) < \infty$, and consider an arbitrary $x \in [0, \infty)$ for which there exists a pair $(C, \pi) \in \mathcal{A}(x; K)$ whose associated wealth process satisfies $X^{x,C,\pi}(T) \geq B$ almost surely. Let $\nu(\cdot) \in \mathcal{D}$ be given, and define

$$\begin{aligned}
C_\nu(t) &= C(t) + \int_0^t \frac{1}{S_0(s)} \left[X^{x,C,\pi}(s) \zeta(\nu(s)) + \pi'(s)\nu(s) \right] ds \\
&= C(t) + \int_0^t \frac{X^{x,C,\pi}(s)}{S_0(s)} [\zeta(\nu(s)) + p'(s)\nu(s)] ds, \quad 0 \leq t \leq T,
\end{aligned}$$

where the portfolio-proportion process $p(\cdot)$, given by (2.14), satisfies the constraint (2.16). It follows from (4.5) that $C_\nu(\cdot)$ is nondecreasing, hence a cumulative consumption process. Using the consumption and portfolio process pair (C_ν, π) in the market $\mathcal{M}_\nu$, we generate the wealth process $X_\nu^{x,C_\nu,\pi}(\cdot)$ of (5.16), which is the unique solution of (5.17):

$$\frac{X_\nu^{x,C_\nu,\pi}(t)}{S_0(t)} + \int_{(0,t]} \frac{dC(s)}{S_0(s)} + \int_0^t \frac{1}{S_0(s)} \left[X^{x,C,\pi}(s)\zeta(\nu(s)) + \pi'(s)\nu(s) \right] ds$$

$$= x + \int_0^t \frac{1}{S_0(s)} \left[(X_\nu^{x,C_\nu,\pi}(s)\zeta(\nu(s)) + \pi'(s)\nu(s)) \, ds \right.$$
$$\left. + \pi'(s)\sigma(s) \, dW_0(s) \right].$$

Comparing this equation with (2.12) we see that $X_\nu^{x,C_\nu,\pi}(\cdot) = X^{x,C,\pi}(\cdot)$, and because of the budget constraint (5.20) we obtain

$$x \geq E[H_\nu(T)X_\nu^{x,C_\nu,\pi}(T)] = E[H_\nu(T)X^{x,C,\pi}(T)] \geq E[H_\nu(T)B] = u_\nu.$$

Since $\nu(\cdot) \in \mathcal{D}$ is arbitrary, we conclude that $x \geq \hat{u}$. This implies (6.8).

We turn to the reverse inequality

$$h_{up}(K) \leq \hat{u}, \tag{6.9}$$

which is obvious if $\hat{u} = \infty$. Thus let us assume for the remainder of the section that $\hat{u} < \infty$, and study in some detail the properties of the upper hedging value process (6.7). We need the following technical result.

Proposition 6.5: *Under the assumption $\hat{u} < \infty$, the upper hedging value process $\widehat{X}(\cdot)$ of (6.7) is finite and satisfies the dynamic programming equation*

$$\widehat{X}(s) = ess \; sup_{\nu \in \mathcal{D}} \frac{E[H_\nu(t)\widehat{X}(t)|\mathcal{F}(s)]}{H_\nu(s)}, \quad 0 \leq s \leq t \leq T. \tag{6.10}$$

Furthermore, $\widehat{X}(\cdot)$ has a right-continuous, left-limited (RCLL) modification; choosing this modification, we have that the process $H_\nu(\cdot)\widehat{X}(\cdot)$ is an RCLL supermartingale for every $\nu(\cdot) \in \mathcal{D}$.

PROOF.   To alleviate the notation we write $H_\nu(s,t) \overset{\Delta}{=} H_\nu(t)/H_\nu(s)$ for $0 \leq s \leq t \leq T$ and set $J_\nu(t) \overset{\Delta}{=} E[H_\nu(t,T)B|\mathcal{F}(t)]$, so that

$$\widehat{X}(t) = ess \; sup_{\nu \in \mathcal{D}} J_\nu(t). \tag{6.11}$$

From (6.7) and (6.11), we have

$$\widehat{X}(s) = ess \; sup_{\nu \in \mathcal{D}} E[H_\nu(s,t)J_\nu(t)|\mathcal{F}(s)]$$
$$\leq ess \; sup_{\nu \in \mathcal{D}} E[H_\nu(s,t)\widehat{X}(t)|\mathcal{F}(s)].$$

To prove the reverse inequality, and thus (6.10), it suffices to fix an arbitrary process $\nu(\cdot) \in \mathcal{D}$ and show that

$$\widehat{X}(s) \geq E[H_\nu(s,t)\widehat{X}(t)|\mathcal{F}(s)] \tag{6.12}$$

almost surely. This is the supermartingale property for $H_\nu(\cdot)\widehat{X}(\cdot)$. With $\nu(\cdot) \in \mathcal{D}$ fixed, we denote by $\mathcal{D}_{t,\nu}$ the set of all processes $\mu(\cdot) \in \mathcal{D}$ that agree with $\nu(\cdot)$ on $[0,t] \times \Omega$. Since $H_\mu(t,T)$ depends only on the values of $\mu(v)$ of $\mu(\cdot)$ for $t \leq v \leq T$, we may rewrite (6.11) as

$$\widehat{X}(t) = \text{ess sup}_{\mu \in \mathcal{D}_{t,\nu}} J_\mu(t).$$

But the collection $\{J_\mu(t)\}_{\mu \in \mathcal{D}_{t,\nu}}$ is closed under pairwise maximization. Indeed, for any two given processes $\mu_1(\cdot)$ and $\mu_2(\cdot)$ in $\mathcal{D}_{t,\nu}$, and setting $A \triangleq \{\omega \in \Omega; J_{\mu_1}(t,\omega) \geq J_{\mu_2}(t,\omega)\}$ and $\mu(v,\omega) \triangleq \mu_1(v,\omega)1_A(\omega) + \mu_2(v,\omega)1_{A^c(\omega)} \in \mathcal{D}_{t,\nu}$, we have

$$\begin{aligned}
J_\mu(t) &= E[H_\mu(t,T)B|\mathcal{F}(t)] \\
&= E\left[(1_A H_{\mu_1}(t,T) + 1_{A^c} H_{\mu_2}(t,T))\, B|\, \mathcal{F}(t)\right] \\
&= 1_A E[H_{\mu_1}(t,T)B|\mathcal{F}(t)] + 1_{A^c} E[H_{\mu_2}(t,T)B|\mathcal{F}(t)] \\
&= J_{\mu_1}(t) \vee J_{\mu_2}(t).
\end{aligned}$$

It follows from Theorem A.3 of Appendix A that there is a sequence $\{\mu_k(\cdot)\}_{k=1}^\infty$ in $\mathcal{D}_{t,\nu}$ such that $\{J_{\mu_k}(t)\}_{k=1}^\infty$ is nondecreasing and

$$\widehat{X}(t) = \lim_{k \to \infty} J_{\mu_k}(t). \tag{6.13}$$

The monotone convergence theorem now implies

$$\begin{aligned}
E[H_\nu(s,t)\widehat{X}(t)|\mathcal{F}(s)] &= \lim_{k \to \infty} E[H_\nu(s,t)J_{\mu_k}(t)|\mathcal{F}(s)] \\
&= \lim_{k \to \infty} E[H_{\mu_k}(s,t)E[H_{\mu_k}(t,T)B|\mathcal{F}(t)]|\mathcal{F}(s)] \\
&= \lim_{k \to \infty} E[H_{\mu_k}(s,T)B|\mathcal{F}(s)] \\
&= \lim_{k \to \infty} J_{\mu_k}(s) \\
&\leq \text{ess sup}_{\mu \in \mathcal{D}} J_\mu(s) \\
&= \widehat{X}(s),
\end{aligned}$$

and (6.12) is established. Setting $s = 0$, we obtain

$$E[H_\nu(t)\widehat{X}(t)] \leq \widehat{X}(0) = \hat{u} < \infty,$$

which shows that $\widehat{X}(t)$ is finite for all $t \in [0,T]$ almost surely.

It remains to show that $\widehat{X}(\cdot)$ has an RCLL modification. From the supermartingale property of $H_\nu(\cdot)\widehat{X}(\cdot)$ for fixed $\nu \in \mathcal{D}$, and from the right-continuity of the filtration $\{\mathcal{F}(t)\}$, we know that the right-hand limit

$$\widehat{X}_+(t,\omega) \triangleq \begin{cases} \lim_{s\downarrow t, s\in\mathbb{Q}} \widehat{X}(s,\omega), & 0 \leq t < T, \\ \widehat{X}(T,\omega) = B(\omega), & t = T \end{cases}$$

is defined and finite for every ω in some set $\Omega^* \in \mathcal{F}(T)$ with $P(\Omega^*) = 1$ (Karatzas and Shreve (1991), Proposition 1.3.14). Here $\mathbb{Q}$ is the set of rational numbers. Furthermore, $H_\nu(\cdot)\widehat{X}_+(\cdot)$ is an RCLL supermartingale, and

$$\widehat{X}_+(t) \geq E[H_\nu(t,T)B|\mathcal{F}(t)], \quad 0 \leq t \leq T$$

almost surely. This last inequality holds for every $\nu(\cdot) \in \mathcal{D}$, which implies $\widehat{X}_+(t) \geq \widehat{X}(t)$ for every $t \in [0,T]$. On the other hand, the right-continuity of $\widehat{X}_+(\cdot)$, Fatou's lemma, and (6.12) show that for a fixed $\nu(\cdot) \in \mathcal{D}$, for any $t \in [0,T)$, and for any sequence of rationals $\{t_n\}_{n=1}^\infty$ converging down to t,

$$\widehat{X}_+(t) = E\left[\lim_{n\to\infty} H_\nu(t,t_n)\widehat{X}(t_n)\bigg| \mathcal{F}(t)\right]$$
$$\leq \liminf_{n\to\infty} E[H_\nu(t,t_n)\widehat{X}(t_n)|\mathcal{F}(t)]$$
$$\leq \widehat{X}(t)$$

almost surely. Thus $\widehat{X}(\cdot)$ and $\widehat{X}_+(\cdot)$ are modifications of one another.   □

Remark 6.6: The supermartingale property for the nonnegative RCLL process $H_\nu(\cdot)\widehat{X}(\cdot)$ implies that we have

$$\widehat{X}(t) = 0, \quad \forall t \in [\hat{\tau}, T]$$

almost surely on $\{\hat{\tau} < T\}$, where

$$\hat{\tau} \overset{\Delta}{=} \inf\{t \in [0,T); \widehat{X}(t) = 0\} \wedge T \tag{6.14}$$

and $\widehat{X}(\cdot)$ is defined by (6.7). Of course, if $P(B > 0) = 1$, then $\widehat{X}(\cdot)$ is strictly positive on $[0,T]$ and $\hat{\tau} = T$ almost surely.

Remark 6.7: Let B be a contingent claim. For each $\{\mathcal{F}(t)\}$-stopping time τ taking values in $[0,T]$, let us define

$$\widetilde{X}(\tau) \overset{\Delta}{=} \text{ess sup}_{\nu \in \mathcal{D}} \frac{E[H_\nu(T)B|\mathcal{F}(\tau)]}{H_\nu(\tau)}.$$

For constant $\tau \equiv t \in [0,T]$, the random variable $\widehat{X}(t)$ of (6.7) agrees with $\widetilde{X}(t)$. However, if the stopping time τ is not constant, then $\widehat{X}(\tau(\omega),\omega)$ obtained from substitution of $\tau(\omega)$ for t in $\widehat{X}(t,\omega)$ is defined differently from $\widetilde{X}(\tau,\omega)$. Nonetheless, when we take a right-continuous modification of $\widehat{X}(\cdot)$, we have $\widehat{X}(\tau) = \widetilde{X}(\tau)$ almost surely. A similar result in a more difficult context receives a detailed treatment in Appendix D. In the present setting we provide a simpler proof.

Let $\nu(\cdot) \in \mathcal{D}$ be given. Because $H_\nu(\cdot)\widehat{X}(\cdot)$ is a supermartingale with $H_\nu(T)\widehat{X}(T) = H_\nu(T)B$, we have from Doob's optional sampling theorem that for any stopping time τ taking values in $[0,T]$,

$$H_\nu(\tau)\widehat{X}(\tau) \geq E[H_\nu(T)B|\mathcal{F}(\tau)].$$

Dividing by $H_\nu(\tau)$ and taking essential supremum over $\nu(\cdot) \in \mathcal{D}$, we conclude that

$$\widehat{X}(\tau) \geq \widetilde{X}(\tau) \tag{6.15}$$

holds almost surely.

On the other hand, a straightforward modification of the proof of Proposition 6.5 shows that whenever ρ and τ are stopping times satisfying $0 \leq \rho \leq \tau \leq T$ almost surely, then $\widetilde{X}(\cdot)$ satisfies the dynamic programming equation

$$\widetilde{X}(\rho) = \text{ess sup}_{\nu \in \mathcal{D}} \frac{E[H_\nu(\tau)\widetilde{X}(\tau)|\mathcal{F}(\rho)]}{H_\nu(\rho)}, \tag{6.16}$$

and, in particular,

$$H_\nu(\rho)\widetilde{X}(\rho) \geq E[H_\nu(\tau)\widetilde{X}(\tau)|(\rho)], \quad \forall \nu(\cdot) \in \mathcal{D}. \tag{6.17}$$

Because $\widehat{X}(t) = \widetilde{X}(t)$ almost surely for each deterministic time t, we also have $\widehat{X}(\tau) = \widetilde{X}(\tau)$ almost surely for each stopping time τ taking only finitely many values. Let τ be an arbitrary stopping time with values in $[0, T]$, and construct a sequence of stopping times $\{\tau_n\}_{n=1}^\infty$, each of which takes only finitely many values, and such that $\tau_n \downarrow \tau$ almost surely as $n \to \infty$ (Karatzas and Shreve (1991), Problem 1.2.24). For each set $A \in \mathcal{F}(\tau)$, the right continuity of $\widehat{X}(\cdot)$, Fatou's lemma, and (6.17) imply

$$\int_A H_\nu(\tau)\widehat{X}(\tau)\,dP \leq \lim_{n\to\infty} \int_A H_\nu(\tau_n)\widehat{X}(\tau_n)\,dP$$

$$= \lim_{n\to\infty} \int_A H_\nu(\tau_n)\widetilde{X}(\tau_n)\,dP$$

$$\leq \int_A H_\nu(\tau)\widetilde{X}(\tau)\,dP.$$

This gives us the reverse of inequality (6.15).

To complete the proof of Theorem 6.2 it remains to show that when $\hat{u} < \infty$, there exists a pair $(\widehat{C}, \hat{\pi}) \in \mathcal{A}(\hat{u}; K)$ such that the corresponding wealth process $X^{\hat{u},\widehat{C},\hat{\pi}}(\cdot)$ satisfies almost surely

$$X^{\hat{u},\widehat{C},\hat{\pi}}(t) = \widehat{X}(t), \quad 0 \leq t \leq T, \tag{6.18}$$

with $\widehat{X}(\cdot)$ given by (6.7). This will imply (6.9).

Fix $\nu(\cdot) \in \mathcal{D}$. For the nonnegative supermartingale $H_\nu(\cdot)\widehat{X}(\cdot)$, define the sequence of stopping times

$$\rho_n \overset{\Delta}{=} \inf\{t \in [0, T); H_\nu(t)\widehat{X}(t) = n\} \wedge T, \quad n = 1, 2, \ldots. \tag{6.19}$$

Since the paths of $H_\nu(\cdot)\widehat{X}(\cdot)$ are almost surely right continuous with left-hand limits, the paths are almost surely bounded on $[0, T]$, and $\rho_n \uparrow T$ as

$n \to \infty$. Each stopped supermartingale $H_\nu(t \wedge \rho_n)\widehat{X}(t \wedge \rho_n)$, $0 \le t \le T$, is bounded and has a unique Doob–Meyer decomposition (Karatzas and Shreve (1991), pp. 24–27 or Protter (1990), p. 94), and this leads to a unique Doob–Meyer decomposition of the nonstopped supermartingale as

$$H_\nu(t)\widehat{X}(t) = \hat{u} + \int_0^t \psi_\nu'(s)\, dW(s) - A_\nu(t), \quad 0 \le t \le T, \qquad (6.20)$$

almost surely. Here

(i) $A_\nu(\cdot)$ is an adapted, natural process with nondecreasing, right-continuous paths almost surely, $EA_\nu(T) < \infty$, $A(0) = 0$;

(ii) $\psi_\nu(\cdot)$ is a progressively measurable, $\mathbb{R}^N$-valued process satisfying the square-integrability condition $\int_0^T \|\psi_\nu(t)\|^2\, dt < \infty$ almost surely.

Remark 6.6 implies that we have, almost surely,

$$A_\nu(t) = A_\nu(\hat{\tau}) \quad \text{and} \quad \psi_\nu(t) = 0 \text{ for Lebesgue-a.e. } t \in [\hat{\tau}, T]. \qquad (6.21)$$

Remark 6.8: The filtration $\{\mathcal{F}(t)\}$ is generated by the d-dimensional Brownian motion $W(\cdot)$, and every right-continuous martingale of this filtration must be continuous because it has a stochastic integral representation with respect to $W(\cdot)$. Therefore, *every adapted, nondecreasing and right-continuous process is natural* (Karatzas and Shreve (1991), Definition 4.5, p. 23). Furthermore, if $Z_\nu(\cdot)$ is a martingale (so that the market $\mathcal{M}_\nu$ is complete and standard), then every P_ν-martingale has a stochastic integral representation with respect to $W_\nu(\cdot)$ and thus is continuous (Lemma 1.6.7). Hence, every adapted nondecreasing process is natural under P_ν as well as under P.

Now let $\mu(\cdot)$ be another process in $\mathcal{D}$, and compute

$$d\left(\frac{H_\mu(t)}{H_\nu(t)}\right) = \frac{H_\mu(t)}{H_\nu(t)} \left[(\theta_\nu(t) - \theta_\mu(t))'\, dW(t) \right.$$
$$\left. + (\theta_\nu(t) - \theta_\mu(t))'\theta_\nu(t)\, dt + (\zeta(\nu(t)) - \zeta(\mu(t)))\, dt \right],$$

which implies

$$d(H_\mu(t)\widehat{X}(t)) = d\left[\frac{H_\mu(t)}{H_\nu(t)} \cdot H_\nu(t)\widehat{X}(t) \right]$$
$$= H_\mu(t)\widehat{X}(t)\left[(\theta_\nu(t) - \theta_\mu(t))'\, dW(t) \right.$$
$$\left. + (\theta_\nu(t) - \theta_\mu(t))'\theta_\nu(t)\, dt + (\zeta(\nu(t)) - \zeta(\mu(t)))\, dt \right]$$
$$+ \frac{H_\mu(t)}{H_\nu(t)}(\psi_\nu'(t)\, dW(t) - dA_\nu(t))$$
$$+ \frac{H_\mu(t)}{H_\nu(t)}(\theta_\nu(t) - \theta_\mu(t))'\psi_\nu(t)\, dt$$

in conjunction with (6.20). But we have also $d(H_\mu(t)\widehat{X}(t)) = \psi'_\mu(t)\,dW(t) - dA_\mu(t)$. Thus, equating the local martingale terms, we obtain

$$H_\mu(t)\widehat{X}(t)(\theta_\nu(t) - \theta_\mu(t)) + \frac{H_\mu(t)}{H_\nu(t)}\psi_\nu(t) = \psi_\mu(t),$$

or equivalently,

$$\varphi(t) \triangleq \frac{\psi_\nu(t)}{H_\nu(t)} + \widehat{X}(t)\theta_\nu(t) = \frac{\psi_\mu(t)}{H_\mu(t)} + \widehat{X}(t)\theta_\mu(t). \qquad (6.22)$$

Equating terms of finite variation, we obtain

$$\widehat{X}(t)(\theta_\nu(t) - \theta_\mu(t))'\theta_\nu(t) + (\theta_\nu(t) - \theta_\mu(t))'\frac{\psi_\nu(t)}{H_\nu(t)}$$

$$= \widehat{X}(t)(\zeta(\mu(t)) - \zeta(\nu(t))) + \frac{dA_\nu(t)}{H_\nu(t)} - \frac{dA_\mu(t)}{H_\mu(t)},$$

which implies

$$\widehat{C}(t) \triangleq \int_{(0,t]} \frac{dA_\nu(s)}{H_\nu(s)} - \int_0^t [\widehat{X}(s)\zeta(\nu(s)) + \varphi'(s)\sigma^{-1}(s)\nu(s)]\,ds$$

$$= \int_{(0,t]} \frac{dA_\mu(s)}{H_\mu(s)} - \int_0^t [\widehat{X}(s)\zeta(\mu(s)) + \varphi'(s)\sigma^{-1}(s)\mu(s)]\,ds. \qquad (6.23)$$

In particular, the processes $\varphi(\cdot)$ and $\widehat{C}(\cdot)$ defined in (6.22) and (6.23) *do not depend on* $\nu(\cdot) \in \mathcal{D}$, and satisfy almost surely

$$\widehat{C}_\nu(t) = \widehat{C}_\nu(\hat{\tau}) \quad \text{and} \quad \varphi(t) = 0, \text{ for Lebesgue-a.e. } t \in [\hat{\tau}, T]. \qquad (6.24)$$

Finally, we have $\int_0^T \|\varphi(t)\|^2\,dt < \infty$ almost surely.

The process $\widehat{C}(\cdot)$ of (6.23) is adapted, with RCLL paths. Writing (6.23) with $\nu(\cdot) \equiv 0$, we obtain

$$\widehat{C}(t) = \int_{(0,t]} \frac{dA_0(s)}{H_0(s)}, \qquad (6.25)$$

which shows that $\widehat{C}(\cdot)$ is nondecreasing; $\widehat{C}(\cdot)$ will play the role of cumulative consumption process in (6.18). The role of portfolio process will be played by

$$\hat{\pi}(t) \triangleq (\sigma'(t))^{-1}\varphi(t), \quad 0 \leq t \leq T. \qquad (6.26)$$

In terms of this process, we may rewrite (6.23) as

$$\widehat{C}(t) = \int_{(0,t]} \frac{dA_\nu(s)}{H_\nu(s)} - \int_0^t [\widehat{X}(s)\zeta(\nu(s)) + \hat{\pi}'(s)\nu(s)]\,ds, \quad 0 \leq t \leq T. \qquad (6.27)$$

From (6.20), (6.22), (6.26), and (6.25) with $\nu(\cdot) \equiv 0$, we have

$$H_0(t)\widehat{X}(t) = \hat{u} + \int_0^t \psi_0'(s)\,dW(s) - A_0(t)$$

$$= \hat{u} + \int_0^t H_0(s)[\sigma'(s)\hat{\pi}(s) - \widehat{X}(s)\theta(s)]'\,dW(s)$$

$$- \int_{(0,t]} H_0(s)\,d\widehat{C}(s). \tag{6.28}$$

Comparing this with (2.13), we conclude that (6.18) holds.

Finally, we define the portfolio-proportion process

$$\hat{p}(t) \overset{\Delta}{=} \begin{cases} \dfrac{\hat{\pi}(t)}{\widehat{X}(t)}, & \text{if } \widehat{X}(t) \neq 0, \\[2mm] p_*, & \text{if } \widehat{X}(t) = 0, \end{cases} \tag{6.29}$$

where p_* is an arbitrary but fixed vector in K. In order to conclude the proof of Theorem 6.2, we need to show that

$$\hat{p}(t) \in K \text{ for Lebesgue-a.e. } t \in [0, \hat{\tau}] \tag{6.30}$$

holds almost surely. To this end, consider the process $\nu(\cdot) \in \mathcal{D}$ given by Lemma 4.2. For any positive integer k, the process $k\nu(\cdot)$ is also in $\mathcal{D}$, and (6.27) gives

$$0 \leq \int_{(0,\hat{\tau}]} \frac{dA_{k\nu}(s)}{H_{k\nu}(s)}$$

$$= \widehat{C}(\hat{\tau}) + k\int_0^{\hat{\tau}} \widehat{X}(s)[\zeta(\nu(s)) + \hat{p}'(s)\nu(s)]\,ds$$

almost surely. Because $\nu(\cdot)$ satisfies (4.8), the integrand on the right-hand side of this inequality is nonpositive, and by choosing k sufficiently large the right-hand side can be made negative with positive probability, unless

$$\zeta(\nu(t)) + \hat{p}'(t)\nu(t) = 0 \text{ for Lebesgue-a.e. } t \in [0, \hat{\tau}] \tag{6.31}$$

holds almost surely. Thus (6.31) must hold, and with it, (6.30) must hold as well. The proof of Theorem 6.2 is complete.

Definition 6.9: Assume $\hat{u} < \infty$, let $\hat{\tau}$ be the stopping time (6.14), let $(\widehat{C}, \hat{\pi}) \in \mathcal{A}(\hat{u}; K)$ be the pair of processes (6.26) and (6.27) constructed in the proof of Theorem 6.2, and define $\hat{p}(\cdot)$ by (6.29). The set of dual processes satisfying the *complementarity condition* is

$$\mathcal{D}^{(c)} \overset{\Delta}{=} \{\nu \in \mathcal{D};\ \zeta(\nu(t)) + \hat{p}'(t)\nu(t) = 0 \text{ Lebesgue-a.e. } t \in [0, \hat{\tau}],\ a.s.\}. \tag{6.32}$$

Remark 6.10: We may rewrite (6.27) as

$$\widehat{C}(t) = \int_{(0,t]} \frac{dA_\nu(s)}{H_\nu(s)} - \int_0^t \widehat{X}(s)\,[\zeta(\nu(s)) + \hat{p}'(s)\nu(s)]\,ds, \quad 0 \leq t \leq T,$$

and (6.30), (4.1) imply that the process $\int_0^t \widehat{X}(s)[\zeta(\nu(s)) + \hat{p}'(s)\nu(s)] \, ds$ is nondecreasing.

Remark 6.11: In the proof of Theorem 6.2 one may replace $\mathcal{D}$ by $\mathcal{D}^{(b)}$, the set of *bounded* processes in $\mathcal{D}$. One can thus show

$$h_{up}(K) = \sup_{\nu \in \mathcal{D}} u_\nu = \sup_{\nu \in \mathcal{D}^{(b)}} u_\nu. \tag{6.33}$$

5.7 Upper Hedging with Constant Coefficients

Throughout this section we assume that

$$r(\cdot) \equiv r, \sigma(\cdot) \equiv \sigma \text{ are constant}, \tag{7.1}$$

$$A(\cdot) \equiv 0, \ \delta(\cdot) \equiv 0, \tag{7.2}$$

$$\theta_0(\cdot) \text{ is bounded}. \tag{7.3}$$

We imposed similar assumptions in Section 2.4, except that there we did not require the dividend rate vector to be zero. Just as in that section, we have here that the money market price process is $S_0(t) = e^{rt}$, the standard martingale measure P_0 is defined, and with $S(t) \triangleq (S_1(t), \ldots, S_N(t))'$ and $\varphi \colon (0, \infty)^N \to [0, \infty)$ a Borel-measurable function, the value at time $t \in [0, T]$ of the contingent claim $B = \varphi(S(T))$ in the unconstrained market $\mathcal{M}(K)$ is given by $u(T - t, S(t))$, where $u(T - t, x)$ is defined by (2.4.6). In the present context it is convenient to write the function $u(T - t, x)$ as

$$u(T - t, x; \varphi) \triangleq e^{-r(T-t)} E_0 \varphi(x_1 Y_1(t, T), \ldots, x_N Y_N(t, T)),$$
$$0 \le t \le T, \quad x \in (0, \infty)^N, \tag{7.4}$$

where

$$Y_n(t, T)$$
$$\triangleq \exp \left\{ \sum_{d=1}^{D} \sigma_{nd} \left(W_0^{(d)}(T) - W_0^{(d)}(t) \right) - \frac{1}{2}(T - t) \sum_{d=1}^{D} \sigma_{nd}^2 + r(T - t) \right\}. \tag{7.5}$$

In the constrained market $\mathcal{M}(K)$, Theorem 6.2 and Remark 6.11 assert that the upper-hedging value process (Definition 6.4) for $\varphi(S(T))$ is

$$\widehat{X}(t) \triangleq \text{ess sup}_{\nu \in \mathcal{D}} \frac{E[H_\nu(T)\varphi(S(T))|\mathcal{F}(t)]}{H_\nu(t)}$$
$$= \sup_{\nu \in \mathcal{D}^{(b)}} u_\nu(T - t, S(t); \varphi), \quad 0 \le t \le T, \tag{7.6}$$

where

$$u_\nu(T - t, x; \varphi)$$
$$\triangleq e^{-r(T-t)} E_\nu \left[e^{-\int_t^T \zeta(\nu(s))\, ds} \cdot \varphi(x_1 Y_1(t, T), \ldots, x_N Y_N(t, T)) \right],$$

$$= e^{-r(T-t)} E_\nu \left[e^{-\int_t^T \zeta(\nu(s))\, ds} \right.$$
$$\left. \cdot \varphi \left(x_1 e^{-\int_t^T \nu_1(s)\, ds} Y_1^{(\nu)}(t, T), \ldots, x_N e^{-\int_t^T \nu_N(s)\, ds} Y_N^{(\nu)}(t, T) \right) \right] \quad (7.7)$$

$$Y_n^{(\nu)}(t, T)$$
$$\triangleq \exp \left\{ \sum_{d=1}^D \sigma_{nd} \left(W_\nu^{(d)}(T) - W_\nu^{(d)}(t) \right) - \frac{T - t}{2} \sum_{d=1}^D \sigma_{nd}^2 + r(T - t) \right\},$$
$$\quad (7.8)$$

and P_ν is defined by (5.15).

The computations of this section exploit the fact that $W_\nu(\cdot)$ is a Brownian motion under P_ν, so that the distribution of the N-dimensional random process $\{(Y_1^{(\nu)}(t, T), \ldots, Y_N^{(\nu)}(t, T)); 0 \le t \le T\}$ under P_ν is the same as the distribution of the random process $\{(Y_1(t, T), \ldots, Y_N(t, T)); 0 \le t \le T\}$ under P_0. This would still be true if we assumed instead of (7.1) that $r(\cdot)$ and $\sigma(\cdot)$ are nonrandom but not necessarily constant; with minor changes, the results of this section hold under this weaker assumption.

Theorem 7.1: *Assume (7.1)–(7.3) and let $B = \varphi(S(T))$ be a contingent claim, where $\varphi: (0, \infty)^N \to [0, \infty)$ is a lower-semicontinuous function satisfying the polynomial growth condition*

$$0 \le \varphi(x) \le C_1 + C_2 \|x\|^\gamma, \quad \forall x \in (0, \infty)^N \quad (7.9)$$

for some positive constants C_1, C_2, and γ. Define the nonnegative function

$$\widehat{\varphi}(x) \triangleq \sup_{\nu \in \widetilde{K}} \left[e^{-\zeta(\nu)} \varphi(x_1 e^{-\nu_1}, \ldots, x_N e^{-\nu_N}) \right], \quad x \in (0, \infty)^N. \quad (7.10)$$

Then the hypotheses of Theorem 6.2 are satisfied, and the upper-hedging value process $\widehat{X}(\cdot)$ for the contingent claim $\varphi(S(T))$ is given by

$$\widehat{X}(t) = e^{-r(T-t)} E_0[\widehat{\varphi}(S(T)) | \mathcal{F}(t)]$$
$$= u(T - t, S(t); \widehat{\varphi}), \quad 0 \le t < T, \quad (7.11)$$

almost surely. In other words, the upper-hedging value process for the contingent claim $\varphi(S(T))$ in the market $\mathcal{M}(K)$ with constraint set K is the same as the value of the contingent claim $\widehat{\varphi}(S(T))$ in the unconstrained market $\mathcal{M}(\mathbb{R}^N)$.

PROOF. We first show that

$$
\sup_{\nu \in \mathcal{D}^{(b)}} E_\nu \left[e^{-\int_t^T \zeta(\nu(s))\,ds} \right.
$$

$$
\left. \cdot \varphi \left(x_1 e^{-\int_t^T \nu_1(s)\,ds} Y_1^{(\nu)}(t,T), \ldots, x_N e^{-\int_t^T \nu_N(s)\,ds} Y_N^{(\nu)}(t,T) \right) \right]
$$

$$
\le E_0 \widehat{\varphi}(x_1 Y_1(t,T), \ldots, x_N Y_N(t,T)), \quad 0 \le t \le T. \tag{7.12}
$$

In light of (7.4)–(7.8), this will imply

$$
\widehat{X}(t) \le u(T-t, S(t); \widehat{\varphi}), \quad 0 \le t < T, \tag{7.13}
$$

almost surely. Because $\widetilde{K}$ is a convex cone, $\nu(\cdot) \in \mathcal{D}^{(b)}$ implies that the vector $(\int_t^T \nu_1(s)\,ds, \ldots, \int_t^T \nu_N(s)\,ds)$ is in $\widetilde{K}$. The definition (4.1) of ζ yields

$$
\zeta \left(\int_t^T \nu_1(s)\,ds, \ldots, \int_t^T \nu_N(s)\,ds \right) \overset{\triangle}{=} \sup_{p \in K} \left(- \int_t^T \sum_{n=1}^N p_n \nu_n(s)\,ds \right)
$$

$$
\le \int_t^T \sup_{p \in K} \left(- \sum_{n=1}^N p_n \nu_n(s) \right) ds
$$

$$
= \int_t^T \zeta(\nu(s))\,ds. \tag{7.14}
$$

Therefore, for an arbitrary process $\nu(\cdot) \in \mathcal{D}^{(b)}$, we have

$$
E_\nu \left[e^{-\int_t^T \zeta(\nu(s))\,ds} \right.
$$

$$
\left. \cdot \varphi \left(x_1 e^{-\int_t^T \nu_1(s)\,ds} Y_1^{(\nu)}(t,T), \ldots, x_N e^{-\int_t^T \nu_N(s)\,ds} Y_N^{(\nu)}(t,T) \right) \right]
$$

$$
\le E_\nu \left[e^{-\zeta \left(\int_t^T \nu_1(s)\,ds, \ldots, \int_t^T \nu_N(s)\,ds \right)} \right.
$$

$$
\left. \cdot \varphi \left(x_1 e^{-\int_t^T \nu_1(s)\,ds} Y_1^{(\nu)}(t,T), \ldots, x_N e^{-\int_t^T \nu(s)\,ds} Y_N^{(\nu)}(t,T) \right) \right]
$$

$$
\le E_\nu \widehat{\varphi} \left(x_1 Y_1^{(\nu)}(t,T), \ldots, x_N Y_N^{(\nu)}(t,T) \right)
$$

$$
= E_0 \widehat{\varphi}(x_1 Y_1(t,T), \ldots, x_N Y_N(t,T)).
$$

Multiplying by $e^{-r(T-t)}$ and taking the supremum over $\nu(\cdot) \in \mathcal{D}^{(b)}$, we see that (7.13) holds.

It remains to show the reverse of inequality (7.13). Let us fix $x \in (0, \infty)$ and choose a sequence $\{\nu^{(m)}\}_{m=1}^\infty$ of vectors in $\widetilde{K}$ such that

$$
\sup_m \left[e^{-\zeta(\nu^{(m)})} \varphi(x_1 e^{-\nu_1^{(m)}}, \ldots, x_N e^{-\nu_N^{(m)}}) \right] = \widehat{\varphi}(x).
$$

For a fixed m, we define a process $\nu(\cdot)$ in $\mathcal{D}^{(b)}$ by setting $\nu(s) = 0$ for $0 \le s < t$ and $\nu(s) = \frac{1}{T-t}\nu^{(m)}$ for $t \le s \le T$. Let $x(\cdot): [0,T] \to (0,\infty)^N$ be a continuous (nonrandom) function, with $x(T) = x$. Then

$$u_\nu(T - t, x(t); \varphi) \tag{7.15}$$

$$= e^{-r(T-t)} \cdot E_\nu \left[e^{- \int_t^T \zeta(\nu(s))\, ds} \right.$$

$$\left. \cdot \varphi \left(x_1(t) e^{- \int_t^T \nu_1(s)\, ds} Y_1^{(\nu)}(t, T), \dots, x_N(t) e^{- \int_t^T \nu_N(s)\, ds} Y_N^{(\nu)}(t, T) \right) \right] \tag{7.16}$$

$$= e^{-r(T-t)}$$

$$\cdot E_\nu \left[e^{-\zeta(\nu^{(m)})} \varphi \left(x_1(t) e^{-\nu_1^{(m)}} Y_1^{(\nu)}(t, T), \dots, x_N(t) e^{-\nu_N^{(m)}} Y_N^{(\nu)}(t, T) \right) \right] \tag{7.17}$$

$$= e^{-r(T-t)}$$

$$\cdot E_0 \left[e^{-\zeta(\nu^{(m)})} \varphi \left(x_1(t) e^{-\nu_1^{(m)}} Y_1(t, T), \dots, x_N(t) e^{-\nu_N^{(m)}} Y_N(t, T) \right) \right]. \tag{7.18}$$

Because of the polynomial growth condition (7.9), we have

$$E_0 \left[\sup_{0 \le t \le T} \varphi \left(x_1(t) e^{-\nu_1^{(m)}} Y_1(t, T), \dots, x_N(t) e^{-\nu_N^{(m)}} Y_N(t, T) \right) \right] < \infty,$$

and using the dominated convergence theorem in (7.15), we obtain from the lower semicontinuity of φ that

$$\liminf_{t \uparrow T} \sup_{\mu \in \mathcal{D}^{(b)}} u_\mu(T - t, x(t); \varphi) \ge \lim_{t \uparrow T} u_\nu(T - t, x(t); \varphi)$$

$$\ge e^{-\zeta(\nu^{(m)})} \varphi \left(x_1 e^{-\nu_1^{(m)}}, \dots, x_N e^{-\nu_N^{(m)}} \right).$$

Taking the supremum over m and recalling (7.10) and (7.6), we see that

$$\liminf_{t \uparrow T} \widehat{X}(t) \ge \widehat{\varphi}(S(T)). \tag{7.19}$$

According to Proposition 6.5, $H_0(\cdot)\widehat{X}(\cdot)$ is a supermartingale, i.e.,

$$H_0(t)\widehat{X}(t) \ge E[H_0(u)\widehat{X}(u)|\mathcal{F}(t)], \quad 0 \le t \le u < T.$$

Letting $u \uparrow T$ and using (7.19) along with Fatou's lemma for conditional expectations, we obtain

$$\widehat{X}(t) \ge \frac{1}{H_0(t)} E[H_0(T)\widehat{\varphi}(S(T))|\mathcal{F}(t)]$$

$$= e^{-r(T-t)} E_0[\widehat{\varphi}(S(T))|\mathcal{F}(t)]$$

$$= u(T - t, S(t); \widehat{\varphi}), \quad 0 \le t < T$$

almost surely, and the proof is complete. □

Under the conditions of Theorem 7.1, Theorem 6.2 guarantees the existence of $(\widehat{C}, \widehat{\pi}) \in \mathcal{A}(\widehat{u}, K)$ such that

$$X^{\hat{u},\widehat{C},\hat{\pi}}(t) = e^{-r(T-t)} E_0[\widehat{\varphi}(S(T))|\mathcal{F}(t)], \quad 0 \le t < T, \qquad (7.20)$$

$$X^{\hat{u},\widehat{C},\hat{\pi}}(T) = \varphi(S(T)). \qquad (7.21)$$

In particular,

$$H_0(t)X^{\hat{u},\widehat{C},\hat{\pi}}(t) = E[H_0(T)\widehat{\varphi}(S(T))|\mathcal{F}(t)], \quad 0 \le t < T$$

is a martingale, hence a local martingale. From (2.13) we see further that

$$H_0(t)X^{\hat{u},\widehat{C},\hat{\pi}}(t) + \int_{(0,t]} H_0(s)\, d\widehat{C}(s), \quad 0 \le t \le T$$

is also a local martingale. This implies that

$$\widehat{C}(t) = 0, \quad 0 \le t < T, \qquad (7.22)$$

almost surely. From (7.20), (7.21) it is apparent that

$$\widehat{C}(T) = \widehat{\varphi}(S(T)) - \varphi(S(T)) \qquad (7.23)$$

almost surely, and this quantity is typically positive with positive probability. When $P[\widehat{C}(T) > 0] > 0$, the contingent claim $B = \varphi(S(T))$ is not attainable in the sense of Definition 3.1(iii).

Of course, the portfolio process $\hat{\pi}(\cdot)$ in (7.20) is just the process used to hedge the contingent claim $\widehat{\varphi}(S(T))$. This is given by the usual formula

$$\hat{\pi}(t) = \begin{bmatrix} S_1(t)\frac{\partial}{\partial x_1}u(T-t, S(t); \widehat{\varphi}) \\ \vdots \\ S_N(t)\frac{\partial}{\partial x_N}u(T-t, S(t); \widehat{\varphi}) \end{bmatrix}, \quad 0 \le t \le T, \qquad (7.24)$$

of (2.4.9). We summarize the preceding discussion.

Corollary 7.2: *Under the assumptions of Theorem 7.1, the portfolio process $\hat{\pi}(\cdot)$ of (7.24) and the cumulative consumption process $\widehat{C}(\cdot)$ of (7.22), (7.23) satisfy $(\widehat{C}, \hat{\pi}) \in \mathcal{A}(\hat{u}, K)$ and $X^{\hat{u},\widehat{C},\hat{\pi}}(\cdot) = \widehat{X}(\cdot)$. In other words, $\hat{\pi}(\cdot)$ is a superreplicating portfolio process.*

Example 7.3 *(European call option):* We consider one stock $S(\cdot) = S_1(\cdot)$ driven by a single Brownian motion, we assume (7.1)–(7.3), and we denote σ_{11} by σ. A European call option corresponds to $\varphi(x) = (x - q)^+$, where $q \ge 0$ is the exercise price. We consider $K = [\alpha, \beta]$ as in Example 4.1(ix), with $-\infty \le \alpha \le 0 \le \beta \le \infty$. It is tedious but straightforward to verify that $\widehat{\varphi} \equiv \infty$ if $0 \le \beta < 1$, $\widehat{\varphi}(x) = x$ if $\beta = 1$, $\widehat{\varphi}(x) = (x - q)^+$ if $\beta = \infty$, and for $1 < \beta < \infty$,

$$\widehat{\varphi}(x) = \begin{cases} \left(\dfrac{\beta-1}{q}\right)^{\beta-1} \left(\dfrac{x}{\beta}\right)^\beta, & \text{if } 0 < x \le \dfrac{\beta q}{\beta-1}, \\[3mm] x - q, & \text{if } x \ge \dfrac{\beta q}{\beta-1}. \end{cases} \qquad (7.25)$$

The function $\widehat{\varphi}$ does not depend on α.

From the above formulas and Theorems 6.2, 7.1, we see immediately that $h_{up}([\alpha, \beta]) = \infty$ if $0 \leq \beta < 1$; no matter how large the (finite) initial wealth, the European call cannot be hedged if the fraction of total wealth invested in the stock is bounded above by a number strictly less than 1. If $\beta = 1$, (7.11) implies

$$h_{up}([\alpha, 1]) = \widehat{X}(0) = E_0[e^{-rT} S(T)] = S(0).$$

The hedging portfolio is to *buy and hold one share of stock* $(\pi(\cdot) = S(\cdot))$, and to consume $S(T) - (S(T) - q)^+ = S(T) \wedge q$ at time $t = T$. If $\beta = \infty$, the portfolio constraint is never active; $h_{up}([\alpha, \infty))$ is the usual Black–Scholes value. If $1 < \beta < \infty$,

$$h_{up}([\alpha, \beta]) = e^{-rT} E_0 \left[\widehat{\varphi}(S(0) \exp\{\sigma W_0(T) + (r - \sigma^2/2)T\}) \right], \qquad (7.26)$$

where $\widehat{\varphi}$ is given by (7.25). This value, and the corresponding hedging portfolio of (7.24), can be written down explicitly in terms of the cumulative normal distribution.

Example 7.4 *(European put option):* We assume again (7.1)–(7.3) and consider one stock. A European put option corresponds to $\varphi(x) = (q - x)^+$, where $q \geq 0$. We consider again $K = [\alpha, \beta]$ with $-\infty \leq \alpha \leq 0 \leq \beta \leq \infty$. It turns out that $\widehat{\varphi}$ does not depend on β. If $\alpha = -\infty$, then $\widehat{\varphi}(x) = (x - q)^+$; the portfolio constraint is not active. If $\alpha = 0$, then $\widehat{\varphi}(x) = q$; the cheapest way to hedge the put, when short-selling is prohibited, is to begin with initial capital $e^{-rT} q$ and keep all wealth in the money market. Finally, for $-\infty < \alpha < 0$,

$$\widehat{\varphi}(x) = \begin{cases} q - x, & \text{if } 0 < x \leq \dfrac{\alpha q}{\alpha - 1}, \\ \left(\dfrac{|\alpha - 1|}{q}\right)^{\alpha - 1} \left(\dfrac{x}{|\alpha|}\right)^{\alpha}, & \text{if } x \geq \dfrac{\alpha q}{\alpha - 1}; \end{cases} \qquad (7.27)$$

the value of the put is given by (7.26) with (7.27) substituted for $\widehat{\varphi}$.

5.8   Optimal Dual Processes

In Section 6 we constructed the upper-hedging value process $\widehat{X}(\cdot) = X^{\hat{u}, \widehat{C}, \hat{\pi}}(\cdot)$ of (6.5), (6.7), whose final value is almost surely equal to a given contingent claim B, and which can be generated by a cumulative consumption process $\widehat{C}(\cdot)$ and a portfolio process $\hat{\pi}(\cdot)$ such that the corresponding portfolio-proportion process $\hat{p}(\cdot)$ takes values only in the constraint set K. The initial value $\hat{u}$ of this wealth process is the upper-hedging price of the contingent claim B in the constrained market $\mathcal{M}(K)$.

In the first part of this section we examine the question of when we can take the consumption process $\widehat{C}(\cdot)$ to be identically zero, so that the superreplicating portfolio process $\hat{\pi}(\cdot)$ constructed in Theorem 6.2 is in fact a *replicating* portfolio process. This is intimately connected with the existence of a dual process $\hat{\nu}(\cdot)$ that attains the supremum in (6.4); we call such a process an *optimal dual process*. As we see in Theorem 8.1, any such optimal dual process must be in the set $\mathcal{D}^{(c)}$ of Definition 6.9.

The analysis of the first part of this section was motivated by the incomplete market of Example 4.1(iii). In this example, stocks $M+1, \ldots, N$ are unavailable for investment in the constrained market $\mathcal{M}(K)$. In the unconstrained market $\mathcal{M}_\nu$, all stocks are available for investment, but the mean rate of return of the nth stock for $n = M+1, \ldots, N$ is $b_n(\cdot) + \nu_n(\cdot)$, rather than $b_n(\cdot)$. Contingent claims, however, are still defined in terms of the original stocks with mean rates of return $(b_1(\cdot), \ldots, b_N(\cdot))$, so the change in the mean rates of return for the investment opportunities does affect contingent claim pricing and hedging. The essence of Theorem 8.1 and its Corollary 8.3 is that the optimal dual process $\hat{\nu}(\cdot)$ makes this adjustment to the mean rates of return in such a way that the unconstrained hedging portfolio in the market $\mathcal{M}_{\hat{\nu}}$ satisfies the constraint in the market $\mathcal{M}(K)$. This reduction of a constrained problem to an unconstrained one is the traditional role of Lagrange multipliers.

In the second part of this section, Theorem 8.9 and its proof, an optimal dual process is posited for the problem of a contingent claim paying off at intermediate times as well as at the final time. Under this condition, we show the existence of a hedging portfolio whose proportion process $p(\cdot)$ takes values in the constraint set K.

The results of this section will not be used in subsequent developments.

Theorem 8.1: *Let B be a contingent claim and assume that $\hat{u}$ defined by (6.1) is finite. Let $\widehat{X}(\cdot)$ be the upper hedging value process defined by (6.7) and let $(\widehat{C}, \hat{\pi}) \in \mathcal{A}(\hat{u}; K)$ be as in Theorem 6.2, so that $X^{\hat{u},\widehat{C},\hat{\pi}}(\cdot) = \widehat{X}(\cdot)$. For a given process $\hat{\nu}(\cdot) \in \mathcal{D}$, the following conditions are equivalent:*

$$\hat{\nu}(\cdot) \text{ is optimal, i.e., } \hat{u} = u_{\hat{\nu}}, \tag{8.1}$$

$$H_{\hat{\nu}}(t)\widehat{X}(t), \quad 0 \le t \le T \text{ is a martingale}, \tag{8.2}$$

$$\left\{ \begin{array}{l} B \text{ is } K\text{-attainable, and for the associated} \\ \text{wealth process } X^{\hat{u},0,\pi}(\cdot), \text{ the product} \\ \text{process } H_{\hat{\nu}}(\cdot)X^{\hat{u},0,\pi}(\cdot) \text{ is a martingale.} \end{array} \right\} \tag{8.3}$$

Any of the above conditions implies that

$$\hat{\nu}(\cdot) \in \mathcal{D}^{(c)} \quad and \quad P[\widehat{C}(t) = 0, \quad \forall 0 \le t \le T] = 1. \tag{8.4}$$

PROOF.   In view of Propositon 6.5, $H_{\hat{\nu}}(\cdot)\widehat{X}(\cdot)$ is a supermartingale. It is a martingale if and only if $E[H_{\hat{\nu}}(T)\widehat{X}(T)] = \widehat{X}(0)$, i.e, if and only if $u_{\hat{\nu}} = \hat{u}$. Hence, (8.1) and (8.2) are equivalent.

Now suppose that the equivalent conditions (8.2) and (8.1) hold. Because $H_{\hat{\nu}}(\cdot)\widehat{X}(\cdot)$ is a martingale, the nondecreasing process $A_{\hat{\nu}}(\cdot)$ of the Doob–Meyer decomposition (6.20) is identically equal to zero, and Remark 6.10

shows that the nondecreasing process $\widehat{C}(\cdot)$ of (6.25) is also nonincreasing and hence identically zero. We have $\hat{\nu}(\cdot) \in \mathcal{D}^{(c)}$, and (8.4) follows. Condition (8.3) holds with $\pi(\cdot) = \hat{\pi}(\cdot)$.

Finally, let us suppose that (8.3) holds. This implies

$$\hat{u} = E\left[H_{\hat{\nu}}(T)X^{\hat{u},0,\pi}(T)\right] = E\left[H_{\hat{\nu}}(T)B\right] = u_{\hat{\nu}},$$

which is (8.1). $\qquad\qquad\qquad\qquad\qquad\qquad\qquad\qquad\qquad\qquad\qquad\square$

Remark 8.2: From Remark 6.10 we see that (8.4) is equivalent to

$$A_{\hat{\nu}}(t) = 0, \quad 0 \le t \le T, \tag{8.5}$$

almost surely.

Corollary 8.3: *Suppose that there exists a process $\hat{\nu}(\cdot) \in \mathcal{D}$ such that $u_{\hat{\nu}} < \infty$. Suppose also that with $\pi_{\hat{\nu}}(\cdot)$ defined to be the unconstrained hedging portfolio process in Definition 6.1 and with*

$$p_{\hat{\nu}}(t) \triangleq \begin{cases} \dfrac{\pi_{\hat{\nu}}(t)}{X^{u_{\hat{\nu}},0,\pi_{\hat{\nu}}}(t)}, & \text{if } X^{u_{\hat{\nu}},0,\pi_{\hat{\nu}}}(t) \ne 0, \\[2ex] p_*, & \text{if } X^{u_{\hat{\nu}},0,\pi_{\hat{\nu}}}(t) = 0, \end{cases}$$

where p_* is an arbitrary but fixed vector in K, we have almost surely

$$p_{\hat{\nu}}(t) \in K, \ \zeta(\hat{\nu}(t)) + p'_{\hat{\nu}}(t)\hat{\nu}(t) = 0 \ \text{for Lebesgue-a.e. } t \in [0,T]. \tag{8.6}$$

Then the equivalent conditions (8.1)–(8.3) hold, and $\hat{\pi}(\cdot)$ in Theorem 8.1 is $\pi_{\hat{\nu}}(\cdot)$. Conversely, suppose that the equivalent conditions of Theorem 8.1 hold; then (8.6) holds as well.

PROOF. Let $X_{\hat{\nu}}(\cdot) = X_{\hat{\nu}}^{u_{\hat{\nu}},0,\pi_{\hat{\nu}}}(\cdot)$ denote the wealth process generated by $(0, \pi_{\hat{\nu}})$ in the market $\mathcal{M}_{\hat{\nu}}$ given by (5.16) and (6.2). Then $X_{\hat{\nu}}(T) = B$ almost surely. Comparing (5.17) in the form

$$\frac{X_{\hat{\nu}}(t)}{S_0(t)} = u_{\hat{\nu}} + \int_0^t \frac{X_{\hat{\nu}}(s)}{S_0(s)}\left[(\zeta(\nu(s)) + p'_{\hat{\nu}}(s)\nu(s))\,ds + p'_{\hat{\nu}}(s)\sigma(s)\,dW_0(s)\right]$$

with (2.12), we see from (8.6) that $X_{\hat{\nu}}(\cdot)$ agrees with $X^{u_{\hat{\nu}},0,\pi_{\hat{\nu}}}(\cdot)$, the wealth process in $\mathcal{M}$ given by (2.12). Furthermore, $(0, \pi_{\hat{\nu}}) \in \mathcal{A}(u_{\hat{\nu}}; K)$, and Definition 3.1(ii) shows that $u_{\hat{\nu}} \ge h_{up}(K)$. On the other hand, Theorem 6.2 implies $u_{\hat{\nu}} \le \hat{u} = h_{up}(K)$. We have $u_{\hat{\nu}} = \hat{u}$, and the remaining equivalent conditions of Theorem 8.1 follow.

For the converse, we assume that the equivalent conditions of Theorem 8.1 hold. Then there is a portfolio process $\pi(\cdot)$ in the class $\mathcal{A}(\hat{u}; K)$ such that $X^{\hat{u},0,\pi}(T) = B$ almost surely, and $H_{\hat{\nu}}(\cdot)X^{\hat{u},0,\pi}(\cdot)$ is a martingale. The unconstrained portfolio process $\pi_{\hat{\nu}}(\cdot) \in \mathcal{A}_{\hat{\nu}}(u_{\hat{\nu}})$ also satisfies $X_{\hat{\nu}}^{\hat{u},0,\pi_{\hat{\nu}}}(T) = B$, and $H_{\hat{\nu}}(\cdot)X_{\hat{\nu}}^{\hat{u},0,\pi_{\hat{\nu}}}(\cdot)$ is a supermartingale (Remark 5.4).

But

$$X_{\hat{\nu}}^{\hat{u},0,\pi_{\hat{\nu}}}(0) = \hat{u} = E\left[H_{\hat{\nu}}(T)X^{\hat{u},0,\pi}(T)\right]$$
$$= E\left[H_{\hat{\nu}}(T)B\right]$$
$$= E\left[H_{\hat{\nu}}(T)X_{\hat{\nu}}^{\hat{u},0,\pi_{\hat{\nu}}}(T)\right],$$

which shows that $H_{\hat{\nu}}(\cdot)X_{\hat{\nu}}^{\hat{u},0,\pi_{\hat{\nu}}}(\cdot)$ is actually a martingale, and must thus coincide with $H_{\hat{\nu}}(\cdot)X^{\hat{u},0,\pi}(\cdot)$. Therefore,

$$X^{\hat{u},0,\pi}(t) = X_{\hat{\nu}}^{\hat{u},0,\pi_{\hat{\nu}}}(t), \quad 0 \le t \le T,$$

almost surely. Comparison of (2.12) and (5.17) shows that

$$X_{\nu}^{\hat{u},0,\pi_{\hat{\nu}}}(s)\left[\zeta(\hat{\nu}(s)) + p_{\hat{\nu}}'(s)\nu(s)\right] = 0, \quad 0 \le s \le T,$$

almost surely. Further comparison of (2.12) and (5.17) shows that $\pi(\cdot) = \pi_{\hat{\nu}}(\cdot)$. Because $\pi(\cdot) \in \mathcal{A}(\hat{u}; K)$, we have $p_{\hat{\nu}}(t) \in K$ for Lebesgue-almost-every $t \in [0,T]$ almost surely. $\square$

The next result provides conditions under which $\nu(\cdot) \equiv 0$ is an optimal dual process. In such a case, the unconstrained hedging price in the original market $\mathcal{M}$ is the upper-hedging price. To state the result, we need to introduce some notation. We denote by $\mathcal{S}$ the set of all stopping times taking values in $[0,T]$, and we say that an $\{\mathcal{F}(t)\}$-adapted process $Y(\cdot)$ is of class $\mathcal{D}[0,T]$ if the family of random variables $\{Y(\rho)\}_{\rho \in \mathcal{S}}$ is uniformly integrable. A local martingale of class $\mathcal{D}[0,T]$ is in fact a martingale. Finally, recall the notation $\mathcal{D}^{(m)}$ of Definition 5.1.

Theorem 8.4: *Let B be a contingent claim and assume that $\hat{u} \triangleq \sup_{\nu \in \mathcal{D}} u_{\nu}$ is finite. Assume further that*

$$H_{\nu}(\cdot)\widehat{X}(\cdot) \text{ is of class } \mathcal{D}[0,T], \quad \forall \nu(\cdot) \in \mathcal{D}^{(m)}. \tag{8.7}$$

Then, for a given $\hat{\nu}(\cdot) \in \mathcal{D}^{(m)}$, all four conditions (8.1)–(8.4) are equivalent and imply

$$\left\{ \begin{array}{l} B \text{ is } K\text{-attainable, and for the associated wealth} \\ \text{process } X^{\hat{u},0,\pi}(\cdot), \text{ the product process} \\ H_0(\cdot)X^{\hat{u},0,\pi}(\cdot) \text{ is a martingale.} \end{array} \right\} \tag{8.8}$$

In particular, if there exists any $\hat{\nu}(\cdot) \in \mathcal{D}^{(m)}$ satisfying the equivalent conditions (8.1)–(8.4), then $\nu(\cdot) \equiv 0$ is an optimal dual process, i.e., $\hat{u} = u_0$. Conversely, if (8.8) holds, then (8.1)–(8.4) are satisfied for all $\hat{\nu}(\cdot) \in \mathcal{D}^{(m)} \cap \mathcal{D}^{(c)}$.

PROOF.   The equivalence of conditions (8.1)–(8.3) has already been established, as has the implication (8.1)$\Rightarrow$(8.4). Let us assume that (8.4) holds for some $\hat{\nu}(\cdot) \in \mathcal{D}^{(m)}$. In light of Remark 8.2, we know that $A_{\hat{\nu}}(\cdot) \equiv 0$, and (6.20) shows that $H_{\hat{\nu}}(\cdot)\widehat{X}(\cdot)$ is a local martingale. Condition (8.7) allows us to conclude that (8.2) holds.

To obtain (8.8) from (8.1)–(8.4) we observe that if (8.4) is satisfied for some $\hat{\nu}(\cdot) \in \mathcal{D}^{(m)}$, then it is satisfied by $\hat{\nu}(\cdot) \equiv 0$. (Note that the process $\widehat{C}(\cdot)$ appearing in (8.4), given by (6.23), does not depend on $\hat{\nu}(\cdot)$.) With the choice $\hat{\nu}(\cdot) \equiv 0$, (8.3) becomes (8.8). From (8.8) we have immediately that

$$u_0 \stackrel{\Delta}{=} E\left[H_0(T)B\right] = E\left[H_0(T)X^{\hat{u},0,\pi}(T)\right] = X^{\hat{u},0,\pi}(0) = \hat{u}.$$

Finally, let us assume only that (8.7) and (8.8) hold. The equivalence just proved shows that (8.1)–(8.4) all hold with $\hat{\nu}(\cdot) \equiv 0$. But then (8.4) holds for every $\hat{\nu}(\cdot) \in \mathcal{D}^{(c)}$, and the first paragraph of this proof shows that (8.1)–(8.3) hold for all $\hat{\nu}(\cdot) \in \mathcal{D}^{(m)} \cap \mathcal{D}^{(c)}$.   □

The next two propositions provide conditions on the contingent claim B that guarantee that the upper hedging price is finite and that condition (8.7) is satisfied.

Proposition 8.5: *If the contingent claim B is almost surely bounded from above, i.e., $P[0 \le B \le \beta] = 1$ for some $\beta \in (0, \infty)$, then $\hat{u} \stackrel{\Delta}{=} \sup_{\nu \in \mathcal{D}} u_\nu$ is finite and (8.7) holds.*

PROOF.   From (5.13) we have

$$0 \le H_\nu(T)B \le \frac{\beta}{s_0} e^{-\zeta_0 T} Z_\nu(T)$$

almost surely. But $Z_\nu(\cdot)$ is a supermartingale and $EZ_\nu(T) \le 1$; hence $u_\nu = EH_\nu(T)B \le \frac{\beta}{s_0} e^{-\zeta_0 T}$ for all $\nu(\cdot) \in \mathcal{D}$. This proves the finiteness of $\hat{u}$.

Now let $\nu(\cdot) \in \mathcal{D}^{(m)}$ and $\tau \in \mathcal{S}$ be given. Define $\mathcal{D}_{\tau,\nu}$ to be the set of processes $\mu(\cdot) \in \mathcal{D}$ that agree with $\nu(\cdot)$ up to time τ. According to Remark 6.7, we may write

$$
\begin{aligned}
0 &\le H_\nu(\tau)\widehat{X}(\tau) \\
&= H_\nu(\tau) \cdot \text{ess sup}_{\mu \in \mathcal{D}} E\left[\frac{H_\mu(T)B}{H_\mu(\tau)} \Big| \mathcal{F}(\tau)\right] \\
&= H_\nu(\tau) \cdot \text{ess sup}_{\mu \in \mathcal{D}_{\tau,\nu}} E\left[\frac{H_\mu(T)B}{H_\mu(\tau)} \Big| \mathcal{F}(\tau)\right] \\
&= \text{ess sup}_{\mu \in \mathcal{D}_{\tau,\nu}} E[H_\mu(T)B | \mathcal{F}(\tau)] \qquad (8.9)
\end{aligned}
$$

because $H_\nu(\tau) = H_\mu(\tau)$ for all $\mu(\cdot) \in \mathcal{D}_{\tau,\nu}$. But

$$
\begin{aligned}
E[H_\mu(T)B | \mathcal{F}(\tau)] &\le \frac{\beta}{s_0} e^{-\zeta_0 T} E[Z_\mu(T) | \mathcal{F}(\tau)] \\
&\le \frac{\beta}{s_0} e^{-\zeta_0 T} Z_\mu(\tau) = \frac{\beta}{s_0} e^{-\zeta_0 T} Z_\nu(\tau).
\end{aligned}
$$

Because $Z_\nu(\cdot)$ is a martingale, this last expression is $\frac{\beta}{s_0} e^{-\zeta_0 T} E[Z_\nu(T) | \mathcal{F}(\tau)]$. We conclude that

$$0 \le H_\nu(\tau)\widehat{X}(\tau) \le \frac{\beta}{s_0} e^{-\zeta_0 T} E[Z_\nu(T) | \mathcal{F}(\tau)] \qquad (8.10)$$

for all stopping times $\tau \in \mathcal{S}$, and this implies the uniform integrability of the collection of random variables $\{H_\nu(\tau)\widehat{X}(\tau)\}_{\tau \in \mathcal{S}}$.   □

Proposition 8.6: *Suppose that for some $n \in \{1, \ldots, N\}$ the dividend rate process $\delta_n(\cdot)$ is bounded from below. Suppose also that with the notation $\nu = (\nu_1, \ldots, \nu_N)$ we have*

$$\nu \mapsto \zeta(\nu) + \nu_n \text{ is bounded from below on } \widetilde{K}, \tag{8.11}$$

and the contingent claim B satisfies

$$0 \leq B \leq \alpha S_n(T) + \beta \tag{8.12}$$

almost surely for some $\alpha > 0$, $\beta > 0$. Then $\hat{u} \triangleq \sup_{\nu \in \mathcal{D}} u_\nu$ is finite and (8.7) holds.

Conversely, if $\theta(\cdot)$ is bounded, $\delta_n(\cdot)$ is bounded from above, $B = (S_n(T) - q)^+$ is a European call option with exercise price $q \geq 0$, and (8.11) fails, then $\hat{u} = \infty$.

PROOF.   Let $\sigma_n(\cdot) = (\sigma_{n1}(\cdot), \ldots, \sigma_{nN}(\cdot))$ be the nth row of the volatility matrix $\sigma(\cdot)$. For $\nu(\cdot) \in \mathcal{D}$, define

$$F_\nu(t) = H_\nu(t)S_n(t)\exp\left\{\int_0^t [\delta_n(s) + \zeta(\nu(s)) + \nu_n(s)]\, ds\right\}.$$

Using (5.7), (5.9), and (1.5.18), we may rewrite this as

$$F_\nu(t) = \frac{S_n(t)Z_\nu(t)}{S_0(t)}\exp\left\{\int_0^t [\delta_n(s) + \nu_n(s)]\, ds\right\}$$

$$= S_n(0)Z_\nu(t)\exp\left\{\int_0^t \sigma_n(s)\, dW_0(s) + \int_0^t \left[\nu_n(s) - \frac{1}{2}\|\sigma_n'(s)\|^2\right]ds\right\}.$$

Finally, recalling (5.10), (5.11), we conclude that

$$F_\nu(t)$$

$$= S_n(0)Z_\nu(t)\exp\left\{\int_0^t \sigma_n(s)\, dW_\nu(s) - \frac{1}{2}\int_0^t \|\sigma_n(s)\|^2\, ds\right\}$$

$$= S_n(0)\exp\left\{\int_0^t (\sigma_n(v) - \theta_\nu'(v))\, dW(v) - \frac{1}{2}\int_0^t \|\sigma_n'(v) - \theta_\nu(v)\|^2 dv\right\},$$

which is a nonnegative local martingale, and hence a supermartingale, under P. If $\nu(\cdot) \in \mathcal{D}^{(m)}$, then $W_\nu(\cdot)$ is a Brownian motion under the probability measure P_ν of Remark 5.2. Because of the assumption (2.4) of boundedness of $\sigma_n(\cdot)$, the process

$$G_\nu(t) \triangleq \exp\left\{\int_0^t \sigma_n(v)\, dW_\nu(v) - \frac{1}{2}\int_0^t \|\sigma_n'(v)\|^2 dv\right\}$$

is a martingale under P_ν. For $0 \leq s \leq t \leq T$, Bayes's rule (Karatzas and Shreve (1991), p. 193) implies

$$E[F_\nu(t)|\mathcal{F}(t)] = S_n(0)E[Z_\nu(t)G_\nu(t)|\mathcal{F}(s)]$$
$$= S_n(0)Z_\nu(s)E_\nu[G_\nu(t)|\mathcal{F}(s)]$$
$$= S_n(0)Z_\nu(s)G_\nu(s)$$
$$= F_\nu(s).$$

In other words, $F_\nu(\cdot)$ is a P-martingale for $\nu(\cdot) \in \mathcal{D}^{(m)}$.

Let $-\gamma \in \mathbb{R}$ be a lower bound on $\delta_n(\cdot) + \zeta(\nu(\cdot)) + \nu_n(\cdot)$. For B as in (8.12) and $\nu(\cdot) \in \mathcal{D}$, we have

$$u_\nu = E[H_\nu(T)B]$$
$$\leq \alpha E[H_\nu(T)S_n(T)] + \beta EH_\nu(T)$$
$$\leq \alpha e^{\gamma T} EF_\nu(T) + \frac{\beta}{s_0}e^{-\zeta_0 T}EZ_\nu(T)$$
$$\leq \alpha e^{\gamma T} S_n(0) + \frac{\beta}{s_0}e^{-\zeta_0 T}.$$

It follows that $\hat{u}$ is finite.

To obtain (8.7), fix $\nu(\cdot) \in \mathcal{D}^{(m)}$, let $\tau \in \mathcal{S}$ be given, and let $\mathcal{D}_{\tau,\nu}$ be the set of processes $\mu(\cdot) \in \mathcal{D}$ that agree with $\nu(\cdot)$ up to time τ. According to Remark 6.7,

$$0 \leq H_\nu(\tau)\widehat{X}(\tau)$$
$$= H_\nu(\tau) \text{ ess sup}_{\mu \in \mathcal{D}} \frac{E[H_\mu(T)B|\mathcal{F}(\tau)]}{H_\mu(\tau)}$$
$$= \text{ess sup}_{\mu \in \mathcal{D}_{\tau,\nu}} E[H_\mu(T)B|\mathcal{F}(\tau)]$$
$$\leq \alpha e^{\gamma T} \text{ess sup}_{\mu \in \mathcal{D}_{\tau,\nu}} E[F_\mu(T)|\mathcal{F}(\tau)]$$
$$+ \frac{\beta}{s_0}e^{-\zeta_0 T} \text{ess sup}_{\mu \in \mathcal{D}_{\tau,\nu}} E[Z_\mu(T)|\mathcal{F}(\tau)]$$
$$\leq \alpha e^{\gamma T} F_\nu(\tau) + \frac{\beta}{s_0}e^{-\zeta_0 T} Z_\nu(\tau)$$
$$= E\left[\alpha e^{\gamma T} F_\nu(T) + \frac{\beta}{s_0}e^{-\zeta_0 T} Z_\nu(T) \,\Big|\, \mathcal{F}(\tau)\right].$$

Here, the last inequality holds because $Z_\mu(\cdot)$ is a supermartingale for every $\mu(\cdot) \in \mathcal{D}_{\tau,\mu}$, and the last equality holds because both $F_\nu(\cdot)$ and $Z_\nu(\cdot)$ are martingales when $\nu(\cdot) \in \mathcal{D}^{(m)}$. This shows that the family of random variables $\{H_\nu(\tau)\widehat{X}(\tau)\}_{\tau \in \mathcal{S}}$ is uniformly integrable.

For the second part of the proposition, we assume that $B = (S_n(T) - q)^+$, $\theta(\cdot)$ is bounded, and $\delta_n(\cdot)$ is bounded above by some constant Δ_0, but (8.11) fails. For every $\nu \in \widetilde{K}$ the process $\nu(\cdot) \equiv \nu$ is in $\mathcal{D}^{(m)}$; therefore,

$$E[H_\nu(T)S_n(T)] \geq EF_\nu(T) \cdot \exp\{-\Delta_0 T - (\zeta(\nu) + \nu_n)T\}$$
$$= S_n(0) \exp\{-\Delta_0 T - (\zeta(\nu) + \nu_n)T\},$$

and from Jensen's inequality,

$$
\begin{aligned}
u_\nu &= E\left[H_\nu(T)(S_n(T) - q)^+\right] \\
&\geq (E[H_\nu(T)S_n(T)] - qEH_\nu(T))^+ \\
&\geq S_n(0)\exp\{-\Delta_0 T - (\zeta(\nu) + \nu_n)T\} - \frac{q}{s_0}e^{-\zeta_0 T}.
\end{aligned}
$$

Since (8.11) fails, this last quantity can be made arbitrarily large by choice of $\nu \in \widetilde{K}$.   □

Remark 8.7: Condition (8.11) is satisfied if the convex set K contains both the origin and the nth unit vector, and thus also the entire line segment adjoining these two points. In this case

$$
\nu_n + \zeta(\nu) \geq \nu_n + \sup_{0 \leq \alpha \leq 1}(-\alpha\nu_n) = \sup_{0 \leq a \leq 1}(a\nu_n) = \nu_n^+ \geq 0, \quad \forall \nu \in \widetilde{K}.
$$

This condition holds in Examples 4.1(i), (ii), (vi), (vii), and (viii) for all choices of the index n. In particular, under prohibition or constraints on either borrowing or short-selling, it is possible to find an initial wealth large enough to permit the construction of a portfolio whose final value almost surely dominates that of any contingent claim B satisfying (8.12). This is also the case in Examples 4.1(iii) and (iv) for $n \in \{1, \ldots, M\}$. For $n \in \{M+1, \ldots, N\}$ in Examples 4.1(iii), (iv), condition (8.11) is violated. Theorem 6.2 and the second part of Proposition 8.6 applied to these examples show that *when a European call is written on a stock that cannot be held by the hedging portfolio, the upper hedging price is infinite*; in other words, no matter how large the initial wealth, it is not possible to construct a portfolio whose final value dominates almost surely the payoff of the option.

Example 8.8 *(Incomplete market):* Consider the case $K = \{p \in \mathbb{R}^N; p_{M+1} = \cdots = p_N = 0\}$ of Example 4.1(iii), where there are only M stocks available for investment, but these are driven by the N-dimensional Brownian motion $W(\cdot)$ with $N > M$. Then $\zeta(\cdot) \equiv 0$ on the barrier cone $\widetilde{K} = \{\nu \in \mathbb{R}^N; \nu_1 = \cdots = \nu_M = 0\}$ of $-K$, and thus

$$
\zeta(\nu) + p'\nu = 0, \quad \forall p \in K, \quad \nu \in \widetilde{K}.
$$

In particular, $\mathcal{D}^{(c)}$ of Definition 6.9 agrees with $\mathcal{D}$.

Consider now a contingent claim B for which $\hat{u} \stackrel{\Delta}{=} \sup_{\nu \in \mathcal{D}} u_\nu$ is finite and for which (8.7) holds, as is the case under the conditions of Proposition 8.5 or 8.6. Theorem 8.4 shows that if the supremum in the definition of $\hat{u}$ is attained by *some* $\nu(\cdot) \in \mathcal{D}^{(m)}$, then it is attained by *every* $\nu(\cdot) \in \mathcal{D}^{(m)}$. This verifies a conjecture of Harrison and Pliska (1981), p. 257; see also Jacka (1992) and Ansel and Stricker (1994) for related results.

We conclude this section with a result that generalizes both Theorem 8.1 (as it allows for European contingent claims that make payments prior to expiration, somewhat like the claims introduced in Definition 2.2.1) and Theorem 2.2 (as it deals with constrained portfolios). For simplicity, we deal only with the case that the payment at the final time is strictly positive.

Theorem 8.9:   *Let $C(\cdot)$ be a cumulative consumption process and $B \colon \Omega \to (0, \infty)$ an $\mathcal{F}(T)$-measurable random variable such that*

$$x \stackrel{\Delta}{=} \sup_{\mu \in \mathcal{D}} u(\mu) = u(\hat{\nu}) < \infty \qquad (8.13)$$

for some $\hat{\nu}(\cdot) \in \mathcal{D}$, where

$$u(\mu) \stackrel{\Delta}{=} E\left[\int_{(0,T]} H_\mu(s)\, dC(s) + H_\mu(T)B \right], \quad \mu(\cdot) \in \mathcal{D}. \qquad (8.14)$$

Then there exists a portfolio process $\pi(\cdot)$ such that $X^{x,C,\pi}(\cdot)$, the wealth process in the original market $\mathcal{M}(\mathbb{R}^N)$, satisfies

$$X^{x,C,\pi}(t) = X_{\hat{\nu}}(t) \stackrel{\Delta}{=} \frac{1}{H_{\hat{\nu}}(t)} E\left[\int_{(t,T]} H_{\hat{\nu}}(s)\, dC(s) + H_{\hat{\nu}}(T)B \,\middle|\, \mathcal{F}(t) \right],$$

$$(8.15)$$

for $0 \le t \le T$. Furthermore, with the portfolio-proportion process $p(\cdot)$ defined by

$$p(t) \stackrel{\Delta}{=} \begin{cases} \dfrac{\pi(t)}{X_{\hat{\nu}}(t)}, & \text{if } X_{\hat{\nu}}(t) \ne 0, \\ p_*, & \text{if } X_{\hat{\nu}}(t) = 0, \end{cases} \qquad (8.16)$$

where p_* is an arbitrary but fixed element of K (cf. (2.14)), we have

$$p(t) \in K, \quad \text{for Lebesgue-a.e. } t \in [0, T], \qquad (8.17)$$

almost surely and

$$\zeta(\hat{\nu}(t)) + p'(t)\hat{\nu}(t) = 0, \quad \text{for Lebesgue-a.e. } t \in [0, T]. \qquad (8.18)$$

In particular, $(C, \pi) \in \mathcal{A}(x; K)$, $X^{x,C,\pi}(T) = B$ almost surely, and

$$H_{\hat{\nu}}(t) X^{x,C,\pi}(t) + \int_{(0,t]} H_{\hat{\nu}}(s)\, dC(s), \quad 0 \le t \le T, \qquad (8.19)$$

is a martingale.

This result can be established by arguments similar to those used in Theorems 6.2 and 8.1. We opt here for a different proof, based on a variational principle that we shall find useful in Chapter 6.

PROOF OF THEOREM 8.9.   From Theorem 5.5 we know that there exists a portfolio process $\pi(\cdot)$ such that $X^{x,C,\pi}_{\hat{\nu}}(\cdot)$, the wealth process in the auxiliary market $\mathcal{M}_{\hat{\nu}}$ corresponding to initial wealth $x \ge 0$ and $(C, \pi) \in \mathcal{A}_{\hat{\nu}}(x)$, is given as $X_{\hat{\nu}}(\cdot) \equiv X^{x,C,\pi}_{\hat{\nu}}(\cdot)$ by (8.15) and satisfies

$$\frac{X_{\hat{\nu}}(t)}{S_0(t)} + \int_{(0,t]} \frac{dC(s)}{S_0(s)}$$

$$= x + \int_0^t \frac{1}{S_0(s)} \left[(X_{\hat{\nu}}(s)\zeta(\hat{\nu}(s)) + \pi'(s)\hat{\nu}(s)) \, ds + \pi'(s)\sigma(s) \, dW_0(s) \right]$$

$$= x + \int_0^t \frac{X_{\hat{\nu}}(s)}{S_0(s)} [(\zeta(\hat{\nu}(s)) + p'(s)\hat{\nu}(s)) \, ds + p'(s)\sigma(s) \, dW_0(s)], \quad (5.17)$$

where $p(\cdot)$ is defined by (8.16). In order to prove the theorem, we must show both (8.17) and (8.18). From (8.18) and the comparison of (5.17) with (2.12), we can conclude that $X_{\hat{\nu}}^{x,C,\pi}(\cdot)$, the wealth process in the auxiliary market $\mathcal{M}_{\hat{\nu}}$, agrees with $X^{x,C,\pi}(\cdot)$, the wealth process in the original market $\mathcal{M}(\mathbb{R}^N)$. Moreover, (8.17) shows that $(C, \pi) \in \mathcal{M}(K)$, so that $X^{x,C,\pi}(\cdot)$ is in fact a wealth process in the constrained market $\mathcal{M}(K)$.

Step 1: For any $\mu(\cdot) \in \mathcal{D}$ and any $\epsilon \in (0,1)$, the convex combination $(1-\epsilon)\hat{\nu}(\cdot) + \epsilon\mu(\cdot)$ is in $\mathcal{D}$, because of the convexity of $\widetilde{K}$ and the positive homogeneity and subadditivity of ζ (see (4.3), (4.4)), which guarantee that $(1-\epsilon)\hat{\nu}(\cdot) + \epsilon\mu(\cdot)$ satisfies (5.2). We shall be interested in two particular choices of $\mu(\cdot)$. The first is $\mu(\cdot) \equiv 0$, which is an element of $\mathcal{D}$ because $0 \in \widetilde{K}$ and $\zeta(0) = 0$. The other is $\mu(\cdot) = \hat{\nu}(\cdot) + \lambda(\cdot)$ for some $\lambda(\cdot) \in \mathcal{D}$; this $\mu(\cdot)$ is in $\mathcal{D}$ because $\widetilde{K}$ is a convex cone and thus closed under addition, and ζ is subadditive.

Let $\{\tau_n\}_{n=1}^\infty$ be a nondecreasing sequence of stopping times converging up to T, and consider the random perturbation of $\hat{\nu}$ given by

$$\nu_{\epsilon,n}(t) \triangleq \begin{cases} (1-\epsilon)\hat{\nu}(t) + \epsilon\mu(t), & 0 \le t \le \tau_n, \\ \hat{\nu}(t), & \tau_n < t \le T, \end{cases} \quad (8.20)$$

$$= \hat{\nu}(t) + \epsilon(\mu(t) - \hat{\nu}(t))1_{\{t \le \tau_n\}}, \quad 0 \le t \le T.$$

Because $\nu_{\epsilon,n}(\cdot) \in \mathcal{D}$ and $\hat{\nu}(\cdot)$ maximizes $u(\hat{\nu})$ over $\mathcal{D}$, we must have

$$0 \le EY_{\epsilon,n} = \frac{u(\hat{\nu}) - u(\nu_{\epsilon,n})}{\epsilon} \quad (8.21)$$

for every $\epsilon \in (0,1)$ and $n = 1, 2, \ldots$, where

$$Y_{\epsilon,n} \triangleq \frac{H_{\hat{\nu}}(T)B}{\epsilon}\left(1 - \frac{H_{\nu_{\epsilon,n}}(T)}{H_{\hat{\nu}}(T)}\right) + \int_{(0,T]} \frac{H_{\hat{\nu}}(t)}{\epsilon}\left(1 - \frac{H_{\nu_{\epsilon,n}}(t)}{H_{\hat{\nu}}(t)}\right) dC(t).$$

$$(8.22)$$

Step 2. A straightforward computation using (5.9)–(5.12) shows that

$$\Lambda_{\epsilon,n}(t) \triangleq \frac{H_{\nu_{\epsilon,n}}(t)}{H_{\hat{\nu}}(t)}$$

$$= \exp\left[-\epsilon N(t \wedge \tau_n) - \frac{\epsilon^2}{2}\langle N\rangle(t \wedge \tau_n) \right.$$

$$-\int_0^{t\wedge\tau_n}\left(\zeta\big((1-\epsilon)\hat{\nu}(s)+\epsilon\mu(s)\big)-\zeta\big(\hat{\nu}(s)\big)\right)ds\Bigg],\quad (8.23)$$

where

$$N(t)\overset{\Delta}{=}\int_0^t\left(\sigma^{-1}(s)(\mu(s)-\hat{\nu}(s))\right)'dW_{\hat{\nu}}(s),\qquad (8.24)$$

$$\langle N\rangle(t)\overset{\Delta}{=}\int_0^t\|\sigma^{-1}(s)(\mu(s)-\hat{\nu}(s))\|^2\,ds.\qquad (8.25)$$

In the case that $\mu(\cdot)=0$, we have

$$\zeta\big((1-\epsilon)\hat{\nu}(s)+\epsilon\mu(s)\big)-\zeta\big(\hat{\nu}(s)\big)=\zeta\big((1-\epsilon)\hat{\nu}(s)\big)-\zeta\big(\hat{\nu}(s)\big)$$
$$=-\epsilon\zeta(\hat{\nu}(s)),$$

whereas in the case that $\mu(\cdot)=\hat{\nu}(\cdot)+\lambda(\cdot)$ for some $\lambda(\cdot)\in\mathcal{D}$, we have

$$\zeta\big((1-\epsilon)\hat{\nu}(s)+\epsilon\mu(s)\big)-\zeta\big(\hat{\nu}(s)\big)=\zeta\big(\hat{\nu}(s)+\epsilon\lambda(s)\big)-\zeta\big(\hat{\nu}(s)\big)$$
$$\leq\epsilon\zeta(\lambda(s)).$$

We define

$$\xi(s)=\begin{cases}-\zeta(\hat{\nu}(s)), & \text{if }\mu(\cdot)=0,\\\zeta(\lambda(s)), & \text{if }\mu=\hat{\nu}(\cdot)+\lambda(\cdot)\text{ for some }\lambda(\cdot)\in\mathcal{D},\end{cases}$$

and $L(t)\overset{\Delta}{=}\int_0^t\xi(s)\,ds$. With this notation, we may rewrite (8.23) as

$$\Lambda_{\epsilon,n}(t)\geq Q_{\epsilon,n}(t)\qquad (8.26)$$
$$\overset{\Delta}{=}\exp\left\{-\epsilon(N(t\wedge\tau_n)+L(t\wedge\tau_n))-\frac{\epsilon^2}{2}\langle N\rangle(t\wedge\tau_n)\right\}.$$

Step 3. For each positive integer n, we define the stopping time

$$\tau_n\overset{\Delta}{=}\inf\Bigg\{t\in[0,T];\ |N(t)|+\langle N\rangle(t)+|L(t)|\geq n$$

$$\text{or }\int_0^t\|\theta_{\hat{\nu}}(s)\|^2\,ds\geq n$$

$$\text{or }\int_0^t\left(\frac{X_{\hat{\nu}}(s)}{S_0^{(\hat{\nu})}(s)}\right)^2\|\sigma^{-1}(s)(\mu(s)-\hat{\nu}(s))\|^2\,ds\geq n$$

$$\text{or }\int_0^t(L(s)+N(s))^2\|\sigma'(s)\pi(s)\|^2\,ds\geq n\Bigg\}\wedge T.$$

Clearly, $\tau_n\uparrow T$ almost surely as $n\to\infty$. According to the Girsanov and Novikov theorems (e.g., Karatzas and Shreve (1991), §3.5), the process

$$W_{\hat{\nu},n}(t)\overset{\Delta}{=}W(t)+\int_0^{t\wedge\tau_n}\theta_{\hat{\nu}}(s)\,ds,\quad 0\leq t\leq T,\qquad (8.27)$$

is a Brownian motion under the probability measure

$$\widetilde{P}_{\hat{\nu},n}(A) \stackrel{\Delta}{=} E[Z_{\hat{\nu}}(\tau_n) \cdot 1_A], \quad A \in \mathcal{F}(T). \tag{8.28}$$

Step 4. With τ_n the stopping time defined in Step 3, the process $Q_{\epsilon,n}(\cdot)$ of (8.26) has the lower bound

$$Q_{\epsilon,n}(t) \geq e^{-\epsilon n}, \quad 0 \leq t \leq T, \tag{8.29}$$

and consequently

$$\frac{1}{\epsilon}\left(1 - \frac{H_{\nu_{\epsilon,n}}(t)}{H_{\hat{\nu}}(t)}\right) = \frac{1 - \Lambda_{\epsilon,n}(t)}{\epsilon} \leq K_n, \quad 0 \leq t \leq T, \tag{8.30}$$

almost surely, where $K_n \stackrel{\Delta}{=} \sup_{0 < \epsilon < 1} \frac{1}{\epsilon}(1 - e^{-\epsilon n})$ is finite. Furthermore,

$$\varlimsup_{\epsilon \downarrow 0} \frac{1 - \Lambda_{\epsilon,n}(t)}{\epsilon} \leq N(t \wedge \tau_n) + L(t \wedge \tau_n),$$

and it follows from Fatou's lemma that

$$\varlimsup_{\epsilon \downarrow 0} Y_{\epsilon,n} \leq H_{\hat{\nu}}(T)B(N(\tau_n) + L(\tau_n))$$

$$+ \int_{(0,T]} H_{\hat{\nu}}(t)(N(t \wedge \tau_n) + L(t \wedge \tau_n))\, dC(t).$$

In addition, each $Y_{\epsilon,n}$ is bounded from above by

$$Y_n \stackrel{\Delta}{=} K_n\left[H_{\hat{\nu}}(T)B + \int_{(0,T]} H_{\hat{\nu}}(t)\, dC(t)\right],$$

which is integrable because $EY_n = K_n u(\hat{\nu}) < \infty$. Another application of Fatou's lemma yields

$$0 \leq \varlimsup_{\epsilon \downarrow 0} \frac{u(\hat{\nu}) - u(\nu_{\epsilon,n})}{\epsilon} = \varlimsup_{\epsilon \downarrow 0} EY_{\epsilon,n} \leq E\left(\varlimsup_{\epsilon \downarrow 0} Y_{\epsilon,n}\right)$$

$$\leq E\left[H_{\hat{\nu}}(T)B(N(\tau_n) + L(\tau_n))\right.$$

$$\left. + \int_{(0,T]} H_{\hat{\nu}}(t)(N(t \wedge \tau_n) + L(t \wedge \tau_n))\, dC(t)\right]. \tag{8.31}$$

Step 5. We next prove that (8.31) leads to

$$E\int_0^{\tau_n} H_{\hat{\nu}}(t)X_{\hat{\nu}}(t)[p'(t)(\mu(t) - \hat{\nu}(t) + \xi(t))]\, dt \geq 0, \quad n \in \mathbb{N}. \tag{8.32}$$

To see this, we first recall from (5.16) that

$$d\left(\frac{X_{\hat{\nu}}(t)}{S_0^{(\hat{\nu})}(t)}\right) = -\frac{dC(t)}{S_0^{(\hat{\nu})}(t)} + \frac{X_{\hat{\nu}}(t)}{S_0^{(\hat{\nu})}(t)}p'(t)\sigma(t)\, dW_{\hat{\nu}}(t),$$

and so

$$
d\left(\frac{X_{\hat{\nu}}(t)}{S_0^{(\hat{\nu})}(t)}(L(t)+N(t))\right)
$$

$$
=\frac{X_{\hat{\nu}}(t)}{S_0^{(\hat{\nu})}(t)}(dL(t)+dN(t))+(L(t)+N(t))d\left(\frac{X_{\hat{\nu}}(t)}{S_0^{(\hat{\nu})}(t)}\right)
$$

$$
+\frac{X_{\hat{\nu}}(t)}{S_0^{(\hat{\nu})}(t)}p'(t)(\mu(t)-\hat{\nu}(t))\,dt.
$$

Integration of this equation yields

$$
\frac{X_{\hat{\nu}}(\tau_n)}{S_0^{(\hat{\nu})}(\tau_n)}(L(\tau_n)+N(\tau_n))+\int_{(0,\tau_n]}\frac{L(t)+N(t)}{S_0^{(\hat{\nu})}(t)}dC(t)
$$

$$
=\int_0^{\tau_n}\frac{X_{\hat{\nu}}(t)}{S_0^{(\hat{\nu})}(t)}[p'(t)(\mu(t)-\hat{\nu}(t)+\xi(t)]\,dt
$$

$$
+\int_0^{\tau_n}\frac{X_{\hat{\nu}}(t)}{S_0^{(\hat{\nu})}(t)}[\sigma^{-1}(t)(\mu(t)-\hat{\nu}(t))
$$

$$
+(L(t)+N(t))\sigma'(t)p(t)]'\,dW_{\hat{\nu}}(t).
$$

By the choice of τ_n, the integrand of the Itô integral in this last expression is square-integrable, and thus has expectation zero under $P_{\hat{\nu},n}$. Taking expectations under this probability measure, we obtain

$$
E\int_0^{\tau_n}H_{\hat{\nu}}(t)X_{\hat{\nu}}(t)[p'(t)(\mu(t)-\hat{\nu}(t))+\xi(t)]\,dt
$$

$$
=E\left[H_{\hat{\nu}}(\tau_n)X(\tau_n)+\int_{(0,\tau_n]}H_{\hat{\nu}}(t)(L(t)+N(t))\,dC(t)\right].\quad(8.33)
$$

Applying the optional sampling theorem to the martingale in (8.19), we see that (8.15) is still valid if we replace t in that equation by the stopping time τ_n. Using this fact, we rewrite (8.33) as

$$
E\int_0^{\tau_n}H_{\hat{\nu}}(t)X_{\hat{\nu}}(t)[p'(t)(\mu(t)-\hat{\nu}(t))+\xi(t)]\,dt
$$

$$
=E\left[(L(\tau_n)+N(\tau_n))\left(H_{\hat{\nu}}(T)B+\int_{(\tau_n,T]}H_{\hat{\nu}}(t)\,dC(t)\right)\right.
$$

$$
\left.+\int_{(0,\tau_n]}H_{\hat{\nu}}(t)(L(t)+N(t))\,dC(t)]\right],
$$

which is the right-hand side of (8.31), hence a nonnegative quantity. This completes the proof of (8.32).

Step 6. We invoke Lemma 4.2 to obtain a process $\lambda(\cdot)\in\mathcal{D}$ satisfying (4.8), and we take $\mu(\cdot)=\hat{\nu}(\cdot)+\lambda(\cdot)$, so that $\xi(t)=\zeta(\lambda(t))$. Equation

(8.32) becomes

$$E \int_0^{\tau_n} H_{\hat{\nu}}(t) X_{\hat{\nu}}(t) [p'(t)\lambda(t) + \zeta(\lambda(t))] \, dt \geq 0, \quad n \in \mathbb{N},$$

which, together with (4.8), implies (8.17). From (4.5) we have immediately that

$$p'(t)\hat{\nu}(t) + \zeta(\hat{\nu}(t)) \geq 0 \quad \text{for Lebesgue-a.e. } t \in [0, T]$$

holds almosts surely. We next take $\mu(\cdot) \equiv 0$ in (8.32), so that $\xi(t) = -\zeta(\hat{\nu}(t))$, and (8.32) becomes

$$E \int_0^{\tau_n} H_{\hat{\nu}}(t) X_{\hat{\nu}}(t) [p'(t)\hat{\nu}(t) + \zeta(\hat{\nu}(t))] \, dt \leq 0, \quad n \in \mathbb{N}$$

which proves (8.18).   □

5.9   Lower Hedging Price

In addition to (2.3)–(2.6), we impose throughout this section the assumption that

$$\text{the process } \theta(\cdot) \text{ of (2.7) is bounded.} \tag{9.1}$$

This implies, in particular, that $Z_0(\cdot)$ of (2.8) is a martingale, that P_0 defined by

$$P_0(A) \triangleq \int_A Z_0(T) \, dP, \quad A \in \mathcal{F}(T), \tag{9.2}$$

is a probability measure on $\mathcal{F}(T)$, and thus $\mathcal{M} = (r(\cdot), b(\cdot), \delta(\cdot), \sigma(\cdot), S(0), A(\cdot))$ of Section 5.2 is a *complete, standard financial market*.

The seller of a contingent claim is interested in the *upper-hedging price* of Section 5.3. This is the amount of money he needs to receive at time $t = 0$ in order to invest in the constrained market in a way that ensures that his final wealth will almost surely dominate his obligation to pay off the contingent claim at the final time. The buyer, on the other hand, is interested in the *lower-hedging price*, which we shall discuss in this section. We imagine that the buyer takes a net negative wealth position, which must include either borrowing from the money market or short-selling of stock, in order to buy the contingent claim. The buyer finances this debt by trading in a constrained market, and desires at the final time to have the payment received from owning the contingent claim be sufficient to pay off the debt without risk (i.e., with probability one). Since the payoffs of the contingent claims in this chapter are nonnegative, the buyer's wealth will always be nonpositive prior to the final time. We define below the lower-hedging price to be the maximum amount of debt the buyer can acquire initially and still be certain that the payoff of the contingent claim will cover his debt at the final time.

We have modeled the constraint on the seller's portfolio-proportion process by a nonempty closed convex set K. For example, if there is only one stock and borrowing in the money market is prohibited, we set $K = (-\infty, 1]$, which enforces the no-borrowing condition by requiring that the investor's stock holdings can never exceed his wealth. To prevent the buyer of the contingent claim, whose wealth is always negative or zero, from borrowing in the money market, we should require that his short position in the stock be always at least as great as his net debt. Thus, the proportion of his (negative) wealth invested in the stock should always lie in the closed convex set $K_- \triangleq [1, \infty)$.

We generalize this situation by requiring that the seller's portfolio-proportion process lie in a nonempty, closed, convex set $K \subset \mathbb{R}^N$ and the buyer's portfolio-proportion process lie in a companion nonempty, closed, convex set $K_- \subset \mathbb{R}^N$. These sets are related by the two conditions

$$K \cap K_- \neq \emptyset, \tag{9.3}$$

$$\forall p_+ \in K, \forall p_- \in K_-, \quad \lambda p_+ + (1 - \lambda)p_- \in \begin{cases} K, & \text{if } \lambda \geq 1, \\ K_-, & \text{if } \lambda \leq 0. \end{cases} \tag{9.4}$$

We impose the assumptions (9.3) and (9.4) because they cover the examples of interest (Examples 9.7) and result in several simplifications of notation and exposition, such as Lemma 9.3 below. See, however, Remark 9.2 below.

We introduce the analogue

$$\zeta_-(\nu) \triangleq \inf_{p \in K_-} (-p'\nu), \quad \nu \in \mathbb{R}^N, \tag{9.5}$$

of the support function $\zeta(\cdot)$ in (4.1). The function $\zeta_-(\cdot)$ maps $\mathbb{R}^N$ into $\mathbb{R} \cup \{-\infty\}$, and $-\zeta_-(\nu) = \sup_{p \in K_-} (p'\nu)$, the support function of K_-, is a closed, proper, positively homogeneous, subadditive, convex function with effective domain

$$\widetilde{K}_- \triangleq \{\nu \in \mathbb{R}^N; \zeta_-(\nu) > -\infty\}. \tag{9.6}$$

By analogy with (4.5), the function $\zeta_-(\cdot)$ satisfies

$$p \in K_- \iff \zeta_-(\nu) + p'\nu \leq 0, \quad \forall \nu \in \widetilde{K}_-. \tag{9.7}$$

The counterpart to Lemma 4.2 for $\zeta_-(\cdot)$ is the following.

Lemma 9.1: *For any given $\{\mathcal{F}(t)\}$-progressively measurable process $p: [0, T] \times \Omega \to \mathbb{R}^N$, there exists a $\widetilde{K}_-$-valued, $\mathcal{F}(t)$-progressively measurable process $\nu(\cdot)$ such that*

$$\|\nu(t)\| \leq 1, \ |\zeta(\nu(t))| \leq 1, \quad 0 \leq t \leq T,$$

and for every $t \in [0, T]$, we have

$$\begin{aligned} p(t) \in K_- &\iff \nu(t) = 0, \\ p(t) \notin K_- &\iff \zeta(\nu(t)) + p'(t)\nu(t) > 0 \end{aligned} \tag{9.8}$$

almost surely.

Remark 9.2: The next lemma relates $\zeta(\cdot)$ to $\zeta_-(\cdot)$ and $\widetilde{K}$ to $\widetilde{K}_-$. We have imposed conditions (9.3), (9.4) in order to obtain this simplifying result. However, the only consequence of Lemma 9.3 below that is actually used in this section is the bound

$$\zeta_-(\nu) \geq \zeta_0, \quad \forall \nu \in \widetilde{K}_- \tag{9.9}$$

for some $\zeta_0 \in \mathbb{R}$. If K_- leads to a function ζ_- satisfying (9.9), then the lower-hedging price results of this section hold, even if there is no nonempty closed, convex set K related to K_- by (9.3), (9.4).

Lemma 9.3: *Under conditions (9.3) and (9.4), we have $\widetilde{K}_- = \widetilde{K}$ and $\zeta_-(\cdot) = \zeta(\cdot)$ on $\widetilde{K}$. In particular, (4.6) implies (9.9).*

PROOF. Clearly, (9.3) gives $\zeta_-(\nu) \leq \zeta(\nu)$ for all $\nu \in \mathbb{R}^N$. Consider an arbitrary $\nu \in \widetilde{K}$; then $-\infty < \zeta(\nu) < \infty$. For fixed $\lambda > 1$ and arbitrary $p_+ \in K$ and $p_- \in K_-$, we have, thanks to (9.4),

$$-\lambda p'_+\nu + (\lambda - 1)p'_-\nu = -(\lambda p_+ + (1-\lambda)p_-)'\nu \leq \sup_{p \in K}(-p'\nu) = \zeta(\nu).$$

Taking the supremum of the left-hand side over $p_+ \in K$ and $p_- \in K_-$, we deduce

$$\lambda\zeta(\nu) - (\lambda - 1)\zeta_-(\nu) \leq \zeta(\nu). \tag{9.10}$$

The inequality $\zeta_-(\nu) > -\infty$ follows from the finiteness of $\zeta(\nu)$, which gives $\widetilde{K} \subseteq \widetilde{K}_-$. The inequality (9.10) also implies $\zeta(\nu) \leq \zeta_-(\nu)$ for all $\nu \in \widetilde{K}$.

We next consider an arbitrary $\nu \in \widetilde{K}_-$, so that $\zeta_-(\nu)$ is finite. Fix $\lambda < 0$, and let $p_+ \in \widetilde{K}$ and $p_- \in \widetilde{K}_-$ be arbitrary. From (9.4) we now have

$$\lambda p'_+\nu + (1-\lambda)p'\nu = (\lambda p_+ + (1-\lambda)p_-)'\nu \leq \sup_{p \in K_-}(p'\nu) = -\zeta_-(\nu).$$

Taking the supremum of the left-hand side over $p_+ \in K$ and $p_- \in K_-$ yields

$$-\lambda\zeta(\nu) - (1-\lambda)\zeta_-(\nu) \leq -\zeta_-(\nu). \tag{9.11}$$

The inequality $\zeta(\nu) < \infty$ follows from the finiteness of $\zeta_-(\nu)$, which gives $\widetilde{K}_- \subseteq \widetilde{K}$, and thus also $\widetilde{K}_- = \widetilde{K}$. The inequality (9.11) also implies $\zeta(\nu) \leq \zeta_-(\nu)$ for all $\nu \in \widetilde{K}_- = \widetilde{K}$.   □

In light of Lemma 9.3, processes $\nu(\cdot)$ mapping $[0, T] \times \Omega$ into $\widetilde{K}$ also map into $\widetilde{K}_-$. In this section, we shall consider the set of processes $\mathcal{D}$ of Definition 5.1 and more particularly, the set $\mathcal{D}^{(b)}$ of bounded processes in $\mathcal{D}$. According to Remark 5.2, $Z_\nu(\cdot)$ is a martingale for every $\nu(\cdot) \in \mathcal{D}^{(b)}$, and the standard martingale measure P_ν for the market $\mathcal{M}_\nu$ is defined by (5.15).

We shall write henceforth $\widetilde{K}$ and $\zeta(\cdot)$ rather than $\widetilde{K}_-$ and $\zeta_-(\cdot)$, in order to simplify notation. However, in accordance with Remark 9.2, there need not actually be a set K and function $\zeta(\cdot)$; in such a case, the set $\widetilde{K}_-$ and the function $\zeta_-(\cdot)$ should be substituted for $\tilde{K}$ and $\zeta(\cdot)$ throughout the remainder of this section and the next.

Consider now an agent with cumulative income process $\Gamma(t) = x - C(t)$, $0 \leq t \leq T$, where $x \leq 0$ is his *initial wealth* and $C(\cdot)$ is his *cumulative consumption*: a nondecreasing, $\{\mathcal{F}(t)\}$-adapted process with nondecreasing, right-continuous paths and $C(0) = 0$, $C(T) < \infty$ almost surely. A *portfolio process* $\pi(\cdot)$ is an $\mathbb{R}^N$-valued, $\{\mathcal{F}(t)\}$-progressively measurable process satisfying (2.11).

Definition 9.4: Given $x \leq 0$ and a pair (C, π) of a cumulative consumption process and a portfolio process, the corresponding wealth process is given by

$$\frac{X^{x,\pi,C}(t)}{S_0(t)} + \int_{(0,t]} \frac{dC(v)}{S_0(v)} = x + \int_0^t \frac{1}{S_0(v)} \pi'(v)\sigma(v)\,dW_0(v), \quad 0 \leq t \leq T,$$
(2.12)

which can also be written as (2.13), and the corresponding *portfolio-proportion process* is defined by

$$p(t) \triangleq \begin{cases} \dfrac{\pi(t)}{X^{x,C,\pi}(t)}, & \text{if } X^{x,C,\pi}(t) \neq 0, \\ p_*, & \text{if } X^{x,C,\pi}(t) = 0, \end{cases}$$
(2.14)

where p_* is an arbitrary but fixed vector in K_-.

We say that a pair (C, π) consisting of a cumulative consumption process and a portfolio process is *admissible in the market* $\mathcal{M}(K_-)$ for the initial wealth $x \leq 0$, and we write $(C, \pi) \in \mathcal{A}(x; K_-)$, if $p(\cdot)$ satisfies the *buyer's constraint*

$$p(t) \in K_- \text{ for Lebesgue-a.e. } t \in [0, T]$$
(9.12)

almost surely, and

$$X^{x,C,\pi}(t) \leq 0, \quad \forall t \in [0, T] \text{ a.s.,}$$
(9.13)

$$E\left(\max_{0 \leq t \leq T} \frac{|X^{x,C,\pi}(t)|}{S_0(t)}\right)^\gamma < \infty, \quad \text{for some } \gamma > 1.$$
(9.14)

Remark 9.5: Consider the random variable

$$\Lambda \triangleq \max_{0 \leq t \leq T} \frac{|X^{x,C,\pi}(t)|}{S_0(t)}$$
(9.15)

in (9.14). Because of (9.1) we have $EZ_0^q(T) < \infty$ for every $q \in \mathbb{R}$, and (9.14) coupled with Hölder's inequality implies

$$E_0\Lambda^{\frac{\gamma+1}{2}} = E\left[Z_0(T)\Lambda^{\frac{\gamma+1}{2}}\right] \leq \left(EZ_0^{\frac{2\gamma}{\gamma-1}}(T)\right)^{\frac{\gamma-1}{2\gamma}} (E\Lambda^\gamma)^{\frac{\gamma+1}{2\gamma}} < \infty.$$

Hence, (9.14) implies

$$E_0\Lambda^{\gamma_0} < \infty \text{ for some } \gamma_0 > 1.$$
(9.16)

Similarly, (9.16) coupled with Hölder's inequality implies (9.14). Indeed, the condition

$$E_\nu \Lambda^{\gamma_\nu} < \infty \text{ for some } \gamma_\nu > 1 \qquad (9.17)$$

for *some* $\nu(\cdot) \in \mathcal{D}^{(b)}$ implies the validity of (9.17) for *every* $\nu(\cdot) \in \mathcal{D}^{(b)}$ and also the validity of (9.14).

Remark 9.6: Suppose (C, π) is admissible in $\mathcal{M}(K)$ for $x \leq 0$. Let $X^{x,C,\pi}(\cdot)$ be the wealth process with initial condition x generated by (C, π), and define Λ by (9.15), so that $E\Lambda^\gamma < \infty$ for some $\gamma > 1$. Then condition (9.16) holds. We have

$$-\Lambda \leq \frac{X^{x,C,\pi}(t)}{S_0(t)} + \int_{(0,t]} \frac{dC(v)}{S_0(v)} = x + \int_0^t \frac{1}{S_0(v)} \pi'(v)\sigma(v)\, dW_0(v), \quad (9.18)$$

and Fatou's lemma shows that the P_0-local martingale

$$M(t) \triangleq \int_0^t \frac{1}{S_0(v)} \pi'(v)\sigma(u)\, dW_0(v)$$

is a P_0-supermartingale. We have then the *budget constraint*

$$E_0 \left[\frac{X^{x,C,\pi}(T)}{S_0(T)} + \int_{(0,T]} \frac{dC(v)}{S_0(v)} \right] \leq x, \quad \forall (C, \pi) \in \mathcal{A}(x; K_-), \qquad (9.19)$$

which implies

$$E_0 \left[\int_{(0,T]} \frac{dC(v)}{S_0(v)} \right] \leq x - E_0 \left[\frac{X^{x,C,\pi}(T)}{S_0(T)} \right] \leq x + E_0\Lambda.$$

Returning to (9.18), we conclude that

$$-\Lambda \leq x + M(t) = \frac{X^{x,C,\pi}(t)}{S_0(t)} + \int_{(0,t]} \frac{dC(v)}{S_0(v)} \leq \int_{(0,T]} \frac{dC(v)}{S_0(v)}, \quad 0 \leq t \leq T.$$

Being bounded both from below and from above by P_0-integrable random variables, the P_0-local martingale $M(\cdot)$ is in fact a P_0-*martingale*. Moreover, the budget constraint (9.19) holds with equality:

$$E_0 \left[\frac{X^{x,C,\pi}(T)}{S_0(T)} + \int_{(0,T]} \frac{dC(v)}{S_0(v)} \right] = x, \quad \forall (C, \pi) \in \mathcal{A}(x; K_-). \qquad (9.20)$$

Example 9.7: The following pairs of nonempty closed, convex sets K and K_- satisfy the conditions of (9.3) and (9.4). The corresponding convex cone $\widetilde{K} = \widetilde{K}_-$ and support function $\zeta(\cdot) = \zeta_-(\cdot)$, given in Examples 4.1, are repeated here for reference.

(i) *Unconstrained case:* $K = K_- = \mathbb{R}^N$, $\widetilde{K} = \{0\}$, and $\zeta \equiv 0$ on $\widetilde{K}$.

(ii) *Prohibition of short-selling:* $K = [0, \infty)^N$, $K_- = (-\infty, 0]^N$, $\widetilde{K} = [0, \infty)^N$, and $\zeta \equiv 0$ on $\widetilde{K}$. In this case, in both the market $\mathcal{M}(K)$

with $X(\cdot) \geq 0$ and in the market $\mathcal{M}(K_-)$ with $X(\cdot) \leq 0$, the amount of wealth $\pi_n(t) = X(t)p_n(t)$ invested in the nth stock is nonnegative for Lebesgue-almost-all $t \in [0,T]$ almost surely, for $n = 1, \ldots, N$.

(iii) *Incomplete market:* $K = K_- = \{p \in \mathbb{R}^N; p_{M+1} = \cdots = p_N = 0\}$ for some $M \in \{1, \ldots, N-1\}$, $\widetilde{K} = \{\nu \in \mathbb{R}^N; \nu_1 = \cdots = \nu_M = 0\}$, and $\zeta \equiv 0$ on $\widetilde{K}$.

(iv) *Incomplete market with prohibition of short-selling:* In this case, $K = \{p \in \mathbb{R}^N; p_1 \geq 0, \ldots, p_M \geq 0, p_{M+1} = \cdots = p_N = 0\}$ and $K_- = \{p \in \mathbb{R}^N; p_1 \leq 0, \ldots, p_M \leq 0, p_{M+1} = \cdots = p_N = 0\}$. Then $\widetilde{K} = \{\nu \in \mathbb{R}^N; \nu_1 \geq 0, \ldots, \nu_M \geq 0\}$ and $\zeta \equiv 0$ on $\widetilde{K}$.

(v) K is a nonempty, closed convex cone in $\mathbb{R}^N$ and $K_- = -K$. Then $\widetilde{K} = \{\nu \in \mathbb{R}^N; p'\nu \leq 0, \quad \forall p \in K_-\}$ and $\zeta \equiv 0$ on $\widetilde{K}$. This generalizes examples (i)–(iv).

(vi) *Prohibition of borrowing:* $K = \{p \in \mathbb{R}^N; \sum_{n=1}^{N} p_n \leq 1\}$ and $K_- = \{p \in \mathbb{R}^N; \sum_{n=1}^{N} p_n \geq 1\}$, so that the amount of money

$$\pi_0(t) \overset{\Delta}{=} X(t) - \sum_{n=1}^{N} \pi_n(t) = X(t)\left(1 - \sum_{n=1}^{N} p_n(t)\right)$$

invested in the money market is nonnegative for Lebesgue-almost-all $t \in [0,T]$ almost surely, regardless of whether $X(t) \geq 0$ or $X(t) \leq 0$. In this case $\widetilde{K} = \{\nu \in \mathbb{R}^N; \nu_1 = \cdots = \nu_N \leq 0\}$ and $\zeta(\nu) = -\nu_1$ on $\widetilde{K}$.

(vii) *Constraints on short-selling:* $K = [-\kappa, \infty)^N$ for some $\kappa > 0$. It is not clear how one should place a short-selling constraint on an agent whose total wealth is negative, but the only set K_- corresponding to the K of this example and satisfying (9.3) and (9.4) is $K_- = (-\infty, -\kappa]^N$. Regardless of the sign of $X(t)$, the wealth $\pi_n(t) = X(t)p_n(t)$ invested in the nth stock is at least $-\kappa X(t)$, a nonnegative quantity when $X(t) \leq 0$. In this case $\widetilde{K} = [0, \infty)^N$ and $\zeta(\nu) = \kappa \sum_{n=1}^{N} \nu_n$ on $\widetilde{K}$.

(viii) *Constraints on borrowing:* $K = \{p \in \mathbb{R}^N; \sum_{n=1}^{N} p_n \leq \kappa\}$ for some $\kappa > 1$. It is not clear how one should place a borrowing constraint on an agent whose total wealth is negative, but the only set K_- corresponding to the K of this example and satisfying (9.3) and (9.4) is $K_- = \{p \in \mathbb{R}^N; \sum_{n=1}^{N} p_n \geq \kappa\}$. The amount $\pi_0(t)$ invested in the money market, defined in (vi), satisfies $\pi_0(t) \geq (1 - \kappa)X(t)$ regardless of the sign of $X(t)$. In this case, $\widetilde{K} = \{\nu \in \mathbb{R}^N; \nu_1 = \cdots = \nu_N \leq 0\}$ and $\zeta(\nu) = -\kappa \nu_1$ on $\widetilde{K}$.

Definition 9.8:

(i) A *contingent claim* B is a nonnegative, $\mathcal{F}(T)$-measurable random variable. We call

$$u_0 \overset{\Delta}{=} E[H_0(T)B] = E_0 \left[\frac{B}{S_0(T)} \right] \tag{3.1}$$

the *unconstrained hedging price* of B. (This is a repetition of Definition 3.1(i).)

(ii) The *lower-hedging price in* $\mathcal{M}(K_-)$ of the contingent claim B is defined to be

$$h_{low}(K_-) \tag{9.21}$$
$$\overset{\Delta}{=} -\inf\{x \leq 0; \exists(C,\pi) \in \mathcal{A}(x;K_-) \text{ with } X^{x,C,\pi}(T) + B \geq 0 \text{ a.s.}\}.$$

(iii) We say that B is K_--*attainable* if there exists a portfolio process $\pi(\cdot)$ such that

$$(0,\pi) \in \mathcal{A}(x;K_-) \quad \text{and} \quad X^{x,0,\pi}(T) + B = 0 \text{ a.s.} \tag{9.22}$$

with $x = -h_{low}(K_-)$.

In other words, if the infimum in (9.21) is attained, then $h_{low}(K_-)$ is the maximal amount of initial debt the buyer of the contingent claim can afford to acquire and still be sure that by investment in the constrained market $\mathcal{M}(K_-)$, he can be guaranteed (with probability one) to have nonnegative wealth at the terminal time $t = T$, once the payoff of the contingent claim has been received.

Lemma 9.9: *Let B be a contingent claim. We have*

$$0 \leq h_{low}(K_-) \leq u_0 \leq h_{up}(K). \tag{9.23}$$

PROOF. Taking $x = 0$, $\pi(\cdot) \equiv 0$, and $C(\cdot) \equiv 0$, we have $X^{x,C,\pi}(\cdot) \equiv 0$, and $X^{x,C,\pi}(T) + B \geq 0$ almost surely. Hence, the infimum in (9.21) is nonpositive and $h_{low}(K_-) \geq 0$.

The inequality $u_0 \leq h_{up}(K)$ follows immediately from Theorem 6.2, but here is an elementary argument. The inequality holds trivially if $h_{up}(K) = \infty$; if $h_{up}(K) < \infty$, then for any $x \geq 0$ for which there exists $(C,\pi) \in \mathcal{A}(x;K_-)$ with $X^{x,C,\pi}(T) \geq B$ almost surely, the budget constraint (2.17) gives

$$x \geq E\left[H_0(T)X^{x,C,\pi}(T) + \int_{(0,T]} H_0(v)\, dC(v)\right] \geq E[H_0(T)B] = u_0,$$

whence $h_{up}(K) \geq u_0$.

It remains to show that

$$h_{low}(K_-) \leq u_0. \tag{9.24}$$

Let $x \leq 0$ be such that there exists $(C,\pi) \in \mathcal{A}(x;K_-)$ with $X^{x,C,\pi}(T) + B \geq 0$ almost surely. According to the budget constraint (9.20),

$$x = E_0 \left[\frac{X^{x,C,\pi}(T)}{S_0(T)} + \int_{(0,T]} \frac{dC(v)}{S_0(v)} \right] \geq -E_0 \left[\frac{B}{S_0(T)} \right] = -u_0.$$

The desired inequality (9.24) follows.   □

We observe that in the presence of condition (9.1), the finiteness of u_0 is implied by the condition

$$E \left[\frac{B}{S_0(T)} \right]^\gamma < \infty \quad \text{for some } \gamma > 1. \tag{9.25}$$

Indeed, when (9.25) holds, we may proceed as in Remark 9.5 to obtain, for every process $\nu(\cdot) \in \mathcal{D}^{(b)}$,

$$E_\nu \left[\frac{B}{S_0(T)} \right]^{\gamma_\nu} < \infty \quad \text{for some } \gamma_\nu > 1. \tag{9.26}$$

For $\nu(\cdot) \in \mathcal{D}^{(b)}$, all moments of $Z_\nu(T)$ are finite under both P and P_ν. Another application of Hölder's inequality shows that (9.26) is equivalent to

$$E_\nu \left[Z_\nu^{q_\nu}(T) \left(\frac{B}{S_0(T)} \right)^{\gamma_\nu} \right] < \infty \quad \text{for every } q_\nu \in \mathbb{R} \tag{9.27}$$

for some $\gamma_\nu > 1$.

It turns out that condition (9.25) actually leads to a much stronger result: *a characterization of the lower-hedging price $h_{low}(K_-)$ in terms of a stochastic control problem*, in the spirit of Theorem 6.2. We recall the Definition 6.1 of u_ν, which can now be written in terms of P_ν as

$$u_\nu \triangleq E[H_\nu(T)B] = E_\nu \left[\frac{B}{S_0^{(\nu)}(T)} \right]. \tag{9.28}$$

Theorem 9.10: *Assume (9.1) and (9.9), and let B be a contingent claim satisfying (9.25). The lower-hedging price of B is given by*

$$h_{low}(K_-) = \check{u} \triangleq \inf_{\nu \in \mathcal{D}^{(b)}} u_\nu. \tag{9.29}$$

Furthermore, there exists a pair $(\check{C}, \check{\pi}) \in \mathcal{A}(-\check{u}, K_-)$ with corresponding wealth process

$$X^{-\check{u},\check{C},\check{\pi}}(t) = -\text{ess inf}_{\nu \in \mathcal{D}^{(b)}} \frac{E[H_\nu(T)B|\mathcal{F}(t)]}{H_\nu(t)}, \quad 0 \leq t \leq T, \tag{9.30}$$

and in particular,

$$X^{-\check{u},\check{C},\check{\pi}}(T) + B = 0 \tag{9.31}$$

almost surely.

The proof of this theorem is provided after Proposition 9.13 below.

Definition 9.11: We call the process

$$\check{X}(t) \overset{\Delta}{=} -\operatorname{ess\,inf}_{\nu \in \mathcal{D}^{(b)}} \frac{E[H_\nu(T)B|\mathcal{F}(t)]}{H_\nu(t)}, \quad 0 \leq t \leq T, \tag{9.32}$$

on the right-hand side of (9.30), the *lower-hedging value process* for the contingent claim B. We shall always take a right-continuous, left-limited modification (see Proposition 9.13 below) of this process.

Remark 9.12: In the *unconstrained case* $K = K_- = \mathbb{R}^N$ of Example 9.5(i) we have $\widetilde{K} = \widetilde{K}_- = \{0\}$, so that $\mathcal{D} = \mathcal{D}^{(b)}$ contains only the zero process $\nu(\cdot) \equiv 0$. Thus, in Theorems 6.2 and 9.10,

$$\check{u} = \hat{u} = u_0, \ h_{low}(K_-) = u_0 = h_{up}(K), \ \check{C}(\cdot) = \widehat{C}(\cdot) \equiv 0,$$

and

$$-X^{-u_0, \check{C}, \check{\pi}}(t) = X^{u_0, \widehat{C}, \hat{\pi}}(t) = \frac{E[H_0(T)B|\mathcal{F}(t)]}{H_0(t)} = S_0(t)E_0\left[\left.\frac{B}{S_0(T)}\right|\mathcal{F}(t)\right]$$

coincides with the process $V^{ECC}(\cdot)$ of (2.2.13) on $[0,T)$, whereas $\check{\pi}(\cdot) = \hat{\pi}(\cdot)$ is the hedging portfolio of Definition 2.2.6.

Proposition 9.13: *Assume (9.1) and (9.9). Under the additional assumption (9.25), the lower-hedging value process*

$$\check{X}(t) \overset{\Delta}{=} -\operatorname{ess\,inf}_{\nu \in \mathcal{D}^{(b)}} \frac{E[H_\nu(T)B|\mathcal{F}(t)]}{H_\nu(t)}, \quad 0 \leq t \leq T, \tag{9.33}$$

is finite and satisfies the dynamic programming equation

$$\check{X}(s) = \operatorname{ess\,sup}_{\nu \in \mathcal{D}^{(b)}} \frac{E[H_\nu(t)\check{X}(t)|\mathcal{F}(s)]}{H_\nu(s)}, \quad 0 \leq s \leq t \leq T. \tag{9.34}$$

In particular, $\check{X}(0) = -\check{u}$ *and* $\check{X}(T) + B = 0$ *almost surely.*

The process $\check{X}(\cdot)$ has a right-continuous, left-limited (RCLL) modification, which we shall always choose. With this choice, and for every process $\nu(\cdot) \in \mathcal{D}^{(b)}$, the process $H_\nu(\cdot)\check{X}(\cdot)$ is a uniformly integrable RCLL supermartingale under P. Furthermore, for each $\nu(\cdot) \in \mathcal{D}^{(b)}$, the process $\frac{\check{X}(\cdot)}{S_0^{(\nu)}(\cdot)}$ is a supermartingale of class $\mathcal{D}[0,T]$ under P_ν; i.e., the collection of random variables $\left\{\frac{\check{X}(\tau)}{S_0^{(\nu)}(\tau)}\right\}_{\tau \in \mathcal{S}}$ is uniformly integrable under P_ν, where $\mathcal{S}$ is the set of all stopping times taking values in $[0,T]$.

PROOF.   We prove first that for each $\nu(\cdot) \in \mathcal{D}^{(b)}$, the process $H_\nu(\cdot)\check{X}(\cdot)$ is uniformly integrable under P. Let $\nu(\cdot) \in \mathcal{D}^{(b)}$ and $t \in [0,T]$ be given, and denote by $\mathcal{D}_{t,\nu}^{(b)}$ the set of all processes $\mu(\cdot) \in \mathcal{D}^{(b)}$ that agree with $\nu(\cdot)$ on $[0,t] \times \Omega$. Since $H_\mu(s,t) \overset{\Delta}{=} H_\mu(t)/H_\mu(s)$ depends only on the values of $\mu(v)$

for $s \leq v \leq t$, we may rewrite (9.33) as

$$\check{X}(t) = \text{ess sup}_{\mu \in \mathcal{D}_{t,\nu}^{(b)}} \frac{E[-H_\mu(T)B|\mathcal{F}(t)]}{H_\mu(t)}$$

$$= \frac{1}{H_\nu(t)} \cdot \text{ess sup}_{\mu \in \mathcal{D}_{t,\nu}^{(b)}} E[-H_\mu(T)B|\mathcal{F}(t)]$$

$$\geq -\frac{1}{H_\nu(t)} E[H_\nu(T)B|\mathcal{F}(t)],$$

whence

$$|H_\nu(t)\check{X}(t)| \leq E[H_\nu(T)B|\mathcal{F}(t)], \quad 0 \leq t \leq T.$$

It follows from Jensen's inequality and Doob's maximal martingale inequality that for every $\gamma_\nu > 1$,

$$E\left[\sup_{0 \leq t \leq T} |H_\nu(t)\check{X}(t)|^{\gamma_\nu}\right] \leq \left(\frac{\gamma_\nu}{\gamma_\nu - 1}\right)^{\gamma_\nu} \cdot E\left[H_\nu(T)B\right]^{\gamma_\nu}, \qquad (9.35)$$

and the uniform integrability of $H_\nu(\cdot)\check{X}(\cdot)$ under P follows from the inequality

$$E[H_\nu(T)B]^{\gamma_\nu} \leq e^{-\gamma_\nu \zeta_0 T} E_\nu\left[Z_\nu^{\gamma_\nu - 1}(T)\left(\frac{B}{S_0(T)}\right)^{\gamma_\nu}\right]$$

and (9.27).

We now imitate the proof of Proposition 6.5, with $-B$ replacing B, and using the uniform integrability of $H_\nu(\cdot)\check{X}(\cdot)$ under P instead of Fatou's lemma in the last step. The proof of Proposition 6.5 also involves random variables, now defined by $J_\mu(t) \triangleq E[-H_\mu(t,T)B|\mathcal{F}(t)]$, which have the property that for each process $\nu(\cdot) \in \mathcal{D}^{(b)}$ and $t \in [0,T]$, the collection $\{J_\mu(t)\}_{\mu \in \mathcal{D}_{t,\nu}^{(b)}}$ is closed under pairwise maximization. It follows that the collection

$$\{(J_\mu(t) \vee J_\nu(t)) + E[H_\nu(t,T)B|\mathcal{F}(t)]\}_{\mu \in \mathcal{D}_{t,\nu}^{(b)}}$$

of nonnegative random variables is closed under pairwise maximization, and Theorem A.3 of Appendix A can be applied to this collection to extract a sequence $\{\mu_k(\cdot)\}_{k=1}^\infty$ such that $\{J_{\mu_k}(t)\}_{k=1}^\infty$ is nondecreasing and

$$\check{X}(t) = \lim_{k \to \infty} J_{\mu_k}(t)$$

(cf. (6.13)). With these modifications, we obtain from the proof of Proposition 6.5 all the assertions of the present proposition, except the last.

For the final assertion of Proposition 9.13 we let $\nu(\cdot) \in \mathcal{D}^{(b)}$ be given and start by verifying the P_ν-supermartingale property for $0 \leq s \leq t \leq T$:

$$E_\nu\left[\left.\frac{\check{X}(t)}{S_0^{(\nu)}(t)}\right|\mathcal{F}(s)\right] = \frac{1}{Z_\nu(s)}E[H_\nu(t)\check{X}(t)|\mathcal{F}(s)]$$

$$\le \frac{1}{Z_\nu(s)}H_\nu(s)\check{X}(s) = \frac{\check{X}(s)}{S_0^{(\nu)}(s)}.$$

Because $-\frac{\check{X}(\cdot)}{S_0^{(\nu)}(\cdot)}$ is a nonnegative P_ν-submartingale, so is $\left|\frac{\check{X}(\cdot)}{S_0^{(\nu)}(\cdot)}\right|^{\gamma_\nu}$ with $\gamma_\nu > 1$ chosen to satisfy (9.26), thanks to which this submartingale has a P_ν-integrable last element

$$\left|\frac{\check{X}(T)}{S_0^{(\nu)}(T)}\right|^{\gamma_\nu} \le e^{-\gamma_\nu\zeta_0 T}\left(\frac{B}{S_0(T)}\right)^{\gamma_\nu}.$$

This establishes the uniform integrability under P_ν of the collection of random variables $\left\{\frac{\check{X}(\tau)}{S_0^{(\nu)}(\tau)}\right\}_{\tau\in\mathcal{S}}$.   □

PROOF OF THEOREM 9.10. Let $x \le 0$ and $(C,\pi) \in \mathcal{A}(x;K_-)$ be such that $X^{x,C,\pi}(T)+B \ge 0$ almost surely. In differential notation, with $X(\cdot) = X^{x,C,\pi}(\cdot)$ and $p(\cdot)$ defined by (2.12) and (2.14), we have

$$d\left(\frac{X(t)}{S_0(t)}\right) = -\frac{dC(t)}{S_0(t)} + \frac{X(t)}{S_0(t)}p'(t)\sigma(t)\,dW_0(t),$$

and thus

$$d\left(\frac{X(t)}{S_0^{(\nu)}(t)}\right) = d\left(e^{-\int_0^t \zeta(\nu(s))\,ds}\frac{X(t)}{S_0(t)}\right)$$

$$= -\zeta(\nu(t))e^{-\int_0^t \zeta(\nu(s))\,ds}\frac{X(t)}{S_0(t)}\,dt + e^{-\int_0^t \zeta(\nu(s))\,ds}d\left(\frac{X(t)}{S_0(t)}\right)$$

$$= \frac{X(t)}{S_0^{(\nu)}(t)}[-\zeta(\nu(t))\,dt + p'(t)\sigma(t)\,dW_0(t)] - \frac{dC(t)}{S_0^{(\nu)}(t)}$$

$$= -\frac{X(t)}{S_0^{(\nu)}(t)}[\zeta(\nu(t)) + p'(t)\nu(t)]\,dt - \frac{dC(t)}{S_0^{(\nu)}(t)}$$

$$\quad + \frac{X(t)}{S_0^{(\nu)}(t)}p'(t)\sigma(t)\,dW_\nu(t)$$

for any $\nu(\cdot) \in \mathcal{D}^{(b)}$. Integration yields

$$\frac{X^{x,C,\pi}(t)}{S_0^{(\nu)}(t)} + \int_0^t \frac{X^{x,C,\pi}(s)}{S_0^{(\nu)}(s)}[\zeta(\nu(s)) + p'(s)\nu(s)]\,ds + \int_{(0,t]}\frac{dC(s)}{S_0^{(\nu)}(s)}$$

$$= x + \int_0^t \frac{X^{x,C,\pi}(s)}{S_0^{(\nu)}(s)}p'(v)\sigma(s)\,dW_\nu(s). \qquad (9.36)$$

From (9.5) and Lemma 9.3, we see that $\zeta(\nu(s)) + p'(s)\nu(s)$ is nonpositive, as is $X^{x,C,\pi}(s)$; thus the process on the left-hand side of (9.36) is bounded from below by

$$\frac{X^{x,C,\pi}(t)}{S_0^{(\nu)}(t)} \geq -\Lambda \exp\left\{-\int_0^t \zeta(\nu(s))\,ds\right\} \geq -\Lambda e^{-\zeta_0 T}, \quad 0 \leq t \leq T,$$

where the random variable Λ of (9.15) satisfies (9.14) as well as (9.17). The P_ν-local martingale (9.36), being bounded from below by a P_ν-integrable random variable, must be a P_ν-supermartingale. Hence

$$x \geq E_\nu\left[\frac{X^{x,C,\pi}(T)}{S_0^{(\nu)}(T)} + \int_0^T \frac{X^{x,C,\pi}(s)}{S_0^{(\nu)}(s)}[\zeta(\nu(s)) + p'(s)\nu(s)]\,ds\right.$$

$$\left. + \int_{(0,T]} \frac{dC(v)}{S_0^{(\nu)}(v)}\right]$$

$$\geq E_\nu\left[-\frac{B}{S_0^{(\nu)}(T)}\right] = -E[H_\nu(T)B] = -u_\nu, \quad \forall \nu \in \mathcal{D}^{(b)}.$$

Taking successively the infimum over x, and then over $\nu(\cdot) \in \mathcal{D}^{(b)}$, we obtain first $h_{low}(K_-) \leq u_\nu$ for every $\nu(\cdot) \in \mathcal{D}^{(b)}$, and then

$$h_{low}(K_-) \leq \check{u}, \tag{9.37}$$

respectively. □

To prove the reverse inequality, we consider the lower-hedging value process $\check{X}(\cdot)$ of Proposition 9.13. For each $\nu(\cdot) \in \mathcal{D}^{(b)}$, the P_ν-supermartingale $\frac{\check{X}(\cdot)}{S_0^{(\nu)}(\cdot)}$ is of class $\mathcal{D}[0,T]$; thus, it admits a *unique* Doob–Meyer decomposition of the form

$$\frac{\check{X}(t)}{S_0^{(\nu)}(t)} = -\check{u} + M_\nu(t) - A_\nu(t), \quad 0 \leq t \leq T. \tag{9.38}$$

Here $M_\nu(\cdot)$ is a P_ν-martingale, representable as a stochastic integral in the form

$$M_\nu(t) = \int_0^t \psi_\nu'(s)\,dW_\nu(s), \quad 0 \leq t \leq T,$$

for some $\{\mathcal{F}(t)\}$-progressively measurable process $\psi_\nu : [0,T] \times \Omega \to \mathbb{R}^N$ with $\int_0^T \|\psi_\nu(t)\|^2\,dt < \infty$ almost surely (Lemma 1.6.7, applied to the market $\mathcal{M}_\nu$), and $A_\nu(\cdot)$ is an $\{\mathcal{F}(t)\}$-adapted, natural (see Remark 6.8), nondecreasing process with right-continuous paths, $E_\nu A_\nu(T) < \infty$, and $A_\nu(0) = 0$.

For any two processes $\mu(\cdot)$, $\nu(\cdot) \in \mathcal{D}^{(b)}$ we have

$$d\left(\frac{S_0^{(\nu)}(t)}{S_0^{(\mu)}(t)}\right) = [\zeta(\nu(t)) - \zeta(\mu(t))]\frac{S_0^{(\nu)}(t)}{S_0^{(\mu)}(t)}\,dt$$

and hence

$$
\begin{aligned}
d\left(\frac{\check{X}(t)}{S_0^{(\mu)}(t)}\right) &= d\left(\frac{\check{X}(t)}{S_0^{(\nu)}(t)} \cdot \frac{S_0^{(\nu)}(t)}{S_0^{(\mu)}(t)}\right) \\
&= \frac{S_0^{(\nu)}(t)}{S_0^{(\mu)}(t)}\psi_\nu'(t)\,dW_\nu(t) - \frac{S_0^{(\nu)}(t)}{S_0^{(\mu)}(t)}dA_\nu(t) \\
&\quad + \frac{\check{X}(t)}{S_0^{(\mu)}(t)}[\zeta(\nu(t)) - \zeta(\mu(t))]\,dt \\
&= \frac{S_0^{(\nu)}(t)}{S_0^{(\mu)}(t)}\psi_\nu'(t)\,dW_\mu(t) + \frac{S_0^{(\nu)}(t)}{S_0^{(\mu)}(t)}\psi_\nu'(t)\sigma^{-1}(t)(\nu(t) - \mu(t))\,dt \\
&\quad - \frac{S_0^{(\nu)}(t)}{S_0^{(\mu)}(t)}dA_\nu(t) + \frac{\check{X}(t)}{S_0^{(\mu)}(t)}[\zeta(\nu(t)) - \zeta(\mu(t))]\,dt.
\end{aligned}
$$

But (9.38) with $\nu(\cdot)$ replaced by $\mu(\cdot)$ implies

$$
d\left(\frac{\check{X}(t)}{S_0^{(\mu)}(t)}\right) = \psi_\mu'(t)\,dW_\mu(t) - dA_\mu(t).
$$

From the uniqueness of the Doob–Meyer decomposition, we deduce then $\frac{S_0^{(\nu)}(t)}{S_0^{(\mu)}(t)}\psi_\nu(t) = \psi_\mu(t)$, i.e., that

$$
\varphi(t) \triangleq S_0^{(\nu)}(t)\psi_\nu(t) = S_0^{(\mu)}(t)\psi_\mu(t), \quad 0 \le t \le T, \tag{9.39}
$$

does not depend on the process $\nu(\cdot) \in \mathcal{D}^{(b)}$, and that

$$
\begin{aligned}
\check{C}(t) &\triangleq \int_{(0,t]} S_0^{(\nu)}(s)dA_\nu(s) - \int_0^t [\check{X}(s)\zeta(\nu(s)) + \varphi'(s)\sigma^{-1}(s)\nu(s)]\,ds \\
&= \int_{(0,t]} S_0^{(\mu)}(s)dA_\mu(s) - \int_0^t [\check{X}(s)\zeta(\mu(s)) + \varphi'(s)\sigma^{-1}(s)\mu(s)]\,ds
\end{aligned}
\tag{9.40}
$$

does not depend on $\nu(\cdot) \in \mathcal{D}^{(b)}$ either. We will take $\check{C}(\cdot)$ in (9.40) to be the consumption process in Theorem 9.10; setting $\nu(\cdot) \equiv 0$ in (9.40), we see that this process $\check{C}(t) = \int_{(0,t]} S_0^{(0)}(s)\,dA_0(s)$ is nondecreasing. We set

$$
\check{\pi}(t) \triangleq (\sigma'(t))^{-1}\varphi(t), \tag{9.41}
$$

and define $\check{p}(\cdot)$ by

$$
\check{p}(t) \triangleq \begin{cases} \dfrac{\check{\pi}(t)}{\check{X}(t)}, & \text{if } \check{X}(t) \ne 0, \\ p_*, & \text{if } \check{X}(t) = 0 \end{cases} \tag{9.42}
$$

(cf. (2.14)), where p_* is an arbitrary vector in K_-.

With $\check{C}(\cdot)$, $\check{\pi}(\cdot)$, and $\check{p}(\cdot)$ as defined above, we have from (9.38) that for every $\nu(\cdot) \in \mathcal{D}^{(b)}$,

$$\frac{\check{X}(t)}{S_0^{(\nu)}(t)} + \int_0^t \frac{\check{X}(s)}{S_0^{(\nu)}(s)}[\zeta(\nu(s)) + \check{p}'(s)\nu(s)]\,ds + \int_{(0,t]} \frac{d\check{C}(s)}{S_0^{(\nu)}(s)}$$

$$= -\check{u} + \int_0^t \frac{1}{S_0^{(\nu)}(s)}\check{\pi}'(s)\sigma(s)\,dW_\nu(s), \quad 0 \le t \le T. \tag{9.43}$$

With $\nu(\cdot) \equiv 0$, this equation reduces to (2.12), which means that $X^{-\check{u},\check{C},\check{\pi}}(\cdot) = \check{X}(\cdot)$ and (9.30) holds.

It remains to verify that $(\check{C}, \check{\pi}) \in \mathcal{A}(-\check{u}; K_-)$. We first check that (9.14) holds. From (9.33) we have

$$\frac{\check{X}(t)}{S_0(t)} \ge -\frac{1}{Z_0(t)}E[H_0(T)B|\mathcal{F}(t)] = -E_0\left[\frac{B}{S_0(T)}\bigg|\mathcal{F}(t)\right], \tag{9.44}$$

which leads us to consider the random variable

$$\Lambda_B \overset{\Delta}{=} \sup_{0 \le t \le T} E_0\left[\frac{B}{S_0(T)}\bigg|\mathcal{F}(t)\right]. \tag{9.45}$$

With $\gamma_0 > 1$ as in (9.26), Doob's maximal martingale inequality implies

$$E_0\Lambda^{\gamma_0} \le \frac{\gamma_0}{\gamma_0 - 1}E_0\left[\frac{B}{S_0(T)}\right]^{\gamma_0} < \infty.$$

Inequality (9.14) follows from Remark 9.5.

Finally, we must show that we have

$$\check{p}(t) \in K_- \text{ for Lebesgue-a.e. } t \in [0, T] \tag{9.46}$$

almost surely. Let $\nu(\cdot) \in \mathcal{D}^{(b)}$ be the process corresponding to $\check{p}(\cdot)$ given by Lemma 9.1. For each positive integer k, $k\nu(\cdot) \in \mathcal{D}^{(b)}$. Using (9.41), (9.42), we may rewrite (9.40) as

$$0 \le \int_{(0,T]} S_0^{(\nu)}(s)\,dA_{k\nu}(s)$$

$$= \check{C}(T) + k\int_0^T \check{X}(s)[\zeta(\nu(s)) + \check{p}'(s)\nu(s)]\,ds.$$

The integrand on the right-hand side of this inequality is nonpositive (because $\check{X}(\cdot)$ is nonpositive and $\nu(\cdot)$ and $\check{\pi}(\cdot)$ are related by (9.8)); choosing k sufficiently large, the right-hand side can be made negative with positive probability, unless

$$\zeta(\nu(t)) + \check{p}(t)\nu(t) = 0 \text{ for Lebesgue-a.e. } t \in [0, T] \tag{9.47}$$

holds almost surely. Thus (9.47) must hold, and with it, (9.46) as well. $\quad\square$

The following two propositions provide some conditions in general markets under which the lower-hedging price is either zero or positive. In the next section, we take up more specific computations of the lower-hedging price in markets with constant coefficients.

Proposition 9.14: *Assume (9.1) and (9.9). We have*

$$h_{low}(K_-) = \check{u} \overset{\Delta}{=} \inf_{\nu \in \mathcal{D}^{(b)}} u_\nu = 0$$

in either of the following cases:

(i) $0 \le B \le \beta$ *almost surely for some* $\beta > 0$, *and*

$$\nu \mapsto \zeta(\nu) \text{ is unbounded from above on } \widetilde{K}; \qquad (9.48)$$

(ii) $0 \le B \le \alpha S_n(T) + \beta$ *almost surely for some* $\alpha > 0$, $\beta \ge 0$, *and some* $n = 1, \dots, N$ *for which the dividend rate process* $\delta_n(\cdot)$ *is bounded from below;* $ES_n^\gamma(T) < \infty$ *for some* $\gamma > 1$;

$$\nu \mapsto \zeta(\nu) + \nu_n \text{ is unbounded from above on } \widetilde{K}; \qquad (9.49)$$

and either $\beta = 0$ *or else (9.48) holds.*

PROOF.   For case (i), the validity of (9.25) is immediate from (2.5), and thus Theorem 9.10 applies. We let $\nu(\cdot) \equiv \nu \in \widetilde{K}$ be a constant process in $\mathcal{D}^{(b)}$, so that

$$u_\nu \overset{\Delta}{=} E_\nu \left[\frac{B}{S_0^{(\nu)}(T)} \right] \le \frac{\beta}{s_0} e^{-\zeta(\nu)T}.$$

Taking the infimum over $\nu \in \widetilde{K}$, we see that $\check{u} = 0$.

In case (ii), we first note that

$$E \left[\frac{B}{S_0(T)} \right]^\gamma \le s_0^{-\gamma} E \left[2(\beta \vee \alpha S_n(T)) \right]^\gamma \le \left(\frac{2}{s_0} \right)^\gamma (\beta^\gamma + \alpha^\gamma ES_n^\gamma(T)) < \infty,$$

and again (9.25) is satisfied. We next use (1.1.10) and (5.11) to verify that

$$\frac{S_n(t)}{S_0(t)} = S_n(0) \exp \left\{ \int_0^t \sum_{d=1}^N \sigma_{nd}(s)\, dW_\nu^{(d)}(s) - \frac{1}{2} \int_0^t \sum_{d=1}^N \sigma_{nd}^2(s)\, ds \right.$$

$$\left. - \int_0^t (\nu_n(s) + \delta_n(s))\, ds \right\}, \quad 0 \le t \le T, \qquad (9.50)$$

is valid for every $\nu(\cdot) \in \mathcal{D}$. Again taking $\nu(\cdot) = \nu \in \widetilde{K}$ to be constant, and replacing $\delta_n(\cdot)$ by its lower bound Δ, we have

$$E_\nu \left(\frac{S_n(T)}{S_0(T)} \right)$$

$$\le S_n(0) e^{-(\nu_n + \Delta)T} E_\nu \exp \left\{ \int_0^t \sum_{d=1}^N \sigma_{nd}(s)\, dW_\nu^{(d)}(s) - \frac{1}{2} \int_0^t \sum_{d=1}^N \sigma_{nd}^2(s)\, ds \right\}$$

$$= S_n(0) e^{-(\nu_n + \Delta)T}.$$

It follows that

$$u_\nu \stackrel{\Delta}{=} E_\nu \left[\frac{B}{S_0^{(\nu)}(T)} \right] = e^{-\zeta(\nu)T} E_\nu \left[\frac{B}{S_0(T)} \right]$$

$$\leq e^{-\zeta(\nu)T} \frac{\beta}{s_0} + \alpha S_n(0) e^{-(\zeta(\nu)+\nu_n+\Delta)T}.$$

Taking the infimum over $\nu \in \widetilde{K}$, we obtain $\check{u} = 0$. □

The condition on the contingent claim in Proposition 9.14(i) is satisfied by a European put option $B = (q - S_n(T))^+$, and the condition of Proposition 9.14(ii) is satisfied with $\alpha = 1$ and $\beta = 0$ by a European call option $B = (S_n(T) - q)^+$, where $q \geq 0$. The validity of (9.48) or (9.49) depends on the nature of the portfolio constraints. In Examples 9.7(ii), (iv), (vii), and (viii), condition (9.49) is satisfied for every n, and in Example 9.7(iii) it is satisfied if $n \in \{M+1, \ldots, N\}$, i.e., *for a European call option written on a stock in which trading is not allowed*. Condition 9.48 is satisfied in Examples 9.7(vi), (vii), and (viii).

The following proposition provides sufficient conditions for a positive lower-hedging price for a European call option.

Proposition 9.15: *Assume (9.1) and (9.9), and suppose $B = (S_n(T) - q)^+$ for some $q > 0$ and $n \in \{1, \ldots, N\}$ for which the dividend rate process $\delta_n(\cdot)$ is bounded from above, $ES_n^\gamma(T) < \infty$ for some $\gamma > 1$, and*

$$\nu \mapsto \zeta(\nu) + \nu_n \text{ is bounded from above on } \widetilde{K}. \tag{9.51}$$

Then

$$h_{low}(K) = \check{u} \stackrel{\Delta}{=} \inf_{\nu \in \mathcal{D}^{(b)}} u_\nu > 0$$

for $q > 0$ sufficiently small.

PROOF. From (9.50) and the formula $S_0^{(\nu)}(T) = S_0(T) e^{-\int_0^T \zeta(\nu(t))\, dt}$, we have

$$u_\nu = E_\nu \left(\frac{S_n(T) - q}{S_0^{(\nu)}(T)} \right)^+$$

$$= E_\nu \left(\exp \left\{ -\int_0^T (\zeta(\nu(t)) + \nu_n(t) + \delta_n(t))\, dt \right\} \right.$$

$$\cdot \exp \left\{ \int_0^T \sum_{d=1}^N \sigma_{nd}(t)\, dW_\nu^{(d)}(t) - \frac{1}{2} \int_0^T \sum_{d=1}^N \sigma_{nd}^2(s)\, ds \right\}$$

$$\left. - \frac{q}{S_0(T)} \exp \left\{ -\int_0^T \zeta(\nu(t))\, dt \right\} \right)^+.$$

Letting Γ denote an upper bound on $\zeta(\nu) + \nu_n + \delta_n(t)$, valid for all $\nu \in \widetilde{K}$ and $t \in [0, T]$, almost surely, we may use Jensen's inequality to write

$$
u_\nu \geq \left(e^{-\Gamma T} E_\nu \exp \left\{ \int_0^T \sum_{d=1}^N \sigma_{nd}(t) \, dW_\nu^{(d)}(t) - \frac{1}{2} \int_0^T \sum_{d=1}^N \sigma_{nd}^2(s) \, ds \right\} \right.
$$
$$
\left. - \frac{q}{s_0} e^{-\zeta_0 T} \right)^+
$$
$$
= \left(e^{-\Gamma T} - \frac{q}{s_0} e^{-\zeta_0 T} \right)^+, \quad \forall \nu(\cdot) \in \mathcal{D}^{(b)}.
$$

For $q > 0$ sufficiently small, this last expression is positive. $\square$

5.10   Lower Hedging with Constant Coefficients

This section is the counterpart to Section 7 for the case of lower-hedging. We show here that the lower-hedging price of a contingent claim defined in terms of the final stock prices is the value in the unconstrained market of a related contingent claim. The section concludes with European call and put option examples.

As in Section 7, we assume here that (7.1)–(7.3) hold. We consider a contingent claim of the form $B = \varphi(S(T))$, where $S(T) = (S_1(T), \ldots, S_N(T))'$ is the vector of final stock prices, and $\varphi : (0, \infty)^N \to [0, \infty)$ is an upper semicontinuous function satisfying the polynomial growth condition (7.9). The value at time $t \in [0, T]$ of the contingent claim $\varphi(S(T))$ in the unconstrained market $\mathcal{M}(\mathbb{R}^N)$ is $u(T - t, S(t))$, where $u(T - t, x)$ is defined by (2.4.6). In the constrained market $\mathcal{M}(K_-)$, Theorem 9.10 asserts that the lower-hedging value process of Definition 9.11 for $\varphi(S(T))$ is

$$
\check{X}(t) \triangleq -\operatorname{ess\,inf}_{\nu \in \mathcal{D}^{(b)}} \frac{E[H_\nu(T) \varphi(S(T)) | \mathcal{F}(t)]}{H_\nu(t)} \tag{10.1}
$$
$$
= - \inf_{\nu \in \mathcal{D}^{(b)}} u_\nu(T - t, S(t); \varphi),
$$

where $u_\nu(T - t, x; \varphi)$ is defined by (7.7).

Theorem 10.1: *Assume (7.1)–(7.3), (9.9), and let $B = \varphi(S(T))$ be a contingent claim, where $\varphi : (0, \infty)^N \to [0, \infty)$ is an upper semicontinuous function satisfying the polynomial growth condition (7.9). Define the nonnegative function*

$$
\check{\varphi}(x) \triangleq \inf_{\nu \in \widetilde{K}} \left[e^{-\zeta(\nu)} \varphi(x_1 e^{-\nu_1}, \ldots, x_N e^{-\nu_N}) \right], \quad x \in (0, \infty)^N. \tag{10.2}
$$

Then the hypotheses of Theorem 9.10 are satisfied, and the lower-hedging value process $\check{X}(\cdot)$ for the contingent claim $\varphi(S(T))$ is given by

$$\check{X}(t) = -e^{-r(T-t)} E_0[\check{\varphi}(S(T)))|\mathcal{F}(t)] \tag{10.3}$$
$$= -u(T - t, S(t); \check{\varphi}), \quad 0 \le t < T,$$

almost surely, where $u(T - t, x; \varphi)$ is defined by (7.4). In other words, the lower-hedging value process for the contingent claim $\varphi(S(T))$ in the market $\mathcal{M}(K_-)$ with constraint set K_- is the value of the contingent claim $-\check{\varphi}(S(T))$ in the unconstrained market $\mathcal{M}(\mathbb{R}^N)$.

PROOF.   We note first that (7.1)–(7.2) and the polynomial growth condition (7.9) ensure that $B = \varphi(S(T))$ satisfies (9.25). Thus the hypotheses of Theorem 9.10 are satisfied.

The definition (9.5) of $\zeta(\cdot) = \zeta_-(\cdot)$ yields

$$\zeta\left(\int_t^T \nu_1(s)\,ds, \ldots, \int_t^T \nu_N(s)\,ds\right) \triangleq \inf_{p \in K_-}\left(-\int_t^T \sum_{n=1}^N p_n \nu_n(s)\,ds\right)$$

$$\ge \int_t^T \inf_{p \in K_-}\left(-\sum_{n=1}^N p_n \nu_n(s)\right)ds$$

$$= \int_t^T \zeta(\nu(s))\,ds. \tag{10.4}$$

Using this inequality instead of (7.14), we may imitate the first part of the proof of Theorem 7.1 to conclude that

$$\inf_{\nu \in \mathcal{D}^{(b)}} E_\nu\left[e^{-\int_t^T \zeta(\nu(s))\,ds}\right.$$
$$\left.\cdot \varphi\left(x_1 e^{-\int_t^T \nu_1(s)\,ds} Y_1^{(\nu)}(t, T), \ldots, x_N e^{-\int_t^T \nu_N(s)\,ds} Y_N^{(\nu)}(t, T)\right)\right]$$
$$\ge E_0 \check{\varphi}(x_1 Y_1(t, T), \ldots, x_N Y_N(t, T)), \quad 0 \le t \le T, \tag{10.5}$$

which implies the almost sure inequality

$$-\check{X}(t) \ge u(T - t, S(t); \check{\varphi}), \quad 0 \le t < T. \tag{10.6}$$

As in the proof of Theorem 7.1, we fix $x \in (0, \infty)^N$ and let $x(\cdot)$ be a continuous function mapping $[0, T]$ into $(0, \infty)^N$ so that $x(T) = x$. We choose a sequence $\{\nu^{(m)}\}_{m=1}^\infty$ in $\widetilde{K}$ such that

$$\inf_m\left[e^{-\zeta(\nu^{(m)})}\varphi(x_1 e^{-\nu_1^{(m)}}, \ldots, x_N e^{-\nu_N^{(m)}})\right] = \check{\varphi}(x).$$

Using the dominated convergence theorem in (7.15) and the upper semicontinuity of φ, we obtain

$$\limsup_{t \uparrow T} \inf_{\mu \in \mathcal{D}^{(b)}} u_\mu(T - t, x(t); \varphi) \le \lim_{t \uparrow T} u_\nu(T - t, x(t); \varphi)$$

$$\le e^{-\zeta(\nu^{(m)})}\varphi(x_1 e^{-\nu_1^{(m)}}, \ldots, x_N e^{-\nu_N^{(m)}}),$$

whereas taking the infimum over m and recalling (10.1) we see that

$$\limsup_{t \uparrow T}(-\check{X}(t)) \leq \check{\varphi}(S(T)). \tag{10.7}$$

According to Proposition 9.13, $-H_0(\cdot)\check{X}(\cdot)$ is a uniformly integrable P-submartingale. In particular,

$$-H_0(s)\check{X}(s) \leq -E[H_0(t)\check{X}(t)|\mathcal{F}(s)], \quad 0 \leq s \leq t < T.$$

Letting $t \uparrow T$ and using (10.7) and the uniform integrability of $H_0(\cdot)\check{X}(\cdot)$, we obtain

$$\begin{aligned}
-\check{X}(s) &\leq \frac{1}{H_0(s)} E[H_0(T)\check{\varphi}(S(T))|\mathcal{F}(s)] \\
&= e^{-r(T-s)} E_0[\check{\varphi}(S(T))|\mathcal{F}(s)] \\
&= u(T-s, S(s); \check{\varphi}), \quad 0 \leq s < T,
\end{aligned}$$

almost surely. This inequality, combined with (10.6), completes the proof. $\qquad\square$

Example 10.2 *(Prohibition of short-selling):* We consider a market with constant coefficients and one stock, i.e., $N = 1$. When short-selling is prohibited (Example 9.7(ii), $K = [0, \infty)$, $K_- = (-\infty, 0]$), we have $\widetilde{K} = [0, \infty)$, $\zeta(\cdot) \equiv 0$ on $\widetilde{K}$, and the function $\check{\varphi}$ of (10.2) is given by

$$\check{\varphi}(x) = \inf_{\nu \geq 0} \varphi(xe^{-\nu}), \quad \forall x > 0.$$

For a *European call*, we have

$$\check{\varphi}(x) = \inf_{\nu \geq 0}(xe^{-\nu} - q)^+ = 0, \quad \forall x > 0,$$

and the lower hedging value is zero. This is also the conclusion of Proposition 9.14(ii) in a more general context. For a *European put option*,

$$\check{\varphi}(x) = \inf_{\nu \geq 0}(q - xe^{-\nu})^+ = (q - x)^+, \quad \forall x > 0;$$

the hedge of a long position in a European put option does not sell stock short, and is thus unaffected by the prohibition of short-selling.

Example 10.3 *(Prohibition of borrowing):* We consider again a market with constant coefficients and one stock, i.e., $N = 1$. When borrowing from the money market is prohibited (Example 9.7(vi), $K = (-\infty, 1]$, $K_- = [1, \infty)$), we have $\widetilde{K} = (-\infty, 0]$, $\zeta(\nu) = -\nu$ on $\widetilde{K}$, and the function $\check{\varphi}$ of (10.2) is given by

$$\check{\varphi}(x) = \inf_{\nu \leq 0}\left[e^{\nu}\varphi(xe^{-\nu})\right], \quad \forall x > 0.$$

For a *European call option*, we have

$$\check{\varphi}(x) = \inf_{\nu \leq 0}(x - qe^{\nu})^+ = (x - q)^+, \quad \forall x > 0;$$

the hedge of a long position in a European call option does not borrow from the money market and is thus unaffected by prohibition of borrowing. For a *European put*,

$$\check{\varphi}(x) = \inf_{\nu \leq 0} (q e^{\nu} - x)^{+} = 0, \quad \forall x > 0,$$

and the lower hedging value is zero. This is also the conclusion of Proposition 9.14(i) in a more general context.

5.11 Notes

The notions and results on *superreplicating portfolio* and *upper-hedging price* were developed first for incomplete markets, by El Karoui and Quenez (1991, 1995). These authors employed the fictitious-completion and duality approach developed by Karatzas, Lehoczky, Shreve, and Xu (1991) in the context of utility maximization in an incomplete market (see notes to Chapter 6), and derived the formula (6.3) for the upper-hedging price of a contingent claim.

In a parallel development, and in a discrete-time/finite-state setting, Edirisinghe, Naik, and Uppal (1993) noted that in the presence of leverage constraints, superreplication may actually be "cheaper" than exact replication. Naik and Uppal (1994) studied the effects of leverage constraints on the pricing and hedging of stock options by deriving a recursive solution scheme as a *linear programming* formulation for the minimum-cost hedging problem under such constraints. For a very nice exposition of this approach, see Musiela and Rutkowski (1997), Chapter 4.

Sections 2–6, 8: The material here comes from Cvitanić and Karatzas (1993), who extended the approach of El Karoui and Quenez (1991, 1995) to the case of *general convex constraints on portfolio proportions* and derived the stochastic-control-type representation (6.3) for the upper-hedging price. Crucial in this development, and of considerable independent probabililistic interest, is the "simultaneous Doob–Meyer decomposition" of (6.20), valid for all processes $\nu(\cdot) \in \mathcal{D}$. This approach echoes the more general themes of the purely probabilistic treatment for stochastic control problems, based on martingale theory, which was developed in the 1970s; see Chapter 16 of Elliott (1982) and the references therein. The approach has been extended further to more general semimartingale price processes by Kramkov (1996a,b), Föllmer and Kramkov (1998), Föllmer and Kabanov (1998). Related results, for hedging contingent claims under margin requirements and short-sale constraints, appear in Heath and Jarrow (1987) and in Jouini and Kallal (1995a), respectively. In Cvitanić and Karatzas (1993) it is also shown how to modify the approach of this section in order to obtain a stochastic control representation of the type (6.3) for the upper-hedging price of contingent claims *in the presence of a higher interest rate for borrowing than for investing*, and how to specialize this representation to specific contingent claims such as options; see also Barron and Jensen (1990), Korn (1992), and Bergman (1995). A similar development, but in the context of utility maximization rather than hedging, appears in Section 6.8.

Section 9: The material here comes from Karatzas and Kou (1996), who studied in detail the *lower-hedging price* for the buyer of the contingent claim, derived the representation (9.29) for it, and computed several examples. Following an idea of Davis (1994) that seems to go back at least to Lucas (1978), these authors also showed how to select a unique price

$$\hat{p} \equiv u_\lambda \stackrel{\Delta}{=} E[H_\lambda(T)B] \text{ for some process } \lambda(\cdot) \in \mathcal{D} \qquad (11.1)$$

inside the arbitrage-free interval $[h_{low}(K_-), h_{up}(K)]$, which is "fair" in the following sense. If the contingent claim sells at price p at time $t = 0$, and an agent with initial capital x and utility function U diverts an amount $\delta \in (-x, x)$ to buy δ/p units of the contingent claim, then $\hat{p}$ is characterized by the requirement that the marginal maximal expected utility be *zero* at $\delta = 0$:

$$\frac{\partial Q}{\partial \delta}(0, \hat{p}, x) = 0, \qquad (11.2)$$

where

$$Q(\delta, x, p) \stackrel{\Delta}{=} \sup_{(C,p) \in \mathcal{A}(x-\delta; K)} EU\left(X^{x-\delta, C, p}(T) + \frac{\delta}{p}B\right) \qquad (11.3)$$

is the resulting maximal expected utility from terminal wealth. Using notions reminiscent of viscosity solutions (cf. Fleming and Soner (1993)), one can make sense of the requirement (11.2) even when the function $Q(\cdot, p, x)$ is not known a priori to be differentiable; it can then be shown that $\hat{p}$ is uniquely determined by this requirement, and can be represented in the form (11.1) for a suitable process $\lambda(\cdot) \in \mathcal{D}$; see Karatzas and Kou (1996), and Karatzas (1996), Chapter 6. In fact, this *fair price* can be computed explicitly in several interesting cases, for instance if the utility is logarithmic; in the case of constant coefficients and cone constraints, the fair price in fact does not depend on the particular form of the utility function or on the initial capital. There are also connections with relative-entropy minimization, with the *minimal martingale measure* of Föllmer and Sondermann (1986), Föllmer and Schweizer (1991), and Hofman et al. (1992), with the utility-based approach of Barron and Jensen (1990) for the pricing of options with differential interest rates, and with utility-based approaches for pricing in the presence of transaction costs (e.g., Hodges and Neuberger (1989), Panas (1993), Constantinides (1993), Davis, Panas, and Zariphopoulou (1993), Davis and Panas (1994), Davis and Zariphopoulou (1995), Cvitanić and Karatzas (1996), Section 7, and Constantinides and Zariphopoulou (1997)). For an interesting synthesis, see the doctoral dissertation of Mercurio (1996) and the survey article of Jouini (1997).

The interested reader should consult Karatzas and Kou (1996) for the derivations of these results, and Karatzas and Kou (1998) for extensions of the results in this chapter to American contingent claims.

Sections 7, 10: The results of Section 7 are due to Broadie, Cvitanić, and Soner (1998) who also show by example how to extend this methodology to cover exotic (path-dependent) options. Wystup (1998) provides a systematic development for path-dependent options. A similar methodology has been used by Cvitanić, Pham, and Touzi (1997, 1998) to discuss the superreplication of contingent claims in the contexts of stochastic volatility and transaction costs, respectively. The results of Section 10 build on those of Section 7, and are apparently new.

6

Constrained Consumption and Investment

6.1 Introduction

As we saw in Chapter 5, when a financial market is incomplete due to portfolio constraints, it may no longer be possible to construct a perfect hedge for contingent claims. This led to the introduction in that chapter of *super-replicating* portfolios and *upper-hedging prices* for contingent claims. This is a conservative approach to pricing, since it begins from the assumption that agents trade only if their probability of loss is zero.

A more venerable and less conservative approach to pricing in the presence of constraints is based on utilities, or "preferences." In this chapter we consider the problem of optimal consumption and investment in a constrained financial market. The *duality theory* introduced in the previous chapter plays a key role here as well. Indeed, the problem of this chapter is well suited to duality and the related notion of Lagrange multipliers. For each market $\mathcal{M}(K)$ in which portfolio proportions are constrained to lie in a nonempty, closed convex set K, we seek to construct a related market $\mathcal{M}_{\hat{\nu}}$ in which portfolio proportions are unconstrained, but such that the optimal portfolio-proportion process lies in K *of its own accord*, so to speak. This is the fundamental idea of Lagrange multipliers; in this context, the Lagrange multiplier is a process $\hat{\nu}(\cdot)$. Once $\hat{\nu}(\cdot)$ has been determined, and it can be explicitly computed in a variety of nontrivial special cases, the optimal consumption and investment problem in the unconstrained market $\mathcal{M}_{\hat{\nu}}$ is the one solved in Chapter 3.

Just as in Chapter 5, the convex set K can represent prohibition or restriction on short-selling, prohibition or restriction on borrowing from the money market, or *incompleteness* in the sense that some stocks (or other sources of uncertainty) are unavailable for investment. The analysis of this chapter also extends to cover a market in which the interest rate for borrowing is higher than the interest rate for investing.

Section 2 of this chapter sets out the constrained optimal consumption and investment problem. In contrast to Chapter 3, here we consider only the problem of consumption *and* investment. The problems of consumption *or* investment can also be addressed by duality theory, except that there is no satisfactory theory of existence of the optimal dual process in the case of utility from consumption only (see Remark 5.8). Section 3 introduces the related unconstrained problems, parametrized by dual processes $\nu(\cdot)$. The central result of this chapter is Theorem 4.1, which provides four conditions stated in terms of dual processes, equivalent to optimality in the constrained problem. The most useful of these is condition (D), the existence of an optimal dual process; this is adddressed in some detail in Section 5, where examples with explicit computations are provided.

The remaining sections consider further refinements of the general theory in important special cases. When the market coefficients are deterministic and K is a convex cone, the value function for the constrained problem satisfies a *nonlinear* Hamilton–Jacobi–Bellman (HJB) parabolic partial differential equation, and the value function for the dual problem satisfies a *linear* HJB equation. The former provides the optimal consumption and portfolio proportion processes in *feedback form*. Section 6 presents these matters. Section 7 works out special cases of *incompleteness*, when the form of K prevents investment in some of the stocks. If there are more sources of uncertainty than assets that can be traded, the unavailable stocks can represent these sources of uncertainty. Finally, Section 8 alters the basic model to allow for a higher interest rate for borrowing than for investing. The duality theory of Sections 4 and 5 applies to this case, and explicit computations are again possible.

6.2 Utility Maximization with Constraints

In this chapter we return to the Problem 3.5.4 of maximizing expected total utility from both consumption and terminal wealth, but now impose the constraint that the portfolio-proportion process should take values in a given nonempty, closed, convex subset K of $\mathbb{R}^N$. We assume throughout this chapter that K contains the zero vector 0 in $\mathbb{R}^N$. As discussed in Examples 5.4.1, the constraint set K can be used to model a variety of market conditions, including incompleteness. We shall use again the support function

$$\zeta(\nu) \overset{\Delta}{=} \sup_{p \in K}(-p'\nu), \ \nu \in \mathbb{R}^N, \tag{5.4.1}$$

of (5.4.1). Because K contains the origin, we have

$$\zeta(\nu) \geq 0, \ \forall \nu \in \mathbb{R}^N \tag{2.1}$$

(i.e., (5.4.6) holds with $\zeta_0 = 0$).

As in Chapters 2 and 3 we shall begin with a *complete, standard financial market* $\mathcal{M} = (r(\cdot), b(\cdot), \delta(\cdot), \sigma(\cdot), S(0), A(\cdot))$, governed by the stochastic differential equations

$$dS_0(t) = S_0(t)[r(t)dt + dA(t)], \tag{2.2}$$

$$dS_n(t) = S_n(t)\left[(b_n(t) + \delta_n(t))\,dt + dA(t) + \sum_{d=1}^{D} \sigma_{nd}(t)dW^{(d)}(t)\right], \tag{2.3}$$

for $n = 1, \ldots, N$. We assume that $S_0(\cdot)$ is almost surely bounded away from zero, i.e.,

$$S_0(t) \geq s_0, \ 0 \leq t \leq T, \ \text{for some } s_0 > 0, \tag{2.4}$$

so Assumption 3.2.3 holds. The number of stocks N is equal to the dimension D of the driving Brownian motion, the volatility matrix $\sigma(t) = (\sigma_{nd}(t))$ is nonsingular for Lebesgue-almost-every t almost surely, and the exponential local martingale $Z_0(\cdot)$ of (1.5.2) is a martingale, so that the standard martingale measure P_0 of Definition 1.5.1 is defined. We assume further that

$$E \int_0^T \|\theta(t)\|^2 dt < \infty \tag{2.5}$$

and $\sigma(\cdot)$ satisfies (5.2.4). The filtration $\{\mathcal{F}(t)\}_{0 \leq t \leq T}$ is, as always, the augmentation by P-null sets of the filtration generated by the D-dimensional Brownian motion $W(\cdot) = (W^{(1)}(\cdot), \ldots, W^{(D)}(\cdot))$.

As in Chapter 3 and in contrast to Chapter 5, the agent in this chapter must choose a consumption *rate* process $c(\cdot)$ rather than a *cumulative* consumption process $C(\cdot)$. The cumulative consumption process $C(\cdot)$ of Chapter 5 is related to the consumption rate process $c(\cdot)$ of Chapter 3 and this chapter by the formula

$$C(t) = \int_0^t c(s)\,ds, \ 0 \leq t \leq T. \tag{2.6}$$

Of course, not every cumulative consumption process $C(\cdot)$ has such a representation. We shall be able to solve the optimality problems in this chapter within the class of cumulative consumption processes that do.

We have then the following definition, a repeat of Definition 3.3.1.

Definition 2.1: A *consumption process* is an $\{\mathcal{F}(t)\}$-progressively mea-surable, nonnegative process $c(\cdot)$ satisfying $\int_0^T c(t)dt < \infty$ almost surely.

In contrast to previous chapters, here it will be convenient to describe the agent's investment decisions in terms of a *portfolio-proportion pro-cess* rather than a portfolio process. Note that in the following definition there is no square-integrability condition like (1.2.6), which was imposed on portfolio processes.

Definition 2.2: A *portfolio proportion process* is an $\{\mathcal{F}(t)\}$-progressively measurable, $\mathbb{R}^N$-valued process.

For a given initial wealth $x \geq 0$, consumption process $c(\cdot)$, and portfolio-proportion process $p(\cdot)$, we wish to define the corresponding *wealth process* $X^{x,c,p}(t)$, $0 \leq t \leq T$, by

$$\frac{X^{x,c,p}(t)}{S_0(t)} + \int_0^t \frac{c(u)du}{S_0(u)} = x + \int_0^t \frac{X^{x,c,p}(u)}{S_0(u)}p'(u)\sigma(u)\,dW_0(u). \qquad (2.7)$$

This is just equation (3.3.1), but with the portfolio process $\pi(u)$ replaced by $X^{x,c,p}(u)p(u)$. The solution to (2.7) is easily verified to be given by

$$\frac{X^{x,c,p}(t)}{S_0(t)} = I_p(t)\left[x - \int_0^t \frac{c(u)du}{S_0(u)I_p(u)}\right],$$

where

$$I_p(t) \triangleq \exp\left\{\int_0^t p'(u)\sigma(u)\,dW_0(u) - \frac{1}{2}\int_0^t \|\sigma'(u)p(u)\|^2\,du\right\}. \qquad (2.8)$$

However, because we have not assumed the finiteness of $\int_0^T \|\sigma'(t)p(t)\|^2 dt$, we need to elaborate on this construction.

Lemma 2.3: *Let $p(\cdot)$ be a portfolio proportion process, and set*

$$\tau_p \triangleq \inf\left\{t \in [0,T]; \int_0^t \|\sigma'(u)p(u)\|^2\,du = \infty\right\}. \qquad (2.9)$$

Here we follow the convention $\inf \emptyset = \infty$, *which means that* $\tau_p = \infty$ *if* $\int_0^T \|\sigma'(u)p(u)\|^2 du < \infty$. *Then* $I_p(t)$ *given by (2.8) is defined for* $0 \leq t \leq T$ *on the set* $\{\tau_p = \infty\}$, $I_p(t)$ *is defined for* $0 \leq t < \tau_p$ *on the set* $\{\tau_p \leq T\}$, *and on this latter set we have*

$$\lim_{t\uparrow\tau_p} I_p(t) = 0 \qquad (2.10)$$

almost surely.

PROOF. We define the local P_0-martingale

$$M_p(t) \triangleq \int_0^t p'(u)\sigma(u)\,dW_0(u), \quad 0 \leq t < T \wedge \tau_p,$$

with quadratic variation $\langle M \rangle(t) = \int_0^t \|\sigma'(u)p(u)\|^2 du$ and invert the quadratic variation process by setting

$$T(s) \overset{\Delta}{=} \inf\{t \geq 0; \langle M \rangle(t) > s\}, \quad 0 \leq s < \langle M \rangle(T \wedge \tau_p).$$

According to Karatzas and Shreve (1991), Theorem 3.4.6 and Problem 3.4.7 (with solution on p. 232), the process $B(s) \overset{\Delta}{=} M(T(s))$, $0 \leq s < \langle M \rangle(T \wedge \tau_p)$, is a Brownian motion. On the set $\{\tau_p \leq T\}$, we have

$$\lim_{t \uparrow \tau_p} \left[M(t) - \frac{1}{2}\langle M \rangle(t) \right] = \lim_{s \uparrow \infty} \left[B(s) - \frac{1}{2}s \right] = -\infty$$

almost surely because $\lim_{s \uparrow \infty} \frac{B(s)}{s} = 0$ (Karatzas and Shreve (1991), Problem 2.9.3 with solution on p. 124). Because $I_p(t) = \exp\{M(t) - \langle M \rangle(t)\}$, we have (2.10).   □

For any initial wealth $x \geq 0$ and for any consumption and portfolio-proportion process pair (c, p), the wealth process $X^{x,c,p}(t)$ is defined for $0 \leq t < T \wedge \tau_p$. Moreover, if $X^{x,c,p}(t) \geq 0$, $0 \leq t < T \wedge \tau_p$, then Lemma 2.3 implies $\lim_{t \uparrow \tau_p} X^{x,c,p}(t) = 0$ almost surely on the set $\{\tau_p \leq T\}$. Therefore, the stopping time

$$\tau_0 \overset{\Delta}{=} \inf\{t \in [0, T]; X^{x,c,p}(t) = 0\}$$

satisfies $\tau_0 \leq \tau_p$ almost surely, and the inequality might be strict. Again, we follow the convention $\inf \emptyset = \infty$. We shall require that $c(t) = 0$ for Lebesgue-almost-every $t \in [\tau_0, T]$, and then $X^{x,c,p}(t) = 0$ satisfies (2.7) for $\tau_0 \leq t \leq T$. This permits us to give the following definition for admissibility.

Definition 2.4: Given $x \geq 0$, we say that a consumption and portfolio-process pair (c, p) is *admissible at x in the unconstrained market* $\mathcal{M}$, and write $(c, p) \in \mathcal{A}(x)$, if we have

$$c(t) = 0 \quad \text{for Lebesgue-a.e. } t \in [\tau_0, T] \tag{2.11}$$

almost surely. For $(c, p) \in \mathcal{A}(x)$, we understand $X^{x,c,p}(\cdot)$ to be identically zero on $[\tau_0, T]$. We shall say that (c, p) is *admissible at x in the constrained market* $\mathcal{M}(K)$, and write $(c, p) \in \mathcal{A}(x; K)$, if $(c, p) \in \mathcal{A}(x)$ and

$$p(t) \in K \quad \text{for Lebesgue-a.e. } t \in [0, T] \tag{2.12}$$

almost surely.   □

Remark 2.5: The collection of wealth processes corresponding to $(c, p) \in \mathcal{A}(x)$ of Definition 2.4 coincides with the collection of wealth processes corresponding to $(c, \pi) \in \mathcal{A}(x)$ of Definition 3.3.2, where $p(\cdot)$ and $\pi(\cdot)$ are related by

$$\pi(t) = X^{x,c,p}(t)p(t), \ 0 \leq t \leq T,$$

$$p(t) = \begin{cases} \frac{1}{X^{x,c,\pi}(t)}\pi(t), & 0 \leq t < \tau_\pi, \\ 0, & \tau_\pi \leq t \leq T, \end{cases}$$

and $\tau_\pi \overset{\Delta}{=} \inf\{t \in [0, T]; X^{x,c,\pi}(t) = 0\}$. It is clear that each pair $(c, \pi) \in \mathcal{A}(x)$ of Definition 3.3.2 leads to a portfolio-proportion process $p(\cdot)$ for which $(c, p) \in \mathcal{A}(x)$ as in Definition 2.4, but the reverse construction is not obvious because $\pi(\cdot)$ must satisfy the square-integrability condition (1.2.6), whereas no such condition is imposed on $p(\cdot)$. However, if $(c, p) \in \mathcal{A}(x)$ as in Definition 2.4 is given, we note from (2.7) and with τ_p defined by (2.9) that on the set $\{\tau_p \leq T\}$, the limit

$$\lim_{t \uparrow \tau_p} \int_0^t \frac{X^{x,c,p}(u)}{S_0(u)} p'(u)\sigma(u)\, dW_0(u) = -x + \int_0^{\tau_p} \frac{c(u)du}{S_0(u)}$$

is defined. This implies that

$$\int_0^{\tau_p} \left(\frac{X^{x,c,p}(u)}{S_0(u)}\right)^2 \|\sigma'(u)p(u)\|^2\, du < \infty$$

holds almost surely (Karatzas and Shreve (1991), Problem 3.4.11 with solution on page 232), which gives the desired square-integrability property $P\left[\int_0^T \|\sigma'(u)\pi(u)\|^2 du < \infty\right] = 1$.

Consider now an agent endowed with *initial wealth* $x \geq 0$ and with a *preference structure* $U_1: [0, T] \times \mathbb{R} \to [-\infty, \infty)$, $U_2: \mathbb{R} \to [-\infty, \infty)$ as in Definition 3.5.1. We shall assume throughout this chapter that

$$\bar{c}(t) = 0, \quad \forall t \in [0, T], \quad \text{and} \quad \bar{x} = 0 \tag{2.13}$$

in (3.5.1), (3.5.3), meaning that both $U_1(t, \cdot)$ and $U_2(\cdot)$ are real-valued on $(0, \infty)$. We assume also that

$$U_1'(t, 0) = \infty, \quad \forall t \in [0, T], \quad \text{and} \quad U_2'(0) = \infty. \tag{2.14}$$

For such an agent, we can formulate the counterpart of Problem 3.5.4 in the constrained market $\mathcal{M}(K)$.

Problem 2.6: Given $x \geq 0$, find a pair $(\hat{c}, \hat{p})$ in

$$\mathcal{A}_3(x; K) \overset{\Delta}{=} \left\{ (c, p) \in \mathcal{A}(x; K); \ E \int_0^T \min[0, U_1(t, c(t))]\, dt > -\infty, \right.$$

$$\left. E(\min[0, U_2(X^{x,c,p}(T))]) > -\infty \right\} \tag{2.15}$$

which is optimal for the problem

$$V(x; K) \overset{\Delta}{=} \sup_{(c,p) \in \mathcal{A}_3(x;K)} E\left[\int_0^T U_1(t, c(t))\, dt + U_2\left(X^{x,c,p}(T)\right)\right] \tag{2.16}$$

of maximizing the expected total utility from both consumption and terminal wealth, subject to the portfolio constraint K of (2.12).

Problem 2.6 will be the object of study in this chapter. Since we consider neither the problem of maximizing utility from consumption only (the con-

strained version of Problem 3.5.2) nor, except in Example 7.4, the problem
of maximizing utility from terminal wealth only (the constrained version
of Problem 3.5.3), we suppress the subscript 3 that appears on the value
function $V_3(x)$ in Problem 3.5.4. See, however, Remark 5.8.

We shall embed this problem into a family of auxiliary unconstrained
problems, formulated in the auxiliary markets $\{\mathcal{M}_\nu\}_{\nu \in \mathcal{D}}$ of Section 5.5.
In terms of these auxiliary problems (Problem 3.2), we shall obtain neces-
sary and sufficient conditions for optimality in Problem 2.6 (Theorem 4.1),
general existence results based on convex duality and martingale methods (Theorem 5.4), as well as specific computations for the value function
$V(\cdot; K)$ and the optimal pair $(\hat{c}, \hat{\pi})$ that attains the supremum in (2.16)
(Examples 4.2, 6.6, and 6.7 as well as Section 7).

6.3   A Family of Unconstrained Problems

Let us consider now the counterpart of Problem 3.5.4 in the unconstrained
market $\mathcal{M}_\nu$ introduced in Section 5.5. The processes $\nu(\cdot)$ are taken from
the set $\mathcal{D}$ of Definition 5.5.1, and these will play the role of Lagrange multipliers in the constrained Problem 2.6. In the market $\mathcal{M}_\nu$, the wealth
process $X_\nu^{x,c,p}(\cdot)$ corresponding to initial condition $x \geq 0$, consumption
process $c(\cdot)$, and portfolio-proportion process $p(\cdot)$ is given by (5.5.16), or
equivalently (5.5.17), (5.5.18), where now $dC(s)$ in those equations is inter-
preted as $c(s)ds$, and $\pi(t)$ in those equations is replaced by $X^{x,c,p}(t)p(t)$.
We reproduce these equations for reference:

$$\frac{X_\nu^{x,c,p}(t)}{S_0^{(\nu)}(t)} + \int_0^t \frac{c(u)du}{S_0^{(\nu)}(u)} = x + \int_0^t \frac{X_\nu^{x,c,p}(u)}{S_0^{(\nu)}(u)} p'(u)\sigma(u)dW_\nu(u), \quad (3.1)$$

$$\frac{X_\nu^{x,c,p}(t)}{S_0(t)} + \int_0^t \frac{c(u)du}{S_0(u)} = x + \int_0^t \frac{X_\nu^{x,c,p}(u)}{S_0(u)} \cdot$$
$$\cdot \left[(\zeta(\nu(u)) + p'(u)\nu(u)) \, du + p'(u)\sigma(u)dW_0(u) \right], \quad (3.2)$$

$$H_\nu(t)X_\nu^{x,c,p}(t) + \int_0^t H_\nu(u)c(u)du \quad (3.3)$$

$$= x + \int_0^t H_\nu(u)X_\nu^{x,c,p}(u)[\sigma'(u)p(u) - \theta_\nu(u)]'dW(u).$$

Just as in the previous section, we first use (3.1) to define $X_\nu^{x,c,p}(\cdot)$ up
to the stopping time

$$\tau_\nu \triangleq \inf\{t \in [0,T]; X_\nu^{x,c,p}(t) = 0\},$$

which must precede τ_p defined by (2.9). We have the following counterpart
to Definition 2.4.

Definition 3.1: Given $x \geq 0$, we say that the consumption and portfolio-
proportion process pair (c, p) is *admissible at x in the unconstrained market*

$\mathcal{M}_\nu$, and we write $(c, p) \in \mathcal{A}_\nu(x)$, if

$$c(t) = 0 \ \text{ for Lebesgue-a.e. } t \in [\tau_\nu, T] \tag{3.4}$$

holds almost surely. For $(c, p) \in \mathcal{A}_\nu(x)$, we understand $X_\nu^{x,c,p}(\cdot)$ to be identically equal to zero on $[\tau_\nu, T]$.

Problem 3.2: Given a process $\nu(\cdot) \in \mathcal{D}$ and given $x \geq 0$, find a pair (c_ν, p_ν) in

$$A_3^{(\nu)}(x) \triangleq \left\{ (c, p) \in \mathcal{A}_\nu(x); E \int_0^T \min[0, U_1(t, c(t))] \, dt > -\infty, \right.$$

$$\left. E\left(\min[0, U_2(X_\nu^{x,c,p}(T))]\right) > -\infty \right\} \tag{3.5}$$

which is optimal for the problem

$$V_\nu(x) \triangleq \sup_{(c,p) \in A_3^{(\nu)}(x)} E\left[\int_0^T U_1(t, c(t)) \, dt + U_2(X_\nu^{x,c,p}(T)) \right] \tag{3.6}$$

of maximizing the expected total utility from both consumption and terminal wealth without regard to the portfolio constraint K. □

Note that when $\nu(\cdot) \equiv 0$, the function $V_0(x)$ is just the value function for the unconstrained version of Problem 2.6. Consequently,

$$V(x; K) \leq V_0(x), \ x \geq 0. \tag{3.7}$$

Remark 3.3: Suppose $(c, p) \in \mathcal{A}(x; K)$, so that $X^{x,c,p}(\cdot)$ is defined by (2.7). If we choose $\nu(\cdot) \in \mathcal{D}$, then $X_\nu^{x,c,p}(\cdot)$ is defined by (3.2). It turns out that

$$X_\nu^{x,c,p}(t) \geq X^{x,c,p}(t), \ 0 \leq t \leq T, \tag{3.8}$$

as we show below. In particular,

$$\mathcal{A}_3(x; K) \subseteq A_3^{(\nu)}(x), \ x \geq 0, \ \nu(\cdot) \in \mathcal{D}, \tag{3.9}$$

and since the utility function U_2 is increasing on $(0, \infty)$, we also have

$$V(x; K) \leq V_\nu(x), \ x \geq 0, \ \nu(\cdot) \in \mathcal{D}. \tag{3.10}$$

To derive (3.8), we set

$$\Delta(t) \triangleq \frac{X_\nu^{x,c,p}(t) - X^{x,c,p}(t)}{S_0(t)}$$

and subtract (2.7) from (3.2) to obtain

$$\Delta(t) = \int_0^t \Delta(u) \left[(\zeta(\nu(u)) + p'(u)\nu(u)) \, du + p'(u)\sigma(u)dW_0(u) \right]$$

$$+ \int_0^t \frac{X^{x,c,p}(u)}{S_0(u)} \left(\zeta(\nu(u)) + p'(u)\nu(u) \right) du.$$

We next define the nonnegative process

$$J(t) = \exp\left\{-\int_0^t p'(u)\sigma(u)\,dW_0(u) + \frac{1}{2}\int_0^t \|\sigma'(u)p(u)\|^2\,du\right.$$
$$\left. - \int_0^t (\zeta(\nu(u)) + p'(u)\nu(u))\,du\right\}$$

and compute the differential

$$d(\Delta(t)J(t)) = \frac{J(t)X^{x,c,p}(t)}{S_0(t)}\,(\zeta(\nu(t)) + p'(t)\nu(t))\,dt.$$

Integrating this equation and using the fact that $\Delta(0) = 0$, we conclude that

$$\Delta(t) = \frac{1}{J(t)}\int_0^t \frac{J(u)X^{x,c,p}(u)}{S_0(u)}\,(\zeta(\nu(u)) + p'(u)\nu(u))\,du. \qquad (3.11)$$

From the fact that $\nu(\cdot) \in \mathcal{D}$ and $p(t) \in K$ for Lebesgue-almost-every t, we see from (5.4.5) that the integrand in (3.11) is nonnegative and hence (3.8) holds. From (3.11), we see also that

$$\zeta(\nu(t)) + p'(t)\nu(t) = 0 \quad \text{for Lebesgue-a.e. } t \in [0, T] \qquad (3.12)$$

almost surely, implies

$$X_\nu^{x,c,p}(t) = X^{x,c,p}(t), \ 0 \le t \le T, \qquad (3.13)$$

almost surely. □

Because of our assumptions that $S_0(\cdot)$ is bounded away from zero and that K contains the origin in $\mathbb{R}^N$, we have for each $\nu(\cdot) \in \mathcal{D}$ the variant

$$E\left[\int_0^T H_\nu(t)\,dt + H_\nu(T)\right] < \infty \qquad (3.14)$$

of Assumption 3.2.3 for the market $\mathcal{M}_\nu$. Problem 3.2 is just the unconstrained Problem 3.5.4 in the auxiliary market $\mathcal{M}_\nu$; its solution is described in Section 3.6 and is given as follows. For every $\nu(\cdot) \in \mathcal{D}$, introduce the function

$$\mathcal{X}_\nu(y) \overset{\Delta}{=} E\left[H_\nu(T)I_2(yH_\nu(T)) + \int_0^T H_\nu(t)I_1(t, yH_\nu(t))\,dt\right], \ 0 < y < \infty.$$
$$(3.15)$$

For each $\nu(\cdot) \in \mathcal{D}$ satisfying $\mathcal{X}_\nu(y) < \infty$ for all $y > 0$, the function $\mathcal{X}_\nu(\cdot)$ maps $(0, \infty)$ *onto* itself, with $\mathcal{X}_\nu(0) \overset{\Delta}{=} \lim_{y\downarrow 0} \mathcal{X}_\nu(y) = \infty$ and $\mathcal{X}_\nu(\infty) \overset{\Delta}{=} \lim_{y\to\infty} \mathcal{X}_\nu(y) = 0$.

We denote by $\mathcal{Y}_\nu(\cdot)$ the inverse of $\mathcal{X}_\nu(\cdot)$, which maps $[0, \infty]$ onto $[0, \infty]$, with $\mathcal{Y}_\nu(0) = \infty$ and $\mathcal{Y}_\nu(\infty) = 0$. For $x \ge 0$, we introduce the nonnegative random variable

$$B_\nu \stackrel{\Delta}{=} I_2(\mathcal{Y}_\nu(x)H_\nu(T)) \tag{3.16}$$

and the nonnegative random processes

$$c_\nu(t) \stackrel{\Delta}{=} I_1(t, \mathcal{Y}_\nu(x)H_\nu(t)), \tag{3.17}$$

$$X_\nu(t) \stackrel{\Delta}{=} \frac{1}{H_\nu(t)} E\left[\int_t^T H_\nu(u)c_\nu(u)\, du + H_\nu(T)B_\nu \,\middle|\, \mathcal{F}(t) \right], \tag{3.18}$$

$$M_\nu(t) \stackrel{\Delta}{=} \int_0^t H_\nu(u)c_\nu(u)\, du + H_\nu(t)X_\nu(t)$$

$$= E\left[\int_0^T H_\nu(u)c_\nu(u)\, du + H_\nu(T)B_\nu \,\middle|\, \mathcal{F}(t) \right], \tag{3.19}$$

defined for $0 \le t \le T$. In particular, $X_\nu(T) = B_\nu$ almost surely, and the process $M_\nu(\cdot)$ is a P-martingale with expectation

$$M_\nu(0) = EM_\nu(T)$$

$$= E\left[\int_0^T H_\nu(u)I_1(u, \mathcal{Y}_\nu(x)H_\nu(u))\, du + H_\nu(T)I_2(\mathcal{Y}_\nu(x)H_\nu(T)) \right]$$

$$= \mathcal{X}_\nu(\mathcal{Y}_\nu(x)) \;=\; x.$$

According to the martingale representation theorem (Karatzas and Shreve, (1991), Theorem 3.4.15 and Problem 3.4.16), there is a progressively measurable, $\mathbb{R}^N$-valued process $\psi_\nu(\cdot)$, unique up to Lebesgue$\times P$-almost everywhere equivalence, such that $\int_0^T \|\psi_\nu(t)\|^2 dt < \infty$ and

$$M_\nu(t) = x + \int_0^t \psi_\nu'(s)dW(s), \quad 0 \le t \le T, \tag{3.20}$$

almost surely. Letting

$$p_\nu(t) \stackrel{\Delta}{=} (\sigma^{-1}(t))'\left(\frac{\psi_\nu(t)}{H_\nu(t)X_\nu(t)} + \theta_\nu(t) \right) \tag{3.21}$$

we see that $X_\nu(\cdot)$ satisfies (3.3); hence

$$X_\nu(\cdot) = X_\nu^{x,c_\nu,p_\nu}(\cdot), \tag{3.22}$$

and (c_ν, p_ν) attains the supremum in (3.6).

Remark 3.4: The proof of Theorem 3.6.3 applied to the market $\mathcal{M}_\nu$ shows that when $x > 0$ and $\nu(\cdot) \in \mathcal{D}$ satisfies $\mathcal{X}_\nu(y) < \infty$ for all $y > 0$, we have that the pair (c_ν, p_ν) belongs to the class $\mathcal{A}_3^{(\nu)}(x)$ of (3.5), not just to the class $\mathcal{A}(x)$ of Definition 2.4. This condition can actually be stated without reference to the portfolio-proportion process $p_\nu(\cdot)$ as

$$E \int_0^T \min[0, U_1(t, c_\nu(t))]\, dt > -\infty, \quad E\left(\min[0, U_2(B_\nu)]\right) - \infty. \tag{3.23}$$

Remark 3.5: Under the simplifying assumptions (2.13), (2.14) imposed in this chapter on our utility functions, we have $I_1(t, y) > 0$, $I_2(y) > 0$ for all $y > 0$, which implies that the process $c_\nu(\cdot)$ and the random variable B_ν are strictly positive, almost surely, provided that $x > 0$. This in turn implies that $M_\nu(\cdot)$ and $X_\nu(\cdot)$ are likewise strictly positive at all times, almost surely.

Even if $\mathcal{X}_\nu(y) < \infty$ for all $y > 0$, it is possible that $V_\nu(x) = \infty$. We set

$$\mathcal{D}_0 \triangleq \{\nu(\cdot) \in \mathcal{D}; \mathcal{X}_\nu(y) < \infty \;\forall y \in (0, \infty), V_\nu(x) < \infty \;\forall x \in (0, \infty)\}. \tag{3.24}$$

Remark 3.6: Let $\nu(\cdot) \in \mathcal{D}_0$ be given. Then Theorem 3.6.11, applied to the market $\mathcal{M}_\nu$, shows that

$$V_\nu(x) = G_\nu(\mathcal{Y}_\nu(x)), \quad x > 0, \tag{3.25}$$

where

$$G_\nu(y) \triangleq E\left[\int_0^T U_1(I_1(t, yH_\nu(t)))\, dt + U_2(I_2(yH_\nu(T)))\right], \quad 0 < y < \infty. \tag{3.26}$$

Indeed, $\nu(\cdot) \in \mathcal{D}_0$ if and only if $G_\nu(y) < \infty$ for all $y > 0$. The convex dual of V_ν is given by

$$\begin{aligned}
\widetilde{V}_\nu(y) &\triangleq \sup_{x>0}[V_\nu(x) - xy] \\
&= G_\nu(y) - y\mathcal{X}_\nu(y) \\
&= E\left[\int_0^T \widetilde{U}_1(t, yH_\nu(t))\, dt + \widetilde{U}_2(yH_\nu(T))\right] \\
&< \infty, \quad 0 < y < \infty,
\end{aligned} \tag{3.27}$$

in the notation of Definition 3.4.2. From the proof of Theorem 3.6.11 (in particular, the arguments leading to (3.6.29), (3.6.30)), we also know that for any given $x \in (0, \infty)$, $y \in (0, \infty)$, $\nu(\cdot) \in \mathcal{D}_0$, and $(c, p) \in \mathcal{A}_3^{(\nu)}$, equality holds in

$$E\left[\int_0^T U_1(t, c(t))\, dt + U_2(X_\nu^{x,c,p}(T))\right] \le V_\nu(x) \le \widetilde{V}_\nu(y) + xy \tag{3.28}$$

if and only if the equations

$$c(t) = I_1(t, yH_\nu(t)) \quad \text{for Lebesgue-a.e. } t \in [0, T], \tag{3.29}$$

$$X_\nu^{x,c,p}(T) = I_2(yH_\nu(T)), \tag{3.30}$$

$$x = E\left[\int_0^T H_\nu(t)c(t)\, dt + H_\nu(T)X_\nu^{x,c,p}(T)\right] \tag{3.31}$$

hold, the first two in the almost sure sense. Finally, from (3.6.25), we know that the function $\widetilde{V}_\nu$ is continuously differentiable, with

$$\widetilde{V}_\nu'(y) = -\mathcal{X}_\nu(y), \quad 0 < y < \infty. \tag{3.32}$$

In what follows, we shall need to consider $\widetilde{V}_\nu(y)$ when the auxiliary process $\nu(\cdot)$ is in the class $\mathcal{D}$ but not necessarily in the class $\mathcal{D}_0$. We take as our definition

$$\widetilde{V}_\nu(y) \triangleq E\left[\int_0^T \widetilde{U}_1(t, yH_\nu(t))\, dt + \widetilde{U}_2(yH_\nu(T))\right], \quad y > 0, \ \nu(\cdot) \in \mathcal{D} \setminus \mathcal{D}_0, \tag{3.33}$$

which agrees with the definition in (3.27) when $\nu(\cdot) \in \mathcal{D}$. For $\nu(\cdot) \in \mathcal{D}\setminus\mathcal{D}_0$, the hypotheses of Theorem 3.6.11 are not necessarily satisfied, and we may not have the duality representation of $\widetilde{V}_\nu(y)$ given in (3.27). Furthermore, it is not immediately clear that the expectation on the right-hand side of (3.33) is defined. Here is a resolution of these difficulties.

Proposition 3.7: *Let $\nu(\cdot) \in \mathcal{D}$ and $y > 0$ be given. Then we have*

$$E\int_0^T \min[0, \widetilde{U}_1(t, yH_\nu(t))]\, dt > -\infty, \ E\left(\min[0, \widetilde{U}_2(yH_\nu(T))]\right) > -\infty, \tag{3.34}$$

and

$$\sup_{x>0}[V_\nu(x) - xy] \le \widetilde{V}_\nu(y) \triangleq E\left[\int_0^T \widetilde{U}_1(t, yH_\nu(t))\, dt + \widetilde{U}_2(yH_\nu(T))\right]. \tag{3.35}$$

Furthermore, if $\widetilde{V}_\nu(y) < \infty$, then equality holds in (3.35).

PROOF.   Let $x > 0$ be given, and consider $c(t) = \frac{x}{2T}S_0^{(\nu)}(t)$, $p(t) = 0$ for all $t \in [0, T]$. For this choice (3.1), (5.5.7), (2.1), and (2.4) imply

$$2X^{x,c,p}(T) = xS_0^{(\nu)}(T) \ge xs_0 > 0.$$

Similarly,

$$2Tc(t) \ge xs_0 > 0.$$

The inequality $\widetilde{U}_1(t, y) \ge U_1(t, x) - xy$ implies that

$$\min[0, U_1(c(t))] \le \min[0, \widetilde{U}_1(t, yH_\nu(t)) + yH_\nu(t)c(t)]$$
$$\le \min[0, \widetilde{U}_1(t, yH_\nu(t))] + \frac{xy}{2T}Z_\nu(t).$$

Integrating from $t = 0$ to $t = T$, taking expectations, and using the inequality $EZ_\nu(t) \le 1$, we obtain the first part of (3.34). The proof of the second part is similar.

Now, for any $(c,p) \in \mathcal{A}_3^{(\nu)}(x)$, we have

$$U_1(t, c(t)) \leq \widetilde{U}_1(t, yH_\nu(t)) + yH_\nu(t)c(t),$$
$$U_2(X_\nu^{x,c,p}(T)) \leq \widetilde{U}_2(yH_\nu(T)) + yH_\nu(T)X_\nu^{x,c,p}(T).$$

Integrating the first inequality, summing the two, and taking expectations, we obtain

$$E\left[\int_0^T U_1(t, c(t))\,dt + U_2(X_\nu^{x,c,p}(T))\right]$$

$$\leq E\left[\int_0^T \widetilde{U}_1(t, yH_\nu(t))\,dt + \widetilde{U}_2(yH_\nu(T))\right]$$

$$+ yE\left[\int_0^T H_\nu(t)c(t)\,dt + H_\nu(T)X_\nu^{x,c,p}(T)\right]$$

$$\leq E\left[\int_0^T \widetilde{U}_1(t, yH_\nu(t))\,dt + \widetilde{U}_2(yH_\nu(T))\right] + xy.$$

In the last step, we have used the budget constraint (5.5.20) for the market $\mathcal{M}_\nu$. The above inequality implies that

$$V_\nu(x) \leq E\left[\int_0^T \widetilde{U}_1(t, yH_\nu(t))\,dt + \widetilde{U}_2(yH_\nu(T))\right] + xy, \quad x > 0, \ y > 0,$$

and (3.35) follows.

Finally, let us assume $\widetilde{V}_\nu(y) < \infty$. We define $c_\nu(t) = I_1(t, yH_\nu(t))$, $B_\nu = I_2(yH_\nu(T))$ and consider for each positive integer $n \in \mathbb{N}$ the process $c_\nu^{(n)}(t) = c_\nu(t)1_{\{c_\nu(t)\leq n\}} + 1_{\{c_\nu(t)>n\}}$ as well as the random variable $B_\nu^{(n)} = B_\nu 1_{\{B_\nu \leq n\}} + 1_{\{B_\nu > n\}}$. We set

$$x^{(n)} = E\left[\int_0^T H_\nu(t)c_\nu^{(n)}(t)\,dt + H_\nu(T)B_\nu^{(n)}\right],$$

which is finite because of the assumption (3.14). There is a portfolio-proportion process $p_\nu^{(n)}(\cdot)$ for which $X^{x^{(n)},c_\nu^{(n)},p_\nu^{(n)}}(T) = B_\nu^{(n)}$ (Theorem 5.5.5); hence

$$E\left[\int_0^T U_1(t, c_\nu^{(n)}(t))\,dt + U_2(B_\nu^{(n)})\right] - x^{(n)}y \leq V_\nu(x^{(n)}) - x^{(n)}y$$

$$\leq \sup_{x>0}[V_\nu(x) - xy]$$

$$\leq \widetilde{V}_\nu(y).$$

To prove that equality holds in (3.35), it suffices to show that

$$\lim_{n\to\infty}\left(E\left[\int_0^T U_1(t, c_\nu^{(n)}(t))\,dt + U_2(B_\nu^{(n)})\right] - x^{(n)}y\right) = \widetilde{V}_\nu(y). \quad (3.36)$$

From Lemma 3.4.3 we have

$$\int_0^T \left[U_1(t, c_\nu^{(n)}(t)) - yH_\nu(t)c_\nu^{(n)}(t)\right]dt$$

$$= \int_0^T \left[U_1(t, I_1(t, yH_\nu(t))) - yH_\nu(t)I_1(t, yH_\nu(t))\right]1_{\{yH_\nu(t)\ge U_1'(t,n)\}}\,dt$$

$$+ \int_0^T \left[U_1(t, 1) - yH_\nu(t)\right]1_{\{yH_\nu(t)<U_1'(t,n)\}}\,dt$$

$$= \int_0^T \left[\widetilde{U}_1(t, yH_\nu(t))1_{\{yH_\nu(t)\ge U_1'(t,n)\}}\right.$$

$$\left. + (U_1(t, 1) - yH_\nu(t))1_{\{yH_\nu(t)<U_1'(t,n)\}}\right]dt.$$

The last integrand is bounded from above by $\widetilde{U}_1(t, yH_\nu(t)) + |U_1(t, 1)|$, and because $\widetilde{V}_\nu(y) < \infty$ and (3.34) holds, $E\int_0^T[\widetilde{U}_1(t, yH_\nu(t)) + |U_1(t, 1)|]dt < \infty$. The integrand is bounded from below by

$$\min[0, \widetilde{U}_1(t, yH_\nu(t))\,dt] - |U_1(t, 1)| - yH_\nu(t),$$

which also is integrable. The dominated convergence theorem implies that

$$\lim_{n\to\infty} E\int_0^T \left[U_1(t, c_\nu^{(n)}(t)) - yH_\nu(t)c_\nu^{(n)}(t)\right]dt = E\int_0^T \widetilde{U}_1(t, yH_\nu(t))\,dt.$$
$$(3.37)$$

A similar analysis shows that

$$\lim_{n\to\infty} E\left[U_2(B_\nu^{(n)}) - yH_\nu(T)B_\nu^{(n)}\right] = E\widetilde{U}_2(yH_\nu(T)). \quad (3.38)$$

Equation (3.36) is the sum of (3.37) and (3.38). □

Let the initial wealth x be strictly positive. *Our strategy now is to find a process $\hat{\nu}(\cdot) \in \mathcal{D}_0$ for which the optimal pair $(c_{\hat{\nu}}, p_{\hat{\nu}})$ of (3.17), (3.21) for the unconstrained Problem 3.2 is also optimal for the constrained Problem 2.6.* In other words, we seek a process $\hat{\nu}(\cdot) \in \mathcal{D}_0$ that satisfies

$$V(x; K) = E\left[\int_0^T U_1(t, c_{\hat{\nu}}(t))\,dt + U_2(B_{\hat{\nu}})\right] = V_{\hat{\nu}}(x), \quad (3.39)$$

and

$$p_{\hat{\nu}}(t) \in K \quad \text{for Lebesgue-a.e. } t \in [0, T] \quad (3.40)$$

almost surely. Remark 3.3 shows that such a $\hat{\nu}(\cdot)$ should satisfy the "complementary slackness" condition

$$\zeta(\hat{\nu}(t)) + p_{\hat{\nu}}'(t)\hat{\nu}(t) = 0 \quad \text{for Lebesgue-a.e. } t \in [0, T] \quad (3.41)$$

almost surely as well, so that $X^{x,c,p}(T) = X_\nu^{x,c,p}(T)$ almost surely.

Proposition 3.8: *Let $x > 0$ be given, and suppose for some $\hat{\nu}(\cdot) \in \mathcal{D}_0$ that (3.40) and (3.41) are satisfied. Then the pair $(c_{\hat{\nu}}, p_{\hat{\nu}})$ of (3.17), (3.21) is optimal for the constrained Problem 2.6 and $V(x; K) = V_{\hat{\nu}}(x)$. Furthermore, $\hat{\nu}(\cdot)$ minimizes $V_{\nu}(x)$ over $\nu(\cdot) \in \mathcal{D}$.*

PROOF. From (3.40) and Remark 3.3 we see that

$$
V(x; K) \geq E\left[\int_0^T U_1(t, c_{\hat{\nu}}(t))\, dt + U_2(X^{x, c_{\hat{\nu}}, p_{\hat{\nu}}}(T))\right]
$$

$$
= E\left[\int_0^T U_1(t, c_{\hat{\nu}}(t))\, dt + U_2(X_{\nu}^{x, c_{\hat{\nu}}, p_{\hat{\nu}}}(T))\right]
$$

$$
= V_{\hat{\nu}}(x). \tag{3.42}
$$

Remark 3.3 also implies that $V(x; K) \leq V_{\nu}(x)$ for every $\nu(\cdot) \in \mathcal{D}$. Hence equality holds in (3.42), and $\hat{\nu}(\cdot)$ minimizes $V_{\nu}(x)$ over $\nu(\cdot) \in \mathcal{D}$. Since $p_{\hat{\nu}}(\cdot)$ satisfies (3.40), it is optimal for the constrained Problem 2.6.   □

The *necessity* of conditions (3.40), (3.41) for optimality in Problem 2.6, and their precise relationship to (3.39) and other equivalent conditions, will be explored in the next section. We shall see there that if we can find $\hat{\nu}(\cdot) \in \mathcal{D}_0$ that minimizes $V_{\nu}(x)$ over $\nu(\cdot) \in \mathcal{D}$, then the corresponding pair $(c_{\hat{\nu}}, p_{\hat{\nu}})$ is indeed optimal for Problem 2.6; in other words, there is no "duality gap."

Remark 3.9: Recall from Remark 3.6.8 that if the utility functions are given by

$$
U_1(t, x) = U_2(x) = \frac{1}{\beta} x^{\beta} \tag{3.43}
$$

for some $\beta < 1$, $\beta \neq 0$, and if $r(\cdot)$, $\theta(\cdot)$, and $A(\cdot)$ are bounded, then

$$
\mathcal{X}_0(y) = \kappa y^{1/(1-\beta)}, \; y > 0, \tag{3.44}
$$

for some finite positive constant κ. It follows that

$$
V_0(x) = E\left[\int_0^T U_1(I_1(t, \mathcal{Y}_0(x)H_0(t)))\, dt + U_2(I_2(\mathcal{Y}_0(x)H_0(T)))\right] < \infty,
$$

for all $x > 0$, and consequently $V(x; K)$ is finite because of (3.7). Indeed, for this result it suffices to assume only that

$$
U_1(t, x) + U_2(x) \leq \kappa(1 + x^{\beta}), \quad 0 \leq t \leq T, \; x > 0, \tag{3.45}
$$

where $0 < \beta < 1$ and $\kappa > 0$.

If in addition to the above assumptions (which include (3.43) and (5.2.4)) we have $\beta < 0$ and that $\nu(\cdot)$ is in the class $\mathcal{D}^{(b)}$ of bounded processes in $\mathcal{D}$, then $r_{\nu}(\cdot)$ is bounded from below (see (2.1) and (5.5.3)), $\theta_{\nu}(\cdot)$ is bounded (see (5.5.9)), and the argument of Remark 3.6.8 shows that $\mathcal{X}_{\nu}(y)$ has the form (3.44) and $V_{\nu}(x) < \infty$ for all $x > 0$. In other words, $\mathcal{D}^{(b)} \subseteq \mathcal{D}_0$.

6.4   Equivalent Optimality Conditions

For a fixed initial capital $x > 0$, let $(\hat{c}, \hat{p})$ be a consumption and portfolio-proportion process pair in the class $\mathcal{A}_3(x; K)$ of Problem 2.6, and denote by $\widehat{X}(\cdot) = X^{x,\hat{c},\hat{p}}(\cdot)$ the corresponding wealth process in the market $\mathcal{M}(K)$. Consider the statement that this pair is optimal for Problem 2.6:

(A) Optimality of $(\hat{c}, \hat{p})$: We have

$$V(x; K) = E\left[\int_0^T U_1(t, \hat{c}(t))\, dt + U_2(\widehat{X}(T))\right] < \infty. \qquad (4.1)$$

We shall characterize property (A) in terms of the following conditions (B)–(E), which concern a given process $\hat{\nu}(\cdot)$ in the class $\mathcal{D}_0$ of (3.24). The notation of (3.16)–(3.18) will be used freely in what follows.

(B) Financeability of $(c_{\hat{\nu}}(\cdot), B_{\hat{\nu}})$: There exists a portfolio-proportion process $p_{\hat{\nu}}(\cdot)$ such that the pair $(c_{\hat{\nu}}(\cdot), p_{\hat{\nu}}(\cdot))$ is in the class $\mathcal{A}_3(x; K)$ and the properties

$$p_{\hat{\nu}}(t) \in K, \quad \zeta(\hat{\nu}(t)) + p'_{\hat{\nu}}(t)\hat{\nu}(t) = 0 \quad \text{for Lebesgue-a.e. } t \in [0, T], \quad (4.2)$$
$$X^{x,c_{\hat{\nu}},p_{\hat{\nu}}}(\cdot) = X_{\hat{\nu}}(\cdot) \qquad (4.3)$$

are valid almost surely.

(C) Minimality of $\hat{\nu}(\cdot)$: We have

$$V_{\hat{\nu}}(x) \le V_\nu(x), \quad \forall \nu(\cdot) \in \mathcal{D}. \qquad (4.4)$$

(D) Dual optimality of $\hat{\nu}(\cdot)$: With $y = \mathcal{Y}_{\hat{\nu}}(x)$, we have

$$\widetilde{V}_{\hat{\nu}}(y) \le \widetilde{V}_\nu(y), \quad \forall \nu(\cdot) \in \mathcal{D}. \qquad (4.5)$$

(E) Parsimony of $\hat{\nu}(\cdot)$: We have

$$E\left[\int_0^T H_\nu(t)c_{\hat{\nu}}(t)\, dt + H_\nu(T)B_{\hat{\nu}}\right] \le x, \quad \forall \nu(\cdot) \in \mathcal{D}. \qquad (4.6)$$

The portfolio-proportion process $p_{\hat{\nu}}(\cdot)$ in (B) is *not assumed* to be given by (3.21) with $\nu(\cdot) = \hat{\nu}(\cdot)$; this follows from (3.22), (4.3), and Remark 3.3 (see (3.13)).

We have already encountered conditions (A)–(C) in Proposition 3.8 and the preceding discussion. Condition (D) is the "dual" version of (C) in the sense of convex duality. Condition (E) asserts that

$$u(\nu) \triangleq E\left[\int_0^T H_\nu(t)c_{\hat{\nu}}(t)\, dt + H_\nu(T)B_{\hat{\nu}}\right], \qquad (4.7)$$

the "price of the European contingent claim $B_{\hat{\nu}}1_{\{t=T\}} + \int_0^t c_{\hat{\nu}}(s)ds$, $0 \le t \le T$ in the market $\mathcal{M}_\nu$," attains its maximum over $\mathcal{D}$ at $\nu(\cdot) = \hat{\nu}(\cdot)$, and this maximum value is

$$E\left[\int_0^T H_{\hat{\nu}}(t)c_{\hat{\nu}}(t)\,dt + H_{\hat{\nu}}(T)B_{\hat{\nu}}\right] = \mathcal{X}_{\hat{\nu}}(\mathcal{Y}_{\hat{\nu}}(x)) = x. \qquad (4.8)$$

Theorem 4.1: *The conditions (B)–(E) are equivalent, and imply condition (A) with $(\hat{p}, \hat{c}) = (p_{\hat{\nu}}, c_{\hat{\nu}})$. Conversely, condition (A) implies the existence of a process $\hat{\nu}(\cdot) \in \mathcal{D}_0$ that satisfies (B)–(E) with $p_{\hat{\nu}}(\cdot) = \hat{p}(\cdot)$, provided that the utility functions U_1 and U_2 satisfy the conditions (3.4.15), (3.6.17).*

Theorem 4.1 is the central result of this chapter. It provides necessary and sufficient conditions for optimality in Problem 2.6. Its condition (D) is perhaps the most important, as it will provide the cornerstone for our general *existence theory* in Section 5, based on methods from convex duality and martingale theories (Theorem 5.4). Condition (D) also underlies the computations of optimal consumption and portfolio policies in Section 6. Even without such a general existence theory, either condition (C) or (D) is sufficient for a complete treatment of logarithmic utilities (see Examples 4.2, 7.2, and 7.3), and condition (B) suffices for treating the important case of independent coefficients and utilities of power type (Example 7.4). The convex-duality approach also allows us to make connections with the *Hamilton–Jacobi–Bellman theory* of stochastic control in Section 6.

The important special case of *incomplete markets* receives special treatment in Section 7.

PROOF OF THEOREM 4.1: We first prove that (B)⇒(E). Assume (B) and let $\nu(\cdot) \in \mathcal{D}$ be given. From Remark 3.3, we have

$$X_{\nu}^{x,c_{\hat{\nu}},p_{\hat{\nu}}}(t) \geq X^{x,c_{\hat{\nu}},p_{\hat{\nu}}}(t) = X_{\hat{\nu}}(t), \ 0 \leq t \leq T.$$

Condition (E) now follows from Remark 5.5.4 (recalling (2.6)) and the equality $X_{\hat{\nu}}(T) = B_{\hat{\nu}}$.

We next prove (E)⇒(D). To begin, we note that (3.4.13), (3.4.14) imply

$$\widetilde{U}_2(U_2'(\xi)) + \xi U_2'(\xi) = U_2(\xi) \leq \widetilde{U}_2(\eta) + \xi\eta, \quad \forall \xi > 0, \ \forall \eta > 0. \qquad (4.9)$$

Assume that $\hat{\nu}(\cdot)$ satisfies (E), and let $y = \mathcal{Y}_{\hat{\nu}}(x)$. Let $\nu(\cdot) \in \mathcal{D}$ be given. Take $\xi = B_{\hat{\nu}} = I_2(yH_{\hat{\nu}}(T))$ so that $U_2'(\xi) = yH_{\hat{\nu}}(T)$, and take $\eta = yH_{\nu}(T)$ in (4.9), which becomes

$$\widetilde{U}_2(yH_{\hat{\nu}}(T)) + yH_{\hat{\nu}}(T)B_{\hat{\nu}} \leq \widetilde{U}_2(yH_{\nu}(T)) + yH_{\nu}(T)B_{\hat{\nu}}.$$

Similarly,

$$\int_0^T \left[\widetilde{U}_1(t, yH_{\hat{\nu}}(t)) + yH_{\hat{\nu}}(t)c_{\hat{\nu}}(t)\right]dt$$

$$\leq \int_0^T \left[\widetilde{U}_1(t, yH_{\nu}(t)) + yH_{\nu}(t)c_{\hat{\nu}}(t)\right]dt.$$

Summing these two inequalities and taking expectations, we obtain

$$\tilde{V}_{\hat{\nu}}(y) + yE\left[\int_0^T H_{\hat{\nu}}(t)c_{\hat{\nu}}(t)\,dt + H_{\hat{\nu}}(T)B_{\hat{\nu}}\right]$$

$$\leq \tilde{V}_{\nu}(y) + yE\left[\int_0^T H_{\nu}(t)c_{\hat{\nu}}(t)\,dt + H_{\nu}(T)B_{\hat{\nu}}\right].$$

By assumption,

$$E\left[\int_0^T H_{\hat{\nu}}(t)c_{\hat{\nu}}(t)\,dt + H_{\hat{\nu}}(T)B_{\hat{\nu}}\right] \geq E\left[\int_0^T H_{\nu}(t)c_{\hat{\nu}}(t)\,dt + H_{\nu}(T)B_{\hat{\nu}}\right],$$

and we conclude that (4.5) holds.

The implications (B)$\Rightarrow$(A) and (B)$\Rightarrow$(C) are consequences of Proposition 3.8.

(C)$\Rightarrow$(D): With $y = \mathcal{Y}_{\hat{\nu}}(x)$, we have

$$\tilde{V}_{\hat{\nu}}(y) = V_{\hat{\nu}}(\mathcal{X}_{\hat{\nu}}(y)) - y\mathcal{X}_{\hat{\nu}}(y) = V_{\hat{\nu}}(x) - xy$$

$$\leq V_{\nu}(x) - xy \leq \sup_{\xi>0}[V_{\nu}(\xi) - \xi y] \leq \tilde{V}_{\nu}(y)$$

from (3.25), (3.27), and Proposition 3.7.

(D)$\Rightarrow$(B): Assume (D), and let $c_{\hat{\nu}}(\cdot)$, $p_{\hat{\nu}}(\cdot)$ be given by (3.17) and (3.21), respectively. The corresponding wealth process $X_{\hat{\nu}}^{x,c_{\hat{\nu}},p_{\hat{\nu}}}(\cdot)$ in the market $\mathcal{M}_{\hat{\nu}}$ is defined by the equivalent equations (3.1)–(3.3), where we replace $c(\cdot)$ by $c_{\hat{\nu}}(\cdot)$ and $p(\cdot)$ by $p_{\hat{\nu}}(\cdot)$. In light of (3.22), we have

$$X_{\hat{\nu}}^{x,c_{\hat{\nu}},p_{\hat{\nu}}}(\cdot) = X_{\hat{\nu}}(\cdot), \tag{4.10}$$

where the latter is given by (3.18) with $\hat{\nu}(\cdot)$ replacing $\nu(\cdot)$. We divide the remainder of the proof into six steps.

Step 1. For any $\mu(\cdot) \in \mathcal{D}$ and any $\epsilon \in (0,1)$, the convex combination $(1-\epsilon)\hat{\nu}(\cdot) + \epsilon\mu(\cdot)$ is in $\mathcal{D}$, because of the convexity of $\widetilde{K}$ and the positive homogeneity and subadditivity of $\zeta(\cdot)$ (see (5.4.3), (5.4.4)), which guarantee that $(1-\epsilon)\hat{\nu}(\cdot) + \epsilon\mu(\cdot)$ satisfies (5.5.2). We shall be interested in two particular choices of $\mu(\cdot)$. The first is $\mu(\cdot) \equiv 0$, which is an element of $\mathcal{D}$ because $\underline{0} \in \widetilde{K}$ and $\zeta(\underline{0}) = 0$. The other is $\mu(\cdot) = \hat{\nu}(\cdot) + \lambda(\cdot)$ for some $\lambda(\cdot) \in \mathcal{D}$; this process $\mu(\cdot)$ is in $\mathcal{D}$ because $\widetilde{K}$ is a convex cone and thus closed under addition, and $\zeta(\cdot)$ is subadditive.

Let $\{\tau_n\}_{n=1}^{\infty}$ be a nondecreasing sequence of stopping times converging up to T, and consider the small random perturbation of $\hat{\nu}(\cdot)$ given by

$$\nu_{\epsilon,n}(t) \overset{\Delta}{=} \hat{\nu}(t) + \epsilon(\mu(t) - \hat{\nu}(t))1_{\{t\leq\tau_n\}}$$

$$= \begin{cases} (1-\epsilon)\hat{\nu}(t) + \epsilon\mu(t), & 0 \leq t \leq \tau_n, \\ \hat{\nu}(t), & \tau_n < t \leq T. \end{cases}$$

Because $\nu_{\epsilon,n}(\cdot) \in \mathcal{D}$ and $\hat{\nu}(\cdot)$ minimizes $\widetilde{V}_\nu(y)$ over $\mathcal{D}$, we must have

$$0 \leq EY_{\epsilon,n} = \frac{\widetilde{V}_{\nu_{\epsilon,n}}(y) - \widetilde{V}_{\hat{\nu}}(y)}{\epsilon y}, \qquad \forall \epsilon \in (0,1), \; n = 1, 2, \ldots, \qquad (4.11)$$

where

$$Y_{\epsilon,n} \triangleq \frac{1}{\epsilon y}\left[\left\{ \widetilde{U}_2(yH_{\nu_{\epsilon,n}}(T)) - \widetilde{U}_2(yH_{\hat{\nu}}(T)) \right\} \right.$$

$$\left. + \int_0^T \left\{ \widetilde{U}_1(t, yH_{\nu_{\epsilon,n}}(t)) - \widetilde{U}_1(t, yH_{\hat{\nu}}(t)) \right\} dt \right]. \qquad (4.12)$$

Step 2. A straightforward computation using (5.5.7) and (5.5.10)–(5.5.12) shows that

$$\Lambda_{\epsilon,n}(t) \triangleq \frac{H_{\nu_{\epsilon,n}}(t)}{H_{\hat{\nu}}(t)}$$

$$= \exp\left[-\epsilon N(t \wedge \tau_n) - \frac{\epsilon^2}{2} \langle N \rangle (t \wedge \tau_n) \right.$$

$$\left. - \int_0^{t \wedge \tau_n} \left(\zeta((1-\epsilon)\hat{\nu}(s) + \epsilon\mu(s)) - \zeta(\hat{\nu}(s)) \right) ds \right],$$

where

$$N(t) \triangleq \int_0^t \left(\sigma^{-1}(s)(\mu(s) - \hat{\nu}(s)) \right)' dW_{\hat{\nu}}(s),$$

$$\langle N \rangle(t) \triangleq \int_0^t \| \sigma^{-1}(s)(\mu(s) - \hat{\nu}(s)) \|^2 ds.$$

In the case $\mu(\cdot) = 0$, we have

$$\zeta\left((1-\epsilon)\hat{\nu}(s) + \epsilon\mu(s) \right) - \zeta\left(\hat{\nu}(s) \right) = \zeta((1-\epsilon)\hat{\nu}(s)) - \zeta(\hat{\nu}(s))$$

$$= -\epsilon\zeta(\hat{\nu}(s)),$$

whereas in the case $\mu(\cdot) = \hat{\nu}(\cdot) + \lambda(\cdot)$ for some $\lambda(\cdot) \in \mathcal{D}$, we have

$$\zeta\left((1-\epsilon)\hat{\nu}(s) + \epsilon\mu(s) \right) - \zeta\left(\hat{\nu}(s) \right) = \zeta\left(\hat{\nu}(s) + \epsilon\lambda(s) \right) - \zeta\left(\hat{\nu}(s) \right)$$

$$\leq \epsilon\zeta(\lambda(s)).$$

If we define

$$\xi(s) = \begin{cases} -\zeta(\hat{\nu}(s)), & \text{if } \mu(\cdot) = 0, \\ \zeta(\lambda(s)), & \text{if } \mu(\cdot) = \hat{\nu}(\cdot) + \lambda(\cdot) \text{ for some } \lambda(\cdot) \in \mathcal{D}, \end{cases}$$

and $L(t) \triangleq \int_0^t \xi(s)\, ds$, we obtain the lower bound

$$\Lambda_{\epsilon,n}(t) \geq Q_{\epsilon,n}(t)$$

$$\triangleq \exp\left\{ -\epsilon(N(t \wedge \tau_n) + L(t \wedge \tau_n)) - \frac{\epsilon^2}{2} \langle N \rangle (t \wedge \tau_n) \right\}. \qquad (4.13)$$

Step 3. For each positive integer n, we introduce the stopping time

$$\tau_n \overset{\Delta}{=} \inf \left\{ t \in [0, T]; \; |N(t)| + \langle N \rangle(t) + |L(t)| \geq n \right.$$

$$\text{or } \int_0^t \|\theta_{\hat{\nu}}(s)\|^2 \, ds \geq n$$

$$\text{or } \int_0^t \left(\frac{X_{\hat{\nu}}(s)}{S_0^{(\hat{\nu})}(s)} \right)^2 \|\sigma^{-1}(s)(\mu(s) - \hat{\nu}(s))\|^2 \, ds \geq n$$

$$\left. \text{or } \int_0^t (L(s) + N(s))^2 \|\sigma'(s)p_{\hat{\nu}}(s)\|^2 \, ds \geq n \right\} \wedge T.$$

Clearly, $\tau_n \uparrow T$ almost surely as $n \to \infty$. According to the Girsanov and Novikov theorems (e.g., Karatzas and Shreve (1991), §3.5), the process

$$W_{\hat{\nu},n}(t) \overset{\Delta}{=} W(t) + \int_0^{t \wedge \tau_n} \theta_{\hat{\nu}}(s) \, ds, \; 0 \leq t \leq T,$$

is Brownian motion under the probability measure

$$P_{\hat{\nu},n}(A) \overset{\Delta}{=} E[Z_{\hat{\nu}}(\tau_n) 1_A], \; A \in \mathcal{F}(T).$$

With this choice of τ_n, the process $Q_{\epsilon,n}(\cdot)$ of (4.13) admits the lower bound

$$Q_{\epsilon,n}(t) \geq e^{-\epsilon n}, \; 0 \leq t \leq T,$$

and consequently,

$$\frac{1}{\epsilon} \left(1 - \frac{H_{\nu_{\epsilon,n}}(t)}{H_{\hat{\nu}}(t)} \right) = \frac{1 - \Lambda_{\epsilon,n}(t)}{\epsilon} \leq K_n, \; 0 \leq t \leq T \qquad (4.14)$$

holds almost surely, where $K_n \overset{\Delta}{=} \sup_{0 < \epsilon < 1} \frac{1}{\epsilon}(1 - e^{-\epsilon n})$ is finite. Furthermore,

$$\overline{\lim_{\epsilon \downarrow 0}} \frac{1 - \Lambda_{\epsilon,n}(t)}{\epsilon} \leq N(t \wedge \tau_n) + L(t \wedge \tau_n).$$

Step 4. The functions $-\tilde{U}_2(\cdot)$ and $-\tilde{U}_1(t, \cdot)$ have derivatives $I_2(\cdot)$ and $I_1(t, \cdot)$, respectively, and these derivatives are nonincreasing functions (Section 3.4). For $U_2(\cdot)$ we have the inequality

$$\tilde{U}_2(yH_{\nu_{\epsilon,n}}(T)) - \tilde{U}_2(yH_{\hat{\nu}}(T)) \leq yH_{\hat{\nu}}(T)I_2(ye^{-n}H_{\hat{\nu}}(T))(1 - \Lambda_{\epsilon,n}(T))^+.$$

From (4.14), we see that

$$\frac{1}{\epsilon}[\tilde{U}_2(yH_{\nu_{\epsilon,n}}(T)) - \tilde{U}_2(yH_{\hat{\nu}}(T))] \leq Y_{n,2} \overset{\Delta}{=} K_n yH_{\hat{\nu}}(T)I_2(ye^{-n}H_{\hat{\nu}}(T))$$

and also that

$$\overline{\lim_{\epsilon \downarrow 0}} \frac{1}{\epsilon} [\widetilde{U}_2(yH_{\nu_{\epsilon},n}(T)) - \widetilde{U}_2(yH_{\hat{\nu}}(T))] = yH_{\hat{\nu}}(T)I_2(yH_{\hat{\nu}}(T))[N(\tau_n) + L(\tau_n)]$$

$$\leq yH_{\hat{\nu}}(T)B_{\hat{\nu}}[N(\tau_n) + L(\tau_n)]. \quad (4.15)$$

Because $EY_{n,2} \leq K_n y \mathcal{X}_{\hat{\nu}}(ye^{-n})$, which is finite by the assumption $\hat{\nu}(\cdot) \in \mathcal{D}_0$, we may apply Fatou's lemma in (4.15) to conclude that

$$\overline{\lim_{\epsilon \downarrow 0}} E \frac{1}{\epsilon} [\widetilde{U}_2(yH_{\nu_{\epsilon},n}(T)) - \widetilde{U}_2(yH_{\hat{\nu}}(T))] \leq yE \{H_{\hat{\nu}}(T)B_{\hat{\nu}}[N(\tau_n) + L(\tau_n)]\}.$$

$$(4.16)$$

We proceed similarly with the difference

$$\int_0^T [\widetilde{U}_1(t, yH_{\hat{\nu}_{\epsilon},n}(t)) - \widetilde{U}_1(t, yH_{\hat{\nu}}(t))]dt$$

to obtain in the end, by analogy with (4.16), that

$$\overline{\lim_{\epsilon \downarrow 0}} \frac{1}{\epsilon} E \int_0^T [\widetilde{U}_1(t, yH_{\hat{\nu}_{\epsilon},n}(t)) - \widetilde{U}_1(t, yH_{\hat{\nu}}(t))]dt$$

$$\leq yE \left\{ \int_0^T H_{\hat{\nu}}(t)c_{\hat{\nu}}(t)[N(t \wedge \tau_n) + L(t \wedge \tau_n)] \, dt \right\}. \quad (4.17)$$

Finally, from (4.11) and (4.12) we have

$$E \left\{ H_{\hat{\nu}}(T)B_{\hat{\nu}}[N(\tau_n) + L(\tau_n)] + \int_0^T H_{\hat{\nu}}(t)c_{\hat{\nu}}(t)[N(t \wedge \tau_n) + L(t \wedge \tau_n)] \, dt \right\}$$

$$\geq 0. \quad (4.18)$$

Step 5. We next prove that

$$E \int_0^{\tau_n} H_{\hat{\nu}}(t)X_{\hat{\nu}}(t)[p'_{\hat{\nu}}(t)(\mu(t) - \hat{\nu}(t)) + \xi(t)] \, dt \geq 0, \quad n = 1, 2, \ldots. \quad (4.19)$$

To see this, we first recall from (4.10) and (3.1) that

$$d\left(\frac{X_{\hat{\nu}}(t)}{S_0^{(\hat{\nu})}(t)} \right) + \frac{c_{\hat{\nu}}(t)dt}{S_0^{(\hat{\nu})}(t)} = \frac{X_{\hat{\nu}}(t)}{S_0^{(\hat{\nu})}(t)} p'_{\hat{\nu}}(t)\sigma(t) \, dW_{\hat{\nu}}(t),$$

and so

$$d\left(\frac{X_{\hat{\nu}}(t)}{S_0^{(\hat{\nu})}(t)} (L(t) + N(t)) \right)$$

$$= \frac{X_{\hat{\nu}}(t)}{S_0^{(\hat{\nu})}(t)} (dL(t) + dN(t)) + (L(t) + N(t)) \, d\left(\frac{X_{\hat{\nu}}(t)}{S_0^{(\hat{\nu})}(t)} \right)$$

$$+ \frac{X_{\hat{\nu}}(t)}{S_0^{(\hat{\nu})}(t)} p'_{\hat{\nu}}(t)(\mu(t) - \hat{\nu}(t)) \, dt.$$

Integration of this equation yields

$$\frac{X_{\hat{\nu}}(\tau_n)}{S_0^{(\hat{\nu})}(\tau_n)}(L(\tau_n) + N(\tau_n)) + \int_0^{\tau_n} \frac{L(t) + N(t)}{S_0^{(\hat{\nu})}(t)} c_{\hat{\nu}}(t)\, dt$$

$$= \int_0^{\tau_n} \frac{X_{\hat{\nu}}(t)}{S_0^{(\hat{\nu})}(t)} [p_{\hat{\nu}}'(t)(\mu(t) - \hat{\nu}(t)) + \xi(t)]\, dt$$

$$+ \int_0^{\tau_n} \frac{X_{\hat{\nu}}(t)}{S_0^{(\hat{\nu})}(t)} [\sigma^{-1}(t)(\mu(t) - \hat{\nu}(t))$$

$$+ (L(t) + N(t))\sigma'(t)p_{\hat{\nu}}(t)]'\, dW_{\hat{\nu}}(t).$$

By the choice of τ_n, the integrand of the Itô integral in this last expression is square-integrable, and thus has expectation zero under $P_{\hat{\nu},n}$. Taking expectations under this probability measure, we obtain

$$E \int_0^{\tau_n} H_{\hat{\nu}}(t)X_{\hat{\nu}}(t)[p_{\hat{\nu}}'(t)(\mu(t) - \hat{\nu}(t)) + \xi(t)]\, dt \qquad (4.20)$$

$$= E\left[(L(\tau_n) + N(\tau_n))H_{\hat{\nu}}(\tau_n)X_{\hat{\nu}}(\tau_n) + \int_0^{\tau_n} H_{\hat{\nu}}(t)(L(t) + N(t))c_{\hat{\nu}}(t)\, dt\right].$$

An application of the optional sampling theorem to the martingale of (3.19) shows that (3.18) is still valid if we replace t in that equation by the stopping time τ_n. Using this fact, we rewrite (4.20) as

$$E \int_0^{\tau_n} H_{\hat{\nu}}(t)X_{\hat{\nu}}(t)[p_{\hat{\nu}}'(t)(\mu(t) - \hat{\nu}(t)) + \xi(t)]\, dt$$

$$= E\left[(L(\tau_n) + N(\tau_n))\left(H_{\hat{\nu}}(T)B_{\hat{\nu}} + \int_{\tau_n}^T H_{\hat{\nu}}(t)c_{\hat{\nu}}(t)\, dt\right)\right.$$

$$\left. + \int_0^{\tau_n} H_{\hat{\nu}}(t)(L(t) + N(t))c_{\hat{\nu}}(t)\, dt\right],$$

which is the left-hand side of (4.18), hence a nonnegative quantity. This completes the proof of (4.19).

Step 6. We invoke Lemma 5.4.2 to obtain a process $\lambda(\cdot) \in \mathcal{D}$ satisfying (5.4.8) with $p(\cdot) = p_{\hat{\nu}}(\cdot)$, and take $\mu(\cdot) = \hat{\nu}(\cdot) + \lambda(\cdot)$ so that $\xi(\cdot) = \zeta(\lambda(\cdot))$. The inequality (4.19) becomes

$$E \int_0^{\tau_n} H_{\hat{\nu}}(t)X_{\hat{\nu}}(t)[p_{\hat{\nu}}'(t)\lambda(t) + \zeta(\lambda(t))]\, dt \geq 0, \quad n = 1, 2, \ldots,$$

which, together with (5.4.8), implies

$$p_{\hat{\nu}}(t) \in K \text{ for Lebesgue-a.e. } t \in [0, T] \qquad (4.21)$$

almost surely. From (5.4.5) we have also that

$$\zeta(\hat{\nu}(t)) + p_{\hat{\nu}}'(t)\hat{\nu}(t) \geq 0 \text{ for Lebesgue-a.e. } t \in [0, T]$$

almost surely. We next take $\mu(\cdot) \equiv 0$, so that $\xi(\cdot) = -\zeta(\hat{\nu}(\cdot))$, and (4.19) becomes

$$E \int_0^{T_n} H_{\hat{\nu}}(t) X_{\hat{\nu}}(t)[p'_{\hat{\nu}}(t)\hat{\nu}(t) + \zeta(\hat{\nu}(t))]\, dt \leq 0, \quad n = 1, 2, \ldots,$$

which proves

$$\zeta(\hat{\nu}(t)) + p'_{\hat{\nu}}(t)\hat{\nu}(t) = 0 \text{ for Lebesgue-a.e. } t \in [0, T] \qquad (4.22)$$

almost surely. Conditions (4.21) and (4.22) are (4.2) of condition (B). Condition (4.3) follows from (4.2), (4.10), and Remark 3.3.

(A)$\Rightarrow$(B): This implication is proved in Appendix C; the proof may be skipped on first reading without harm, as this implication will not be invoked in the sequel. $\square$

Example 4.2: $U_1(t, x) = U_2(x) = \log x, \ \forall (t, x) \in [0, T] \times (0, \infty)$.

As in Example 3.6.6, we have in this case $I_1(t, y) = I_2(y) = \frac{1}{y}$, and consequently, $\widetilde{U}_1(t, y) = \widetilde{U}_2(y) = -(1 + \log y)$ for $0 < y < \infty$. Furthermore, for all processes $\nu(\cdot) \in \mathcal{D}$, we have $\mathcal{X}_\nu(y) = \frac{T+1}{y}$ for $0 < y < \infty$ and $\mathcal{Y}_\nu(x) = \frac{T+1}{x}$ for $0 < x < \infty$. In particular, $\mathcal{D}_0 = \mathcal{D}$. Direct computations show that for all $\nu(\cdot) \in \mathcal{D}$, $t \in [0, T]$, $x \in (0, \infty)$ and $y \in (0, \infty)$,

$$B_\nu = \frac{x}{(T+1)H_\nu(T)}, \quad c_\nu(t) = \frac{x}{(T+1)H_\nu(t)}, \quad X_\nu(t) = \frac{(T+1-t)x}{(T+1)H_\nu(t)},$$

$$p_\nu(t) = (\sigma^{-1}(t))'\theta_\nu(t) = (\sigma(t)\sigma'(t))^{-1}[b(t) + \delta(t) - r(t)\underline{1} + \nu(t)],$$

$$V_\nu(x) = (T+1) \log\left(\frac{x}{T+1}\right) + f(\nu),$$

$$\widetilde{V}_\nu(y) = -(T+1)(1 + \log y) + f(\nu),$$

where

$$f(\nu) \triangleq -E\left(\int_0^T \log H_\nu(t)\, dt + \log H_\nu(T)\right).$$

For $\nu(\cdot) \in \mathcal{D}$, we have

$$-E \log H_\nu(t) = E\left[A(t) + \int_0^t \left(r(s) + \zeta(\nu(s)) + \frac{1}{2}\|\theta_\nu(s)\|^2\right) ds\right],$$

so that conditions (C) and (D) amount to *pointwise minimization* of the expression

$$\zeta(\nu) + \frac{1}{2}\|\theta(t) + \sigma^{-1}(t)\nu\|^2 \quad \text{over} \quad \nu \in \widetilde{K}, \qquad (4.23)$$

where $\widetilde{K}$, the effective domain of the function $\zeta(\cdot)$, is the barrier cone of $-K$ (see (5.4.2)).

We denote by $L^\times(\mathbb{R}^N \times \mathbb{R}^N)$ the set of nonsingular $N \times N$ matrices. For $\nu \in \widetilde{K}$, $\theta \in \mathbb{R}^N$, and $\sigma \in L^\times(\mathbb{R}^N \times \mathbb{R}^N)$, we define

$$g(\nu, \theta, \sigma) \triangleq \zeta(\nu) + \frac{1}{2}\|\theta + \sigma^{-1}\nu\|^2.$$

For fixed $\theta \in \mathbb{R}^N$ and $\sigma \in L^\times(\mathbb{R}^N \times \mathbb{R}^N)$, the function $\nu \mapsto g(\nu, \theta, \sigma)$ is strictly convex, and thus has a unique minimizer. For each positive integer n, we denote by $\widetilde{K}_n$ the compact set $\widetilde{K} \cap \{\nu \in \mathbb{R}^N; \|\nu\| \leq n\}$, and define g_n to be the restriction of g to $\widetilde{K}_n \times \mathbb{R}^N \times L^\times(\mathbb{R}^N \times \mathbb{R}^N)$. According to measurable selection theorems of Dubins–Savage (1965) type (see Schäl (1974), (1975) or Bertsekas and Shreve (1978), Proposition 7.33), for each n there is a Borel-measurable function $\varphi_n \colon \mathbb{R}^N \times L^\times(\mathbb{R}^N \times \mathbb{R}^N) \to \widetilde{K}_n$ such that

$$g_n(\varphi_n(\theta, \sigma), \theta, \sigma) = \min_{\nu \in \widetilde{K}_n} g(\nu, \theta, \sigma), \quad \forall \theta \in \mathbb{R}^N, \ \sigma \in L^\times(\mathbb{R}^N \times \mathbb{R}^N).$$

Because $g(\nu, \theta, \sigma) \to \infty$ as $\|\nu\| \to \infty$, for fixed $\theta \in \mathbb{R}^N$ and $\sigma \in L^\times(\mathbb{R}^N \times \mathbb{R}^N)$, the function values $\varphi_n(\theta, \sigma)$ are bounded uniformly in n, and hence do not depend on n for n sufficiently large. We define $\varphi(\theta, \sigma) = \lim_{n \to \infty} \varphi_N(\theta, \sigma)$, which satisfies

$$g(\varphi(\theta, \sigma), \theta, \sigma) = \min_{\nu \in \widetilde{K}} g(\nu, \theta, \sigma), \quad \forall \theta \in \mathbb{R}^N, \ \sigma \in L^\times(\mathbb{R}^N \times \mathbb{R}^N).$$

The progressively measurable process

$$\hat{\nu}(t) \overset{\Delta}{=} \varphi\left(\theta(t), \sigma(t)\right), \ 0 \leq t \leq T,$$

is in $\mathcal{D}$ and minimizes the expression (4.23) for all $t \in [0, T]$ almost surely. The inequality

$$g\left(\hat{\nu}(t), \theta(t), \sigma(t)\right) \leq g\left(0, \theta(t), \sigma(t)\right) \tag{4.24}$$

and (2.5) imply that

$$E \int_0^T \zeta(\hat{\nu}(t)) \, dt + \frac{1}{2} E \int_0^T \|\theta(t) + \sigma(t)\hat{\nu}(t)\|^2 \, dt \leq E \int_0^T \|\theta(t)\|^2 \, dt < \infty,$$

and thus $f(\hat{\nu}) < \infty$; in particular, $V_{\hat{\nu}}(x) < \infty$ for all $x > 0$.

The implications (C)$\Rightarrow$(A) or (D)$\Rightarrow$(A) of Theorem 4.1 show that the *optimal consumption, portfolio-proportion, and wealth processes* for Problem 2.6 are given by

$$\hat{c}(t) = \frac{x}{(T+1)H_{\hat{\nu}}(t)}, \quad \widehat{X}(t) = \frac{(T+1-t)x}{(T+1)H_{\hat{\nu}}(t)},$$
$$\hat{p}(t) = (\sigma(t)\sigma'(t))^{-1}[b(t) + \delta(t) - r(t)\mathbf{1} + \hat{\nu}(t)], \quad 0 \leq t \leq T,$$

and the *value function* of (2.16) is

$$V(x; K) = V_{\hat{\nu}}(x) = (T+1) \log\left(\frac{x}{T+1}\right) + f(\hat{\nu}) < \infty, \ 0 < x < \infty.$$

It is perhaps worth noting in this example that if $\theta(\cdot)$ satisfies the Novikov condition $E\left[\exp\left\{\frac{1}{2} \int_0^T \|\theta(t)\|^2 dt\right\}\right] < \infty$ mentioned in Remark 1.5.2, then

(4.24) and (2.1) show that $\theta_{\hat{\nu}}(\cdot)$ also satisfies this condition; thus both financial markets $\mathcal{M}_0$ and $\mathcal{M}_{\hat{\nu}}$ are "standard" (in the sense of Definition 1.5.1) and complete.   □

6.5   Duality and Existence

We now turn to the *dual optimization problem* associated with the "primal" constrained Problem 2.6. The dual problem is, for fixed $y > 0$, to minimize $\widetilde{V}_\nu(y)$ over $\nu(\cdot) \in \mathcal{D}$, and its value function is defined by

$$\widetilde{V}(y) \triangleq \inf_{\nu \in \mathcal{D}} \widetilde{V}_\nu(y), \ 0 < y < \infty, \tag{5.1}$$

in the notation of (3.27), (3.33). This problem is suggested by condition (D) preceding Theorem 4.1, which amounts to $\widetilde{V}(y) = \widetilde{V}_{\hat{\nu}}(y)$ for some $\hat{\nu}(\cdot) \in \mathcal{D}_0$ and $y = \mathcal{Y}_{\hat{\nu}}(x)$. The terminology "dual" comes from the fact that, as we shall show in Propositions 5.1 and 5.2 below, the value function $\widetilde{V}(\cdot)$ of (5.1) is the *convex dual* of the concave function $V(\cdot; K)$ of (2.16), in the sense of Definition 3.4.2.

We shall assume throughout this section that

$$\widetilde{V}(y) < \infty, \ 0 < y < \infty. \tag{5.2}$$

Our plan in this section is to show that under reasonable conditions, for any given $y > 0$, the dual problem (5.1) has a minimizer $\nu_y(\cdot) \in \mathcal{D}_0$; i.e.,

$$\forall y \in (0, \infty), \ \exists \ \nu_y(\cdot) \in \mathcal{D}_0 \text{ such that } \widetilde{V}(y) = \widetilde{V}_{\nu_y}(y) \tag{5.3}$$

(see Theorem 5.3, whose proof occupies a good part of this section). We can then argue that in conjunction with Theorem 4.1, this solution implies the existence of an optimal consumption and portfolio-proportion pair for the "primal" Problem 2.6 (see Theorem 5.4).

Proposition 5.1 (Weak Duality): *Suppose (5.2) and (5.3) hold. Then, for any given $y \in (0, \infty)$ and with $x = \mathcal{X}_{\nu_y}(y)$,*

(i) *there exists an optimal consumption and portfolio-proportion process pair $(\hat{c}, \hat{p}) \in \mathcal{A}_3(x; K)$ for Problem 2.6, and*

(ii) *we have*

$$\widetilde{V}(y) = \sup_{\xi > 0}[V(\xi; K) - \xi y]. \tag{5.4}$$

PROOF.   First, let us note that (5.3), (3.10), and (3.27) imply

$$V(x; K) \le \widetilde{V}(y) + xy, \ \forall y > 0, \ \forall x > 0. \tag{5.5}$$

Now fix $y \in (0, \infty)$, let $x = \mathcal{X}_{\nu_y}(y)$, and note that the assumption $\widetilde{V}_{\nu_y}(y) \le \widetilde{V}_\nu(y)$ for all $\nu(\cdot) \in \mathcal{D}$ of (5.3) amounts to (4.5) with $\hat{\nu}(\cdot) = \nu_y(\cdot)$. The implications (D)$\Rightarrow$(B), (D)$\Rightarrow$(A) in Theorem 4.1 and Proposition 3.8 show that the pair $(c_{\hat{\nu}}(\cdot), p_{\hat{\nu}}(\cdot))$ given by (3.17), (3.21) is in $\mathcal{A}_3(x; K)$ and is optimal for Problem 2.6. By (3.17), (3.22), (3.18), and (3.16),

$$c_{\hat{\nu}}(t) = I_1(t, yH_{\hat{\nu}}(t)),$$
$$X_{\hat{\nu}}^{x, c_{\hat{\nu}}, p_{\hat{\nu}}}(T) = X_{\hat{\nu}}(T) = B_{\hat{\nu}} = I_2(yH_{\hat{\nu}}(T)),$$
$$x = X_{\hat{\nu}}^{x, c_{\hat{\nu}}, p_{\hat{\nu}}}(0) = E\left[\int_0^T H_{\hat{\nu}}(t)c_{\hat{\nu}}(t)\, dt + H_{\hat{\nu}}(T)X_{\hat{\nu}}^{x, c_{\hat{\nu}}, p_{\hat{\nu}}}(T)\right].$$

In other words, equations (3.29)–(3.31) hold and imply equality in (3.28); i.e.,

$$\widetilde{V}(y) = \widetilde{V}_{\hat{\nu}}(y) = V_{\hat{\nu}}(x) - xy. \tag{5.6}$$

But condition (B) also implies, via Proposition 3.8, that

$$V_{\hat{\nu}}(x) - xy = V(x; K) - xy, \tag{5.7}$$

which, of course, is bounded above by $\sup_{\xi>0}[V(\xi; K) - \xi y]$. We have obtained the inequality

$$\widetilde{V}(y) \le \sup_{\xi>0}[V(\xi; K) - \xi y].$$

The reverse inequality is obvious from (5.5), and (5.4) follows.   □

Proposition 5.2 (Strong Duality): *Assume that (5.2), (5.3), and*

$$U_2(\infty) = \infty \tag{5.8}$$

hold. Then, for any given $x \in (0, \infty)$, we have

$$V(x; K) = \inf_{y>0}[\widetilde{V}(y) + xy], \tag{5.9}$$

and the infimum in (5.9) is attained at some $y = y(x) \in (0, \infty)$ that satisfies $x = \mathcal{X}_{\nu_{y(x)}}(y(x))$.

PROOF.   From the nonincrease and convexity of $\widetilde{U}_1(t, \cdot)$ and $\widetilde{U}(\cdot)$, we have, in conjunction with (3.27), (3.33), and Jensen's inequality,

$$\widetilde{V}_\nu(y) \ge \int_0^T \widetilde{U}_1(t, yEH_\nu(t))\, dt + \widetilde{U}_2(yEH_\nu(T))$$
$$\ge \int_0^T \widetilde{U}_1\left(t, \frac{y}{s_0}\right) dt + \widetilde{U}_2\left(\frac{y}{s_0}\right), \quad 0 < y < \infty, \; \nu(\cdot) \in \mathcal{D}, \tag{5.10}$$

where s_0 is as in (2.4) and we have used (2.1). Since $\widetilde{U}_2(0+) = U_2(\infty) = \infty$ (Lemma 3.4.3), it develops that $\widetilde{V}(0+) = \infty$ and the convex function $f_x(\eta) \triangleq \widetilde{V}(\eta) + x\eta$, $0 < \eta < \infty$, satisfies $f_x(0+) = \infty$. But from (5.4),

$$f_x(y) = \sup_{\xi>0}[V(\xi; K) - \xi y] + xy \ge V\left(\frac{x}{2}; K\right) + \frac{xy}{2},$$

which shows that $f_x(\infty) = \infty$. Being convex and finite on $(0, \infty)$, the function f_x is continuous there and must attain its minimum at some $y = y(x) \in (0, \infty)$. We have to show that $x = \mathcal{X}_{\nu_y}(y)$.

To see this last property, observe that

$$\inf_{u>0} \left[\widetilde{V}_{\nu_y}(uy) + xuy \right] = \inf_{\eta>0} \left[\widetilde{V}_{\nu_y}(\eta) + x\eta \right]$$

$$\geq \inf_{\eta>0} \left[\widetilde{V}(\eta) + x\eta \right]$$

$$= \widetilde{V}(y) + xy$$

$$= \widetilde{V}_{\nu_y}(y) + xy,$$

where $\nu_y(\cdot) \in \mathcal{D}_0$ is the process of (5.3). In other words, the function $u \mapsto \widetilde{V}_{\nu_y}(uy) + xuy$ attains its minimum over $(0, \infty)$ at $u = 1$. But from (3.32) this function is continuously differentiable with derivative $xy - y\mathcal{X}_{\nu_y}(uy)$, which has to vanish at $u = 1$. Thus $\mathcal{X}_{\nu_y}(y) = x$, as desired.

It remains to prove (5.9). As in Proposition 5.1 we have (5.6) and (5.7), but now with $\hat{\nu}(\cdot)$ replaced by $\nu_y(\cdot)$, and these equations imply

$$V(x; K) = \widetilde{V}_{\nu_y}(y) + xy = \widetilde{V}(y) + xy \geq \inf_{\eta>0} \left[\widetilde{V}(\eta) + x\eta \right].$$

The reverse inequality is a consequence of (5.5).   □

Our next result provides sufficient conditions for requirement (5.3) to be satisfied.

Theorem 5.3 (Existence in the dual problem): *Suppose that (5.2) holds, and that the utility functions $U_1(t, \cdot)$, $U_2(\cdot)$ satisfy*

$$U_1(t, \infty) = \infty, \quad \inf_{0 \leq t \leq T} U_1(t, 0) > -\infty \quad \forall t \in (0, \infty), \qquad (5.11)$$

$$U_2(\infty) = \infty, \quad U_2(0) > -\infty, \qquad (5.12)$$

as well as (3.4.15), (3.6.17). Then (5.3) holds.

We devote the remainder of this section to the proof of Theorem 5.3. But first, let us combine it with Propositions 5.1 and 5.2 to derive the basic existence result for the primal, constrained Problem 2.6.

Theorem 5.4 (Existence in the primal problem): *Under the assumptions of Theorem 5.3 we have $V(x; K) < \infty$ for every $x \in (0, \infty)$, and there exists an optimal pair $(\hat{c}, \hat{p}) \in \mathcal{A}_3(x; K)$ for Problem 2.6. In other words, condition (A) preceding Theorem 4.1 holds.*

The conditions (5.11), (5.12) exclude logarithmic utility functions; for these, however, we have the direct arguments and explicit computations presented in Example 4.2, and thus need not appeal to general existence results.

In order to proceed with the proof of Theorem 5.3, let us fix $y > 0$ and extend the functional $\nu \mapsto \widetilde{V}_\nu(y)$ given by (3.27), (3.33) for $\nu(\cdot) \in \mathcal{D}$ to the entirety of the space $\mathcal{H}$ of Definition 5.5.1, by setting

$$\widetilde{V}_\nu(y) = E\left[\int_0^T \widetilde{U}_1(t, yH_\nu(t))\,dt + \widetilde{U}_2(yH_\nu(T))\right], \quad \nu(\cdot) \in \mathcal{H}. \qquad (5.13)$$

This definition is possible under the assumptions of Theorem 5.3 because

$$\widetilde{U}_1(t, y) = \sup_{x \geq 0}[U_1(t, x) - xy] \geq U_1(t, 0) \geq \inf_{0 \leq t \leq T} U_1(t, 0) > -\infty, \quad (5.14)$$

$$\widetilde{U}_2(y) = \sup_{x \geq 0}[U_2(x) - xy] \geq U_2(0) > -\infty; \qquad (5.15)$$

thus the expectation in (5.13) may be $+\infty$, but is well-defined.

Remark 5.5: Under the conditions of Theorem 5.3,

$$\widetilde{V}_\nu(y) = \infty, \quad \nu(\cdot) \in \mathcal{H} \setminus \mathcal{D}, \ y > 0. \qquad (5.16)$$

To see this, use Jensen's inequality and the convexity of $z \mapsto \widetilde{U}_2(ye^z)$ from (3.4.15'') to write

$$\widetilde{V}_\nu(y) \geq \kappa + E\widetilde{U}_2(yH_\nu(T))$$
$$\geq \kappa + \widetilde{U}_2(ye^{E\log H_\nu(T)})$$
$$\geq \kappa + \widetilde{U}_2\left(\frac{y}{s_0}e^{-E\int_0^T \|\theta_\nu(t)\|^2 dt - E\int_0^T \zeta(\nu(t))dt}\right)$$

for a suitable constant $\kappa > -\infty$. If $\nu(\cdot) \in \mathcal{H} \setminus \mathcal{D}$, then $E\int_0^T \zeta(\nu(t))\,dt = \infty$. Under the conditions of Theorem 5.3, $\widetilde{U}_2(0) = U_2(\infty) = \infty$, and $\nu(\cdot) \in \mathcal{H} \setminus \mathcal{D}$ implies (5.16).

Lemma 5.6: *Fix $y > 0$. Under the assumptions of Theorem 5.3, the functional $\nu \mapsto \widetilde{V}_\nu(y)$ of (5.13) is convex, coercive, that is,*

$$\widetilde{V}_\nu(y) \to \infty \quad \text{if } [\![\nu]\!] \to \infty, \qquad (5.17)$$

and lower semicontinuous; i.e.,

$$\widetilde{V}_\nu(y) \leq \underline{\lim}_{n\to\infty}\widetilde{V}_{\nu_n}(y) \quad \text{if } [\![\nu_n - \nu]\!] \to 0 \text{ as } n \to \infty.$$

PROOF. Note first that

$$\log H_\nu(t) = -A(t) - \int_0^t r(s)\,ds - \int_0^t \left(\theta(s) + \sigma^{-1}(s)\nu(s)\right)dW(s)$$
$$- \frac{1}{2}\int_0^t \|\theta(s) + \sigma^{-1}(s)\nu(s)\|^2\,ds - \int_0^t \zeta(\nu(s))\,ds$$

is a concave function of $\nu(\cdot) \in \mathcal{D}$. Let $\nu_1(\cdot), \nu_2(\cdot) \in \mathcal{D}$ and $\alpha \in (0, 1)$ be given. We have

$$\log H_{\alpha\nu_1 + (1-\alpha)\nu_2}(t) \geq \alpha \log H_{\nu_1}(t) + (1 - \alpha)\log H_{\nu_2}(t). \qquad (5.18)$$

The functions $z \mapsto \tilde{U}_1(t, ye^z)$ and $z \mapsto \tilde{U}_2(ye^z)$ are nonincreasing (Lemma 3.4.3) and convex (equation (3.4.15'')), and so the inequality (5.18) leads to

$$\tilde{U}_1(t, yH_{\alpha\nu_1+(1-\alpha)\nu_2}(t)) \leq \tilde{U}_1(t, y \exp(\alpha \log H_{\nu_1}(t) + (1-\alpha) \log H_{\nu_2}(t)))$$
$$\leq \alpha\tilde{U}_1(t, yH_{\nu_1}(t)) + (1-\alpha)\tilde{U}_1(t, yH_{\nu_2}(t)),$$

and similarly,

$$\tilde{U}_2(yH_{\alpha\nu_1+(1-\alpha)\nu_2}(T)) \leq \alpha\tilde{U}_2(yH_{\nu_1}(T)) + (1-\alpha)\tilde{U}_2(yH_{\nu_2}(T)).$$

These inequalities imply the convexity of $\nu \mapsto \tilde{V}_\nu(y)$.

For $\nu(\cdot) \in \mathcal{H}$, we have

$$\tilde{V}_\nu(y) \geq E \int_0^T \tilde{U}_1 \left(\frac{y}{s_0} \exp\left\{ -\log \frac{1}{Z_\nu(t)} \right\} \right) dt$$
$$+ E\tilde{U}_2 \left(\frac{y}{s_0} \exp\left\{ -\log \frac{1}{Z_\nu(T)} \right\} \right)$$
$$\geq \int_0^T \tilde{U}_1 \left(\frac{y}{s_0} \exp\left\{ -E\log \frac{1}{Z_\nu(t)} \right\} \right) dt$$
$$+ \tilde{U}_2 \left(\frac{y}{s_0} \exp\left\{ -E\log \frac{1}{Z_\nu(T)} \right\} \right)$$
$$\geq \int_0^T \tilde{U}_1 \left(\frac{y}{s_0} \exp\left\{ -\frac{1}{2}E \int_0^t \|\theta_\nu(s)\|^2 \, ds \right\} \right) dt$$
$$+ \tilde{U}_2 \left(\frac{y}{s_0} \exp\left\{ -\frac{1}{2}E \int_0^T \|\theta_\nu(s)\|^2 \, ds \right\} \right).$$

As $[\![\nu]\!] \to \infty$, condition (5.2.4) implies that $E \int_0^T \|\theta_\nu(s)\|^2 ds \to \infty$. But $\tilde{U}_2(0+) = U_2(\infty) = \infty$, and (5.17) follows.

Finally, let $\{\nu_n(\cdot)\}_{n=1}^\infty$ be a sequence in $\mathcal{H}$ converging to a limit $\nu(\cdot) \in \mathcal{H}$ in the norm of Definition 5.5.1. We may extract a subsequence $\{\nu_{n_k}(\cdot)\}_{k=1}^\infty$ for which

$$\lim_{k\to\infty} \tilde{V}_{\nu_{n_k}}(y) = \varliminf_{n\to\infty} \tilde{V}_{\nu_n}(y),$$

and we seek to show that

$$\tilde{V}_\nu(y) \leq \lim_{k\to\infty} \tilde{V}_{\nu_{n_k}}(y). \tag{5.19}$$

Consider the martingales

$$M_{\nu_n}(t) \triangleq \int_0^t \theta_{\nu_n}(s) \, dW(s), \; 0 \leq t \leq T.$$

According to the Burkholder–Davis–Gundy inequality (Karatzas and Shreve (1991), Theorem 3.3.28),

$$E\left[\sup_{0\le t\le T}|M_{\nu_n}(t)-M_\nu(t)|^2\right]\le \kappa E\int_0^T\|\theta_{\nu_n}(s)-\theta_{\nu(s)}\|^2\,ds$$

for some constant κ, and the right-hand side approaches zero as $n\to\infty$. Therefore, we may choose a further subsequence, also labeled $\{\nu_{n_k}(\cdot)\}_{k=1}^\infty$, along which we have

$$\lim_{k\to\infty}\nu_{n_k}(t)=\nu(t)\quad\text{for Lebesgue-a.e. }t\in[0,T],$$

$$\lim_{k\to\infty}\sup_{0\le t\le T}|Z_{\nu_{n_k}}(t)-Z_\nu(t)|=0$$

almost surely. These equalities and the lower semicontinuity of $\zeta(\cdot)$ imply

$$\overline{\lim}_{k\to\infty}H_{\nu_{n_k}}(t)\le H_\nu(t),\quad 0\le t\le T,$$

almost surely. Because $\widetilde U_1(t,\cdot)$ and $\widetilde U_2(\cdot)$ are nondecreasing and continuous on $(0,\infty)$, we have the almost sure inequalities

$$\underline{\lim}_{k\to\infty}\widetilde U_1(t,yH_{\nu_{n_k}}(t))\ge \widetilde U_1(t,yH_\nu(t)),\quad 0\le t\le T,$$

$$\underline{\lim}_{k\to\infty}\widetilde U_2(yH_{\nu_{n_k}}(T))\ge \widetilde U_2(yH_\nu(T)).$$

Moreover, (5.14), (5.15), and Fatou's lemma give us

$$\lim_{k\to\infty}\widetilde V_{\nu_{n_k}}(y)$$

$$=\lim_{k\to\infty}E\left[\int_0^T\widetilde U_1(t,yH_{\nu_{n_k}}(t))\,dt+\widetilde U_2(yH_{\nu_{n_k}}(T))\right]$$

$$\ge\underline{\lim}_{k\to\infty}E\int_0^T\widetilde U_1(t,yH_{\nu_{n_k}}(t))\,dt+\underline{\lim}_{k\to\infty}E\widetilde U_2(yH_{\nu_{n_k}}(T))$$

$$\ge E\int_0^T\underline{\lim}_{k\to\infty}\widetilde U_1(t,yH_{\nu_{n_k}}(t))\,dt+E\left[\underline{\lim}_{k\to\infty}\widetilde U_2(yH_{\nu_{n_k}}(T))\right]$$

$$\ge E\left[\int_0^T\widetilde U_1(t,yH_\nu(t))\,dt+\widetilde U_2(yH_\nu(T))\right]=\widetilde V_\nu(y).\qquad\square$$

PROOF OF THEOREM 5.3: Let $y>0$ be given. From Lemma 5.6 and Proposition 2.12 in Ekeland and Temam (1976), the convex, coercive, and lower-semicontinuous functional $\mathcal H\ni\nu\mapsto\widetilde V_\nu(y)\in(-\infty,\infty]$ attains its infimum on the Hilbert space $\mathcal H$:

$$\inf_{\nu\in\mathcal H}\widetilde V_\nu(y)=\widetilde V_{\hat\nu}(y)\quad\text{for some }\hat\nu(\cdot)\in\mathcal H.$$

According to Remark 5.5 and (5.2), $\widetilde V_{\hat\nu}(y)=\widetilde V(y)<\infty$, so $\hat\nu(\cdot)$ must be in $\mathcal D$. It remains to show that $\hat\nu(\cdot)$ actually belongs to the class $\mathcal D_0$ of (3.24).

The finiteness of $\widetilde{V}_{\hat{\nu}}(y)$ and the inequality (3.35) show that $V_{\hat{\nu}}(x) < \infty$ for all $x > 0$. It remains to check that

$$\mathcal{X}_{\hat{\nu}}(\eta) < \infty, \ \forall \eta > 0. \tag{5.20}$$

From the decrease of $\widetilde{U}$, (3.4.12), the decrease of I, and (3.6.17) (in particular, (3.4.16″)), we have

$$\widetilde{U}(\eta) - \widetilde{U}(\infty) \geq \widetilde{U}(\eta) - \widetilde{U}\left(\frac{\eta}{\beta}\right) = \int_{\eta}^{\frac{\eta}{\beta}} I(u)\, du$$

$$\geq \left(\frac{\eta}{\beta} - \eta\right) I\left(\frac{\eta}{\beta}\right)$$

$$\geq \frac{1 - \beta}{\beta\gamma} \eta I(\eta), \ 0 < \eta < \infty,$$

where $\beta \in (0,1)$ and $\gamma \in (1,\infty)$ are as in (3.4.16″), and $\widetilde{U}(\cdot)$ stands generically for $\widetilde{U}(t,\cdot)$ and $\widetilde{U}_2(\cdot)$. Consequently,

$$y\mathcal{X}_{\hat{\nu}}(y) = E\left[\int_0^T yH_{\hat{\nu}}(t)I_1(t, yH_{\hat{\nu}}(t))\, dt + yH_{\hat{\nu}}(T)I_2(yH_{\hat{\nu}}(T))\right]$$

$$\leq \frac{\beta\gamma}{1-\beta} E\left[\int_0^T \widetilde{U}_1(t, yH_{\hat{\nu}}(t))\, dt + \widetilde{U}_2(yH_{\hat{\nu}}(T))\right.$$

$$\left. - \left(\int_0^T \widetilde{U}_1(t,\infty) + \widetilde{U}_2(\infty)\right)\right]$$

$$\leq \frac{\beta\gamma}{1-\beta}\left[\widetilde{V}_{\hat{\nu}}(y) - \left(\int_0^T U_1(t,0)\, dt + U_2(0)\right)\right] < \infty,$$

where the last inequality is a consequence of (5.14) and (5.15). This proves (5.20) for $\eta = y$ (but not for all $\eta > 0$, because $\hat{\nu}$ depends on y). Assumption (3.6.17″) (see the sentence following that assumption) now implies that (5.20) holds for all $\eta > 0$.   □

Remark 5.7: It can be checked rather easily that utility functions of the form

$$U_1(t,x) = \frac{1}{\beta} e^{-\alpha t} x^\beta, \ U_2(x) = \frac{1}{\beta} e^{-\alpha T} x^\beta, \ 0 \leq t \leq T, \ x > 0,$$

where $\alpha \geq 0$ and $0 < \beta < 1$, satisfy all the conditions of Theorems 5.3 and 5.4.

Remark 5.8: The theory of Sections 2–5 goes through without change if one sets formally $U_1 \equiv 0$ and admits only the identically zero consumption process throughout. This leads to the problem of *maximizing the expected utility from terminal wealth only*. On the other hand, the condition $U_2(\infty) = \infty$ was used extensively in this section. The situation with regard to existence of optimal solutions when $U_2 \equiv 0$ is not well understood.

6.6   Deterministic Coefficients, Cone Constraints

In this section, we consider Problem 2.6 and its dual under the assumption that $r(\cdot)$, $\theta(\cdot)$, and $\sigma(\cdot)$ are *continuous, deterministic functions*; $A(\cdot) \equiv 0$; and the *constraint set K is a cone*. The analysis proceeds through a study of the Hamilton–Jacobi–Bellman (HJB) equations, and for this we use the conditions placed on utility functions in Section 3.8 rather than conditions (3.4.15), (3.6.17) used in the previous section of this chapter. In particular, all functions $U^{(\beta)}(x)$ given by (3.4.4), (3.4.5) are included, not just those for which $0 < \beta < 1$. (We use β for the exponent in place of p throughout this chapter, reserving the symbol $p(\cdot)$ for portfolio-proportion processes.)

More specifically, we assume in this section that Assumptions 3.8.1, 3.8.2 hold, and, in addition, the following condition is satisfied.

Assumption 6.1:   *The process $\sigma(\cdot)$ is nonrandom, $\sigma(t)$ is nonsingular for every $t \in [0,T]$, and $\sigma(\cdot)$ is Hölder-continuous, i.e., for some $\kappa > 0$ and $\rho \in (0,1)$, we have*

$$\|\sigma(t_1) - \sigma(t_2)\| \leq \kappa |t_1 - t_2|^\rho, \ \forall \ t_1, t_2 \in [0,T].$$

In particular, $\sigma(\cdot)$ and $\sigma^{-1}(\cdot)$ are both bounded, and $\sigma^{-1}(\cdot)$ is also Hölder-continuous.

Assumption 6.2:   *The constraint set K is a nonempty closed convex cone.*

From Assumption 6.2 we see that $\widetilde{K}$ is the polar cone of $-K$, which is necessarily closed. On $\widetilde{K}$, the support function $\zeta(\cdot)$ is identically zero (Rockafellar (1970), Theorem 14.1). From Assumption 3.8.1 we have the validity of (2.4) and (2.5). Conditions (2.13) and (2.14) are assumed throughout this chapter, including in the present section.

Our aim is to study the time-dependent generalization of Problem 2.6 via the time-dependent generalization of Problem 3.2. Toward that end, for $0 \leq t \leq T$, $x \geq 0$ and with $(c(\cdot), p(\cdot))$ a consumption and portfolio-proportion process pair, we define the corresponding wealth process $X^{t,x,c,p}(s)$, $t \leq s \leq T$, by

$$\frac{X^{t,x,c,p}(s)}{S_0(s)} + \int_t^s \frac{c(u)du}{S_0(u)} = x + \int_t^s \frac{X^{t,x,c,p}(u)}{S_0(u)} p'(u)\sigma(u)\, dW_0(u)$$

(cf. (2.7)). The process $X^{t,x,c,p}(\cdot)$ reaches zero no later than the stopping time

$$\tau_{t,p} \triangleq \inf \left\{ s \in [t,T]; \int_t^s \|\sigma'(u)p(u)\|^2 \, du = \infty \right\}$$

(cf.(2.9)). We set

$$\tau_{t,0} \triangleq \inf \left\{ s \in [t,T]; X^{t,x,c,p}(s) = 0 \right\},$$

and define $\mathcal{A}(t, x; K)$ to be the set of all consumption and portfolio-process pairs $(c(\cdot), p(\cdot))$ that satisfy the almost sure conditions

$$c(s) = 0, \quad \text{Lebesgue-a.e.} \quad s \in [\tau_{t,0}, T],$$
$$p(s) \in K, \quad \text{Lebesgue-a.e.} \quad s \in [t, T].$$

Following (2.15), (2.16), we set

$$\mathcal{A}_3(t, x; K) \triangleq \left\{ (c, p) \in \mathcal{A}_3(t, x; K); \quad E \int_t^T \min[0, U_1(s, c(s))] ds > -\infty, \right.$$

$$\left. E\left(\min[0, U_2(X^{t,x,c,p}(T))]\right) > -\infty \right\}$$

and define the *time-dependent value function*

$$V(t, x; K) \triangleq \sup_{(c,p) \in \mathcal{A}(t,x;K)} E\left[\int_t^T U_1(s, c(s)) ds + U_2(X^{t,x,c,p}(T)) \right]. \quad (6.1)$$

We show below that the function $V(\cdot, \cdot; K) : [0, T] \times (0, \infty) \to \mathbb{R}$ of (6.1) is continuous and satisfies the *constrained Hamilton–Jacobi–Bellman* equation (see Theorem 3.8.11 for the unconstrained case)

$$V_t(t, x; K) + \max_{\substack{0 \le c < \infty \\ p \in K}} \left[\frac{1}{2} \|\sigma'(t)p\|^2 x^2 V_{xx}(t, x; K) \right.$$

$$\left. + (r(t)x - c + xp'\sigma(t)\theta(t)) V_x(t, x; K) + U_1(t, c) \right] = 0 \quad (6.2)$$

for $0 \le t \le T$, $x > 0$. By definition, $V(\cdot, \cdot; K)$ satisfies the boundary conditions

$$V(t, x; K) = U_2(x), \quad x > 0, \tag{6.3}$$

$$V(t, 0; K) = \int_t^T U_1(t, 0) \, dt + U_2(0), \quad 0 \le t \le T. \tag{6.4}$$

We approach the constrained HJB equation through the dual problem. For $\nu(\cdot) \in \mathcal{D}$, $t \in [0, T]$, and $x \ge 0$, we determine $X_\nu^{t,x,c,p}(\cdot)$ by the equation (cf. (3.1))

$$\frac{X_\nu^{t,x,c,p}(s)}{S_0^{(\nu)}(s)} + \int_t^s \frac{c(u)du}{S_0^{(\nu)}(u)} = x + \int_t^s \frac{X_\nu^{t,x,c,p}(u)}{S_0^{(\nu)}(u)} p'(u)\sigma(u)dW_\nu(u),$$

$$t \le s \le T,$$

or its equivalent variations similar to (3.2) and (3.3). We define $\mathcal{A}_\nu(t, x)$ to be the set of all consumption and portfolio-process pairs (c, p) for which

$$c(s) = 0, \quad \text{Lebesgue-a.e. } s \in [\tau_{t,\nu}, T]$$

almost surely, where

$$\tau_{t,\nu} = \inf \left\{ s \in [t,T]; X_\nu^{t,x,c,P}(s) = 0 \right\}.$$

Following (3.5), (3.6), we set

$$\mathcal{A}_3^{(\nu)}(t,x) \triangleq \Bigg\{ (c,p) \in \mathcal{A}_\nu(t,x); \quad E \int_t^T \min[0, U_1(s,c(s))]ds > -\infty,$$

$$E\left(\min[0, U_2(X_\nu^{t,x,c,P}(T))]\right) > -\infty \Bigg\},$$

$$V_\nu(t,x) \triangleq \sup_{(c,p)\in\mathcal{A}_3^{(\nu)}(t,x)} E\left[\int_t^T U_1(s,c(s))\,ds + U_2(X_\nu^{t,x,c,P}(T))\right]. \quad (6.5)$$

For $\nu(\cdot) \in \mathcal{D}$ and $t \in [0,T]$, we introduce the processes with time parameter $s \in [t,T]$,

$$Z_\nu(t,s) \triangleq \frac{Z_\nu(s)}{Z_\nu(t)} = \exp\left\{-\int_t^s \theta_\nu'(u)\,dW(u) - \frac{1}{2}\int_t^s \|\theta_\nu(u)\|^2\,du\right\},$$

$$H_\nu(t,s) \triangleq \frac{H_\nu(s)}{H_\nu(t)} = \exp\left\{-\int_t^s r(u)\right\} Z_\nu(t,s).$$

The nonnegative function

$$\mathcal{X}_\nu(t,y) \triangleq E\left[\int_t^T H_\nu(t,s)I_1(s,yH_\nu(t,s))ds + H_\nu(t,T)I_2(yH_\nu(t,T))\right],$$

$$0 \le t \le T, \ y > 0, \quad (6.6)$$

is finite if $\nu(\cdot)$ is in the set $\mathcal{D}_0$ of (3.24). In this case, $\mathcal{X}_\nu(t,\cdot)$ is a strictly decreasing function mapping $(0,\infty)$ onto $(0,\infty)$ and has a strictly decreasing inverse $\mathcal{Y}_\nu(t,\cdot)$, likewise mapping $(0,\infty)$ onto $(0,\infty)$.

For $\nu(\cdot) \in \mathcal{D}_0$ we have, just as in Remark 3.6, that

$$V_\nu(t,x) = G_\nu(t,\mathcal{Y}_\nu(t,x)), \quad 0 \le t \le T, \ x > 0, \quad (6.7)$$

where

$$G_\nu(t,y) \triangleq E\left[\int_t^T U_1(I_1(s,yH_\nu(t,s)))\,ds + U_2(I_2(yH_\nu(t,T)))\right]. \quad (6.8)$$

Furthermore,

$$\widetilde{V}_\nu(t,y) \triangleq \sup_{x>0}[V_\nu(t,x) - xy]$$

$$= G_\nu(t,y) - y\mathcal{X}_\nu(t,y)$$

$$= E\left[\int_t^T \widetilde{U}_1(s,yH_\nu(t,s))ds + \widetilde{U}_2(yH_\nu(t,T))\right]$$

$$< \infty, \quad 0 \le t \le T, \ y > 0 \quad (6.9)$$

(cf. (3.27)), and for $\nu(\cdot) \in \mathcal{D} \setminus \mathcal{D}_0$, we follow (3.33) and define

$$\widetilde{V}_\nu(t, y) \triangleq E\left[\int_t^T \widetilde{U}_1(s, yH_\nu(t, s))\, ds + \widetilde{U}_2(yH_\nu(t, T))\right],$$
$$y > 0, \ \nu(\cdot) \in \mathcal{D} \setminus \mathcal{D}_0. \quad (6.10)$$

Proposition 3.7 can be adapted to the time-dependent case to ensure that

$$E\int_t^T \min[0, \widetilde{U}_1(s, yH_\nu(t, s))]\, ds > -\infty, \ E\left(\min[0, \widetilde{U}_2(yH_\nu(t, T))]\right) > -\infty,$$

and hence

$$E\left[\int_t^T |\widetilde{U}_1(s, yH_\nu(t, s))|\, ds + |\widetilde{U}_2(yH_\nu(t, T))|\right] < \infty \quad (6.11)$$

if $\widetilde{V}_\nu(t, y) < \infty$. Also, as in (3.35), $\widetilde{V}_\nu(t, y) < \infty$ implies

$$\sup_{x>0}[V_\nu(t, x) - xy] = \widetilde{V}_\nu(y). \quad (6.12)$$

From the theory of Section 6.5—in particular, from (5.4)—we expect

$$\widetilde{V}(t, y) \triangleq \inf_{\nu \in \mathcal{D}} \widetilde{V}_\nu(t, y) \quad (6.13)$$

to be the convex dual of the function $V(t, \cdot; K)$ in (6.1), i.e.,

$$\widetilde{V}(t, y) = \sup_{\xi>0}[V(t, \xi; K) - \xi y], \quad 0 \le t \le T, \ y > 0. \quad (6.14)$$

From Theorem 3.8.12, we also expect $\widetilde{V}$ to satisfy a *linear* partial differential equation, which turns out to be (cf. (3.8.44))

$$\widetilde{V}_t(t, y) + \frac{1}{2}\min_{\nu \in K}\|\theta(t) + \sigma^{-1}(t)\nu\|^2 y^2 \widetilde{V}_{yy}(t, y) - r(t)y\widetilde{V}_y(t, y) + \widetilde{U}_1(t, y)$$
$$= 0, \quad 0 \le t \le T, \ y > 0, \quad (6.15)$$

as well as the terminal condition

$$\widetilde{V}(T, y) = \widetilde{U}_2(y), \ y > 0. \quad (6.16)$$

Our program then is to construct $\widetilde{V}$ via (6.15), (6.16) and in the process to obtain the minimizer $\hat{\nu}(\cdot)$ in (6.13). Then, just as in Theorem 4.1, we discover that when $\nu(\cdot) = \hat{\nu}(\cdot)$ in (6.5), the maximizing $(c, p) \in \mathcal{A}_3^{(\hat{\nu})}(t, x)$ is optimal in (6.1). Finally, we prove (6.14), (6.2) and obtain the optimal (c, p) for Problem 2.6 in feedback form.

We begin this program with a study of the minimization appearing in (6.15). Given $\epsilon > 0$, let $L_\epsilon(\mathbb{R}^N; \mathbb{R}^N)$ denote the set of $N \times N$ nonsingular matrices σ whose operator norm

$$\|\sigma\| \triangleq \sup_{x \in \mathbb{R}^N, \|x\|=1} \|\sigma x\|$$

satisfies $\|\sigma\| \geq \epsilon$. Then $\|\sigma^{-1}\| \leq \frac{1}{\epsilon}$ for every $\sigma \in L_\epsilon(\mathbb{R}^N; \mathbb{R}^N)$. Recall from (5.2.4) that we have assumed $\sigma(t) \in L_\epsilon(\mathbb{R}^N; \mathbb{R}^N)$ for all $0 \leq t \leq T$ for sufficiently small $\epsilon > 0$. Define the function

$$h(\theta, \sigma) \triangleq \inf_{\nu \in \widetilde{K}} \|\theta + \sigma^{-1}\nu\|$$

mapping $\mathbb{R}^N \times L_\epsilon(\mathbb{R}^N; \mathbb{R}^N)$ to $[0, \infty)$. Because the mapping $\nu \mapsto \|\theta + \sigma^{-1}\nu\|$ is strictly convex and $\lim_{\|\nu\| \to \infty} \|\theta + \sigma^{-1}\nu\| = \infty$ for each fixed $(\theta, \sigma) \in \mathbb{R}^N \times L_\epsilon(\mathbb{R}^N; \mathbb{R}^N)$, there is a unique minimizer $\nu = \Upsilon(\theta, \sigma)$ in $\widetilde{K}$; i.e.,

$$h(\theta, \sigma) = \|\theta + \sigma^{-1}\Upsilon(\theta, \sigma)\| \quad \text{for } \theta \in \mathbb{R}^N, \ \sigma \in L_\epsilon(\theta, \sigma).$$

Lemma 6.3:   *The function $h: \mathbb{R}^N \times L_\epsilon(\mathbb{R}^N; \mathbb{R}^N) \to [0, \infty)$ is locally Lipschitz continuous.*

PROOF.   For $\theta_1, \theta_2 \in \mathbb{R}^N$ and $\sigma_1, \sigma_2 \in L_\epsilon(\mathbb{R}^N; \mathbb{R}^N)$, we have

$$\begin{aligned}
h(\theta_1, \sigma_1) - h(\theta_2, \sigma_2) &\leq \|\theta_1 + \sigma_1^{-1}\Upsilon(\theta_2, \sigma_2)\| - \|\theta_2 + \sigma_2^{-1}\Upsilon(\theta_2, \sigma_2)\| \\
&\leq \|\theta_1 - \theta_2 + (\sigma_1^{-1} - \sigma_2^{-1})\Upsilon(\theta_2, \sigma_2)\| \\
&\leq \|\theta_1 - \theta_2\| + \|\sigma_1^{-1} - \sigma_2^{-1}\|\|\Upsilon(\theta_2, \sigma_2)\| \\
&\leq \|\theta_1 - \theta_2\| + \|\sigma_1^{-1}\|\|\sigma_2^{-1}\|\|\sigma_1 - \sigma_2\|\|\Upsilon(\theta_2, \sigma_2)\| \\
&\leq \|\theta_1 - \theta_2\| + \frac{1}{\epsilon^2}\|\Upsilon(\theta_2, \sigma_2)\|\|\sigma_1 - \sigma_2\|.
\end{aligned}$$

But

$$\begin{aligned}
\|\Upsilon(\theta, \sigma)\| &\leq \|\sigma\theta + \Upsilon(\theta, \sigma)\| + \|-\sigma\theta\| \\
&\leq \|\sigma\|\|\theta + \sigma^{-1}\Upsilon(\theta, \sigma)\| + \|\sigma\|\|\theta\| \\
&= \|\sigma\|(h(\theta, \sigma) + \|\theta\|) \\
&\leq 2\|\sigma\|\|\theta\|.
\end{aligned}$$

It follows that

$$h(\theta_1, \sigma_1) - h(\theta_2, \sigma_2) \leq \|\theta_1 - \theta_2\| + \frac{2}{\epsilon^2}\|\sigma_2\|\|\theta_2\|\|\sigma_1 - \sigma_2\|,$$

and reversing the roles of θ_1, σ_1 with θ_2, σ_2, we obtain

$$\begin{aligned}
|h(\theta_1, \sigma_1) - h(\theta_2, \sigma_2)| &\leq \|\theta_1 - \theta_2\| \\
&\quad + \frac{2}{\epsilon^2}\max\{\|\sigma_1\|\|\theta_1\|, \|\sigma_2\|\|\theta_2\|\} \cdot \|\sigma_1 - \sigma_2\|. \qquad \square
\end{aligned}$$

In light of Lemma 6.3, we may rewrite (6.15) as

$$\widetilde{V}_t(t, y) + \frac{1}{2}\|\theta_{\hat{\nu}}(t)\|^2 y^2 \widetilde{V}_{yy}(t, y) - r(t)y\widetilde{V}_y(t, y) + \widetilde{U}_1(t, y) = 0,$$
$$0 \leq t \leq T, \ y > 0,$$

where

$$\hat{\nu}(t) = \Upsilon(\theta(t), \sigma(t)), \quad \theta_{\hat{\nu}}(t) = \theta(t) + \sigma^{-1}(t)\hat{\nu}(t). \tag{6.17}$$

Assumptions 3.8.1 and 6.1 guarantee that $\|\theta_{\hat{\nu}}(\cdot)\| = h(\theta(\cdot), \sigma(\cdot))$ is a non-random, Hölder-continuous function, and hence all the results of Section 3.8 apply to the market $\mathcal{M}_{\hat{\nu}}$, as we elaborate below.

Lemmas 3.8.4 and 3.8.10, coupled with Theorem 3.8.11, show that $\hat{\nu}(\cdot) \in \mathcal{D}_0$. According to Theorem 3.8.11, $V_{\hat{\nu}}(t, x)$ is of class $C^{1,2}$ on $[0, T) \times (0, \infty)$, continuous on $[0, T] \times (0, \infty)$, satisfies the boundary conditions

$$V_{\hat{\nu}}(T, x) = U_2(x), \quad x > 0,$$

$$V_{\hat{\nu}}(t, 0) = \int_t^T U_1(t, 0)\, dt + U_2(0), \quad 0 \le t \le T,$$

and solves the HJB equation

$$\frac{\partial}{\partial t} V_{\hat{\nu}}(t, x) + \max_{\substack{0 \le c < \infty \\ p \in \mathbb{R}^N}} \left[\frac{1}{2} \|\sigma'(t)p\|^2 x^2 \frac{\partial^2}{\partial x^2} V_{\hat{\nu}}(t, x) \right. \tag{6.18}$$

$$\left. + (r(t)x - c + xp'\sigma(t)\theta_{\hat{\nu}}(t)) \frac{\partial}{\partial x} V_{\hat{\nu}}(t, x) + U_1(t, c) \right] = 0, \ 0 \le t < T, x > 0,$$

for the optimization problem without constraints on $p(\cdot)$. Theorem 3.8.12 implies that for all $t \in [0, T]$,

$$\frac{\partial}{\partial x} V_{\hat{\nu}}(t, x) = \mathcal{Y}_{\hat{\nu}}(t, x), \quad x > 0, \tag{6.19}$$

$$\widetilde{V}_{\hat{\nu}}(t, y) = G_{\hat{\nu}}(t, y) - y\mathcal{X}_{\hat{\nu}}(t, y) \tag{6.20}$$

$$= E\left[\int_t^T \widetilde{U}_1(s, yH_{\hat{\nu}}(t, s))\, ds + \widetilde{U}_2(yH_{\hat{\nu}}(t, T)) \right], \quad y > 0,$$

$$\frac{\partial}{\partial y} \widetilde{V}_{\hat{\nu}}(t, y) = -\mathcal{X}_{\hat{\nu}}(t, y), \quad y > 0. \tag{6.21}$$

Moreover, the convex dual function $\widetilde{V}_{\hat{\nu}}$ in (6.14) is of class $C^{1,2}$ on $[0, T) \times (0, \infty)$, continuous on $[0, T] \times (0, \infty)$, satisfies the boundary conditions

$$\widetilde{V}_{\hat{\nu}}(T, y) = \widetilde{U}_2(y), \quad y > 0, \tag{6.22}$$

and solves the linear, second-order equation

$$\frac{\partial}{\partial t} \widetilde{V}_{\hat{\nu}}(t, y) + \frac{1}{2} \|\theta_{\hat{\nu}}(t)\|^2 y^2 \frac{\partial^2}{\partial y^2} \widetilde{V}_{\hat{\nu}}(t, y) - r(t)y \frac{\partial}{\partial y} \widetilde{V}_{\hat{\nu}}(t, y) + \widetilde{U}_1(t, y)$$

$$= 0, \quad 0 \le t < T, \ y > 0. \tag{6.23}$$

Finally, every solution to (6.22), (6.23) satisfying the growth condition (3.8.21) must agree with $\widetilde{V}_{\hat{\nu}}$.

Theorem 6.4:   *Under Assumptions 3.8.1, 3.8.2, 6.1, and 6.2, and with*
$\hat{\nu}(\cdot)$ *given by (6.17), we have*

$$V(t, x; K) = \tilde{V}_{\hat{\nu}}(t, x), \quad 0 \le t \le T, \ x > 0, \tag{6.24}$$

$$\tilde{V}(t, y) = \tilde{V}_{\hat{\nu}}(t, y), \quad 0 \le t \le T, \ y > 0. \tag{6.25}$$

*The function $V(\cdot, \cdot; K)$ satisfies the constrained HJB equation (6.2), and
$\tilde{V}(\cdot, \cdot)$ satisfies the linear, second-order equation (6.15). In terms of the
"feedback functions"*

$$C(t, x) \triangleq I_1(t, \mathcal{Y}_{\hat{\nu}}(t, x)), \tag{6.26}$$

$$P(t, x) \triangleq -(\sigma'(t))^{-1}\theta_{\hat{\nu}}(t)\frac{\mathcal{Y}_{\hat{\nu}}(t, x)}{x\frac{\partial}{\partial x}\mathcal{Y}_{\hat{\nu}}(t, x)}, \quad 0 \le t \le T, \ x > 0, \tag{6.27}$$

*the optimal consumption and portfolio-proportion processes for Problem 2.6
are given in "feedback form" as*

$$\hat{c}(t) = C(t, \widehat{X}(t)), \quad \hat{p}(t) = P(t, \widehat{X}(t)), \quad 0 \le t \le T, \tag{6.28}$$

with

$$\widehat{X}(t) \triangleq X_{\hat{\nu}}(t) = \mathcal{X}_{\hat{\nu}}(t, \mathcal{Y}_{\hat{\nu}}(0, x)H_{\hat{\nu}}(t)) = X^{x, \hat{c}, \hat{p}}(t), \quad 0 \le t \le T, \tag{6.29}$$

as in (3.18), (6.6), and (2.7).

The proof of Theorem 6.4 requires the following lemma.

Lemma 6.5:   *Let $M(t), 0 \le t \le T$, be a positive, continuous, local martingale, and let $\varphi: (0, \infty) \to \mathbb{R}$ be a convex nonincreasing function satisfying
$E|\varphi(M(t))| < \infty$ for every $t \in [0, T]$. Then $\varphi(M(t)), 0 \le t \le T$, is a
submartingale.*

PROOF.   Let $\{\tau_n\}_{n=1}^{\infty}$ be a nondecreasing sequence of stopping times with
$\tau_n \uparrow T$ almost surely and such that $M(t \wedge \tau_n), 0 \le t \le T$, is a martingale
for every n. For $\epsilon > 0$ and $\kappa > 0$, we introduce the function

$$\varphi_{\epsilon, \kappa}(x) = \begin{cases} \varphi(\epsilon) + (x - \epsilon)D^+\varphi(\epsilon), & 0 < x < \epsilon, \\ \max\{\varphi(x), \varphi(\epsilon + \kappa)\}, & x \ge \epsilon, \end{cases} \tag{6.30}$$

where $D^+\varphi$ denotes the right-hand derivative of φ. Note that $\varphi_{\epsilon, \kappa}$ is
bounded, convex, and nonincreasing. Jensen's inequality implies that
$\varphi_{\epsilon, \kappa}(M(t \wedge \tau_n)), 0 \le t \le T$, is a bounded submartingale. In other words,
for $0 \le s \le t \le T$ and $A \in \mathcal{F}(s)$,

$$\int_A \varphi_{\epsilon, \kappa}(M(s \wedge \tau_n)) \, dP \le \int_A \varphi_{\epsilon, \kappa}(M(t \wedge \tau_n)) \, dP.$$

Letting $n \to \infty$, we obtain

$$\int_A \varphi_{\epsilon, \kappa}(M(s)) \, dP \le \int_A \varphi_{\epsilon, \kappa}(M(t)) \, dP$$

by the bounded convergence theorem. Letting first $\epsilon \downarrow 0$ and then $\kappa \to \infty$, we obtain from two applications of the monotone convergence theorem that

$$\int_A \varphi(M(s)) \, dP \leq \int_A \varphi(M(t)) \, dP. \qquad \square$$

PROOF OF THEOREM 6.4: We first show (6.25), i.e., for every $\nu(\cdot) \in \mathcal{D}$, $y > 0$, and $0 \leq t \leq T$,

$$V_{\hat{\nu}}(t, y) = E\left[\int_t^T \tilde{U}_1(s, yH_{\hat{\nu}}(t, s)) \, ds + \tilde{U}_2(yH_{\hat{\nu}}(t, T))\right]$$

$$\leq E\left[\int_t^T \tilde{U}(s, yH_\nu(t, s)) \, ds + \tilde{U}_2(yH_\nu(t, T))\right]$$

$$= V_\nu(t, y). \qquad (6.31)$$

Of course, if $V_\nu(t, y) = \infty$, this inequality is trivially true. If $V_\nu(t, y) < \infty$, we have (6.11). We will show below that

$$E\varphi(Z_{\hat{\nu}}(t, s)) \leq E\varphi(Z_\nu(t, s)), \quad 0 \leq t \leq s \leq T, \qquad (6.32)$$

whenever $\varphi: (0, \infty) \to \mathbb{R}$ is a convex nonincreasing function satisfying

$$E|\varphi(Z_{\hat{\nu}}(t, s))| < \infty, \quad E|\varphi(Z_\nu(t, s))| < \infty. \qquad (6.33)$$

Taking first

$$\varphi(z) = \tilde{U}_1\left(s, yz \exp\left\{-\int_t^s r(u) \, du\right\}\right),$$

and then $s = T$ and

$$\varphi(z) = \tilde{U}_2\left(yz \exp\left\{-\int_t^T r(u) \, du\right\}\right),$$

we will obtain (6.31) from (6.32).

To simplify notation we set $t = 0$ in (6.32), i.e., we prove only

$$E\varphi(Z_{\hat{\nu}}(t)) \leq E\varphi(Z_\nu(t)), \quad 0 \leq t \leq T. \qquad (6.34)$$

Recall from (5.5.10) that

$$Z_{\hat{\nu}}(t) = \exp\left\{-\int_0^t \theta'_{\hat{\nu}}(u) \, dW(u) - \frac{1}{2}\int_0^t \|\theta_{\hat{\nu}}(u)\|^2 \, du\right\},$$

$$Z_\nu(t) = \exp\left\{-\int_0^t \theta'_\nu(u) \, dW(u) - \frac{1}{2}\int_0^t \|\theta_\nu(u)\|^2 \, du\right\}.$$

We assume initially that for some $\kappa > 0$, we have

$$\|\theta_{\hat{\nu}}(t)\| \geq \kappa > 0, \quad 0 \leq t \leq T, \qquad (6.35)$$

and define

$$\Lambda_{\hat{\nu}}(t) = \int_0^t \|\theta_{\hat{\nu}}(u)\|^2 \, du, \; \Lambda_\nu(t) = \int_0^t \|\theta_\nu(u)\|^2 \, du. \tag{6.36}$$

Because $\|\theta_{\hat{\nu}}(u)\| \leq \|\theta_\nu(u)\|$, $0 \leq u \leq T$, almost surely, we have $\Lambda_{\hat{\nu}}(t) \leq \Lambda_\nu(t)$, $0 \leq t \leq T$, almost surely. We extend $\theta_{\hat{\nu}}(\cdot)$ and $\theta_\nu(\cdot)$ to (T, ∞) by setting $\theta_\nu(t) = \theta_{\hat{\nu}}(t) = \theta_{\hat{\nu}}(T)$, for $t > T$, and subsequently extend $\Lambda_{\hat{\nu}}(\cdot)$, $\Lambda_\nu(\cdot)$ to (T, ∞) via (6.36).

We next define the nondecreasing, progressively measurable process

$$A(t) \overset{\Delta}{=} \inf\{u \geq 0; \Lambda_{\hat{\nu}}(u) = \Lambda_\nu(t)\} \geq t$$

(recall here that $\Lambda_{\hat{\nu}}(\cdot)$ is not random), so that

$$\Lambda_{\hat{\nu}}(A(t)) = \Lambda_\nu(t), \; t \geq 0.$$

But $\Lambda_{\hat{\nu}}(\cdot)$ is continuously differentiable with derivative bounded from below by $\kappa^2 > 0$, so we may invert the above equation to conclude that $A(\cdot)$ is continuously differentiable. Indeed,

$$A'(t) = \frac{\|\theta_\nu(t)\|^2}{\|\theta_{\hat{\nu}}(A(t))\|^2}, \; t \geq 0.$$

Using the fact that $\theta_{\hat{\nu}}(\cdot)$ is deterministic, we may construct a progressively measurable process $\mathcal{O}(\cdot)$ with values in the set of $N \times N$ orthonormal matrices, and such that

$$\sqrt{A'(t)} \, \mathcal{O}(t)\theta_{\hat{\nu}}(A(t)) = \theta_\nu(t), \; t \geq 0$$

holds almost surely. We define the N-dimensional vector of martingales

$$M(t) = (M_1(t), \ldots, M_N(t))' = \int_0^t \sqrt{A'(u)} \, \mathcal{O}'(u) \, dW(u),$$

and note that $\langle M_i, M_j \rangle(t) = \delta_{ij} A(t)$, $t \geq 0$, almost surely. Thus, there exists an N-dimensional Brownian motion $B(\cdot)$ for which $M(t) = B(A(t))$, $t \geq 0$ (see Karatzas and Shreve (1991), Theorem 3.4.6, for the one-dimensional case, which can be easily extended using Lévy's characterization of multidimensional Brownian motion (*ibid.* (1991), Theorem 3.3.16); alternatively, this is a special case of Knight's theorem, (*ibid.* (1991), Theorem 3.4.13)). Changing variables in the formula for $Z_\nu(\cdot)$ (*ibid.* (1991), Proposition 3.4.8), we have

$$Z_\nu(A^{-1}(t)) = \exp\left\{-\int_0^{A^{-1}(t)} \theta_\nu'(u) \, dW(u) - \frac{1}{2} \int_0^{A^{-1}(t)} \|\theta_\nu(u)\|^2 \, du\right\}$$

$$= \exp\left\{-\int_0^{A^{-1}(t)} \sqrt{A'(u)} \, \theta_{\hat{\nu}}'(A(u))\mathcal{O}'(u) \, dW(u)\right.$$

$$-\frac{1}{2}\int_0^{A^{-1}(t)} \|\theta_{\hat{\nu}}(A(u))\|^2 A'(u)\, du \Bigg\}$$

$$= \exp\left\{-\int_0^{A^{-1}(t)} \theta'_{\hat{\nu}}(A(u))\, dM(u) - \frac{1}{2}\int_0^t \|\theta_{\hat{\nu}}(s)\|^2\, ds\right\}$$

$$= \exp\left\{-\int_0^t \theta'_{\hat{\nu}}(s)\, dB(s) - \frac{1}{2}\int_0^t \|\theta_{\hat{\nu}}(s)\|^2\, ds\right\}.$$

This shows that $Z_\nu(A^{-1}(t))$ has the same distribution as $Z_{\hat{\nu}}(t)$, and thus

$$E\varphi(Z_{\hat{\nu}}(t)) = E\varphi(Z_\nu(A^{-1}(t))).$$

Lemma 6.5 implies that $\varphi(Z_\nu(\cdot))$ is a submartingale. Because $A^{-1}(t)$ is a stopping time, and $A^{-1}(t) \leq t$ almost surely, the optional sampling theorem gives

$$E\varphi(Z_\nu(A^{-1}(t))) \leq E\varphi(Z_\nu(t)).$$

We have established (6.34) under the assumption (6.35).

Indeed, we have proved that whenever $\hat{\nu}(\cdot)$ is nonrandom and satisfies (6.35), $\nu(\cdot) \in \mathcal{D}$, and $\|\theta_{\hat{\nu}}(t)\| \leq \|\theta_\nu(t)\|$, $0 \leq t \leq T$, almost surely, then (6.34) holds. If $\hat{\nu}(\cdot)$ does not satisfy (6.35), we may construct sequences $\{\hat{\nu}_n(\cdot)\}_{n=1}^\infty$, $\{\nu_n(\cdot)\}_{n=1}^\infty$ such that for each n, $\hat{\nu}_n(\cdot)$ is nonrandom, $\|\theta_{\hat{\nu}_n}(t)\| \geq \frac{1}{n}$, $\|\theta_{\hat{\nu}_n}(t)\| \leq \|\theta_{\nu_n}(t)\|$ for $0 \leq t \leq T$, and

$$\lim_{n\to\infty} E\int_0^T \|\theta_{\hat{\nu}_n}(t) - \theta_{\hat{\nu}}(t)\|^2\, dt = \lim_{n\to\infty} E\int_0^T \|\theta_{\nu_n}(t) - \theta_\nu(t)\|^2\, dt = 0.$$

Consequently, along a subsequence (which may depend on t), we have

$$Z_{\hat{\nu}_n}(t) \to Z_{\hat{\nu}}(t) \quad \text{and} \quad Z_{\nu_n}(t) \to Z_\nu(t)$$

almost surely. We have already shown that

$$E\varphi(Z_{\hat{\nu}_n}(t)) \leq E\varphi(Z_{\nu_n}(t)), \quad n = 1, 2, \ldots,$$

and it remains to pass to the limit. If φ is bounded, we may use the bounded convergence theorem to do this. Even if φ is unbounded, the function $\varphi_{\epsilon,\kappa}$ of (6.30) is bounded, so

$$E\varphi_{\epsilon,\kappa}(Z_{\hat{\nu}}(t)) \leq E\varphi_{\epsilon,\kappa}(Z_\nu(t)).$$

Letting first $\epsilon \downarrow 0$ and then $\kappa \to \infty$, we obtain (6.34) from two applications of the monotone convergence theorem, using (6.33). Equation (6.25) is proved.

From (6.25) and (6.23), we see that $\widetilde{V}$ satisfies the linear, second-order partial differential equation (6.15). Because (6.25) is condition (D) of Theorem 4.1, we have immediately that conditions (A), (B), (C), and (E) hold. In particular, the first sentence of Theorem 4.1 asserts that the optimal consumption and portfolio-proportion processes for Problem 2.6 are

$\hat{c}(\cdot) = c_{\hat{\nu}}(\cdot)$, $\hat{p}(\cdot) = p_{\hat{\nu}}(\cdot)$ given by (3.17), (3.21), the optimal processes for Problem 3.2 in the market $\mathcal{M}_{\hat{\nu}}$. Equation (6.24) with $t = 0$ follows; its verification for $t \neq 0$ is a straightforward variation of the above argument. Applying Theorem 3.8.8 and equation (3.8.3) to the market $\mathcal{M}_{\hat{\nu}}$, we obtain (6.26)–(6.29).

It remains only to show that the unconstrained HJB equation (6.18) reduces to the constrained equation (6.2), i.e., to show that

$$R \overset{\Delta}{=} \max_{p \in \mathbb{R}^N} \left[\frac{1}{2} \|\sigma'(t)p\|^2 x^2 \frac{\partial^2}{\partial x^2} V_{\hat{\nu}}(t, x) + (p'\sigma(t)\theta(t) + p'\hat{\nu}(t)) x \frac{\partial}{\partial x} V_{\hat{\nu}}(t, x) \right]$$

agrees with

$$L \overset{\Delta}{=} \max_{p \in K} \left[\frac{1}{2} \|\sigma'(t)p\|^2 x^2 \frac{\partial^2}{\partial x^2} V_{\hat{\nu}}(t, x) + p'\sigma(t)\theta(t) x \frac{\partial}{\partial x} V_{\nu}(t, x) \right].$$

Now, $R \geq L$, since $p'\nu \geq 0$ for every $p \in K$, $\nu \in \widetilde{K}$ (see (5.4.5)) and $\frac{\partial}{\partial x} V_{\hat{\nu}}(t, x) \geq 0$. On the other hand, the maximum in the definition of R is obtained by

$$p = -(\sigma'(t))^{-1}\theta_{\hat{\nu}}(t) \frac{\frac{\partial}{\partial x} V_{\hat{\nu}}(t, x)}{x \frac{\partial^2}{\partial x^2} V_{\hat{\nu}}(t, x)} = P(t, x),$$

according to (6.19), (6.26). But $\hat{\nu}(t)$ minimizes $\|\theta(t) + \sigma^{-1}(t)\nu\|$ over $\nu \in \widetilde{K}$, so $\theta_{\hat{\nu}}(t)$ is orthogonal to $\sigma^{-1}\nu$ for every $\nu \in \widetilde{K}$. In other words,

$$\nu'P(t, x) = 0, \quad \forall \nu \in \widetilde{K},$$

and now (5.4.5) shows that $P(t, x) \in K$ and $\hat{\nu}'(t)P(t, x) = 0$. It follows that $R \leq L$. $\qquad \square$

Example 6.6 *(Constant coefficients):* Consider the case of constant $r(\cdot) = r > 0$, $\theta(\cdot) = \theta$, $\sigma(\cdot) = \sigma$, and $A(\cdot) \equiv 0$. Assume that

$$U_1(t, x) = e^{-\alpha t} u_1(x), \quad U_2(x) = e^{-\alpha T} u_2(x), \quad 0 \leq t \leq T, \ x > 0,$$

where $\alpha \geq 0$ and $u_1: (0, \infty) \to \mathbb{R}$ and $u_2: (0, \infty) \to \mathbb{R}$ are three-times continuously differentiable utility functions. Assume further that (3.8.51) and (3.8.52) are satisfied and $\gamma \overset{\Delta}{=} \frac{1}{2} \|\theta_{\hat{\nu}}\|^2 > 0$. Then $X_{\hat{\nu}}(t, y)$, $G_{\hat{\nu}}(t, y)$, and $\widetilde{V}_{\hat{\nu}}(t, y) = \widetilde{V}(t, y)$ are given by (3.8.53), (3.8.54), and (3.8.55), respectively.

Example 6.7 *(Utility functions of power type):* Fix $\beta \in (-\infty, 1) \setminus \{0\}$ and assume

$$U_1(t, x) = U_2(x) = \frac{1}{\beta} x^\beta, \quad 0 \leq t \leq T, \ x > 0.$$

We know from Example 3.8.13 that

$$X_{\hat{\nu}}(t, y) = k(t) y^{\frac{1}{\beta-1}}, \qquad G_{\hat{\nu}}(t, y) = \frac{1}{\beta} k(t) y^{\frac{\beta}{\beta-1}},$$

$$\widetilde{V}(t, y) = \frac{1 - \beta}{\beta} k(t) y^{\frac{\beta}{\beta-1}}, \qquad V(t, x) = \frac{1}{\beta} k(t) \left(\frac{x}{k(t)} \right)^\beta,$$

where

$$k(t) \triangleq e^{\int_t^T \alpha(s)ds} \left[1 + \int_t^T e^{-\int_s^T \alpha(u)du} ds \right],$$

$$\alpha(t) \triangleq \frac{\beta}{(1-\beta)^2} \left[\frac{1}{2} \|\theta_{\hat{\nu}}(t)\|^2 + r(t)(1-\beta) \right].$$

It is now easily calculated, in the notation of (6.17) and Theorem 6.4, that

$$\widehat{X}(t) = \frac{xk(t)}{k(0)} (H_{\hat{\nu}}(t))^{\frac{1}{\beta-1}},$$

$$\hat{c}(t) = \frac{\widehat{X}(t)}{k(t)},$$

$$\hat{p}(t) = \frac{1}{1-\beta} (\sigma(t)\sigma'(t))^{-1}[b(t) + \delta(t) - r(t)\underline{1} + \hat{\nu}(t)].$$

6.7 Incomplete Markets

We return to the general model of a complete, standard financial market set up in Section 6.2. There are N stocks driven by an N-dimensional Brownian motion, as described by (2.3). We simplify the notation by assuming that $\delta(\cdot) \equiv 0$. We choose an integer $M \in \{1, \ldots, N-1\}$, let $L = N - M$, and consider the case that only the first M stocks are available for investment. This corresponds to

$$K = \{p \in \mathbb{R}^N; \quad p_{M-1} = \cdots = p_N = 0\} \tag{7.1}$$

as in Example 5.4.1(iii). Thus, in place of (2.2), (2.3), the relevant equations are

$$dS_0(t) = S_0(t)[r(t)dt + dA(t)], \tag{7.2}$$

$$dS_n(t) = S_n(t) \left[b_n(t)dt + dA(t) + \sum_{d=1}^N \sigma_{nd}(t)dW^{(d)}(t) \right], \tag{7.3}$$

$$n = 1, \ldots, M.$$

We denote by $\tilde{b}(t)$ the M-dimensional column vector $(b_1(t), \ldots, b_M(t))'$, and by $\tilde{\sigma}(t)$ the bounded $M \times N$ matrix $(\sigma_{nd}(t))_{\substack{n=1,\ldots,M \\ d=1,\ldots,N}}$. By assumption, $\tilde{\sigma}(t)$ has full row rank for Lebesgue-almost-every t, almost surely. In the market $\widetilde{\mathcal{M}}$ consisting of the money market and the first M stocks, the market price of risk process (1.4.9) is the N-dimensional vector process

$$\tilde{\theta}(t) \triangleq \tilde{\sigma}'(t)(\tilde{\sigma}(t)\tilde{\sigma}'(t))^{-1}[\tilde{b}(t) - r(t)\underline{1}_M]. \tag{7.4}$$

Although in this section investment in stocks $M+1, \ldots, N$ is not permitted in the constrained Problem 2.6, it is useful to create the unconstrained Problem 3.2 in which such investment is permitted. We may fashion stocks $M+1, \ldots, N$ however we like, provided that we respect the assumptions of Section 6.2. It is convenient to assume that the rows of the $L \times N$ matrix $\rho(t) = (\sigma_{nd}(t))_{\substack{n=M+1,\ldots,N \\ d=1,\ldots,N}}$ are orthonormal vectors spanning the kernel of $\tilde{\sigma}(t)$; i.e.,

$$\rho(t)\rho'(t) = I_L, \quad \tilde{\sigma}(t)\rho'(t) = 0, \quad 0 \le t \le T, \tag{7.5}$$

almost surely, where I_L is the $L \times L$ identity matrix. It is then easily verified that

$$\sigma^{-1}(t) = [\tilde{\sigma}'(t)(\tilde{\sigma}(t)\tilde{\sigma}'(t))^{-1} \mid \rho'(t)]. \tag{7.6}$$

The boundedness of $\sigma(t)$ and $\sigma^{-1}(t)$ follows from the assumption of boundedness of $\tilde{\sigma}(t)$ and $(\tilde{\sigma}(t)\tilde{\sigma}'(t))^{-1}$, conditions that we impose throughout.

To simplify later notation, we denote by $a(t)$ the L-dimensional vector $(b_{M+1}(t), \ldots, b_N(t))'$ of mean rates of return for the unavailable stocks.

With K given by (7.1) we have

$$\widetilde{K} = \{\nu \in \mathbb{R}^N; \nu_1 = \cdots = \nu_M = 0\},$$

and the class $\mathcal{D}$ of Definition 5.5.1 consists of all process of the form

$$\nu(\cdot) = \begin{bmatrix} 0_M \\ \xi(\cdot) \end{bmatrix}, \tag{7.7}$$

where 0_M is the M-dimensional zero vector and $\xi(\cdot)$ is any $\mathbb{R}^L$-valued progressively measurable process satisfying

$$E \int_0^T \|\xi(t)\|^2 \, dt < \infty. \tag{7.8}$$

The evolution of prices in the auxiliary market $\mathcal{M}_\nu$ is given by (7.2), (7.3), and

$$
dS_n^{(\nu)}(t) = S_n^{(\nu)}(t) \Bigg[(a_{n-M}(t) + \xi_{n-M}(t))\, dt + dA(t)
$$

$$
+ \sum_{d=1}^N \rho_{n-M,d}(t)\, dW^{(d)}(t) \Bigg], \quad n = M+1, \ldots, N. \tag{7.9}
$$

We say that $\mathcal{M}_\nu$ is a *fictitious completion* of $\widetilde{\mathcal{M}}$. The theory of this chapter is about choosing the process $\xi(\cdot)$ so that an agent permitted to invest in the stocks $M+1, \ldots, N$ chooses not to.

Remark 7.1: Using (7.4), (7.6), we see that the market price of risk process $\theta_\nu(\cdot)$ for $\mathcal{M}_\nu$ is

$$\theta_\nu(t) = \sigma^{-1}(t)[b(t) + \nu(t) - r(t)1_N]$$
$$= \tilde{\theta}(t) + \rho'(t)[a(t) + \xi(t) - r(t)1_L]. \tag{7.10}$$

Note that $\rho'(t)[a(t) + \xi(t) - r(t)1_L]$ is orthogonal to $\tilde{\theta}(t)$, because of (7.5).

Example 7.2 *(Logarithmic utility, incomplete market):* $U_1(t, x) = U_2(x) = \log x$ for every $(t, x) \in [0, T] \times (0, \infty)$.

This is Example 4.2, specialized to the case of an *incomplete market*; i.e., K given by (7.1). The expression in (4.23) to be minimized over $\xi \in \mathbb{R}^L$ is

$$\frac{1}{2}\|\tilde{\theta}(t) + \rho'(t)(a(t) + \xi - r(t)1_L)\|^2 = \frac{1}{2}\|\tilde{\theta}(t)\|^2$$
$$+ \frac{1}{2}\|\rho'(t)(a(t) + \xi - r(t)1_L)\|^2,$$

and this is minimized by $\hat{\xi}(t) = r(t)1_L - a(t)$. In other words, $\hat{\nu}(\cdot)$ satisfying the equivalent conditions of Theorem 4.1 is

$$\hat{\nu}(t) = \begin{bmatrix} 0_M \\ r(t)1_L - a(t) \end{bmatrix}$$

and $\theta_{\hat{\nu}}(\cdot) = \tilde{\theta}(\cdot)$. This corresponds to choosing the fictitious completion for which $\frac{S_n^{(\nu)}(\cdot)}{S_0(\cdot)}$ is a martingale, $n = M + 1, \ldots, N$. The formula for the optimal portfolio-proportion process in Example 4.2 becomes

$$\hat{p}(t) = \begin{bmatrix} (\tilde{\sigma}(t)\tilde{\sigma}'(t))^{-1}(\tilde{b}(t) - r(t)1_M) \\ 0_L \end{bmatrix}.$$

Example 7.3 *(Logarithmic utility, incomplete market, short-selling prohibited):* $U_1(t, x) = U_2(x) = \log x$ for every $(t, x) \in [0, T] \times (0, \infty)$. In contrast to Example 7.2, we now take

$$K = \{p \in \mathbb{R}^N; \quad p_1 \geq 0, \ldots, p_M \geq 0, p_{M+1} = 0, \ldots, p_M = 0\}$$

as in Example 5.4.1(iv). Because $\widetilde{K} = \{\nu \in \mathbb{R}^N; \nu_1 \geq 0, \ldots, \nu_M \geq 0\}$, the expression in (4.23) becomes

$$\frac{1}{2}\|\sigma^{-1}(t)(b(t) - r(t)1_N) + \sigma^{-1}(t)\nu\|^2$$
$$= \frac{1}{2}\left\| \tilde{\sigma}'(t) \left[(\tilde{\sigma}(t)\tilde{\sigma}(t))^{-1} | \rho'(t) \right] \begin{bmatrix} b(t) - r(t)1_M \\ a(t) - r(t)1_L \end{bmatrix} \right.$$
$$\left. + \left[\tilde{\sigma}'(t)(\tilde{\sigma}(t)\tilde{\sigma}'(t))^{-1} | \rho'(t) \right] \begin{bmatrix} \eta \\ \xi \end{bmatrix} \right\|^2$$
$$= \frac{1}{2}\|\tilde{\sigma}'(t)(\tilde{\sigma}(t)\tilde{\sigma}'(t))^{-1}(b(t) + \eta - r(t)1_M)$$
$$+ \rho'(t)(a(t) + \xi - r(t)1_L)\|^2$$

$$= \frac{1}{2}\|\tilde{\sigma}'(t)(\tilde{\sigma}(t)\tilde{\sigma}'(t))^{-1}(b(t) + \eta - r(t)\underline{1}_M)\|$$

$$+ \frac{1}{2}\|\rho'(t)(a(t) + \xi - r(t)\underline{1}_L)\|^2,$$

and this is to be minimized over $[\eta'|\xi']' \in \widetilde{K}$. As in the previous example, the minimizing ξ is $\hat{\xi}(t) = r(t)\underline{1}_L - a(t)$. The minimizing $\eta \geq 0$ can be solved from the Kuhn–Tucker conditions

$$(\tilde{\sigma}(t)\tilde{\sigma}'(t))^{-1}(\tilde{b}(t) + \hat{\eta}(t) - r(t)\underline{1}_M) \geq \underline{0}_M,$$

$$\hat{\eta}(t) \geq \underline{0}_M,$$

$$\hat{\eta}'(t)(\tilde{\sigma}(t)\tilde{\sigma}'(t))^{-1}(\tilde{b}(t) + \hat{\eta}(t) - r(t)\underline{1}_M) = 0. \tag{7.11}$$

Then

$$\hat{\nu}(t) = \left[\frac{\hat{\eta}(t)}{r(t)\underline{1}_L - a(t)}\right],$$

and

$$\hat{p}(t) = \left[\frac{(\tilde{\sigma}(t)\tilde{\sigma}'(t))^{-1}(\tilde{b}(t) + \hat{\eta}(t) - r(t)\underline{1}_M)}{\underline{0}_L}\right]$$

takes values in K because of (7.11).

Example 7.4 (*"Totally unhedgeable" coefficients, utility of power type*): We take again

$$K = \{p \in \mathbb{R}^N; \quad p_{M+1} = \cdots = p_N = 0\}. \tag{7.12}$$

Suppose now that the $M \times N$-matrix-valued process $\tilde{\sigma}(t) = (\sigma_{nd}(t))_{\substack{n=1,\ldots,M \\ d=1,\ldots,N}}$ is of the form

$$\tilde{\sigma}(t) = \left[\overset{\circ}{\sigma}(\cdot)\middle|\underline{0}_{M\times L}\right], \tag{7.13}$$

where $\overset{\circ}{\sigma}(\cdot)$ is an $M \times M$ nonsingular matrix. Suppose further that

$$\left\{ \begin{array}{l} r(\cdot),\ b(\cdot),\ \sigma(\cdot) \text{ are adapted to the filtration} \\ \check{\mathcal{F}}(t) \overset{\Delta}{=} \sigma(\check{W}(s), 0 \leq s \leq t),\ 0 \leq t \leq T, \\ \text{generated by the } L\text{-dimensional Brownian motion} \\ \check{W}(\cdot) = (W_{M+1}(\cdot), \ldots, W_N(\cdot))' \end{array} \right\} \tag{7.14}$$

and

$$A(\cdot) \equiv 0. \tag{7.15}$$

Then the market $\widetilde{\mathcal{M}}$ of (7.2), (7.3) takes the form

$$dS_0(t) = r(t)S_0(t)\,dt, \tag{7.16}$$

$$dS_n(t) = S_n(t)\left[b_n(t)\,dt + \sum_{d=1}^{M} \overset{\circ}{\sigma}_{nd}(t)d\,\overset{\circ}{W}^{(d)}(t)\right],\ n = 1, \ldots, M; \tag{7.17}$$

it consists of the money market and of M stocks, driven by the M-dimensional Brownian motion

$$\overset{\circ}{W}(t) = (W_1(\cdot), \ldots, W_M(\cdot))'.$$

But the driving Brownian motion $\overset{\circ}{W}(\cdot)$ in (7.17) is independent of $\check{W}(\cdot)$, to whose natural filtration $\{\check{\mathcal{F}}(t)\}$ the coefficients $r(\cdot)$, $b_n(\cdot)$, and $\overset{\circ}{\sigma}_{nd}(\cdot)$ appearing in (7.16), (7.17) are adapted. We refer to this situation as one of *totally unhedgeable coefficients*.

The risk inherent in the coefficient processes is *undiversifiable*, and economic intuition suggests that an investor should simply "ignore" it. While it is by no means obvious how to implement this maxim in general, in the case

$$U_1(t,x) = 0, \quad U_2(x) = \frac{1}{\beta}x^\beta, \quad \forall\, 0 \leq t \leq T,\ x > 0, \tag{7.18}$$

for some $\beta < 1$, $\beta \neq 0$, one might expect the optimal portfolio-proportion process to be given by

$$\overset{\circ}{p}(t) = \frac{1}{1-\beta}(\overset{\circ}{\sigma}(t)\,\overset{\circ}{\sigma}{}'(t))^{-1}[\tilde{b}(t) - r(t)\underline{1}_M], \tag{7.19}$$

which is the same formula as in a complete market with deterministic coefficients (see (3.6.16) in Example 3.6.7). Formula (7.19) directs investment in stocks $1, \ldots, M$; the corresponding portfolio-proportion process in the constrained market $\mathcal{M}(K)$ is

$$\hat{p}(t) = \begin{bmatrix} \overset{\circ}{p}(t) \\ \underline{0}_L \end{bmatrix}. \tag{7.20}$$

The remainder of this example is a proof that (7.19), (7.20) indeed do provide the optimal portfolio-proportion process under the assumptions (7.12)–(7.15), provided that

$$E\left[\exp\left\{\beta \int_0^T \left(r(t) + \frac{1}{2(1-\beta)}\|\overset{\circ}{\theta}(t)\|^2\right) dt\right\}\right] < \infty, \tag{7.21}$$

where

$$\overset{\circ}{\theta}(t) \triangleq (\overset{\circ}{\sigma}(t))^{-1}[\tilde{b}(t) - r(t)\underline{1}_M],$$

and provided that either $\beta \in (0,1)$ or else $\int_0^T r(t)dt$ and $\int_0^T \|\overset{\circ}{\theta}(t)\|^2 dt$ are bounded.

We begin by choosing $a(t) = r(t)\underline{1}_L$ and $\rho(t) = (\sigma_{nd}(t))_{\substack{n=M+1,\ldots,N \\ d=1,\ldots,N}}$ to be of the form $\rho(t) = [\underline{0}_{L\times M}\,|\,\check{\rho}(t)]$, where the $M \times M$ matrix $\check{\rho}(t)$ is orthonormal: $\check{\rho}(t)\check{\rho}'(t) = I_L$. Then $\sigma(t)$ has the block form

$$\sigma(t) = \begin{bmatrix} \overset{\circ}{\sigma}(t) & |\,\underline{0}_{M\times L} \\ \hline \underline{0}_{L\times M} & |\,\check{\rho}(t) \end{bmatrix},$$

and the market price of risk for $\mathcal{M}(K)$ is

$$\theta(t) = \sigma^{-1}(t)[b(t) - r(t)\underline{1}_N] = \begin{bmatrix} \overset{\circ}{\theta}(t) \\ \hline \underline{0}_M \end{bmatrix}.$$

Processes in $\mathcal{D}$ are of the form (7.7), where (7.8) holds, and for such a process $\nu(\cdot) \in \mathcal{D}$ we have

$$\theta_\nu(t) = \begin{bmatrix} \overset{\circ}{\theta}(t) \\ \hline \check{\nu}(t) \end{bmatrix},$$

where

$$\check{\nu}(t) \overset{\Delta}{=} \check{\rho}'(t)\xi(t). \tag{7.22}$$

It follows that

$$H_\nu(T) = \exp\left\{ -\int_0^T \overset{\circ}{\theta}{}'(t)d\overset{\circ}{W}(t) - \int_0^T \check{\nu}'(t)\,d\check{W}(t) \right.$$
$$\left. -\int_0^T \left(r(t) + \frac{1}{2}\|\overset{\circ}{\theta}(t)\|^2 + \frac{1}{2}\|\check{\nu}(t)\|^2 \right) dt \right\} \tag{7.23}$$

and

$$(H_\nu(T))^{\frac{\beta}{\beta-1}} = m_\nu(T) \overset{\circ}{\Lambda}(T)\check{\Lambda}_\nu(T), \tag{7.24}$$

where

$$m_\nu(T) \overset{\Delta}{=} \exp\left\{ \frac{\beta}{1-\beta}\int_0^T r(t)\,dt \right. \tag{7.25}$$
$$\left. + \frac{\beta}{2(1-\beta)^2} \left(\int_0^T \|\overset{\circ}{\theta}(t)\|^2\,dt + \int_0^T \|\check{\nu}(t)\|^2\,dt \right) \right\},$$

$$\overset{\circ}{\Lambda}(T) = \exp\left\{ \frac{\beta}{1-\beta}\int_0^T \overset{\circ}{\theta}{}'(t)d\overset{\circ}{W}(t) \right. \tag{7.26}$$
$$\left. - \frac{1}{2}\left(\frac{\beta}{1-\beta}\right)^2 \int_0^T \|\overset{\circ}{\theta}(t)\|^2\,dt \right\},$$

$$\check{\Lambda}_\nu(T) \overset{\Delta}{=} \exp\left\{ \frac{\beta}{1-\beta}\int_0^T \check{\nu}'(t)\,d\check{W}(t) \right. \tag{7.27}$$
$$\left. - \frac{1}{2}\left(\frac{\beta}{1-\beta}\right)^2 \int_0^T \|\check{\nu}(t)\|^2\,dt \right\}.$$

The implication (B)$\Rightarrow$(A) in Theorem 4.1 and Remark 5.8 show that in order to prove optimality of $\hat{p}(\cdot)$ in (7.20), it suffices to find a process

$\nu(\cdot) \in \mathcal{D}$ such that

$$X^{x,0,\hat{p}}(T) = I_2(\mathcal{Y}_\nu(x)H_\nu(T)). \tag{7.28}$$

With $\hat{p}(\cdot)$ given by (7.19), (7.20), we have

$$\sigma'(t)\hat{p}(t) = \frac{1}{1-\beta}\begin{bmatrix} \overset{\circ}{\theta}(t) \\ 0_L \end{bmatrix},$$

and the left-hand side of (7.28) is easily computed from (2.7) in the form

$$d\left(\frac{X^{x,0,\hat{p}}(t)}{S_0(t)}\right) = \frac{X^{x,0,\hat{p}}(t)}{(1-\beta)S_0(t)}\overset{\circ}{\theta}'(t)\,(\overset{\circ}{\theta}(t)\,dt + d\overset{\circ}{W}(t)).$$

We obtain

$$X^{x,0,\hat{p}}(T) = x \cdot \exp\left\{\int_0^T \left(r(t) + \frac{1-2\beta}{2(1-\beta)^2}\|\overset{\circ}{\theta}(t)\|\right)dt\right.$$

$$\left. + \frac{1}{1-\beta}\int_0^T \overset{\circ}{\theta}'(t)\,d\overset{\circ}{W}(t)\right\}. \tag{7.29}$$

To compute the right-hand side of (7.28) we observe that $I_2(y) = y^{\frac{1}{\beta-1}}$, so that

$$\mathcal{X}_\nu(y) = E\left[H_\nu(T)I_2(yH_\nu(T))\right]$$

$$= y^{\frac{1}{\beta-1}}E\left[(H_\nu(T))^{\frac{\beta}{\beta-1}}\right]$$

$$= y^{\frac{1}{\beta-1}}\mathcal{X}_\nu(1). \tag{7.30}$$

Consequently, $\mathcal{Y}_\nu(x) = \left(\frac{x}{\mathcal{X}_\nu(1)}\right)^{\beta-1}$ and

$$I_2(\mathcal{Y}_\nu(x)H_\nu(T))$$

$$= \frac{x}{\mathcal{X}_\nu(1)}\exp\left\{\frac{1}{1-\beta}\int_0^T \overset{\circ}{\theta}'(t)\,d\overset{\circ}{W}(t) + \frac{1}{1-\beta}\int_0^T \check{\nu}'(t)\,d\check{W}(t)\right.$$

$$\left. + \frac{1}{1-\beta}\int_0^T \left(r(t) + \frac{1}{2}\|\overset{\circ}{\theta}(t)\|^2 + \frac{1}{2}\|\check{\nu}(t)\|^2\right)dt\right\}. \tag{7.31}$$

Comparing (7.29) and (7.31), we see that it suffices to construct a process $\nu(\cdot) \in \mathcal{D}$ such that

$$\mathcal{X}_\nu(1) = \exp\left\{\frac{\beta}{1-\beta}\int_0^T r(t)\,dt + \frac{\beta}{2(1-\beta)^2}\int_0^T \|\overset{\circ}{\theta}(t)\|^2\,dt\right.$$

$$\left. + \frac{1}{2(1-\beta)}\int_0^T \|\check{\nu}(t)\|^2\,dt + \frac{1}{1-\beta}\int_0^T \check{\nu}'(t)\,d\check{W}(t)\right\} \tag{7.32}$$

almost surely.

Consider now the $\{\check{\mathcal{F}}(t)\}$-martingale

$$Q(t) \overset{\Delta}{=} E\left[\exp\left\{\beta \int_0^T r(u)\,du + \frac{\beta}{2(1-\beta)} \int_0^T \|\overset{\circ}{\theta}(u)\|^2\,du\right\} \,\Big|\, \check{\mathcal{F}}(t)\right],$$

(7.33)

defined for $0 \le t \le T$ and taking values in $[\gamma, \infty)$, where γ is a positive constant. Indeed, if $\beta \in (0,1)$, then we may take $\gamma = s_0^\beta$, whereas we use the assumed boundedness of $\int_0^T r(t)dt$ and $\int_0^T \|\overset{\circ}{\theta}(t)\|^2 dt$ to construct γ if $\beta < 0$. According to the martingale representation theorem,

$$Q(t) = Q(0) + \int_0^t \psi'(u)\,d\check{W}(u), \quad 0 \le t \le T,$$

for some $\{\check{\mathcal{F}}(t)\}$-progressively measurable $\psi(\cdot): [0,T] \times \Omega \to \mathbb{R}^L$ satisfying $\int_0^T \|\psi(u)\|^2 du < \infty$ almost surely. We set $\check{\nu}(t) = -\frac{1}{Q(t)}\psi(t)$. Since $Q(\cdot)$ is bounded away from zero, we have $\int_0^T \|\check{\nu}(t)\|^2 dt < \infty$ almost surely and

$$Q(t) = Q(0)\exp\left\{-\int_0^t \check{\nu}'(u)\,d\check{W}(u) - \frac{1}{2}\int_0^t \|\check{\nu}(u)\|^2\,du\right\}, \quad 0 \le t \le T.$$

(7.34)

With $\tau_n \overset{\Delta}{=} \inf\{t \in [0,T); \int_0^t \|\check{\nu}(u)\|^2 du = n\} \wedge T$, we obtain from (7.34)

$$\frac{1}{2}E\int_0^{\tau_n} \|\check{\nu}(t)\|^2\,dt = \log Q(0) - E\log Q(\tau_n)$$

$$\le \log Q(0) - \log \gamma < \infty.$$

Letting $n \to \infty$, we conclude from the monotone convergence theorem that $E\int_0^T \|\check{\nu}(t)\|^2 dt < \infty$. Now set $\xi(t) = \check{\rho}(t)\check{\nu}(t)$ and $\nu(t) = \begin{bmatrix} 0_M \\ \xi(t) \end{bmatrix}$, in accordance with (7.22) and (7.7), and observe that $\nu(\cdot) \in \mathcal{D}$ because $\check{\rho}(\cdot)$ is bounded. From (7.33), (7.34) we also have

$$\exp\left\{\beta \int_0^T r(t)dt + \frac{\beta}{2(1-\beta)}\int_0^T \|\overset{\circ}{\theta}(t)\|^2\,dt\right\}$$

$$= Q(0)\exp\left\{-\int_0^T \check{\nu}'(t)\,d\check{W}(t) - \frac{1}{2}\int_0^T \|\check{\nu}(t)\|^2\,dt\right\},$$

or equivalently,

$$\exp\left\{\frac{\beta}{1-\beta}\int_0^T r(t)\,dt + \frac{\beta}{2(1-\beta)^2}\int_0^T \|\overset{\circ}{\theta}(t)\|^2 dt\right.$$

(7.35)

$$\left. + \frac{1}{2(1-\beta)}\int_0^T \|\check{\nu}(t)\|^2\,dt + \frac{1}{1-\beta}\int_0^T \check{\nu}'(t)\,d\check{W}(t)\right\} = (Q(0))^{\frac{1}{1-\beta}}.$$

To prove (7.32), it suffices to show that

$$\mathcal{X}_\nu(1) = (Q(0))^{\frac{1}{1-\beta}}. \tag{7.36}$$

From (7.30), (7.24), we have

$$\mathcal{X}_\nu(1) = E\left[(H_\nu(T))^{\frac{\beta}{\beta-1}}\right]$$

$$= E\left[m_\nu(T) \overset{\circ}{\Lambda}(T)\check{\Lambda}_\nu(T)\right]$$

$$= E\left[m_\nu(T)\check{\Lambda}_\nu(T)E\left(\overset{\circ}{\Lambda}(T)\middle| \check{\mathcal{F}}(T)\right)\right],$$

where we have used the $\check{\mathcal{F}}(T)$-measurability of $\hat{\nu}(\cdot)$. Because $\overset{\circ}{\theta}(\cdot)$ is $\{\check{\mathcal{F}}(t)\}$-progressively measurable and $\overset{\circ}{W}(\cdot)$ is independent of $\check{\mathcal{F}}(T)$, we see that $E[\overset{\circ}{\Lambda}(T)|\check{\mathcal{F}}(T)] = 1$, just as would be the case if $\overset{\circ}{\theta}(\cdot)$ were nonrandom. We conclude that

$$\mathcal{X}_\nu(1) = E\left[m_\nu(T)\check{\Lambda}_\nu(T)\right]$$

$$= E\left[\exp\left\{\frac{\beta}{1-\beta}\int_0^T r(t)\,dt + \frac{\beta}{2(1-\beta)^2}\int_0^T \|\overset{\circ}{\theta}(t)\|^2\,dt\right.\right.$$

$$\left.\left. + \frac{\beta}{2(1-\beta)}\int_0^T \|\check{\nu}(t)\|^2\,dt + \frac{\beta}{1-\beta}\int_0^T \check{\nu}'(t)\,d\check{W}(t)\right\}\right]$$

$$= E\left[(Q(T))^{\frac{1}{1-\beta}}\left(\frac{Q(0)}{Q(T)}\right)^{\frac{\beta}{1-\beta}}\right],$$

where we have used (7.33) to evaluate $(Q(T))^{\frac{1}{1-\beta}}$ and (7.34) to evaluate $\left(\frac{Q(0)}{Q(T)}\right)^{\frac{\beta}{1-\beta}}$. But

$$E\left[(Q(T))^{\frac{1}{1-\beta}}\left(\frac{Q(0)}{Q(T)}\right)^{\frac{\beta}{1-\beta}}\right] = (Q(0))^{\frac{\beta}{1-\beta}}EQ(T)$$

$$= (Q(0))^{\frac{\beta}{1-\beta}}Q(0) = (Q(0))^{\frac{1}{1-\beta}},$$

and (7.36) follows.

6.8 Higher Interest Rate for Borrowing Than for Investing

In this section we modify the optimal consumption and investment Problem 3.5.4 to allow the interest rate for borrowing to exceed the interest rate for investing. The solution of this problem can be obtained via a dual problem in the spirit of Theorem 4.1, although not as a direct application of that theorem. We indicate below how to make the appropriate modifications to the arguments in Sections 2–5.

As in Section 2 we begin with a complete, standard financial market with asset prices governed by (2.2), (2.3); we assume that (2.4), (2.5), and (5.2.4) hold, and that $Z_0(\cdot)$ is a martingale, so the standard martingale measure P_0 is defined. Consumption processes and portfolio-proportion processes are as in Definitions 2.1 and 2.2, respectively.

The process $S_0(\cdot)$ records the value of investment in the money market. Borrowing, on the other hand, can occur only at a premium $R(\cdot) - r(\cdot)$ above the money-market rate. More specifically, there is a bounded, nonnegative, progressively measurable process $R(\cdot) \geq r(\cdot)$ such that for initial wealth $x \geq 0$, consumption process $c(\cdot)$, and portfolio-proportion process $p(\cdot)$, the corresponding *wealth process* $X^{x,c,p}(\cdot)$, $0 \leq t \leq T$, satisfies (cf. (2.7)),

$$\frac{X^{x,c,p}(t)}{S_0(t)} + \int_0^t \frac{c(u)du}{S_0(u)} = x + \int_0^T \frac{X^{x,c,p}(u)}{S_0(u)} p'(u)\sigma(u)\,dW_0(u) \tag{8.1}$$
$$- \int_0^t \frac{X^{x,c,p}(u)}{S_0(u)}(R(u) - r(u))(p'(u)\mathbf{1}_N - 1)^+\,du.$$

The last integral in (8.1) accounts for the higher interest paid when the fraction of total wealth borrowed,

$$(p'(u)\mathbf{1}_N - 1)^+ = \max\left\{0, \sum_{n=1}^N p_n(u) - 1\right\},$$

is strictly positive. Equation (8.1) is equivalent to

$$dX^{x,c,p}(t) = -c(t)\,dt + (1 - p'(t)\mathbf{1}_N)^+ X^{x,c,p}(t)(r(t)\,dt + dA(t)) \tag{8.2}$$
$$- (p'(u)\mathbf{1}_N - 1)^+ X^{x,c,p}(t)(R(t)\,dt + dA(t))$$
$$+ X^{x,c,p}(t)p'(t)[b(t)\,dt + \delta(t)\,dt + \mathbf{1}_N\,dA(t) + \sigma(t)\,dW(t)].$$

The solution of (8.1), or equivalently, (8.2), is constructed as follows. Define τ_p by (2.9) and $I_p(t)$, $0 \leq t < \tau_p$, by (2.8), so (2.10) holds. Set

$$K_p(t) \triangleq I_p(t)\exp\left\{-\int_0^t (R(u) - r(u))(p'(u)\mathbf{1}_N - 1)^+du\right\},$$

and then define $X^{x,c,p}(t)$ for $0 \leq t < \tau_p$ by

$$\frac{X^{x,c,p}(t)}{S_0(t)} = K_p(t)\left[x - \int_0^t \frac{c(u)du}{S_p(u)K_p(u)}\right]; \tag{8.3}$$

it is easily verified that (8.1) holds. With $\tau_0 \triangleq \inf\{t \in [0,T]; X^{x,c,p}(t) = 0\}$ we have $\tau_0 \leq \tau_p$ almost surely. As in Definition 2.4 we say that (c,p) is *admissible at x in the market with higher interest rate for borrowing than for investing*, and write $(c,p) \in \mathcal{A}(x)$, if (2.11) holds. For $(c,p) \in \mathcal{A}(x)$, the process $X^{x,c,p}(t)$ defined by (8.3) for $0 \leq t < \tau_0$ and set equal to 0 for $\tau_0 \leq t \leq T$, solves (8.1), or equivalently (8.2).

Here is the problem that we shall study in this section.

Problem 8.1: Given $x \geq 0$, find a pair $(\hat{c}, \hat{p})$ in the class

$$\mathcal{A}_3(x) \triangleq \left\{ (c,p) \in \mathcal{A}(x); \quad E \int_0^T \min[0, U_1(t, c(t))]\, dt > -\infty, \right.$$

$$\left. E\left(\min[0, U_2(X^{x,c,p}(T))]\right) > -\infty \right\}$$

that attains

$$V(x) \triangleq \sup_{(c,p)\in\mathcal{A}_3(x)} E\left[\int_0^T U_1(t, c(t))\, dt + U_2(X^{x,c,p}(T))\right],$$

the maximal expected utility from both consumption and terminal wealth in the market with higher interest rate for borrowing than for investing.

We define the random, set-valued process

$$\widetilde{K}(t) \triangleq \{\nu \in \mathbb{R}^N; -(R(t) - r(t)) \leq \nu_1 = \nu_2 = \cdots = \nu_N \leq 0\}.$$

As in Definition 5.5.1, we denote by $\mathcal{H}$ the Hilbert space of $\{\mathcal{F}(t)\}$-progressively measurable processes $\nu(\cdot): [0,T] \times \Omega \to \mathbb{R}^N$ with $E \int_0^T \|\nu(t)\|^2 dt < \infty$ and by $\mathcal{D}$ the closed subset of $\mathcal{H}$ consisting of processes $\nu(\cdot)$ that satisfy the almost sure condition

$$\nu(t) \in \widetilde{K}(t), \text{ Lebesgue-a.e. } t \in [0,T].$$

For $t \in [0,T]$ and $\nu \in \mathbb{R}^N$, we define the nonnegative random variable

$$\zeta(t,\nu) \triangleq \begin{cases} -\nu_1, & \text{if } \nu \in \widetilde{K}(t), \\ \infty, & \text{if } \nu \notin \widetilde{K}(t). \end{cases}$$

For each ν, $\zeta(\cdot, \nu)$ is $\{\mathcal{F}(t)\}$-progressively measurable. For each (t, ω), the mapping $\nu \mapsto \zeta(t, \nu)$ is convex and lower semicontinuous.

These definitions of $\mathcal{D}$ and ζ are in the spirit of Definition 5.5.1, except that $\widetilde{K}$ depends now on (t, ω) and is not a cone. There is no convex set K corresponding to $\widetilde{K}$ via (5.4.1), (5.4.2).

For $\nu(\cdot) \in \mathcal{D}$, we form the auxiliary market $\mathcal{M}_\nu$ described in Section 5.5.2. In $\mathcal{M}_\nu$ there is a single interest rate for investing and borrowing, namely

$$r_\nu(t) \triangleq r(t) + \zeta(t, \nu(t))$$

(cf. (5.5.3)). Note that

$$0 \leq \zeta(t, \nu(t)) \leq R(t) - r(t), \text{ for Lebesgue-a.e. } t \in [0,T], \tag{8.4}$$

holds almost surely, so $r(t) \leq r_\nu(t) \leq R(t)$. The money market in $\mathcal{M}_\nu$ has price process

$$S_0^{(\nu)}(t) = S_0(t) \exp\left\{ \int_0^t \zeta(s, \nu(s)) \, ds \right\}, \tag{8.5}$$

while the stock prices are given by

$$S_n^{(\nu)}(t) = S_n(t), \quad n = 1, \ldots, N \tag{8.6}$$

(see (5.5.7) and (5.5.8)). The market price of risk in the market $\mathcal{M}_\nu$ is

$$\theta_\nu(t) \overset{\Delta}{=} \theta(t) + \sigma^{-1}(t)\nu(t) = \sigma^{-1}(t)[b(t) + \delta(t) - r(t)\mathbf{1}_N + \nu_1(t)\underline{1}_N]. \tag{8.7}$$

Given $x \geq 0$ and a consumption and portfolio-proportion process pair (c, p), we define the corresponding wealth process $X_\nu^{x,c,p}(\cdot)$ in $\mathcal{M}_\nu$ by the equivalent equations (3.1)–(3.3). Note that in the case at hand, (3.2) may be written as

$$\frac{X_\nu^{x,c,p}(t)}{S_0(t)} + \int_0^T \frac{c(u)du}{S_0(u)} = x + \int_0^t \frac{X_\nu^{x,c,p}(u)}{S_0(u)} p'(u)\sigma(u) \, dW_0(u) \tag{8.8}$$

$$+ \int_0^t \frac{X_\nu^{x,c,p}(u)}{S_0(u)} \zeta(u, \nu(u))(1 - p'(u)\underline{1}_N) \, du.$$

We say that (c, p) is *admissible at x in the unconstrained market* $\mathcal{M}_\nu$, and write $(c, p) \in \mathcal{A}_\nu(x)$, if (3.4) holds. We define $\mathcal{A}_3^{(\nu)}(x)$ and $V_\nu(x)$ as in Problem 3.2.

Remark 8.2: Suppose $(c, p) \in \mathcal{A}(x)$, so that $X^{x,c,p}(\cdot)$ is defined. If we choose $\nu(\cdot) \in \mathcal{D}$, then $X_\nu^{x,c,p}(\cdot)$ is also defined, and we have

$$X_\nu^{x,c,p}(t) \geq X^{x,c,p}(t), \quad 0 \leq t \leq T, \tag{3.8}$$

as we show below. In particular,

$$\mathcal{A}_3(x) \subset \mathcal{A}_3^{(\nu)}(x), \quad \text{for } x \geq 0, \ \nu(\cdot) \in \mathcal{D}, \tag{8.9}$$

and

$$V(x) \leq V_\nu(x), \quad \text{for } x \geq 0, \ \nu(\cdot) \in \mathcal{D}. \tag{8.10}$$

To derive (3.8), we first note that

$$\zeta(t, \nu(t))(1 - p'(t)\underline{1}_N) + (R(t) - r(t))(p'(t)\underline{1}_N - 1)^+$$
$$= \zeta(t, \nu(t))(1 - p'(t)\underline{1}_N)^+ + (R(t) - r(t) - \zeta(t, \nu(t)))(p'(t)\underline{1}_N - 1)^+$$
$$\geq 0. \tag{8.11}$$

Subtraction of (8.1) from (8.8) shows that the process

$$\Delta(t) \overset{\Delta}{=} \frac{X_\nu^{x,c,p}(t) - X^{x,c,p}(t)}{S_0(t)}, \quad 0 \leq t \leq T,$$

satisfies

$$\Delta(t) = \int_0^t \Delta(u)p'(u)\sigma(u) \, dW_0(u) + \int_0^t \Delta(u)\zeta(u, \nu(u))(1 - p'(u)\underline{1}_N) \, du$$

$$+ \int_0^t \frac{X^{x,c,p}(u)}{S_0(u)} \left[\zeta(u, \nu(u))(1 - p'(u)\underline{1}_N) \right.$$
$$\left. + (R(u) - r(u))(p'(u)\underline{1}_N - 1)^+ \right] du.$$

Setting

$$J(t) = \exp\left\{ - \int_0^t p'(u)\sigma(u)dW_0(u) + \frac{1}{2} \int_0^t \|\sigma'(u)p(u)\|^2 \, du \right.$$
$$\left. - \int_0^t \zeta(u, \nu(u))(1 - p'(u)\underline{1}_N) \, du \right\},$$

we compute

$$d(\Delta(t)J(t)) = \frac{J(t)X^{x,c,p}(t)}{S_0(t)} \left[\zeta(t, \nu(t))(1 - p'(t)\underline{1}_N) \right.$$
$$\left. + (R(t) - r(t))(p'(u)\underline{1}_N - 1)^+ \right] dt.$$

We integrate this equation, using (8.11) and the fact that $\Delta(0) = 0$, to conclude that (3.8) holds. Note also that if we have

$$\zeta(t, \nu(t))(1 - p'(t)\underline{1}_N) + (R(t) - r(t))(p'(u)\underline{1}_N - 1)^+ = 0, \quad (8.12)$$
$$\text{for Lebesgue-a.e. } t \in [0, T]$$

almost surely, then

$$X_\nu^{x,c,p}(t) = X^{x,c,p}(t), \quad 0 \le t \le T \tag{8.13}$$

holds almost surely.

The unconstrained market $\mathcal{M}_\nu$ satisfies (3.14), and hence Assumption 3.2.3. We define $\mathcal{X}_\nu(\cdot)$ and $\mathcal{D}_0$ by (3.15) and (3.24), respectively, and for $\nu(\cdot) \in \mathcal{D}$ satisfying $\mathcal{X}_\nu(y) < \infty$ for all $y > 0$, we denote by $\mathcal{Y}_\nu(\cdot)$ the inverse of $\mathcal{X}_\nu(\cdot)$. We define $c_\nu(\cdot)$, $X_\nu(\cdot)$, and $p_\nu(\cdot)$ by (3.16)–(3.21); then (3.22) holds and (c_ν, p_ν) attains the supremum in (3.6). Remarks 3.4, 3.5, and 3.6 apply to $\mathcal{M}_\nu$. For $\nu(\cdot) \in \mathcal{D}_0$ we define $\widetilde{V}_\nu(\cdot)$ by (3.27). We extend this definition to $\nu(\cdot) \in \mathcal{D} \setminus \mathcal{D}_0$ by (3.33), and Proposition 3.7 holds.

Let the initial wealth x be strictly positive. Just as in Section 3, our strategy is to find a process $\hat{\nu}(\cdot) \in \mathcal{D}_0$ for which the optimal pair $(c_{\hat{\nu}}(\cdot), p_{\hat{\nu}}(\cdot))$ for the market $\mathcal{M}_{\hat{\nu}}$ is also optimal for Problem 8.1. In other words, we seek $\nu(\cdot) \in \mathcal{D}_0$ that satisfies

$$V(x) = E\left[\int_0^T U_1(t, c_{\hat{\nu}}(t)) \, dt + U_2(B_{\hat{\nu}}) \right] = V_{\hat{\nu}}(x). \tag{8.14}$$

Remark 8.2 shows that such a $\hat{\nu}(\cdot)$ should satisfy almost surely the *complementary slackness condition*

$$\zeta(t, \hat{\nu}(t))(1 - \hat{p}'(t)\underline{1}_N) + (R(t) - r(t))(p'(u)\underline{1}_N - 1)^+ = 0, \tag{8.15}$$
$$\text{for Lebesgue-a.e. } t \in [0, T].$$

Proposition 8.3: *Let $x > 0$ be given, and suppose that (8.15) is satisfied for some process $\hat{\nu}(\cdot) \in \mathcal{D}_0$. Then the pair $(c_{\hat{\nu}}, p_{\hat{\nu}})$ of (3.17), (3.21) is optimal for Problem 8.1, and $V(x) = V_{\hat{\nu}}(x)$. Furthermore, $\hat{\nu}(\cdot)$ minimizes $V_{\nu}(x)$ over $\nu(\cdot) \in \mathcal{D}$.*

PROOF.   From Remark 8.2 we see that (8.15) implies

$$V(x) \geq E\left[\int_0^T U_1(t, c_{\hat{\nu}}(t))\, dt + U_2(X^{x,c_{\hat{\nu}},p_{\hat{\nu}}}(T))\right]$$

$$= E\left[\int_0^T U_1(t, c_{\hat{\nu}}(t))\, dt + U_2(X_{\hat{\nu}}^{x,c_{\hat{\nu}},p_{\hat{\nu}}}(T))\right]$$

$$= V_{\hat{\nu}}(x). \tag{8.16}$$

Remark 8.2 also gives $V(x) \leq V_{\hat{\nu}}(x)$ for every $\nu(\cdot) \in \mathcal{D}$. Hence equality holds in (8.16), $\hat{\nu}(\cdot)$ minimizes $V_{\nu}(x)$ over $\nu(\cdot) \in \mathcal{D}$, and $(c_{\hat{\nu}}, p_{\hat{\nu}})$ is optimal for Problem 8.1.   □

We alter slightly two of the optimality conditions of Section 6.4. For a fixed initial capital $x > 0$, let $(\hat{c}, \hat{p})$ be a consumption and portfolio-proportion process pair in the class $\mathcal{A}_3(x)$ of Problem 8.1, and denote by $\widehat{X}(\cdot) = X^{x,\hat{c},\hat{p}}(\cdot)$ the corresponding wealth process of (8.1). Consider the statement that this pair is optimal for Problem 8.1:

(A′) Optimality of $(\hat{c}, \hat{p})$: We have

$$V(x) = E\left[\int_0^T U_1(t, \hat{c}(t))\, dt + U_2(\widehat{X}(T))\right] < \infty.$$

Now let $\hat{\nu}(\cdot)$ be a process in $\mathcal{D}_0$. In the next condition, we use the notation (3.16)–(3.18).

(B′) Financeability of $(c_{\hat{\nu}}, B_{\hat{\nu}})$: There exists a portfolio-proportion process $p_{\hat{\nu}}(\cdot)$ such that $(c_{\hat{\nu}}, p_{\hat{\nu}}) \in \mathcal{A}_3(x)$ and

$$\zeta(t, \hat{\nu}(t))(1 - p'_{\hat{\nu}} 1_N) + (R(t) - r(t))(p'_{\hat{\nu}}(t)1_N - 1)^+ = 0, \tag{8.17}$$
$$\text{for Lebesgue-a.e. } t \in [0, T],$$
$$X^{x,c_{\hat{\nu}},p_{\hat{\nu}}}(\cdot) = X_{\hat{\nu}}(\cdot) \tag{8.18}$$

hold almost surely.

As in Section 4.1, the portfolio-proportion process $p_{\hat{\nu}}(\cdot)$ in (B′) is *not assumed* to be given by (3.21) with $\nu(\cdot) = \hat{\nu}(\cdot)$; this follows from (8.17), (8.18), and Remark 8.2 (see (8.13)).

For the present purpose, the other three optimality conditions in Section 6.4 are suitable exactly as stated there. We have the following counterpart to the first assertion in Theorem 4.1.

Theorem 8.4: *The condition (B′) and the conditions (C), (D) and (E) of Section 6.4 are equivalent, and they imply condition (A′) with $(\hat{p}, \hat{c}) = (p_{\hat{\nu}}, c_{\hat{\nu}})$.*

PROOF.   The implications (B′)⇒(E)⇒(D), (B′)⇒(A′), and (B′)⇒ (C)⇒ (D) are proved as in Theorem 4.1. It remains to prove (D)⇒(B′).

We assume (D), define $\mu(t) = (\mu_1(t), \ldots, \mu_1(t))' \in \mathcal{D}$ by

$$\mu_1(t) = \cdots = \mu_N(t) = \begin{cases} 0, & \text{if } 1 - p'_{\hat{\rho}}(t)1_N \geq 0, \\ -(R(t) - r(t)), & \text{if } 1 - p'_{\hat{\rho}}(t)1_N < 0, \end{cases}$$

and proceed through the proof of the implication (D)⇒(B) in Theorem 4.1, taking

$$\xi(s) = \frac{1}{\epsilon}[\zeta(s, (1-\epsilon)\hat{\nu}(s) + \epsilon\mu(s)) - \zeta(s, \hat{\nu}(s))] = \hat{\nu}_1(s) - \mu_1(s),$$

so that (4.19) becomes

$$E \int_0^{T_n} H_{\hat{\rho}}(t) X_{\hat{\rho}}(t) \left[(\hat{\nu}_1(t) - \mu_1(t))(1 - p'_{\hat{\rho}}(t)1_N)\right] dt \geq 0, \quad n = 1, 2, \ldots.$$

Because $-(R(t)-r(t)) \leq \hat{\nu}_1(t) \leq 0$, the integrand is negative unless $\hat{\nu}_1(t) = \mu_1(t)$ or $1 - p'_{\hat{\rho}}(t)1_N = 0$. Because $\tau_n \uparrow T$ almost surely as $n \to \infty$, we must have

$$(\hat{\nu}(t) - \mu(t))(1 - p'_{\hat{\rho}}(t)1_N) = 0, \text{ for Lebesgue-a.e. } t \in [0, T]$$

almost surely. It follows that

$$\begin{aligned} 0 &= \zeta(t, \hat{\nu}(t))(1 - p'_{\hat{\rho}}(t)1_N)^+ + (R(t) - r(t) - \zeta(t, \hat{\nu}(t)))(p'_{\hat{\rho}}(t)1_N - 1)^+ \\ &= \zeta(t, \hat{\nu}(t))(1 - p'_{\hat{\rho}}(t)1_N) + (R(t) - r(t))(p'_{\hat{\nu}(t)}1_N - 1)^+, \\ & \quad \text{for Lebesgue-a.e. } t \in [0, T] \end{aligned}$$

almost surely.   □

The existence theory for the dual problem captured by condition (D) goes through just as in Section 6.5. In particular, Theorem 5.3 holds, and under the assumptions of that theorem, for every $x > 0$, we have $V(x) < \infty$ and there exists an optimal pair $(\hat{c}, \hat{p}) \in \mathcal{A}_3(x)$ for Problem 8.1.

Example 8.5 (*Logarithmic utilities, general coefficients*): In the special case $U_1(t, x) = U_2(x) = \log x$, $0 \leq t \leq T$, $x > 0$, we have $\tilde{U}_1(t, y) = \tilde{U}_2(y) = -(1 + \log y)$, $y > 0$ (see Example 4.2), and

$$\tilde{V}_\nu(y) = -(T+1)(1 + \log y) - \int_0^T E[\log H_\nu(t)]\, dt - E[\log H_\nu(T)].$$

For $\nu(\cdot) \in \mathcal{D}$, we have

$$\begin{aligned} &-E[\log H_\nu(t)] \\ &= EA(t) + \int_0^t E\left[r(u) + \zeta(u, \nu) + \frac{1}{2}\|\theta(u) + \sigma^{-1}\nu(u)\|^2\right] du. \end{aligned}$$

The optimal dual process is thus $\hat{\nu}(t) = \hat{\nu}_1(t)1_N$, where $\hat{\nu}_1(t)$ minimizes

$$-\nu_1 + \frac{1}{2}\|\theta(t) + \nu_1\sigma^{-1}(t)1_N\|^2 \quad \text{over } \nu_1 \in [-(R(t) - r(t)), 0].$$

With

$$B(t) \triangleq \theta'(t)\sigma^{-1}(t)1_N, \quad C(t) \triangleq 1'_N(\sigma(t)\sigma'(t))^{-1}1_N, \qquad (8.19)$$

this minimization is achieved as follows:

$$\hat{\nu}_1(t) = \begin{cases} 0, & \text{if } B(t) - 1 \le 0, \\ \frac{1-B(t)}{C(t)}, & \text{if } 0 < B(t) - 1 < C(t)(R(t) - r(t)), \\ r(t) - R(t), & \text{if } B(t) - 1 \ge C(t)(R(t) - r(t)). \end{cases}$$

Example 3.6.6 applied to the market $\mathcal{M}_{\hat{\nu}}$ shows that the optimal portfolio-proportion process is $p_{\hat{\nu}}(t) = (\sigma'(t))^{-1}\theta_{\hat{\nu}}(t)$; i.e.,

$$p_{\hat{\nu}}(t) = \begin{cases} (\sigma(t)\sigma'(t))^{-1}[b(t) + \delta(t) - r(t)1_N], & \text{if } B(t) - 1 \le 0, \\ (\sigma(t)\sigma'(t))^{-1}\left[b(t) + \delta(t) - r(t) + \frac{1-B(t)}{C(t)}1_N\right], \\ & \text{if } 0 < B(t) - 1 < C(t)(R(t) - r(t)), \\ (\sigma(t)\sigma'(t))^{-1}[b(t) + \delta(t) - R(t)], \\ & \text{if } B(t) - 1 \ge C(t)(R(t) - r(t)). \end{cases}$$

Example 8.6 (*Utilities of power type, deterministic coefficients*): Assume that $U_1(t,x) = U_2(x) = \frac{1}{\beta}x^\beta$, $0 \le t \le T$, $x > 0$, for some $\beta < 0$, $\beta \ne 0$. Assume further that $r(\cdot)$, $R(\cdot)$, $\sigma(\cdot)$, and $\theta(\cdot)$ are deterministic. We have

$$\tilde{U}_1(t,y) = \tilde{U}_2(y) = \frac{1-\beta}{\beta}y^{\frac{\beta}{\beta-1}}, \quad y > 0,$$

and

$$\tilde{V}_\nu(y) = \frac{1-\beta}{\beta}y^{\frac{\beta}{\beta-1}}E\left[\int_0^T (H_\nu(t))^{\frac{\beta}{\beta-1}}dt + (H_\nu(T))^{\frac{\beta}{\beta-1}}\right].$$

We take $\hat{\nu}_1(t)$ to be the (nonrandom) minimizer of

$$-(1-\beta)\nu_1 + \frac{1}{2}\|\theta(t) + \nu_1\sigma^{-1}(t)1_N\|^2 \quad \text{over} \quad \nu_1 \in [-(R(t) - r(t)), 0],$$

i.e.,

$$\nu_1(t) = \begin{cases} 0, & \text{if } B(t) + \beta - 1 \le 0, \\ \frac{1-\beta-B(t)}{C(t)}, & \text{if } 0 < B(t) + \beta - 1 < C(t)(R(t) - r(t)), \\ r(t) - R(t), & \text{if } B(t) + \beta - 1 \ge C(t)(R(t) - r(t)), \end{cases}$$

where $B(t)$ and $C(t)$ are given by (8.19). From Example 3.6.7, we have for all $\nu(\cdot) \in \mathcal{D}$ that $(H_\nu(t))^{\frac{\beta}{\beta-1}} = m_\nu(t)\Lambda_\nu(t)$, where

$$m_\nu(t) \triangleq \exp\left\{\frac{\beta}{1-\beta}\left[A(t) + \int_0^t r(u)du\right] \right.$$

$$+ \frac{\beta}{(1-\beta)^2}\int_0^t \left[-(1-\beta)\nu_1(u)\right.$$

$$\left.\left. + \frac{1}{2}\|\theta(u) + \nu_1(u)\sigma^{-1}(u)1_N\|^2\right]du\right\},$$

$$\Lambda_\nu(t) \stackrel{\Delta}{=} \exp\left\{ \frac{\beta}{1-\beta} \int_0^t \theta_\nu'(u)dW(u) - \frac{\beta^2}{2(1-\beta)^2} \int_0^t \|\theta_\nu(u)\|^2 du \right\}.$$

If $0 < \beta < 1$, then

$$E(H_\nu(t))^{\frac{\beta}{\beta-1}} = E[m_\nu(t)\Lambda_\nu(t)] \geq m_{\hat\nu}(t)E\Lambda_\nu(t) = m_{\hat\nu}(t) = E(H_{\hat\nu}(t))^{\frac{\beta}{\beta-1}},$$

but if $\beta < 0$, then

$$E(H_\nu(t))^{\frac{\beta}{\beta-1}} \leq m_{\hat\nu}(t) = E(H_{\hat\nu}(t))^{\frac{\beta}{\beta-1}}.$$

In either case, we see that $V_{\hat\nu}(y) \leq V_\nu(y)$ holds for all $\nu(\cdot) \in \mathcal{D}$. Having thus solved the dual problem, we appeal to (3.6.16) to obtain the optimal portfolio process in the form $p_{\hat\nu}(t) = \frac{1}{1-\beta}(\sigma'(t))^{-1}\theta_{\hat\nu}(t)$; i.e.,

$$p_{\hat\nu}(t) = \begin{cases} \frac{1}{1-\beta}(\sigma(t)\sigma'(t))^{-1}[b(t) + \delta(t) - r(t)1_N], & \text{if } B(t) + \beta - 1 \leq 0, \\[2mm] \frac{1}{1-\beta}(\sigma(t)\sigma'(t))^{-1}\left[b(t) + \delta(t) - r(t) + \frac{1-\beta-B(t)}{C(t)} \right], \\[2mm] \qquad\qquad \text{if } 0 < B(t) + \beta - 1 < C(t)(R(t) - r(t)), \\[2mm] \frac{1}{1-\beta}(\sigma(t)\sigma'(t))^{-1}[b(t) + \delta(t) - R(t)], \\[2mm] \qquad\qquad \text{if } B(t) + \beta - 1 \geq C(t)(R(t) - r(t)). \end{cases}$$

6.9   Notes

The stochastic duality theory of Bismut (1973) was apparently first employed to study *constrained portfolio optimization problems* in the doctoral dissertation of Xu (1990) (see also Shreve and Xu (1992)), who considered the special case $K = [0, \infty)^N$ of Example 5.4.1(iii) corresponding to *prohibiting the short-sale of stocks*. Xu formulated a dual problem in the spirit of this chapter, whose solution could be shown to exist and to be useful in constructing and characterizing the solution to the original, constrained optimization problem. His methodology was then applied to the more traditional *incomplete market model* of Section 7 by Karatzas, Lehoczky, Shreve, and Xu (1991). He and Pearson (1991b) also dealt with the same problem in a discrete-time, finite-probability-space framework, where they proposed finding the pairs of optimal consumption processes and terminal wealth levels corresponding to each member in a family of equivalent martingale measures and then searching over those pairs for a policy that yields minimal expected utility; they used separating hyperplane theorems to show that the total utility obtained by this two-step "minimax" procedure leads to the value of the optimization problem in the incomplete market. Using techniques of Xu (1990), He and Pearson (1991a) extended He and Pearson (1991b) to continuous time. For earlier work on these matters with Markovian price processes, see also Pagès (1989), who deals with the case of incomplete markets using probabilistic tools, and Zariphopoulou (1989), who discusses constraints using methods from partial differential equations.

For the generality of the model presented in this chapter the reader is referred to the paper by Cvitanić and Karatzas (1992), whose exposition we have followed rather closely. For other expositions, see the lectures by Cvitanić (1997), El Karoui and Quenez (1997), Karatzas (1996). Pliska (1997) has a presentation of these techniques and results in the much simpler setting of a discrete-time model. Work related to the problem of Section 8 (consumption/investment optimization in the presence of a higher interest rate for borrowing than for investing) includes the papers by Brennan (1971) in an equilibrium setting, as well as Fleming and Zariphopoulou (1991), who deal with an infinite-horizon model with discounting, $d = 1$ and constant $R > r$, b_1, σ_{11} and obtain explicit formulae for power-type utility functions. We also refer the reader to Foldes (1992) for related work on infinite-horizon models; to He and Pagès (1993), Browne (1995), Cuoco (1997), El Karoui and Jeanblanc-Picqué (1997) and the references therein, as well as Duffie, Fleming, Soner, and Zariphopoulou (1997), for utility maximization and/or equilibrium problems under various constraints and *random endowment streams*; and to Cuoco and Cvitanić (1996) for generalizations in the context of a "large investor" model. In particular, Cuoco (1997) establishes directly the existence of optimal consumption/investment strategies for fairly general price dynamics and utility functions; his model allows for a random endowment stream and for portfolio constraints that include short sale, borrowing, and incompleteness restrictions, and he characterizes equilibrium risk premia when these exist.

In another very interesting development, Schweizer (1992b) considers an incomplete market with $r(\cdot) \equiv 0$ and *one* stock, with price-per-share process

$$dS(t) = S(t)[b(t)\,dt + \sigma_1(t)\,dW_1(t) + \sigma_2(t)\,dW_2(t)] \qquad (9.1)$$

driven by a *two*-dimensional Brownian motion $W(\cdot) = (W_1(\cdot), W_2(\cdot))$ that generates the filtration $\{\mathcal{F}(t)\}_{0 \le t \le T}$. The coefficients $b(\cdot)$, $\sigma_1(\cdot)$, $\sigma_2(\cdot)$ are bounded and adapted to this filtration, and $\sigma(\cdot) \stackrel{\Delta}{=} \sqrt{\sigma_1^2(\cdot) + \sigma_2^2(\cdot)}$ is bounded away from zero. One considers portfolio processes $\pi : [0, T] \times \Omega \to \mathbb{R}$ in the Hilbert space $\mathcal{H}$ of measurable, $\{\mathcal{F}(t)\}$-adapted processes with $E \int_0^T \pi^2(t)dt < \infty$, and tries to "hedge" with such portfolios a given contingent claim $B \in L^p(\Omega, \mathcal{F}_T, P)$ for some $p > 2$. Since exact duplication in such a setting is typically not possible, Schweizer considers the minimization

$$\inf_{\pi(\cdot) \in \mathcal{H}} E(B - X^{x,\pi}(T))^2 \qquad (9.2)$$

of expected quadratic loss for a given initial capital $x \in \mathbb{R}$; see also Richardson (1989), Duffie and Richardson (1991) for earlier results in this vein. Of course, if the market were complete and $x = E_0(B)$, then the infimum in (9.2) would be zero, since one could employ the hedging porfolio $\hat{\pi}(\cdot)$ of Section 2.2 and achieve the exact hedging $X^{x,\hat{\pi}}(T) = B$ almost surely. In our present setting, such exact duplication is not possible even for $x = E_0(B)$, and (9.2) merely attempts to minimize the L^2-norm of the discrepancy in the hedge at time $t = T$ (or equivalently, to find the "best approximation in L^2 of the random variable B by a stochastic integral").

In the notation of Chapter 1, we have here $X^{x,\pi}(t) = x + G^\pi(t)$ for the gains process $G^\pi(t) = \int_0^t \pi(s)\sigma(s)dW_0(s)$, where we set $\theta(\cdot) \triangleq b(\cdot)/\sigma(\cdot)$, $\rho(\cdot) \triangleq \sigma_1(\cdot)/\sigma(\cdot)$, as well as

$$W(t) \triangleq \int_0^t \rho(u)\,dW_1(u) + \int_0^t \sqrt{1 - \rho^2(u)} \cdot dW_2(u),$$

$$N(t) \triangleq \int_0^t \sqrt{1 - \rho^2(u)} \cdot dW_2(u) - \int_0^t \rho(u)\,dW_2(u),$$

and

$$W_0(t) \triangleq W(t) + \int_0^t \theta(s)\,ds.$$

One notices that $W(\cdot), N(\cdot)$ are Brownian motions with $\langle W, N \rangle \equiv 0$; and that $W_0(\cdot), N(\cdot)$ are Brownian motions with $\langle W_0, N \rangle \equiv 0$ under the so-called *minimal* equivalent martingale measure

$$P_0(A) \triangleq E\left[\exp\left\{-\int_0^T \theta(s)\,dW(s) - \frac{1}{2}\int_0^T \theta^2(s)\,ds\right\} \cdot 1_A\right]$$

on $\mathcal{F}(T)$, introduced by Föllmer and Schweizer (1991). On the other hand, elementary Hilbert-space arguments show that the portfolio $\pi^*(\cdot) \in \mathcal{H}$ which attains the infimum in (9.2) satisfies

$$E[(B - x - G^{\pi^*}(T)) \cdot G^\pi(T)] = 0, \quad \forall \pi(\cdot) \in \mathcal{H}. \tag{9.3}$$

In order to find this optimal portfolio $\pi^*(\cdot)$, one introduces the "intrinsic value process" $V(t) \triangleq E_0[B|\mathcal{F}(t)]$, $0 \le t \le T$; from the martingale representation theorem, this process can be written in the form

$$V(t) = E_0(B) + \int_0^t \xi(u)\sigma(u)\,dW_0(u) + \int_0^t \eta(u)\,dN(u)$$

for suitable processes $\xi(\cdot), \eta(\cdot)$ in $\mathcal{H}$, and (9.3) is implied by the equality

$$E[(V(t) - x - G^{\pi^*}(T)) \cdot G^\pi(T)] = 0, \quad \forall \ 0 \le t \le T. \tag{9.4}$$

Using tools from stochastic analysis, Schweizer (1992b) then produces the optimal portfolio $\pi^*(\cdot)$ in the form

$$\pi^*(t) = \xi(t) + \frac{b(t)}{\sigma^2(t)}\left(V(t) - x - G^*(t)\right), \tag{9.5}$$

under the assumption that the ratio

$$\frac{b(\cdot)}{\sigma(\cdot)} \quad \text{is deterministic}, \tag{9.6}$$

where $G^*(\cdot)$ is the solution of the linear stochastic differential equation

$$dG^*(t) = \left[\sigma(t)\xi(t) + \frac{b(t)}{\sigma(t)}\left(V(t) - x - G^*(t)\right)\right] dW_0(t), \quad G^*(0) = 0. \quad (9.7)$$

This result is further extended to a much more general framework in Schweizer (1994); this paper should also be consulted for the study of additional quadratic-optimization problems such as the optimal choice of both initial capital and hedging strategy, the determination of variance-minimizing strategies and of the "mean-variance frontier," and the approximation of a riskless asset. See also Pham, Rheinländer and Schweizer (1998). For the related notion of "variance-optimal martingale measure," see Schweizer (1996).

Appendix A
Essential Supremum of a Family of Random Variables

The essential infimum of a family of nonnegative random variables is defined and constructed by Dunford and Schwartz (1957). We modify this material to suit our purposes.

Definition A.1: Let $(\Omega, \mathcal{F}, P)$ be a probability space and let $\mathcal{X}$ be a nonempty family of nonnegative random variables defined on $(\Omega, \mathcal{F}, P)$. The *essential supremum* of $\mathcal{X}$, denoted by ess sup $\mathcal{X}$, is a random variable X^* satisfying:

(i) $\forall X \in \mathcal{X}, \ X \leq X^*$ a.s., and
(ii) if Y is a random variable satisfying $X \leq Y$ a.s. for all $X \in \mathcal{X}$, then $X^* \leq Y$ a.s.

Because random variables are defined only up to P-almost sure equivalence, it is in general not meaningful to speak of an "ω by ω" supremum $\sup\{X(\omega); \ X \in \mathcal{X}\}$. The essential supremum substitutes for this concept.

It is apparent from Definition A.1 that if the essential supremum of $\mathcal{X}$ exists, then it is unique. The purpose of this appendix is to establish its existence and a few basic properties.

Given $\mathcal{X}$ as in Definition A.1 and given $A \in \mathcal{F}$, we will say that $\pi = (K; A_1, \ldots, A_K; X_1, \ldots, X_K)$ is an *$\mathcal{X}$-partition of A*, provided that:

(i) K is a positive integer,
(ii) $A_1, \ldots, A_K$ are disjoint sets in $\mathcal{F}$ whose union is A, and
(iii) $X_1, \ldots, X_K$ are random variables in $\mathcal{X}$.

For $\lambda \in (0, \infty]$ we define

$$\mu_\pi^\lambda(A) \triangleq E \sum_{k=1}^K (X_k \wedge \lambda) \cdot 1_{A_k} \,, \tag{A.1}$$

$$\mu^\lambda(A) \triangleq \sup\{\mu_\pi^\lambda(A); \ \pi \text{ is an } \mathcal{X}\text{-partition of } A\}. \tag{A.2}$$

Then μ^λ is a nonnegative set function defined on $\mathcal{F}$, and is easily seen to be finitely additive. By the monotone convergence theorem,

$$\mu^\infty(A) = \sup_\pi \sup_{\lambda \in (0,\infty)} \mu_\pi^\lambda(A) = \sup_{\lambda \in (0,\infty)} \sup_\pi \mu_\pi^\lambda(A) = \sup_{\lambda \in (0,\infty)} \mu^\lambda(A). \tag{A.3}$$

Lemma A.2: *For $\lambda \in (0, \infty]$, μ^λ is countably additive.*

PROOF. We first consider the case $\lambda < \infty$. Let $\{A_j\}_{j=1}^\infty$ be an increasing sequence of sets in $\mathcal{F}$ with $A = \bigcup_{j=1}^\infty A_j$. Then $\mu^\lambda(A) = \mu^\lambda(A_j) + \mu^\lambda(A \backslash A_j) \geq \mu^\lambda(A_j)$, so $\lim_{j \to \infty} \mu^\lambda(A_j) \leq \mu^\lambda(A)$. Given $\varepsilon > 0$, choose j such that $P(A \backslash A_j) < \varepsilon$. From (A.1), (A.2), we have $\mu^\lambda(A \backslash A_j) \leq \lambda \varepsilon$, and thus $\mu^\lambda(A) \geq \mu^\lambda(A_j) - \varepsilon$. It follows that $\lim_{j \to \infty} \mu^\lambda(A_j) \geq \mu^\lambda(A)$.

We finally consider the case $\lambda = \infty$. We have from (A.3) that

$$\lim_{j \to \infty} \mu^\infty(A_j) = \sup_j \sup_{\lambda \in (0,\infty)} \mu^\lambda(A_j) = \sup_{\lambda \in (0,\infty)} \sup_j \mu^\lambda(A_j)$$

$$= \sup_{\lambda \in (0,\infty)} \mu^\lambda(A) = \mu^\infty(A). \qquad \square$$

Theorem A.3: *Let $\mathcal{X}$ be a nonempty family of nonnegative random variables. Then $X^* = \text{ess sup } \mathcal{X}$ exists. Furthermore, if $\mathcal{X}$ is closed under pairwise maximization, i.e., $X, Y \in \mathcal{X}$ implies $X \vee Y \in \mathcal{X}$, then there is a nondecreasing sequence $\{Z_n\}_{n=1}^\infty$ of random variables in $\mathcal{X}$ satisfying $X^* = \lim_{n \to \infty} Z_n$ almost surely.*

PROOF. Note first that μ^∞ defined by (A.1), (A.2) is absolutely continuous with respect to P. Define

$$X^* = \frac{d\mu^\infty}{dP}.$$

For all $X \in \mathcal{X}$ and $A \in \mathcal{F}$, we have $E(1_A X) \leq \mu^\infty(A) = E(1_A X^*)$, so condition (i) of Definition A.1 is satisfied. If Y is as in condition (ii) of Definition A.1, then

$$E(1_A X^*) = \mu^\infty(A) = \sup_\pi \mu_\pi^\infty(A) \leq E(1_A Y),$$

and thus $X^* \leq Y$ almost surely.

Now assume that $\mathcal{X}$ is closed under pairwise maximization. Given a positive integer n, because $\mu^n(\Omega) \leq n < \infty$, we may choose an $\mathcal{X}$-partition

$\pi^{(n)} = (K^{(n)}; A_1^{(n)}, \ldots, A_{K^{(n)}}^{(n)}; X_1^{(n)}, \ldots, X_{K^{(n)}}^{(n)})$ of Ω satisfying

$$\mu^n(\Omega) \leq E \sum_{k=1}^{K^{(n)}} 1_{A_k^{(n)}} \left(X_k^{(n)} \wedge n \right) + \frac{1}{n}.$$

Then $Y_n \overset{\triangle}{=} X_1^{(n)} \vee \cdots \vee X_{K^{(n)}}^{(n)}$ is in $\mathcal{X}$ and $\mu^n(\Omega) \leq EY_n + \frac{1}{n}$. Likewise, $Z_n \overset{\triangle}{=} Y_1 \vee \cdots \vee Y_n$ is in $\mathcal{X}$ and $\mu^n(\Omega) \leq EZ_n + \frac{1}{n}$. Letting $n \to \infty$ in this inequality, we obtain

$$EX^* = \mu^\infty(\Omega) \leq \lim_{n\to\infty} EZ_n = E(\lim_{n\to\infty} Z_n).$$

Because $Z_n \in \mathcal{X}$ for each n, we have $Z^n \leq X^*$ and $\lim_{n\to\infty} Z_n \leq X^*$ almost surely. Hence, $X^* = \lim_{n\to\infty} Z_n$ almost surely. $\square$

Appendix B
On the Model of Section 1.1.

In this appendix we return to the market model of Section 1.1. Recall that $(\Omega, \mathcal{F}, P)$ is a complete probability space; $W(t) = (W^{(1)}(t), \ldots, W^{(D)}(t))'$, $0 \le t \le T$, is a standard D-dimensional Brownian motion on this space; and $\{\mathcal{F}(t)\}_{0 \le t \le T}$ is the augmentation by P-null sets of the filtration generated by $W(\cdot)$. The purpose of this appendix is to verify two assertions made in Section 1.1. We first show that if $S(\cdot)$ is a strictly positive and continuous semimartingale with respect to $\{\mathcal{F}(t)\}_{0 \le t \le T}$, then $S(\cdot)$ has the form

$$dS(t) = S(t) \left[\beta(t)\, dt + d\alpha(t) + \sum_{d=1}^{D} \rho_d(t)\, dW^{(d)}(t) \right], \quad 0 \le t \le T, \quad \text{(B.1)}$$

where $\alpha(\cdot)$ is singularly continuous. Applying this result to each one of the stock-price processes $S_1(\cdot), \ldots, S_N(\cdot)$ of Section 1.1, we conclude that $S_n(\cdot)$ has the form

$$dS_n(t) = S_n(t) \left[b_n(t)\, dt + dA_n(t) + \sum_{d=1}^{D} \sigma_{nd}(t)\, dW^{(d)}(t) \right] \quad \text{(B.2)}$$

for each $n = 1, \ldots, N$. Our second result is that each process $A_n(\cdot)$ of this representation (B.2) must agree with $A(\cdot)$, the process appearing in the representation

$$dS_0(t) = S_0(t)[r(t)\, dt + dA(t)] \quad \text{(B.3)}$$

of the money-market price process $S_0(\cdot)$ as in (1.1.6). If this were not the case, the model would admit arbitrage opportunities.

Proposition B.1:   *Let $S : [0,T] \times \Omega \rightarrow (0,\infty)$ be an $\{\mathcal{F}(t)\}_{0 \leq t \leq T}$-semimartingale with continuous paths. Then (B.1) holds, where $\beta(\cdot)$ and $\rho_d(\cdot), d = 1, \ldots, D$ are suitable $\{\mathcal{F}(t)\}_{0 \leq t \leq T}$-progressively measurable processes with $\int_0^T [|\beta(t)| + \sum_{d=1}^D (\rho_d(t))^2] dt < \infty$ almost surely, and $\alpha(\cdot)$ is an $\{\mathcal{F}(t)\}_{0 \leq t \leq T}$-progressively measurable process with paths that are continuous but singular with respect to Lebesgue measure; i.e., $d\alpha(t)/dt = 0$ for Lebesgue-almost every $t \in [0,T]$ almost surely, and with $\alpha(0) = 0$.*

PROOF.   By the definition of a semimartingale, $S(\cdot)'$ can be written as $S(t) = S(0) + B(t) + M(t)$, $0 \leq t \leq T$, where $B(\cdot)$ is an $\{\mathcal{F}(t)\}_{0 \leq t \leq T}$-adapted process with RCLL paths of finite variation on $[0,T], M(\cdot)$ is an $\{\mathcal{F}(t)\}_{0 \leq t \leq T}$-local martingale, and $B(0) = M(0) = 0$. From the martingale representation theorem for the augmented Brownian filtration $\{\mathcal{F}(t)\}_{0 \leq t \leq T}$ (e.g. Karatzas and Shreve (1991), pp. 182–184), we have the integral representation $M(t) = \sum_{d=1}^D \int_0^t \psi_d(s) \, dW^{(d)}(s)$, $0 \leq t \leq T$, for suitable $\{\mathcal{F}(t)\}_{0 \leq t \leq T}$-progressively measurable processes $\psi_1(\cdot), \ldots, \psi_D(\cdot)$ with $\sum_{d=1}^D \int_0^T (\psi_d(s))^2 \, ds < \infty$ almost surely. In particular, $M(\cdot)$ has continuous paths, and since this is true for $S(\cdot)$, it is also true for $B(\cdot)$. We can then decompose $B(\cdot)$ into its absolutely continuous part $B^{ac}(t) = \int_0^t \frac{d}{ds}B(s) \, ds$ and its singularly continuous part $B^s(t) = B(t) - B^{ac}(t)$. Noting that each path of the continuous, positive process $S(\cdot)$ must be bounded away from zero, we obtain the desired representation, with

$$\beta(t) = \frac{1}{S(t)} \cdot \frac{dB(t)}{dt}, \alpha(t) = \int_0^t \frac{1}{S(u)} dB^s(u), \rho_d(t) = \frac{\psi_d(t)}{S(t)}, \quad d = 1, \ldots, D.$$

$\square$

Consider now a financial market $\mathcal{M}$, consisting in part of a *money market* with price per share $S_0(\cdot)$ governed by (B.3), where $r(\cdot)$ and $A(\cdot)$ are $\{\mathcal{F}(t)\}_{0 \leq t \leq T}$-progressively measurable, $A(\cdot)$ is singularly continuous with $A(0) = 0$, and $\int_0^T |r(t)| dt < \infty$ holds almost surely. The remainder of $\mathcal{M}$ consists of N stocks with prices per share $S_n(\cdot)$ that are continuous, strictly positive $\{\mathcal{F}(t)\}_{0 \leq t \leq T}$-semimartingales, $n = 1, \ldots, N$. According to Proposition B.1 these prices admit the representation (B.2), where all the processes $b_n(\cdot), A_n(\cdot), \sigma_{nd}(\cdot)$ appearing in (B.2) are $\{\mathcal{F}(t)\}_{0 \leq t \leq T}$-progressively measurable, $\int_0^T \sum_{n=1}^N [|b_n(t)| + \sum_{d=1}^D (\sigma_{nd}(t))^2] dt < \infty$ holds almost surely, each $A_n(\cdot)$ is singularly continuous, and $A_n(0) = 0$. This is the model of Definition 1.3.1, except that here we permit the different stocks to have singularly continuous parts that differ from $A(\cdot)$, and for the sake of simplicity we are setting $\delta_n(\cdot) \equiv 0$ for all $n = 1, \ldots, N$. As in Definition 1.3.1, we set $S_0(0) = 1$, but allow $S_1(0), \ldots, S_N(0)$ to be arbitrary positive constants.

Following Definition 1.2.1, we define the *gains process* $G^\pi(\cdot)$ corresponding to a self-financed portfolio $\pi(\cdot) = (\pi_1(\cdot), \ldots, \pi_N(\cdot))'$ to be the solution to the stochastic differential equation

$$dG^\pi(t) = \left(G^\pi(t) - \sum_{n=1}^{N} \pi_n(t)\right)\frac{dS_0(t)}{S_0(t)} + \sum_{n=1}^{N} \pi_n(t)\frac{dS_n(t)}{S_n(t)}$$
$$= G^\pi(t)[r(t)\,dt + dA(t)] + \pi'(t)[(b(t) - r(t)\underline{1})\,dt + dB(t)$$
$$+ \sigma(t)\,dW(t)] \tag{B.4}$$

with the initial condition $G^\pi(0) = 0$, where

$$B_n(\cdot) \overset{\triangle}{=} A_n(\cdot) - A(\cdot), \quad B(\cdot) \overset{\triangle}{=} (B_1(\cdot), \dots, B_N(\cdot))'. \tag{B.5}$$

By analogy with Definitions 1.2.4 and 1.4.1, a self-financed portfolio $\pi(\cdot)$ is called

(i) *tame* if the process $G_\pi(\cdot)/S_0(\cdot)$ is almost surely bounded from below;

(ii) an *arbitrage opportunity* if it is tame and satisfies

$$P[G^\pi(T) \geq 0] = 1, \quad P[G^\pi(T) > 0] > 0. \tag{B.6}$$

Theorem B.2: *In order to exclude arbitrage opportunities from the model $\mathcal{M}$ of (B.2), it is necessary that*

$$A_n(\cdot) = A(\cdot), \quad \forall n = 1, \dots, N, \tag{B.7}$$

hold almost surely.

The proof of Theorem B.2, which occupies the remainder of this section, uses a random time change. To set up the notation, we rewrite (B.4) in the more compact form

$$\frac{G^\pi(t)}{S_0(t)} = \int_0^t \frac{\pi'(u)}{S_0(t)}[dF(u) + \sigma(u)\,dW(u)], \quad 0 \leq t \leq T, \tag{B.8}$$

where

$$F_n(t) \overset{\triangle}{=} B_n(t) + \int_0^t (b_n(u) - r(u))\,du, \quad F(t) \overset{\triangle}{=} (F_1(t), \dots, F_N(t))'. \tag{B.9}$$

We denote by $\check{F}_n(t)$ the total variation of $F_n(\cdot)$ on the interval $[0,t]$.

Lemma B.3: *Introduce the continuous, strictly increasing process*

$$C(t) \overset{\triangle}{=} t + \sum_{n=1}^{N} \check{F}_n(t), \quad 0 \leq t \leq T, \tag{B.10}$$

and its continuous, nondecreasing inverse

$$\tau(s) \overset{\triangle}{=} \inf\{t \in [0,T]; C(t) > s\} \wedge T, \quad 0 \leq s < \infty. \tag{B.11}$$

(i) *For every $s \in [0,\infty)$, $\tau(s)$ is a stopping time of the filtration $\{\mathcal{F}(t)\}_{0 \leq t \leq T}$.*

(ii) *The time-changed filtration*

$$\mathcal{H}(s) \stackrel{\triangle}{=} \mathcal{F}(\tau(s)), \quad 0 \le s < \infty,$$

satisfies the "usual conditions" of being right continuous and containing all the P-null sets of $\mathcal{H}(\infty) = \mathcal{F}(T)$.

(iii) *If $\{X(t); 0 \le t \le T\}$ is an $\{\mathcal{F}(t)\}_{0 \le t \le T}$-progressively measurable process, then $\{\tilde{X}(s) = X(\tau(s)); 0 \le s < \infty\}$ is $\{\mathcal{H}(s)\}_{0 \le s < \infty}$-progressively measurable.*

(iv) *The processes $\tau(\cdot)$ and*

$$H_n(s) \stackrel{\triangle}{=} F_n(\tau(s)), \quad 0 \le s < \infty, \quad n = 1, \ldots, N, \qquad \text{(B.12)}$$

are Lipschitz continuous, with Lipschitz constant 1. In particular,

$$0 \le \dot{\tau}(s) \le 1, \quad \tau(s) = \int_0^s \dot{\tau}(v)\, dv, \quad 0 \le s < \infty, \qquad \text{(B.13)}$$

$$|\dot{H}_n(s)| \le 1, H_n(s) = \int_0^s \dot{H}_n(v)\, dv, \ 0 \le s < \infty, \ n = 1, \ldots, N, \text{(B.14)}$$

where the derivatives $\dot{\tau}(\cdot)$ and $\dot{H}_n(\cdot)$ are $\{\mathcal{H}(s)\}_{0 \le s < \infty}$-progressively measurable.

PROOF. For every $s \ge 0$ and $0 \le t \le T$, we have $\{\tau(s) < t\} = \{C(t) > s\} \in \mathcal{F}(t)$. Therefore, $\{\tau(s) \le t\} = \bigcap_{n=k}^\infty \{\tau(s) < t + \frac{1}{n}\} \in \mathcal{F}((t + \frac{1}{k}) \wedge T)$ for every positive integer k, so $\{\tau(s) \le t\} \in \mathcal{F}(t+) = \mathcal{F}(t)$ (cf. Remark 1.1.1), and (i) is established.

The σ-algebra $\mathcal{H}(s)$ is defined as the collection of sets $A \in \mathcal{F}(T)$ such that $A \cap \{\tau(s) \le t\} \in \mathcal{F}(t), \forall t \in [0, T]$; but because the filtration $\{\mathcal{F}(t)\}_{0 \le t \le T}$ is right continuous, this condition is easily seen to be equivalent to

$$A \cap \{\tau(s) < t\} \in \mathcal{F}(t), \quad \forall t \in [0, T].$$

Since $\mathcal{F}(t)$ contains all P-null sets of $\mathcal{F}(T)$ for $0 \le t \le T$, so does $\mathcal{H}(s)$. Furthermore, $\mathcal{H}(s) \subset \mathcal{F}(T)$ for every $s \ge 0$, so $\mathcal{H}(\infty) \stackrel{\triangle}{=} \sigma\left(\bigcup_{0 \le s < \infty} \mathcal{H}(s)\right)$ is also contained in $\mathcal{F}(T)$. On the other hand, any set $A \in \mathcal{F}(T)$ is of the form $A = \cup_{n=1}^\infty A_n$, where $A_n \stackrel{\triangle}{=} A \cap \{C(T) \le n\}$, and for $0 \le t \le T$ we have

$$A_n \cap \{\tau(n) < t\} = A_n \cap \{C(t) > n\} = \emptyset \in \mathcal{F}(t).$$

It follows that $A_n \in \mathcal{H}(n)$, and hence $A \in \mathcal{H}(\infty)$. We have shown that $\mathcal{F}(T) = \mathcal{H}(\infty)$.

To complete (ii), it remains to show that $\{\mathcal{H}(s)\}_{0 \le s < \infty}$ is right continuous. Let $s \ge 0$ be given, and let $s_n \downarrow s$ with $s_n > s$ for every n. Let $A \in \mathcal{H}(s+) = \bigcap_{n=1}^\infty \mathcal{F}(\tau(s_n))$ be given. Then $A \cap \{\tau(s_n) < t\} \in \mathcal{F}(t)$ for every $t \in [0, T]$ and every positive integer n. In particular, each $\mathcal{F}(t)$

contains the set

$$\bigcup_{n=1}^{\infty} \Big(A \cap \{\tau(s_n) < t\} \Big) = \bigcup_{n=1}^{\infty} \Big(A \cap \{C(t) > s_n\} \Big)$$
$$= A \cap \{C(t) > s\}$$
$$= A \cap \{\tau(s) < t\},$$

which shows that $A \in \mathcal{H}(s)$.

We turn next to (iii). Let $X(\cdot)$ and $\tilde{X}(\cdot)$ be as in the statement of the lemma. For each $s \in [0, \infty), \tau(s)$ is $\mathcal{H}(s)$-measurable. An adapted, continuous process is progressively measurable (cf. Karatzas and Shreve (1991), Chapter 1, Proposition 1.13), so $\{\tau(s); 0 \leq s < \infty\}$ is $\{\mathcal{H}(s)\}_{0 \leq s < \infty}$-progressively measurable. Consider the mapping $\varphi : [0, \infty) \times \Omega \to [0, T] \times \Omega$ defined by

$$\varphi(s, \omega) \stackrel{\triangle}{=} (\tau(s, \omega), \omega), \quad \forall \, (s, \omega) \in [0, \infty) \times \Omega.$$

Fix $r \in (0, \infty)$, and set $\Omega_t \stackrel{\triangle}{=} \{\omega \in \Omega; \tau(r, \omega) \leq t\}, 0 \leq t \leq T$. Then $\varphi|_{[0,r] \times \Omega_t}$ is $\mathcal{B}([0, r]) \otimes \mathcal{F}(t) \,/\, \mathcal{B}([0, t]) \otimes \mathcal{F}(t)$-measurable. By assumption, $X(\cdot, \cdot)|_{[0,t] \times \Omega}$ is $\mathcal{B}([0, t]) \otimes \mathcal{F}(t) \,/\, \mathcal{B}(\mathbb{R})$-measurable. Therefore,

$$(X \circ \varphi)(s, \omega) = X(\tau(s, \omega), \omega),$$

when restricted to $[0, r] \times \Omega_t$, is $\mathcal{B}([0, t]) \otimes \mathcal{F}(t) \,/\, \mathcal{B}(\mathbb{R})$-measurable; i.e.,

$$\{(s, \omega) \in [0, r] \times \Omega; \; X(\tau(s, \omega), \omega) \in B, \; \tau(r, \omega) \leq t\}$$
$$\in \mathcal{B}([0, r]) \otimes \mathcal{F}(t), \quad \forall B \in \mathcal{B}(\mathbb{R}).$$

In particular, for every $B \in \mathcal{B}(\mathbb{R})$, we have

$$\{\omega \in \Omega; X(\tau(r, \omega), \omega) \in B, \tau(r, \omega) \leq t\} \in \mathcal{F}(t), \quad \forall t \in [0, T],$$

which shows that $X(\tau(r))$ is $\mathcal{H}(r)$-measurable. Now let $\epsilon \in (0, r)$ be given, and set

$$A_{n,k} \stackrel{\triangle}{=} \{\omega \in \Omega; (k - 1)2^{-n} < \tau(r - \epsilon, \omega) \leq k2^{-n} < \tau(r, \omega)\}$$

for each pair (n, k) of positive integers. We have $A_{n,k} \in \mathcal{H}(r) \cap \mathcal{F}(k2^{-n})$, and φ restricted to $[0, r - \epsilon] \times A_{n,k}$ is $\mathcal{B}([0, r - \epsilon]) \otimes \mathcal{H}(r) / \mathcal{B}([0, k2^{-n}]) \otimes \mathcal{F}(k2^{-n})$-measurable. It follows that $(s, \omega) \mapsto X(\tau(s, \omega), \omega)$ restricted to $[0, r - \epsilon] \times A_{n,k}$ is $\mathcal{B}([0, r - \epsilon]) \otimes \mathcal{H}(r) / \mathcal{B}(\mathbb{R})$-measurable. But $\bigcup_{n=1}^{\infty} \bigcup_{k=1}^{\infty} A_{n,k} = \Omega$, and we see that $X(\tau(s, \omega), \omega)$ is a $\mathcal{B}([0, r - \epsilon]) \otimes \mathcal{H}(r) / \mathcal{B}(\mathbb{R})$-measurable mapping from $[0, r - \epsilon] \times \Omega$ to $\mathbb{R}$. Since $\epsilon \in [0, r)$ is arbitrary, the mapping $(s, \omega) \mapsto X(\tau(s, \omega), \omega)$ is $\mathcal{B}([0, r)) \otimes \mathcal{H}(r) / \mathcal{B}(\mathbb{R})$-measurable on $[0, r) \times \Omega$. We have already shown that $\omega \mapsto X(\tau(r, \omega), \omega)$ is $\mathcal{H}(r) / \mathcal{B}([0, r])$-measurable, and we conclude that $\tilde{X}(\cdot, \cdot) \,|_{[0,r] \times \Omega}$ is $\mathcal{B}([0, r]) \otimes \mathcal{H}(r) / \mathcal{B}(\mathbb{R})$-measurable. This is the requirement of progressive measurability.

To prove (iv), we note that (B.10) gives, almost surely,

$$C(t_2) - C(t_1) = (t_2 - t_1) + \sum_{n=1}^{N} (\check{F}_n(t_2) - \check{F}_n(t_1)), \quad 0 \leq t_1 \leq t_2 \leq T,$$

whence also

$$s_2 - s_1 \geq C(\tau(s_2)) - C(\tau(s_1))$$
$$= \tau(s_2) - \tau(s_1) + \sum_{n=1}^{N} (\check{F}_n(\tau(s_2)) - \check{F}_n(\tau(s_1)))$$
$$\geq \tau(s_2) - \tau(s_1) + \sum_{n=1}^{N} |H_n(s_2) - H_n(s_1)|, \quad 0 \leq s_1 \leq s_2 < \infty.$$

This proves the Lipschitz continuity of $\tau(\cdot)$ and $H_1(\cdot), \ldots, H_N(\cdot)$, with Lipschitz constant equal to 1. □

We note that for proving Theorem B.2, it is enough to show that the process

$$\left\{ \begin{array}{c} B_n(\cdot) = A_n(\cdot) - A(\cdot) \text{ of (B.5) is absolutely} \\ \text{continuous with respect to Lebesgue measure} \\ \text{almost surely, for every } n = 1, \ldots, N. \end{array} \right\} \qquad (B.15)$$

PROOF OF THEOREM B.2. Following the notation of Lemma B.3 (iii), whenever the process $\{X(t); 0 \leq t \leq T\}$ is $\{\mathcal{F}(t)\}_{0 \leq t \leq T}$-progressively measurable, we denote by $\{\tilde{X}(s) = X(\tau(s)); 0 \leq s < \infty\}$ the corresponding time-changed process. (The sole exception to this convention is the use of $H_n(\cdot)$ to denote the time-change of $F_n(\cdot)$; see (B.12).) With this notation, equation (B.8) becomes (cf. Karatzas and Shreve (1991), Problem 3.4.5(vi) and Proposition 3.4.8)

$$\frac{\tilde{G}^{\pi}(s)}{\tilde{S}_0(s)} = \sum_{n=1}^{N} \int_0^s \frac{1}{\tilde{S}_0(v)} \tilde{\pi}_n(v) \left[\dot{H}_n(v) \, dv + \sum_{d=1}^{D} \tilde{\sigma}_{nd}(v) \, d\tilde{W}^{(d)}(v) \right],$$
$$0 \leq s \leq C(T), \qquad (B.16)$$

in view of (B.14). Each component of the D-dimensional process $\{\tilde{W}(s) \stackrel{\triangle}{=} W(\tau(s)), 0 \leq s < \infty\}$ is an $\{\mathcal{H}(s)\}_{0 \leq s < \infty}$-martingale, with

$$\langle \tilde{W}^{(k)}, \tilde{W}^{(l)} \rangle(s) = \delta_{kl} \cdot \tau(s) = \delta_{kl} \cdot \int_0^s \dot{\tau}(v) \, dv, \qquad (B.17)$$

from (B.13). Therefore, there exists a D-dimensional Brownian motion $\tilde{B}(\cdot)$ on the probability space $(\Omega, \mathcal{F}, P)$ (possibly extended to accommodate additional independent Brownian motions; see Karatzas and Shreve (1991), pp. 170–173), such that

$$\tilde{W}^{(d)}(s) = \int_0^s \sqrt{\dot{\tau}(v)} \, d\tilde{B}^{(d)}(v), \quad 0 \leq s < \infty, \quad d = 1, \ldots, D, \qquad (B.18)$$

almost surely.

Now consider the self-financed portfolio process $\pi(\cdot)$ given as

$$\pi_n(t) = \tilde{\pi}_n(C(t)), \quad 0 \le t \le T, \quad n = 1, \ldots, N, \tag{B.19}$$

where

$$\tilde{\pi}_n(s) \stackrel{\triangle}{=} \operatorname{sgn}(\dot{H}_n(s)) \cdot 1_{\{\dot{\tau}(s)=0\}}, \quad 0 \le s < \infty, \quad n = 1, \ldots, N. \tag{B.20}$$

In light of (B.16), (B.18) we have then

$$\frac{\tilde{G}^\pi(s)}{\tilde{S}_0(s)} = \sum_{n=1}^{N} \int_0^s \frac{1}{\tilde{S}_0(v)} |\dot{H}_n(v)| \cdot 1_{\{\dot{\tau}(v)=0\}} \, dv, \quad 0 \le s \le C(T),$$

or equivalently, inverting the time change,

$$\frac{G^\pi(t)}{S_0(t)} = \sum_{n=1}^{N} \int_0^{C(t)} \frac{1}{\tilde{S}_0(v)} |\dot{H}_n(v)| \cdot 1_{\{\dot{\tau}(v)=0\}} \, dv \ge 0, \quad 0 \le t \le T. \tag{B.21}$$

In particular, the self-financed portfolio $\pi(\cdot)$ is *tame*. Suppose now that

$$\left\{ \begin{array}{c} \text{for some } n = 1, \ldots, N \text{ and for some } \Omega_o \in \mathcal{F} \text{ with} \\ P(\Omega_o) > 0, \text{ we have for every } \omega \in \Omega_o, \\ \operatorname{meas}\{s \in [0, C(T, \omega)]; \dot{H}_n(s, \omega) \ne 0, \dot{\tau}(s, \omega) = 0\} > 0. \end{array} \right\} \tag{B.22}$$

Then from (B.21), (B.22), the random variable $G^\pi(T)$ is almost surely nonnegative, and is positive with positive probability (i.e., satisfies the conditions (B.6)); consequently, the tame portfolio $\pi(\cdot)$ of (B.19), (B.20) is an arbitrage opportunity in $\mathcal{M}$. It follows that *viability* for $\mathcal{M}$, that is, absence of arbitrage opportunities, implies the negation of (B.22), namely,

$$\left\{ \begin{array}{c} \operatorname{meas}\{s \in [0, C(T, \omega)]; \dot{H}_n(s, \omega) \ne 0, \ \dot{\tau}(s, \omega) = 0\} = 0 \\ \text{holds for all } n = 1, \ldots, N, \forall \omega \in \Omega^*, \\ \text{for some event } \Omega^* \in \mathcal{F} \text{ with } P(\Omega^*) = 1. \end{array} \right\} \tag{B.23}$$

The following result then completes the proof of Theorem B.2.   □

Lemma B.4: *Condition (B.23) implies (B.15).*

PROOF.   Assume (B.23); from (B.9) it suffices to show that for each $n = 1, \ldots, N$ and $\omega \in \Omega^*$, we have

$$F_n(\cdot, \omega) \text{ is absolutely continuous on } [0, T] \tag{B.24}$$

with respect to Lebesgue measure. Fix $n = 1, \ldots, N$, $\omega \in \Omega^*$, $\epsilon > 0$, and choose $\delta > 0$ such that

$$\sum_{j=1}^{m} [C^{ac}(t'_j, \omega) - C^{ac}(t_j, \omega)] < \epsilon$$

holds for every finite collection $\{(t_j, t'_j)\}_{j=1}^{m}$ of nonoverlapping intervals in $[0, T]$ with $\sum_{j=1}^{m}(t'_j - t_j) < \delta$. Here $C^{ac}(\cdot, \omega)$ is the absolutely continuous part of the continuous function $C(\cdot, \omega)$ of (B.10). For every $n = 1, \ldots, N$, it follows from Lemma B.3(iv), (B.12), and (B.23) that

$$\sum_{j=1}^{m} |F_n(t'_j, \omega) - F_n(t_j, \omega)| = \sum_{j=1}^{m} |H_n(C(t'_j, \omega), \omega) - H_n(C(t_j, \omega), \omega)|$$

$$\leq \sum_{j=1}^{m} \int_{C(t_j, \omega)}^{C(t'_j, \omega)} |\dot{H}_n(v, \omega)| \, dv$$

$$= \sum_{j=1}^{m} \int_{C(t_j, \omega)}^{C(t'_j, \omega)} |\dot{H}_n(v, \omega)| \cdot 1_{\{\dot{\tau}(v, \omega) > 0\}} \, dv$$

$$\leq \sum_{j=1}^{m} \int_{C(t_j, \omega)}^{C(t'_j, \omega)} \dot{C}(\tau(v, \omega), \omega)\dot{\tau}(v, \omega) \, dv$$

$$= \sum_{j=1}^{m} \int_{t_j}^{t'_j} \dot{C}(\theta, \omega) \, d\theta$$

$$= \sum_{j=1}^{m} [C^{ac}(t'_j, \omega) - C^{ac}(t_j, \omega)] < \epsilon. \qquad \square$$

Appendix C
On Theorem 6.4.1

The purpose of this appendix is to establish the implication (A)$\Rightarrow$(B), and thereby complete the proof of Theorem 6.4.1. For this we assume throughout this appendix the conditions (3.4.15), (3.6.17) as well as (6.2.13) and (6.2.14), which are in force throughout Chapter 6. We assume also that there exists a pair $(\hat{c}, \hat{p}) \in \mathcal{A}_3(x; K)$ that is optimal for the consumption/investment Problem 6.2.6 in the constrained market $\mathcal{M}(K)$, and that $V(x; K) < \infty$. We denote by $\widehat{X}(\cdot) = X^{x,\hat{c},\hat{p}}(\cdot)$ the corresponding wealth process, which satisfies

$$\frac{\widehat{X}(t)}{S_0(t)} + \int_0^t \frac{\hat{c}(u)\,du}{S_0(u)} = x + \int_0^t \frac{\widehat{X}(u)}{S_0(u)}\hat{p}'(u)\sigma(u)\,dW_0(u), \quad 0 \le t \le T, \quad \text{(C.1)}$$

and

$$H_0(t)\widehat{X}(t) + \int_0^t H_0(u)\hat{c}(u)\,du = x + \int_0^t H_0(u)\widehat{X}(u)\lambda'(u)\,dW(u),$$
$$0 \le t \le T, \quad \text{(C.2)}$$

according to (6.3.1), (6.3.2), and (6.3.3), with

$$\lambda(t) \triangleq \sigma'(t)\hat{p}(t) - \theta(t), \quad 0 \le t \le T. \quad \text{(C.3)}$$

Our program is as follows. We shall construct a process $\hat{\nu}(\cdot) \in \mathcal{D}_0$ for which the positive process $\widehat{X}(\cdot)$ can be represented as

$$\widehat{X}(t) = \frac{1}{H_{\hat{\nu}}(t)} E\left[\int_t^T H_{\hat{\nu}}(u) c_{\hat{\nu}}(u)\, du + H_{\hat{\nu}}(T) B_{\hat{\nu}} \,\bigg|\, \mathcal{F}(t) \right], \quad 0 \le t \le T,$$
(C.4)

and the requirements

$$\hat{c}(t) = c_{\hat{\nu}}(t), \text{ Lebesgue-a.e. } t \in [0,T], \tag{C.5}$$
$$\zeta(\hat{\nu}(t)) + \hat{p}'(t)\hat{\nu}(t) = 0, \text{ Lebesgue-a.e. } t \in [0,T] \tag{C.6}$$

hold almost surely. Then we shall have $X_{\hat{\nu}}(\cdot) = \widehat{X}(\cdot) = X^{x,\hat{c},\hat{p}}(\cdot)$ from (6.3.18), (C.4), and Remark 6.3.3. We may also then take $p_{\hat{\nu}}(\cdot) = \hat{p}(\cdot)$, and (B) will be established. We prove the above facts in a series of lemmas.

Lemma C.1: *Suppose that for every $\epsilon \in (0,1)$, there is a consumption and portfolio-proportion process pair $(c_\epsilon, p_\epsilon) \in \mathcal{A}(x;K)$ satisfying*

$$c_\epsilon(t) \ge \alpha(\epsilon)\hat{c}(t), \text{ Lebesgue-a.e. } t \in [0,T] \text{ and } X^{x,c_\epsilon,p_\epsilon}(T) \ge \beta(\epsilon)\widehat{X}(T)$$
(C.7)

almost surely, where $\alpha(\epsilon) > 0$ and $\beta(\epsilon) > 0$ satisfy

$$0 < \lim_{\epsilon \downarrow 0} \frac{1-\alpha(\epsilon)}{\epsilon} < \infty, \quad 0 < \lim_{\epsilon \downarrow 0} \frac{1-\beta(\epsilon)}{\epsilon} < \infty.$$

Then we have $(c_\epsilon, p_\epsilon) \in \mathcal{A}_3(x;K)$ for each $\epsilon \in (0,1)$, and

$$E\left[\int_0^T \overline{\lim_{\epsilon \downarrow 0}}\, \frac{U_1(t,\hat{c}(t)) - U_1(t,c_\epsilon(t))}{\epsilon}\, dt \right.$$
$$\left. + \overline{\lim_{\epsilon \downarrow 0}}\, \frac{U_2(\widehat{X}(T)) - U_2(X^{x,c_\epsilon,p_\epsilon}(T))}{\epsilon} \right] \ge 0. \tag{C.8}$$

PROOF.   The convexity of U_2 implies the inequality

$$U_2(\xi_2) - U_2(\xi_1) \le (\xi_2 - \xi_1)U_2'(\xi_1) \tag{C.9}$$

for all $\xi_1, \xi_2 > 0$. Taking $\xi_1 = \xi \ge 1$ and $\xi_2 = 1$, we obtain

$$\xi U_2'(\xi) \le U_2(\xi) - U_2(1) + U_2'(\xi) \le U_2(\xi) - U_2(1) + U_2'(1), \quad \xi \ge 1.$$

On the other hand,

$$\xi U_2'(\xi) \le U_2'(1), \quad 0 < \xi \le 1,$$

follows from (3.4.15). Similarly, for all $t \in [0,T]$,

$$\xi U_1'(t,\xi) \le \begin{cases} U_1(t,\xi) - U_1(t,1) + U_1'(t,1), & \xi \ge 1, \\ U_1'(t,1), & 0 < \xi \le 1. \end{cases}$$

These inequalities and the finiteness of $V(x; K)$ imply

$$E\left[\int_0^T \hat{c}(t)U_1'(t,\hat{c}(t))1_{\{\hat{c}(t)>0\}}\,dt + \widehat{X}(T)U_2'(\widehat{X}(T))1_{\{\widehat{X}(T)>0\}}\right] \tag{C.10}$$

$$\leq \kappa + E\left[\int_0^T U_1(t,\hat{c}(t))1_{\{\hat{c}(t)\geq 1\}}\,dt + U_2(\widehat{X}(T))1_{\{\widehat{X}(T)\geq 1\}}\right] < \infty$$

for some real constant κ.

For sufficiently small $\epsilon > 0$, we have $0 < \beta(\epsilon) < 1$. From (C.9), the monotonicity of U_2', and (3.4.15), we have also

$$\begin{aligned}
U_2(\widehat{X}(T)) - U_2(X^{x,c_\epsilon,p_\epsilon}(T)) &\leq (\widehat{X}(T) - X^{x,c_\epsilon,p_\epsilon}(T))U_2'(X^{x,c_\epsilon,p_\epsilon}(T)) \\
&\leq \frac{1-\beta(\epsilon)}{\beta(\epsilon)}\beta(\epsilon)\widehat{X}(T)U_2'(\beta(\epsilon)\widehat{X}(T)) \\
&\leq \frac{1-\beta(\epsilon)}{\beta(\epsilon)}\widehat{X}(T)U_2'(\widehat{X}(T)). \tag{C.11}
\end{aligned}$$

The finiteness of $V(x; K)$ implies $E|U_2(\widehat{X}(T))| < \infty$, which, in conjunction with (C.10) and (C.11), leads to

$$E\left(\min[0, U_2(X^{x,c_\epsilon,p_\epsilon}(T))]\right) > -\infty.$$

A similar argument shows that

$$U_1(t,\hat{c}(t)) - U_1(t,c_\epsilon(t)) \leq \frac{1-\alpha(\epsilon)}{\alpha(\epsilon)}\hat{c}(t)U_1'(t,\hat{c}(t)), \tag{C.12}$$

from which we conclude

$$E\int_0^T \min[0, U_1(t,c_\epsilon(t))]\,dt > -\infty.$$

Hence, $(c_\epsilon, p_\epsilon) \in \mathcal{A}_3(x; K)$ for all sufficiently small $\epsilon \in (0,1)$.

Finally, the optimality of $(\hat{c}, \hat{p})$ implies

$$E\left[\int_0^T \frac{U_1(t,\hat{c}(t)) - U_1(t,c_\epsilon(t))}{\epsilon}\,dt + \frac{U_2(\widehat{X}(T)) - U_2(X^{x,c_\epsilon,p_\epsilon}(T))}{\epsilon}\right] \geq 0.$$

The bounds (C.10), (C.11), and (C.12) permit us to use Fatou's lemma to obtain (C.8).   □

Lemma C.2: *We have the almost sure inequalities*

$$\hat{c}(t) > 0, \text{ for Lebesgue-a.e. } t \in [0,T], \tag{C.13}$$
$$\widehat{X}(t) > 0, \text{ for all } t \in [0,T]. \tag{C.14}$$

PROOF.   We set

$$c_1(t) = \frac{x}{2T}S_0(t), \quad p_1(t) = 0 \quad \text{for } 0 \leq t \leq T,$$

so that the corresponding wealth process satisfies

$$\frac{X_1(t)}{S_0(t)} = x\left(\frac{2T-t}{2T}\right).$$

In particular, $(c_1, p_1) \in \mathcal{A}(x; K)$. For $\epsilon \in (0, 1)$ we define

$$c_\epsilon(t) \overset{\triangle}{=} \epsilon c_1(t) + (1-\epsilon)\hat{c}(t), \quad p_\epsilon(t) \overset{\triangle}{=} \frac{(1-\epsilon)\widehat{X}(t)\hat{p}(t)}{\epsilon X_1(t) + (1-\epsilon)\widehat{X}(t)}, \quad 0 \le t \le T,$$

and note that because K is a convex set containing $\hat{p}(t)$ and 0, it also contains $p_\epsilon(t)$. We set $X_\epsilon(t) \overset{\triangle}{=} \epsilon X_1(t) + (1-\epsilon)\widehat{X}(t)$, and note that

$$x - \int_0^t \frac{c_\epsilon(u)}{S_0(u)}\,du + \int_0^t \frac{X_\epsilon(u)}{S_0(u)}p'_\epsilon(u)\sigma(u)\,dW_0(u)$$

$$= \epsilon\left[x - \int_0^t \frac{c_1(u)}{S_0(u)}\,du\right]$$

$$+ (1-\epsilon)\left[x - \int_0^t \frac{\hat{c}(u)}{S_0(u)}\,du + \int_0^t \frac{\widehat{X}(u)}{S_0(u)}\hat{p}(u)\sigma(u)\,dW_0(u)\right]$$

$$= \frac{X_\epsilon(t)}{S_0(t)}, \tag{C.15}$$

so that $X_\epsilon(\cdot) = X^{x,c_\epsilon,p_\epsilon}(\cdot)$. Thus, we have the situation described in Lemma C.1 with $\alpha(\epsilon) = \beta(\epsilon) = 1 - \epsilon$, and this implies

$$0 \le E\left[\int_0^T \overline{\lim_{\epsilon\downarrow 0}}\frac{U_1(t,\hat{c}(t)) - U_1(t,c_\epsilon(t))}{\epsilon}\,dt + \overline{\lim_{\epsilon\downarrow 0}}\frac{U_2(\widehat{X}(T)) - U_2(X_\epsilon(T))}{\epsilon}\right]$$

$$\le E\left[\int_0^T \overline{\lim_{\epsilon\downarrow 0}}(\hat{c}(t) - c_1(t))U_1(t,c_\epsilon(t))\,dt\right.$$

$$\left. + \overline{\lim_{\epsilon\downarrow 0}}(\widehat{X}(T) - X_1(T))U'_2(X_\epsilon(T))\right]$$

$$\le E\left[\int_0^T \hat{c}(t)U'_1(t,\hat{c}(t))1_{\{\hat{c}(t)>0\}}\,dt + \widehat{X}(T)U'_2(\widehat{X}(T))1_{\{\widehat{X}(T)>0\}}\right.$$

$$\left. - \int_0^T c_1(t)U'_1(t,\hat{c}(t))\,dt - X_1(T)U'_2(\widehat{X}(T))\right]$$

in conjunction with (C.9). Since $X_1(T) > 0$ and $c_1(t) > 0$ for all $t \in [0,T]$ almost surely, (C.10) and (6.2.14) imply (C.13) and

$$P[\widehat{X}(T) > 0] = 1. \tag{C.16}$$

The left-hand side of (C.2) is nonnegative, and the right-hand side is a local martingale under P and thus also a supermartingale. Solving (C.2) for

$H_0(t)\widehat{X}(t)$, we see that this process is also a supermartingale. But a continuous nonnegative supermartingale is absorbed when it hits zero (Karatzas & Shreve (1991), Problem 1.3.29), and thus (C.14) follows from (C.16).   □

We introduce the continuous, nondecreasing process

$$A(t) \triangleq \int_0^t \hat{c}(s)U_1'(s, \hat{c}(s)) \, ds, \quad 0 \leq t \leq T, \tag{C.17}$$

and the continuous martingale

$$M(t) \triangleq E[A(T) + \widehat{X}(T)U_2'(\widehat{X}(T))|\mathcal{F}(t)]$$
$$= EM(T) + \int_0^t \psi'(s) \, dW(s), \quad 0 \leq t \leq T. \tag{C.18}$$

Here $EA(T) \leq EM(T) < \infty$ from (C.10), and the progressively measurable process $\psi: [0, T] \times \Omega \to \mathbb{R}^N$, with $\int_0^T \|\psi(t)\|^2 dt < \infty$ almost surely, is uniquely determined from (C.18) by the martingale representation theorem.

Lemma C.3:   *The consumption rate process $\hat{c}(\cdot)$ is given by*

$$\hat{c}(t) = I_1\left(t, \frac{M(t) - A(t)}{\widehat{X}(t)}\right), \quad \textit{for Lebesgue-a.e. } t \in [0, T] \tag{C.19}$$

almost surely.

PROOF.   From (C.1) we see that

$$d\left(\frac{\widehat{X}(t)}{S_0(t)}\right) = -\frac{\hat{c}(t)}{S_0(t)} + \frac{\widehat{X}(t)}{S_0(t)}\hat{p}'(t)\sigma(t) \, dW_0(t),$$

and the solution to this stochastic differential equation with initial condition $\widehat{X}(0) = x$ is given by

$$\frac{\widehat{X}(t)}{S_0(t)} = x \exp\left\{-\int_0^t \left(f(s) + \frac{1}{2}\|\sigma'(s)\hat{p}(s)\|^2\right) ds + \int_0^t \hat{p}'(s)\sigma(s) \, dW_0(s)\right\}, \tag{C.20}$$

where

$$f(t) \triangleq \frac{\hat{c}(t)}{\widehat{X}(t)} > 0 \quad \text{for Lebesgue-a.e. } t \in [0, T] \tag{C.21}$$

almost surely. We introduce a small random perturbation of the process $f(\cdot)$ as follows:

$$f_\epsilon(t) \triangleq f(t) + \epsilon\rho(t), \quad 0 \leq t \leq T,$$

where $\rho: [0, T] \times \Omega \to \mathbb{R}$ is an arbitrary but fixed progressively measurable process with $|\rho(t)| \leq 1 \wedge f(t)$ and $\epsilon \in (0, 1)$. We further define

$$X_\epsilon(t) \triangleq \widehat{X}(t)e^{-\epsilon \int_0^t \rho(s) \, ds};$$

$$c_\epsilon(t) \triangleq X_\epsilon(t) f_\epsilon(t) = \left[\hat{c}(t) + \epsilon\rho(t)\widehat{X}(t)\right] e^{-\epsilon \int_0^t \rho(s)\, ds}.$$

Clearly, $X^{x, c_\epsilon, \hat{p}}(t) = X_\epsilon(t) > 0$, and thus $(c_\epsilon, \hat{p}) \in \mathcal{A}(x; K)$.

We are now in the situation described by Lemma C.1 with $\alpha(\epsilon) = \beta(\epsilon) = e^{-\epsilon T}$, and conclude that

$$E \int_0^T U_1'(t, \hat{c}(t)) \left[\hat{c}(t) \int_0^t \rho(s)\, ds - \rho(t)\widehat{X}(t)\right] dt$$

$$+ E \left[U_2'(\widehat{X}(T))\widehat{X}(T) \int_0^T \rho(t)\, dt\right] \geq 0. \tag{C.22}$$

We examine further the terms appearing in (C.22). Note first that

$$E \int_0^T \left[U_1'(t, \hat{c}(t))\hat{c}(t) \int_0^t \rho(s)\, ds\right] dt$$

$$= E \int_0^T \int_0^t \rho(s)\, ds\, dA(t) = E \int_0^T \int_s^T dA(t) \rho(s)\, ds$$

$$= E \int_0^T [A(T) - A(s)]\rho(s)\, ds = E \int_0^T [A(T) - A(t)]\rho(t)\, dt. \tag{C.23}$$

Furthermore,

$$E \left[U_2'(\widehat{X}(T))\widehat{X}(T) \int_0^T \rho(t)\, dt\right]$$

$$= E \left[(M(T) - A(T)) \int_0^T \rho(t)\, dt\right]$$

$$= \int_0^T E\left[E[M(T)\rho(t)|\mathcal{F}(t)]\right] dt - \int_0^T E\left[A(T)\rho(t)\right] dt$$

$$= E \int_0^T [M(t) - A(T)]\rho(t)\, dt. \tag{C.24}$$

Substitution of (C.23) and (C.24) into (C.22) yields

$$E \int_0^T \rho(t) \left[M(t) - A(t) - \widehat{X}(t)U_1'(t, \hat{c}(t))\right] dt \geq 0. \tag{C.25}$$

But except for the condition $|\rho(t)| \leq 1 \wedge f(t)$, the process $\rho(\cdot)$ is arbitrary. Because $f(t)$ is strictly positive (see (C.21)), the expression

$$M(t) - A(t) - \widehat{X}(t)U_1'(t, \hat{c}(t))$$

must be zero for Lebesgue-almost-every $t \in [0, T]$ almost surely; otherwise, we could choose $\rho(t)$ such that (C.25) is violated. This leads immediately to (C.19).   □

Lemma C.4: *With $\lambda(\cdot)$ and $\psi(\cdot)$ defined by (C.3) and (C.18), respectively, the process*

$$\hat{\nu}(t) \triangleq \sigma(t) \left[\lambda(t) - \frac{\psi(t)}{M(t) - A(t)} \right], \quad 0 \le t \le T, \qquad (C.26)$$

satisfies (C.6) and

$$\int_0^T \left(\zeta(\hat{\nu}(t)) + \|\hat{\nu}(t)\|^2 \right) dt < \infty, \qquad (C.27)$$

$$\hat{\nu}(t) \in \widetilde{K}, \text{ for Lebesgue-a.e. } t \in [0, T] \qquad (C.28)$$

almost surely.

PROOF. Here again the idea is to start with the expression (C.20) for $\widehat{X}(\cdot)$ and introduce a small random perturbation. This time we perturb the optimal portfolio-proportion process $\hat{p}(\cdot)$ by defining

$$p_\epsilon(t) \triangleq \begin{cases} (1 - \epsilon)\hat{p}(t) + \epsilon\eta(t), & 0 \le t \le \tau_n, \\ \hat{p}(t), & \tau_n < t \le T \end{cases}$$

for some arbitrary but fixed progressively measurable process $\eta \colon [0, T] \times \Omega \to K$ with $\int_0^T \|\eta(t)\|^2 dt < \infty$ almost surely. Here $\epsilon \in (0, 1)$, and $\{\tau_n\}_{n=1}^\infty$, to be chosen later, is a nondecreasing sequence of stopping times satisfying $\lim_{n \to \infty} \tau_n = T$ almost surely. Clearly, $p_\epsilon(\cdot)$ takes values in K. With $f(t) \triangleq \frac{\hat{c}(t)}{\widehat{X}(t)}$ as in (C.21), we introduce

$$\frac{X_\epsilon(t)}{S_0(t)} \triangleq x \exp \left\{ -\int_0^t \left(f(s) + \frac{1}{2} \|\sigma'(s)p_\epsilon(s)\|^2 \right) ds \right. \\ \left. + \int_0^t p_\epsilon'(s)\sigma(s) \, dW_0(s) \right\},$$

$$c_\epsilon(t) \triangleq f(t)X_\epsilon(t),$$

and notice that $X^{x,c_\epsilon,p_\epsilon}(t) = X_\epsilon(t) > 0$, so that $(c_\epsilon, p_\epsilon) \in \mathcal{A}(x; K)$. The processes $p_\epsilon(\cdot)$, $c_\epsilon(\cdot)$ and $X_\epsilon(\cdot)$ depend on n, although we do not indicate this explicitly. We write

$$X_\epsilon(t) = \widehat{X}(t) \exp \left[\epsilon N(t \wedge \tau_n) - \frac{1}{2}\epsilon^2 \langle N \rangle(t \wedge \tau_n) \right],$$

where

$$N(t) \triangleq \int_0^t (\eta(s) - \hat{p}(s))'\sigma(s)\widehat{W}(s),$$

$$\langle N \rangle(t) = \int_0^t \|\sigma'(s)(\eta(s) - \hat{p}(s))\|^2 \, ds,$$

$$\widehat{W}(t) \triangleq W(t) - \int_0^t \lambda(s) \, ds.$$

We can define now the nondecreasing sequence of stopping times

$$\tau_n \overset{\triangle}{=} \inf \left\{ t \in [0, T]; \quad |N(t)| + \langle N \rangle(t) + A(t) + |M(t)| \right.$$
$$\left. + \int_0^t \|\psi(s)\|^2 ds \geq n \right\} \wedge T, \quad n = 1, 2, \ldots,$$

and observe that $\lim_{n \to \infty} \tau_n = T$ almost surely, as well as

$$\frac{c_\epsilon(t)}{\hat{c}(t)} = \frac{X_\epsilon(t)}{\widehat{X}(t)} \geq e^{-\frac{3}{2}\epsilon n}, \quad 0 \leq t \leq T.$$

Thus, for each fixed n, we are in the situation described by Lemma C.1 with $\alpha(\epsilon) = \beta(\epsilon) = e^{-\frac{3}{2}\epsilon n}$. The conclusion of that lemma is

$$0 \geq E\left[\int_0^T N(t \wedge \tau_n)\hat{c}(t)U_1'(t, \hat{c}(t))\, dt \right.$$
$$\left. + N(T \wedge \tau_n)\widehat{X}(T)U_2'(\widehat{X}(T)) \right]$$
$$= E\left[\int_0^T N_n(t)\, dA(t) + N_n(T)(M(T) - A(T)) \right], \qquad (C.29)$$

in the notation (C.17), (C.18) and with $N_n(t) = N(t \wedge \tau_n)$. The product rule of stochastic calculus gives now

$$E\big(M(T)N_n(T)\big) = E\big(M(\tau_n)N(\tau_n)\big)$$
$$= E \int_0^{\tau_n} \big(\eta(t) - \hat{p}(t)\big)' \sigma(t)\big(\psi(t) - M(t)\lambda(t)\big)\, dt,$$

$$E\left[\int_0^T N_n(t)\, dA(t) - A(T)N_n(T) \right]$$
$$= -E\left[\int_0^T A(t)\, dN_n(t) \right]$$
$$= E \int_0^{\tau_n} A(t)\big(\eta(t) - \hat{p}(t)\big)' \sigma(t)\big(-dW(t) + \lambda(t)\, dt\big)$$
$$= E \int_0^{\tau_n} A(t)\big(\eta(t) - \hat{p}(t)\big)' \sigma(t)\lambda(t)\, dt.$$

Substituting back into (C.29), we obtain

$$E \int_0^{\tau_n} \big(M(t) - A(t)\big)\big(\eta(t) - \hat{p}(t)\big)' \hat{\nu}(t)\, dt \geq 0, \qquad (C.30)$$

where $\hat{\nu}(\cdot)$ is defined by (C.26).

The process $M(t) - A(t) = \widehat{X}(t)U_1'(t, \hat{c}(t))$ (see (C.19)) is continuous in t and strictly positive almost surely. Letting p be a fixed vector in K and substituting the process $\eta(t) \overset{\triangle}{=} p1_{\{(p-\hat{p}(t))'\hat{\nu}(t)<0\}} + \hat{p}(t)1_{\{(p-\hat{p}(t))'\hat{\nu}(t)\geq 0\}}$ into (C.30), we see that the set

$$F_p \overset{\triangle}{=} \{(t,\omega) \in [0,T] \times \Omega; \quad (p - \hat{p}(t))'\hat{\nu}(t) < 0\}$$

has Lebesgue $\times$ P-measure zero. Likewise, the set $F \overset{\triangle}{=} \cup_{p\in K\cap\mathbb{Q}^N} F_p$ has product measure zero. (Here $\mathbb{Q}^N$ denotes the set of vectors in $\mathbb{R}^N$ with rational coordinates.) We have

$$p'\hat{\nu}(t,\omega) \geq \hat{p}'(t,\omega)\hat{\nu}(t,\omega), \text{ for all } (t,\omega) \notin F. \tag{C.31}$$

For $(t,\omega) \notin F$, (C.31) implies

$$\zeta(\hat{\nu}(t)) = \sup_{p\in K} \left(-p'\hat{\nu}(t) \right) \leq -\hat{p}'(t)\hat{\nu}(t),$$

and since $\hat{p}'(t) \in K$, we in fact have $\zeta(\hat{\nu}(t)) = -\hat{p}'(t)\hat{\nu}(t)$ for Lebesgue-almost-every $t \in [0,T]$ almost surely. This is (C.6).

From (C.6) it follows that $\zeta(\hat{\nu}(t)) < \infty$, and hence $\hat{\nu}(t) \in \widetilde{K}$ for Lebesgue-almost-every $t \in [0,T]$ almost surely. This is (C.28).

We next note that $\widehat{X}(T) > 0$ implies $\int_0^T \|\hat{p}(t)\|^2 dt < \infty$ almost surely (Definition 6.2.4). The uniform boundedness assumed for $\sigma(\cdot)$ implies that $\lambda(\cdot)$ defined by (C.3) satisfies $\int_0^T \|\lambda(t)\|^2 dt < \infty$ almost surely. For fixed ω, the positive continuous process $M(\cdot) - A(\cdot)$ is bounded away from zero uniformly in t, and $\int_0^T \|\psi(t)\|^2 dt < \infty$ holds almost surely. It follows from these facts that $\hat{\nu}(\cdot)$ defined by (C.26) satisfies $\int_0^T \|\hat{\nu}(t)\|^2 dt < \infty$ almost surely. Furthermore,

$$\int_0^T \zeta(\hat{\nu}(t))\, dt \leq \int_0^T |\hat{p}'(t)\hat{\nu}(t)|dt \leq \left(\int_0^T \|\hat{p}(t)\|^2 dt \right)^{\frac{1}{2}} \left(\int_0^T \|\hat{\nu}(t)\|^2 dt \right)^{\frac{1}{2}},$$

which is almost surely finite. Condition (C.27) is satisfied, and the lemma is established. $\qquad \square$

It remains to prove that $\hat{\nu}(\cdot) \in \mathcal{D}_0$ and (C.4), (C.5) are satisfied. We first prove (C.4), (C.5).

Lemma C.5: *Equations (C.4) and (C.5) hold.*

PROOF.   Before beginning the computations, we recall the equations

$$\lambda(t) = \sigma'(t)\hat{p}(t) - \theta(t), \tag{C.3}$$

$$\zeta(\hat{\nu}(t)) = -\hat{p}'(t)\hat{\nu}(t), \tag{C.6}$$

$$A(t) = \int_0^t \hat{c}(s)U_1'(s, \hat{c}(s))\, ds, \tag{C.17}$$

$$M(t) = E[A(T) + \widehat{X}(T)U_2'(\widehat{X}(T))|\mathcal{F}(t)]$$

$$= EM(T) + \int_0^t \psi'(s)\, dW(s), \tag{C.18}$$

$$f(t) = \frac{\hat{c}(t)}{\widehat{X}(t)}, \tag{C.21}$$

$$\hat{\nu}(t) = \sigma(t)\left[\lambda(t) - \frac{\psi(t)}{M(t) - A(t)}\right], \tag{C.26}$$

$$S^{(\hat{\nu})}(t) = S_0(t)\exp\left[\int_0^t \zeta(\hat{\nu}(s))\, ds\right], \tag{5.5.7}$$

$$\theta_{\hat{\nu}}(t) = \theta(t) + \sigma^{-1}(t)\hat{\nu}(t). \tag{5.5.9}$$

We note also that the conclusion of Lemma C.3 can be written as

$$M(t) - A(t) = \widehat{X}(t)U_1'(t, \hat{c}(t)). \tag{C.32}$$

It follows that

$$
\begin{aligned}
d\Big(&\widehat{X}(t)U_1'(t, \hat{c}(t))\Big) \\
&= dM(t) - dA(t) \\
&= \psi'(t)\, dW(t) - \hat{c}(t)U_1'(t, \hat{c}(t))\, dt \\
&= (M(t) - A(t))(\lambda(t) - \sigma^{-1}(t)\hat{\nu}(t))'dW(t) \\
&\quad - f(t)\widehat{X}(t)U_1'(t, \hat{c}(t))\, dt \\
&= \widehat{X}(t)U_1'(t, \hat{c}(t))[(\lambda(t) - \sigma^{-1}(t)\hat{\nu}(t))'dW(t) - f(t)\, dt].
\end{aligned}
$$

From (C.20) we have

$$
\begin{aligned}
\frac{\widehat{X}(t)}{S_0^{(\hat{\nu})}(t)} = x\exp\Bigg\{&-\int_0^t \left(f(s) + \zeta(\hat{\nu}(s)) + \frac{1}{2}\|\sigma'(s)\hat{p}(s)\|^2\right)\, ds \\
&+ \int_0^t \hat{p}'(s)\sigma(s)\, dW_0(s)\Bigg\},
\end{aligned}
$$

which yields the differential formula

$$
\begin{aligned}
d\left(\frac{S_0^{(\hat{\nu})}(t)}{\widehat{X}(t)}\right) = \frac{S_0^{(\hat{\nu})}(t)}{\widehat{X}(t)}&\left[(f(t) + \zeta(\hat{\nu}(t)) + \|\sigma'(t)\hat{p}(t)\|^2)\, dt\right. \\
&\left. - \hat{p}'(t)\sigma(t)\, dW_0(t)\right].
\end{aligned}
$$

The product rule for stochastic integration and the above formulas give

$$d\left(S_0^{(\hat{\nu})}(t)U_1(t, \hat{c}(t))\right) = -\theta_{\hat{\nu}}(t)S_0^{(\hat{\nu})}(t)U_1(t, \hat{c}(t))\, dW(t),$$

and hence $S_0^{(\hat{\nu})}(t)U_1(t, \hat{c}(t)) = y_* Z_{\hat{\nu}}(t)$ (see (5.5.10)), for some $y_* \in (0, \infty)$. In other words,

$$U_1(t, \hat{c}(t)) = y_* H_{\hat{\nu}}(t) \quad \text{for Lebesgue-a.e. } t \in [0, T] \tag{C.33}$$

almost surely. Without loss of generality, we can modify $\hat{c}(t)$ as necessary so that (C.33) holds for *every* $t \in [0,T]$ almost surely. Using (C.33), (C.32), (C.17), and (C.18), we conclude that

$$
\begin{aligned}
y_*\widehat{X}(t)H_{\hat{\nu}}(t)) &= \widehat{X}(t)U_1'(t,\hat{c}(t)) \\
&= M(t) - A(t) \\
&= E\left[\left.\int_t^T y_*H_{\hat{\nu}}(s)\hat{c}(s)\,ds + \widehat{X}(T)U_2'(\widehat{X}(T))\right|\mathcal{F}(t)\right]
\end{aligned}
$$

for $0 \le t \le T$. Evaluating this equation at $t = T$, we see that

$$
y_*H_{\hat{\nu}}(T) = U_2'(\widehat{X}(T)), \tag{C.34}
$$

and hence

$$
\widehat{X}(t) = \frac{1}{H_{\hat{\nu}}(t)}E\left[\left.\int_t^T H_{\hat{\nu}}(s)\hat{c}(s)\,ds + H_{\hat{\nu}}(T)\widehat{X}(T)\right|\mathcal{F}(t)\right]. \tag{C.35}
$$

From (C.33), (C.34), we have

$$
\hat{c}(t) = I_1(t, y_*H_{\hat{\nu}}(t)), \quad \widehat{X}(T) = I_2(y_*H_{\hat{\nu}}(T)), \tag{C.36}
$$

and substitution into (C.35), evaluated at $t = 0$, yields

$$
\begin{aligned}
x &= E\left[\int_0^T H_{\hat{\nu}}(s)I_1(s, y_*H_{\hat{\nu}}(s))\,ds + H_{\hat{\nu}}(T)I_2(y_*H_{\hat{\nu}}(T))\right] \\
&= \mathcal{X}_{\hat{\nu}}(y_*), \tag{C.37}
\end{aligned}
$$

which is thus seen to be finite. Therefore, $y_* = \mathcal{Y}_{\hat{\nu}}(x)$, and $B_{\hat{\nu}}$ defined by (6.3.16) is $\widehat{X}(T)$, while $c_{\hat{\nu}}(\cdot)$ defined by (6.3.17) is $\hat{c}(\cdot)$. Thus, equation (C.5) holds. Equation (C.4) now follows from (C.35). $\qquad\square$

The following lemma concludes the proof of Theorem 6.4.1.

Lemma C.6: *The process $\hat{\nu}(\cdot)$ defined by (C.26) is in the set $\mathcal{D}_0$ of (6.3.24).*

PROOF.   We note first from (6.2.1) and (6.2.4) that

$$
EH_{\hat{\nu}}(t) \le \frac{1}{s_0}EZ_{\hat{\nu}}(t) \le \frac{1}{s_0}e^{\zeta_0 t},
$$

and hence

$$
E\left[\int_0^T H_{\hat{\nu}}(t)\,dt + H_{\hat{\nu}}(T)\right] < \infty.
$$

With $y_* = \mathcal{Y}_{\hat{\nu}}(x)$ as in (C.37), we have

$$
\begin{aligned}
U_1(t, c_{\hat{\nu}}(t)) - y_*H_{\hat{\nu}}(t)c_{\hat{\nu}}(t) &= \tilde{U}_1(t, y_*H_{\hat{\nu}}(t)) \\
&\ge U_1(t, 1) - y_*H_{\hat{\nu}}(t),
\end{aligned}
$$

$$U_2(t, \widehat{X}(T)) - y_* H_{\hat{p}}(T)\widehat{X}(T) = \widetilde{U}_2(y_* H_{\hat{p}}(T))$$
$$\geq U_2(1) - y_* H_{\hat{p}}(T)$$

from the definition of convex dual functions and (C.36). It follows that

$$E\left\{\int_0^T \min[0, U_1(t, c_{\hat{p}}(t))]\, dt + \min[0, U_2(\widehat{X}(T))]\right\}$$
$$\geq \int_0^T \min[0, U_1(t, 1)]\, dt + \min[0, U_2(1)]$$
$$- y_* E\left[\int_0^T H_{\hat{p}}(t)\, dt + H_{\hat{p}}(T)\right] > -\infty. \qquad (C.38)$$

We began this appendix with the assumption that

$$V(x; K) = E\left[\int_0^T U_1(t, \hat{c}(t))\, dt + U_2(\widehat{X}(T))\right],$$

and we have subsequently shown that (C.4) and (C.5) hold. Hence,

$$V(x; K) = E\left[\int_0^T U_1(t, c_{\hat{p}}(t))\, dt + U_2(\widehat{X}(T))\right] = V_{\hat{p}}(x).$$

The left-hand side is finite by assumption, and consequently

$$E\left[\int_0^T |U_1(t, c_{\hat{p}}(t))|\, dt + |U_2(\widehat{X}(T))|\right] < \infty$$

and $V_{\hat{p}}(x)$ is finite.

For each positive integer n, set

$$\tau_n = \inf\left\{t \in [0, T); \int_0^t \|\theta_{\hat{p}}(s)\|^2\, ds = n\right\} \wedge T,$$

so that $E \int_0^{\tau_n} \theta_{\hat{p}}'(s)\, dW(s) = 0$. Using the fact that the mapping $z \mapsto \widetilde{U}_2(e^z)$ is convex (see (3.4.15'')), the fact that $\widetilde{U}_2$ is nonincreasing and convex, and Jensen's inequality, we may write

$$\widetilde{U}_2\left(\frac{y_*}{s_0}\exp\left\{-\frac{1}{2}E\int_0^{\tau_n}\|\theta_{\hat{p}}(s)\|^2\, ds\right\}\right)$$
$$= \widetilde{U}_2\left(\frac{y_*}{s_0}\exp\left\{E\left(\int_0^{\tau_n}\theta_{\hat{p}}'(s)\, dW(s) - \frac{1}{2}\int_0^{\tau_n}\|\theta_{\hat{p}}(s)\|^2 ds\right)\right\}\right)$$
$$\leq E\widetilde{U}_2\left(\frac{y_*}{s_0}\exp\left\{-\int_0^{\tau_n}\theta_{\hat{p}}'(s)\, dW(s) - \frac{1}{2}\int_0^{\tau_n}\|\theta_{\hat{p}}(s)\|^2 ds\right\}\right)$$
$$= E\widetilde{U}_2\left(\frac{y_*}{s_0}Z_{\hat{p}}\mathcal{F}(\tau_n)\right) \leq E\left[\widetilde{U}_2\left(\frac{y_*}{s_0}E[Z_{\hat{p}}(T)|\mathcal{F}(\tau_n)]\right)\right]$$

$$\leq E\left[E\left[\tilde{U}_2\left(\frac{y_*}{s_0}Z_{\hat{\nu}}(T)\right)\Big|\mathcal{F}(\tau_n)\right]\right] = E\tilde{U}_2\left(\frac{y_*}{s_0}Z_{\hat{\nu}}(T)\right)$$

$$\leq E\tilde{U}_2(y_*H_{\hat{\nu}}(T)) = E\tilde{U}_2(U'_2(\hat{X}(T)))$$

$$= E\left[U_2(\hat{X}(T)) - \hat{X}(T)U'_2(\hat{X}(T))\right] \leq EU_2(\hat{X}(T)) < \infty.$$

Letting $n \to \infty$, we obtain

$$\tilde{U}_2\left(\frac{y_*}{s_0}\exp\left\{-\frac{1}{2}E\int_0^T\|\theta_{\hat{\nu}}(s)\|^2\,ds\right\}\right) < \infty.$$

Because $\tilde{U}_2(0) = \infty$, we must have $E\int_0^T\|\theta_{\hat{\nu}}(s)\|^2ds < \infty$. But

$$\hat{\nu}(t) = \sigma(t)\theta_{\hat{\nu}}(t) - \sigma(t)\theta(t),$$

$\sigma(\cdot)$ is bounded, and $\theta(\cdot)$ satisfies (6.2.5). We conclude that $E\int_0^T\|\hat{\nu}(t)\|^2dt$ $< \infty$; i.e., (5.5.1) is satisfied.

Again using Jensen's inequality, we compute

$$\tilde{U}_2\left(\frac{y_*}{s_0}\exp\left\{-E\int_0^T\zeta(\hat{\nu}(s))\,ds - \frac{1}{2}E\int_0^T\|\theta_{\hat{\nu}}(s)\|^2\,ds\right\}\right)$$

$$\leq E\tilde{U}_2\left(\frac{y_*}{s_0}\exp\left\{-\int_0^T\zeta(\hat{\nu}(s))\,ds - \int_0^T\theta'_{\hat{\nu}}(s)\,dW(s)\right.\right.$$

$$\left.\left.-\frac{1}{2}\int_0^T\|\theta(s)\|^2\,ds\right\}\right)$$

$$\leq E\tilde{U}_2(y_*H_{\hat{\nu}}(T)) < \infty.$$

This time we conclude that $E\int_0^T\zeta(\hat{\nu}(t))\,dt < \infty$; i.e., (5.5.2) is satisfied. We have established that $\hat{\nu}(\cdot)$ is in $\mathcal{D}$.

We next show that $\hat{\nu}(\cdot) \in \mathcal{D}_0$. The condition $\mathcal{X}_{\hat{\nu}}(y) < \infty$ for all $y > 0$ follows from the finiteness of $\mathcal{X}_{\hat{\nu}}(y_*)$ in (C.37) and assumption (3.6.17) (see the sentence following (3.6.17)).

We must also show that $V_{\hat{\nu}}(\xi) < \infty$ for all $\xi > 0$. We have already shown that $V_{\hat{\nu}}(x)$ is finite. The function $V_{\hat{\nu}}$ is concave on $[0, \infty)$, which one can show using the argument in the first part of the proof of Theorem 3.6.11. We next show that $V_{\hat{\nu}}$ is not $-\infty$ at any point in $(0, \infty)$. Like $U_1(t, \cdot)$ and U_2, the function $V_{\hat{\nu}}$ is nondecreasing, and hence $V_{\hat{\nu}}(\xi) > -\infty$ for all $\xi \geq x$. For $\epsilon \in (0, 1)$, we set $c_\epsilon(\cdot) = (1 - \epsilon)\hat{c}(\cdot)$. It follows immediately from Definition 6.3.1 that $(c_\epsilon, \hat{p}) \in \mathcal{A}_{\hat{\nu}}((1-\epsilon)x)$. The proof of Lemma C.1 shows that $(c_\epsilon, p) \in \mathcal{A}_3^{(\hat{\nu})}((1 - \epsilon)x)$. The existence of a consumption/portfolio-proportion process pair in $\mathcal{A}_3^{(\hat{\nu})}((1 - \epsilon)x)$ implies that $V_{\hat{\nu}}((1-\epsilon)x) > -\infty$. The concavity of $V_{\hat{\nu}}$ on $(0, \infty)$ and the inequalities $V_{\hat{\nu}}(\xi) > -\infty$ for all $\xi \in (0, \infty)$, $V_{\hat{\nu}}(x) < \infty$ imply $V_{\hat{\nu}}(\xi) < \infty$ for all $\xi \in (0, \infty)$. $\qquad\square$

Appendix D
Optimal Stopping for Continuous-Parameter Processes

We discuss briefly in this appendix the problem of optimal stopping for random processes in continuous time. (A similar treatment for random sequences appears in Chapter VI of Neveu (1975).) The results presented here are not the most general one can obtain, but will suffice for the purposes of Chapter 2, where they are used rather extensively. For more general treatments we refer the reader to El Karoui (1981), as well as Bismut and Skalli (1977), Fakeev (1970, 1971), Xue (1984), and Shiryaev (1978) for the Markovian case.

Throughout this appendix we shall consider a *nonnegative process* $Y = \{Y(t), \mathcal{F}(t); 0 \leq t \leq T\}$ *with right-continuous paths* and $Y(T) \leq \overline{\lim}_{t \uparrow T} Y(t)$ *a.s.*, defined on a probability space $(\Omega, \mathcal{F}, P)$ and adapted to a filtration $\{\mathcal{F}(t)\}_{0 \leq t \leq T}$ satisfying the usual conditions of right continuity and augmentation by the null sets of $\mathcal{F} = \mathcal{F}(T)$. We suppose that $\mathcal{F}(0)$ contains only sets of probability zero or one. The time horizon $T \in (0, \infty]$ is a fixed constant; if $T = \infty$, we interpret $\mathcal{F}(\infty) \stackrel{\triangle}{=} \sigma\left(\bigcup_{0 \leq t < \infty} \mathcal{F}(t)\right)$ and $Y(\infty) \stackrel{\triangle}{=} \overline{\lim}_{t \to \infty} Y(t)$. We denote by $\mathcal{S}$ the class of $\{\mathcal{F}(t)\}$-stopping times with values in $[0, T]$. For any stopping time $v \in \mathcal{S}$, we set $\mathcal{S}_v \stackrel{\triangle}{=} \{\tau \in \mathcal{S}; \tau \geq v \ a.s.\}$.

The *optimal stopping problem* consists in

(i) computing the maximal "expected reward"

$$Z(0) \stackrel{\triangle}{=} \sup_{\tau \in \mathcal{S}} EY(\tau), \tag{D.1}$$

(ii) finding necessary and/or sufficient conditions for the existence of a stopping time τ_* that is "optimal," i.e., its expected reward $EY(\tau_*)$ attains the supremum in (D.1), and then

(iii) characterizing such a τ_* and studying its properties.

Throughout this appendix, we assume

$$0 < Z(0) < \infty. \tag{D.2}$$

Under this assumption, we construct the so-called *Snell envelope* $Z^0(\cdot)$ of $Y(\cdot)$, namely, the smallest RCLL supermartingale that dominates $Y(\cdot)$. A stopping time τ_* is optimal if and only if $\{Z^0(t \wedge \tau_*), \mathcal{F}(t); 0 \le t \le T\}$ is a martingale and $Z^0(\tau_*) = Y(\tau_*)$ a.s. (Theorem D.9). In order to prove the *existence of an optimal stopping time*, we shall impose the stronger condition $E[\sup_{0 \le t \le T} Y(t)] < \infty$, under which the stopping time

$$D_*(0) = \inf\{t \in [0, T]; \quad Z^0(t) = Y(t)\}$$

can be shown to be optimal if, in addition, the paths of $Y(\cdot)$ are continuous (Theorem D.12). Following this result, we decompose the Snell envelope into the difference of a martingale and a nondecreasing process, and use this decomposition to characterize optimal stopping times.

The key to our study is provided by the family $\{Z(v)\}_{v \in \mathcal{S}}$ of random variables

$$Z(v) \overset{\triangle}{=} \text{ess sup}_{\tau \in \mathcal{S}_v} E[Y(\tau)|\mathcal{F}(v)], \quad v \in \mathcal{S}. \tag{D.3}$$

(See Appendix A for the definition and properties of essential supremum.) We shall see in Proposition D.2 that the assumption (D.2) guarantees $EZ(v) < \infty$ for all $v \in \mathcal{S}$. The random variable $Z(v)$ is the optimal conditional expected reward for stopping at time v or later. Since each deterministic time $t \in [0, T]$ is also a stopping time, (D.3) defines a nonnegative, adapted process $\{Z(t), \mathcal{F}(t); 0 \le t \le T\}$. For $v \in \mathcal{S}$, it is tempting to regard $Z(v)$ as the process $\{Z(t); 0 \le t \le T\}$ evaluated at the stopping time v. We shall see in Theorem D.7 that there is indeed a modification $\{Z^0(t), \mathcal{F}(t); 0 \le t \le T\}$ of the process $\{Z(t), \mathcal{F}(t); 0 \le t \le T\}$, i.e., $P[Z^0(t) = Z(t)] = 1$ for all $t \in [0, T]$, such that for each $v \in \mathcal{S}$ we have

$$Z(v)(\omega) = Z^0(v(\omega), \omega), \quad P\text{-a.e. } \omega \in \Omega.$$

This process $Z^0(\cdot)$ is the Snell envelope of $Y(\cdot)$.

We now begin an examination of the family of random variables $\{Z(v)\}_{v \in \mathcal{S}}$.

Lemma D.1: *For any $v \in \mathcal{S}$ and $\tau \in \mathcal{S}_v$, the family $\{E[Y(\rho)|\mathcal{F}(v)]\}_{\rho \in \mathcal{S}_\tau}$ is closed under pairwise maximization. Furthermore, there is a sequence $\{\rho_n\}_{n=1}^\infty$ of stopping times in $\mathcal{S}_\tau$ such that the sequence $\{E[Y(\rho_n)|\mathcal{F}(v)]\}_{n=1}^\infty$ is nondecreasing and*

$$\text{ess sup}_{\rho \in \mathcal{S}_\tau} E[Y(\rho)|\mathcal{F}(v)] = \lim_{n \to \infty} E[Y(\rho_n)|\mathcal{F}(v)].$$

PROOF.   Let ρ_1 and ρ_2 be in $\mathcal{S}_\tau$ and set

$$A = \{E[Y(\rho_1)|\mathcal{F}(v)] \geq E[Y(\rho_2)|\mathcal{F}(v)]\}, \quad \rho_3 = \rho_1 1_A + \rho_2 1_{A^c}.$$

Because $A \in \mathcal{F}_v$, the random time ρ_3 is a stopping time. In particular, $\rho_3 \in \mathcal{S}_\tau$ and

$$E[Y(\rho_3)|\mathcal{F}(v)] = 1_A E[Y(\rho_1)|\mathcal{F}(v)] + 1_{A^c} E[Y(\rho_2)|\mathcal{F}(v)]$$
$$= E[Y(\rho_1)|\mathcal{F}(v)] \vee E[Y(\rho_2)|\mathcal{F}(v)].$$

The remainder of the lemma follows from Theorem A.3 in Appendix A.   □

Proposition D.2:   *For any $v \in \mathcal{S}, \sigma \in \mathcal{S}$, and $\tau \in \mathcal{S}_v$, we have*

$$Z(v) = Z(\sigma) \quad a.s. \quad on \; \{\sigma = v\}, \tag{D.4}$$

$$E[Z(\tau)|\mathcal{F}(v)] = \operatorname{ess\,sup}_{\rho \in \mathcal{S}_\tau} E[Y(\rho)|\mathcal{F}(v)] \quad a.s., \tag{D.5}$$

$$E[Z(\tau)|\mathcal{F}(v)] \leq Z(v) \quad a.s., \tag{D.6}$$

$$EZ(\tau) = \sup_{\rho \in \mathcal{S}_\tau} EY(\rho) \leq Z(0) < \infty. \tag{D.7}$$

PROOF.   To prove (D.4), note that the event $B \stackrel{\triangle}{=} \{v = \sigma\}$ belongs to $\mathcal{F}_v \cap \mathcal{F}_\sigma = \mathcal{F}_{v \wedge \sigma}$ (Karatzas and Shreve (1991), Lemma 1.2.16). Given $\tau \in \mathcal{S}_v$, we define $\tau_B = \tau \cdot 1_B + T \cdot 1_{B^c}$, so $\tau_B \in \mathcal{S}_\sigma$. From the definition of conditional expectation and from Karatzas and Shreve (1988), Problem 2.17(i) and solution (p. 39), we have

$$1_B E[Y(\tau)|\mathcal{F}(v)] = 1_B E[Y(\tau_B)|\mathcal{F}(v)] = 1_B E[Y(\tau_B)|\mathcal{F}(v \wedge \sigma)]$$
$$= 1_B E[Y(\tau_B)|\mathcal{F}(\sigma)] \leq 1_B Z(\sigma) \quad a.s.$$

for every $\tau \in \mathcal{S}_v$. Therefore, $Z(v) \leq Z(\sigma)$ almost surely on B. Reversing the roles of v and σ, we obtain (D.4).

   For (D.5), use Lemma D.1 to choose a sequence $\{\rho_n\}_{n=1}^\infty$ in $\mathcal{S}_\tau$ such that $\{E[Y(\rho_n)|\mathcal{F}(\tau)]\}_{n=1}^\infty$ is nondecreasing and $Z(\tau) = \lim_{n \to \infty} E[Y(\rho_n)|\mathcal{F}(\tau)]$. By the monotone convergence theorem for conditional expectations, we have

$$E[Z(\tau)|\mathcal{F}(v)] = \lim_{n \to \infty} E[E\{Y(\rho_n)|\mathcal{F}(\tau)\}|\mathcal{F}(v)]$$
$$= \lim_{n \to \infty} E[Y(\rho_n)|\mathcal{F}(v)]$$
$$\leq \operatorname{ess\,sup}_{\rho \in \mathcal{S}_\tau} E[Y(\rho)|\mathcal{F}(v)].$$

On the other hand, $Z(\tau) \geq E[Y(\rho)|\mathcal{F}(\tau)]$ holds for all $\rho \in \mathcal{S}_\tau$, and taking conditional expectations on both sides of this inequality we obtain $E[Z(\tau)|\mathcal{F}(v)] \geq E[Y(\rho)|\mathcal{F}(v)]$. This implies

$$E[Z(\tau)|\mathcal{F}(v)] \geq \operatorname{ess\,sup}_{\rho \in \mathcal{S}_\tau} E[Y(\rho)|\mathcal{F}(v)],$$

proving (D.5). Since $\operatorname{ess\,sup}_{\rho \in \mathcal{S}_\tau} E[Y(\rho)|\mathcal{F}(v)] \leq \operatorname{ess\,sup}_{\rho \in \mathcal{S}_v} E[Y(\rho)|\mathcal{F}(v)] = Z(v)$, (D.6) follows from (D.5). We obtain (D.7) by setting $v \equiv 0$ in (D.5) and (D.6).   □

For $v \in \mathcal{S}$, let us now define $\mathcal{S}_v^* \triangleq \{\tau \in \mathcal{S}_v; \tau > v \text{ a.s. on } \{v < T\}\}$ and introduce the family

$$Z^*(v) \triangleq \operatorname{ess\,sup}_{\tau \in \mathcal{S}_v^*} E[Y(\tau)|\mathcal{F}(v)], \quad v \in \mathcal{S}. \tag{D.8}$$

The random variable $Z^*(v)$ is the optimal conditional expected reward for stopping strictly later than v (except that on the event $\{v = T\}$, stopping must be done at time T and $Z^*(v) = Y(T)$). Since $Y(\cdot)$ has *right-continuous paths*, it is no surprise that $Z(v) = Z^*(v)$ holds almost surely (Corollary D.4 below). Before proving this equality, let us observe that the proof of Proposition D.2 can be trivially altered to show that the family $\{Z^*(v)\}_{v \in \mathcal{S}}$ possesses analogues of the properties established in that proposition for the family $\{Z(v)\}_{v \in \mathcal{S}}$. Furthermore, $\{Z^*(v)\}_{v \in \mathcal{S}}$ has the right-continuity property (D.10) below.

Proposition D.3: *For any $v \in \mathcal{S}$, and for any decreasing sequence $\{v_n\}_{n=1}^\infty$ in $\mathcal{S}_v^*$ with $v = \lim_{n \to \infty} v_n$ a.s., we have*

$$Z(v) = Z^*(v) \vee Y(v) \quad a.s., \tag{D.9}$$

$$\int_A Z^*(v)\,dP = \lim_{n \to \infty} \int_A Z^*(v_n)\,dP, \quad \forall A \in \mathcal{F}(v). \tag{D.10}$$

PROOF.   Obviously, $Z(v) \geq Z^*(v) \vee Y(v)$, from the definitions (D.3) and (D.8). For the reverse inequality, take an arbitrary $\sigma \in \mathcal{S}_v$ and note that the sets $\{\sigma = v\}$ and $\{\sigma > v\}$ are in $\mathcal{F}(v)$ (Karatzas and Shreve (1991), Lemma 1.2.16). Therefore,

$$E[Y(\sigma)|\mathcal{F}(v)] = E[Y(v)1_{\{\sigma=v\}} + Y(\sigma)1_{\{\sigma>v\}}|\mathcal{F}(v)]$$
$$\leq Y(v)1_{\{\sigma=v\}} + Z^*(v)1_{\{\sigma>v\}} \leq Z^*(v) \vee Y(v).$$

Taking the essential supremum over $\sigma \in \mathcal{S}_v$, we obtain (D.9).

For (D.10), observe from the analogue of (D.5) for the family $\{Z^*(v)\}_{v \in \mathcal{S}}$ that $E[Z^*(v_n)|\mathcal{F}(v)] = \operatorname{ess\,sup}_{\rho \in \mathcal{S}_{v_n}^*} E[Y(\rho)|\mathcal{F}(v)]$. Thus for $A \in \mathcal{F}(v)$ the sequence of random variables $\{\int_A Z^*(v_n)\,dP\}_{n=1}^\infty$ is nondecreasing, and

$$\lim_{n \to \infty} \int_A Z^*(v_n)\,dP = \lim_{n \to \infty} \int_A E[Z^*(v_n)|\mathcal{F}(v)]\,dP \leq \int_A Z^*(v)\,dP. \tag{D.11}$$

For the reverse inequality, let $\tau \in \mathcal{S}_v^*$ be given and define

$$\tau_n = \begin{cases} \tau, & \text{if } v_n < \tau, \\ T, & \text{if } v_n \geq \tau, \end{cases}$$

so that $\tau_n \in \mathcal{S}_{v_n}^*$ for all n. For $A \in \mathcal{F}(v)$, we have

$$\int_{\{v_n<T\}\cap A} Y(\tau)\,dP - \int_{\{\tau \leq v_n<T\}\cap A} Y(T)\,dP = \int_{\{v_n<\tau\}\cap A} Y(\tau_n)\,dP$$

$$= \int_{\{v_n<\tau\}\cap A} E[Y(\tau_n)|\mathcal{F}(v_n)]\,dP$$

$$\le \int_{\{v_n < \tau\} \cap A} Z^*(v_n)\, dP \le \int_{\{v < T\} \cap A} Z^*(v_n)\, dP.$$

Now $1_{\{v_n < T\} \cap A} \uparrow 1_{\{v < T\} \cap A}$ and $1_{\{\tau \le v_n < T\} \cap A} \downarrow 0$ almost surely, so letting $n \to \infty$ in the previous inequality, we obtain

$$\int_{\{v < T\} \cap A} Y(\tau)\, dP \le \lim_{n \to \infty} \int_{\{v < T\} \cap A} Z^*(v_n)\, dP. \qquad (D.12)$$

On the set $\{v = T\}$, we have $Y(\tau) = Y(T) = Z^*(v_n)$, and thus

$$\int_{\{v = T\} \cap A} Y(\tau)\, dP = \int_{\{v = T\} \cap A} Z^*(v_n)\, dP. \qquad (D.13)$$

Summing (D.12) and (D.13), we conclude that

$$\int_A Y(\tau)\, dP \le \lim_{n \to \infty} \int_A Z^*(v_n)\, dP; \quad \forall \tau \in \mathcal{S}_v^*, \quad \forall A \in \mathcal{F}(v). \qquad (D.14)$$

Using the analogue of Lemma D.1 for $\mathcal{S}_v^*$, we choose a sequence $\{\rho_k\}$ of stopping times in $\mathcal{S}_v^*$ such that $\{E[Y(\rho_k)|\mathcal{F}(v)]\}_{k=1}^{\infty}$ is nondecreasing and

$$Z^*(v) = \operatorname{ess\,sup}_{\tau \in \mathcal{S}_v^*} E[Y(\tau)|\mathcal{F}(v)] = \lim_{k \to \infty} E[Y(\rho_k)|\mathcal{F}(v)] \quad \text{a.s.}$$

The monotone convergence theorem and (D.14) imply

$$\int_A Z^*(v)\, dP = \lim_{k \to \infty} \int_A E[Y(\rho_k)|\mathcal{F}(v)]\, dP$$

$$= \lim_{k \to \infty} \int_A Y(\rho_k)\, dP \le \lim_{n \to \infty} \int_A Z^*(v_n)\, dP. \qquad (D.15)$$

This, combined with (D.11), yields (D.10). $\qquad \square$

Corollary D.4: *For any* $v \in \mathcal{S}$, *we have* $Z^*(v) = Z(v)$ *a.s.*

PROOF.   For any decreasing sequence $\{v_n\}_{n=1}^{\infty}$ in $\mathcal{S}_v^*$ with $\lim_{n \to \infty} v_n = v$, the right continuity of $Y(\cdot)$ and Fatou's lemma for conditional expectations imply

$$Z^*(v) \ge \varlimsup_{n \to \infty} E[Y(v_n)|\mathcal{F}(v)] \ge E[\lim_{n \to \infty} Y(v_n)|\mathcal{F}(v)] = Y(v) \quad \text{a.s.}$$

The conclusion follows from (D.9). $\qquad \square$

As mentioned earlier, we may take the stopping times v in (D.3) to be deterministic and thereby obtain a nonnegative, adapted process $\{Z(t), \mathcal{F}(t); 0 \le t \le T\}$. From (D.6) we see that this process is a supermartingale. Because the function $t \mapsto EZ(t)$ is right continuous (from (D.10) and Corollary D.4), there exists a supermartingale $\{Z^0(t), \mathcal{F}(t); 0 \le t \le T\}$ with RCLL paths that satisfies

$$P[Z(t) = Z^0(t)] = 1, \quad \forall t \in [0, T] \qquad (D.16)$$

(cf. Karatzas and Shreve (1991), Theorem 1.3.13). Recall that $Y(T) \triangleq \overline{\lim}_{t\to\infty} Y(t)$ if $T = \infty$, and thus

$$Z^0(\infty) = Z(\infty) = \overline{\lim_{t\to\infty}} Y(t) \quad \text{a.s.}$$

The nonnegative RCLL supermartingale $Z^0(\cdot)$ is called the *Snell envelope* of $Y(\cdot)$. It is the smallest supermartingale that dominates $Y(\cdot)$ in the sense of the following definition.

Definition D.5: Let $X_1(\cdot) = \{X_1(t); 0 \le t \le T\}$ and $X_2(\cdot) = \{X_2(t); 0 \le t \le T\}$ be arbitrary processes. We say that $X_1(\cdot)$ *dominates* $X_2(\cdot)$ if $P[X_1(t) \ge X_2(t), \forall 0 \le t \le T] = 1$.

Remark D.6: If $X_1(\cdot)$ and $X_2(\cdot)$ are right-continuous processes and $P[X_1(t) \ge X_2(t)] = 1$ for all t in $[0, T]$ (or even in a countable, dense subset of $[0, T]$), then $X_1(\cdot)$ dominates $X_2(\cdot)$.

The following theorem permits us henceforth to focus our attention on the RCLL supermartingale $Z^0(\cdot)$ rather than the more awkward family of random variables $\{Z(v)\}_{v\in\mathcal{S}}$. In (D.17), $Z^0(v)$ denotes the process $Z^0(\cdot)$ evaluated at the stopping time v, whereas $Z(v)$ stands for the random variable of (D.3).

Theorem D.7: *The Snell envelope $Z^0(\cdot)$ of $Y(\cdot)$ satisfies*

$$Z^0(v) = Z(v) \quad \text{a.s.} \tag{D.17}$$

for every $v \in \mathcal{S}$. Moreover, $Z^0(\cdot)$ dominates $Y(\cdot)$, and if $X(\cdot)$ is another RCLL supermartingale dominating $Y(\cdot)$, then $X(\cdot)$ also dominates $Z^0(\cdot)$.

PROOF.   For any $v \in \mathcal{S}$ there is a decreasing sequence $\{v_n\}_{n=1}^{\infty}$ of stopping times in S_v^* with values in $\{T\} \cup \mathbf{D}$ (where $\mathbf{D}$ is the set of dyadic rationals) and with $v = \lim_{n\to\infty} v_n$ a.s. (Karatzas and Shreve (1991), Problem 1.2.24). From Doob's optional sampling theorem (*ibid.*, Theorem 1.3.22), we have

$$E[Z^0(v_n)|\mathcal{F}(v)] \le E[Z^0(v_{n+k})|\mathcal{F}(v)] \le Z^0(v) \quad \text{a.s.}$$

for all positive integers n and k. Therefore, for $A \in \mathcal{F}(v)$, the sequence $\{\int_A Z^0(v_n)\,dP\}_{n=1}^{\infty}$ is nondecreasing and bounded above by $\int_A Z^0(v)\,dP$; hence $\lim_{n\to\infty} \int_A Z^0(v_n)\,dP \le \int_A Z^0(v)\,dP$. The reverse inequality follows from Fatou's lemma and the right continuity of $Z^0(\cdot)$. Coupling these observations with (D.10), Corollary D.4, and (D.16), we deduce

$$\int_A Z(v)\,dP = \lim_{n\to\infty} \int_A Z(v_n)\,dP = \lim_{n\to\infty} \int_A Z^0(v_n)\,dP = \int_A Z^0(v)\,dP.$$

Since $A \in \mathcal{F}(v)$ is arbitrary, (D.17) must hold.

Finally, let $X(\cdot)$ be an RCLL supermartingale dominating $Y(\cdot)$. For $t \in [0, T]$ and $\tau \in \mathcal{S}_t$, the optional sampling theorem implies $E[Y(\tau)|\mathcal{F}(t)] \le E[X(\tau)|\mathcal{F}(t)] \le X(t)$ almost surely. Therefore, for each $t \in [0, T]$, we have

$$Z^0(t) = Z(t) \triangleq \text{ess sup}_{\tau\in\mathcal{S}_t} E[Y(\tau)|\mathcal{F}(t)] \le X(t) \quad \text{a.s.} \qquad \square$$

Remark D.8: If $\{X(t), \mathcal{F}(t); 0 \leq t < T\}$ is an RCLL supermartingale satisfying

$$P[X(t) \geq Y(t)] = 1, \quad \forall\, t \in [0, T),$$

then it is still true that $X(\cdot)$ dominates $Z^0(\cdot)$. To see this, first use the right continuity of $X(\cdot)$ and $Y(\cdot)$ to see that $P[X(t) \geq Y(t), \forall t \in [0, T)] = 1$. Then use the submartingale convergence theorem (Karatzas and Shreve (1991), Theorem 1.3.15) to establish the almost sure existence of $X(T) \stackrel{\triangle}{=} \lim_{t \uparrow T} X(t)$. For $0 \leq s < T$ and with $T_n \uparrow T$, we have

$$E[X(T)|\mathcal{F}(s)] \leq \lim_{n \to \infty} E[X(T_n)|\mathcal{F}(s)] \leq X(s) \quad \text{a.s.}$$

from Fatou's lemma for conditional expectations, so that $\{X(t); 0 \leq t \leq T\}$ is an RCLL supermartingale. Moreover $X(T) = \lim_{t \uparrow T} X(t) \geq \overline{\lim}_{t \uparrow T} Y(t) \geq Y(T)$. Thus the supermartingale with last element $\{X(t), \mathcal{F}(t); 0 \leq t \leq T\}$ dominates $\{Y(t), \mathcal{F}(t); 0 \leq t \leq T\}$. Now apply Theorem D.7.

Theorem D.9: *A stopping time τ_* is optimal, i.e.,*

$$EY(\tau_*) = Z^0(0) = \sup_{\rho \in S} EY(\rho), \tag{D.18}$$

if and only if

$$Z^0(\tau_*) = Y(\tau_*) \quad a.s. \tag{D.19}$$

and the stopped supermartingale

$$\{Z^0(t \wedge \tau_*), \mathcal{F}(t); 0 \leq t \leq T\} \text{ is a martingale.} \tag{D.20}$$

PROOF. Suppose that τ_* is optimal. From (D.7) we have $EY(\tau_*) = \sup_{\rho \in S_{\tau_*}} EY(\rho) = EZ^0(\tau_*)$, as well as $EY(\tau_*) = \sup_{\rho \in S_{\sigma \wedge \tau_*}} EY(\rho) = EZ^0(\sigma \wedge \tau_*)$, for any $\sigma \in S$. Now, $P[Z^0(t) \geq Y(t), \forall t \in [0, T]] = 1$ and $EY(\tau_*) = EZ^0(\tau_*)$ give (D.19), whereas the fact that $EZ^0(\sigma \wedge \tau_*)$ does not depend on $\sigma \in S$ yields (D.20) thanks to Problem 1.3.26 in Karatzas and Shreve (1991). Conversely, (D.19) and (D.20) give $EY(\tau_*) = EZ^0(\tau_*) = Z^0(0) = \sup_{\rho \in S} EY(\rho)$. □

Having characterized optimal stopping times in Theorem D.9, we now seek to establish their existence. We begin by constructing a family of stopping times that are "approximately optimal." For $\lambda \in (0, 1)$ and $v \in S$, define the stopping time

$$D^\lambda(v) \stackrel{\triangle}{=} \inf\{t \in (v, T]; \quad \lambda Z^0(t) \leq Y(t)\} \wedge T \tag{D.21}$$

in S_v. Because of the right continuity of $Y(\cdot)$ and $Z^0(\cdot)$, the process $\{D^\lambda(t); 0 \leq t \leq T\}$ is *right continuous*. Furthermore, for any $v \in S$ we have the inequality

$$\lambda Z^0(D^\lambda(v)) \leq Y(D^\lambda(v)) \quad \text{a.s.} \tag{D.22}$$

(This inequality holds even on the set $\{D^\lambda(v) = T\}$, because $Z^0(T) = Y(T)$.)

Proposition D.10:   *For $0 < \lambda < 1$ and every $v \in \mathcal{S}$, we have*

$$Z^0(v) = E[Z^0(D^\lambda(v))|\mathcal{F}(v)] \quad a.s. \tag{D.23}$$

PROOF.   Define the family of nonnegative random variables $J(v) \overset{\triangle}{=} E[Z^0(D^\lambda(v))|\mathcal{F}(v)]$, $v \in \mathcal{S}$. From (D.6) and the fact that $D^\lambda(v) \in \mathcal{S}_v$, we note that

$$J(v) \le Z^0(v) \quad \text{a.s.,} \quad \forall v \in \mathcal{S}. \tag{D.24}$$

To prove the reverse inequality, we shall show that $J(\cdot)$ *has an RCLL supermartingale modification* $J^0(\cdot)$ *that dominates* $Z^0(\cdot)$. For $v \in \mathcal{S}$ and $\tau \in \mathcal{S}_v$, we have from (D.6) that

$$
\begin{aligned}
E[J(\tau)|\mathcal{F}(v)] &= E[E\{Z^0(D^\lambda(\tau))|\mathcal{F}(\tau)\}|\mathcal{F}(v)] = E[Z^0(D^\lambda(\tau))|\mathcal{F}(v)] \\
&= E[E\{Z^0(D^\lambda(\tau))|\mathcal{F}(D^\lambda(v))\}|\mathcal{F}(v)] \\
&\le E[Z^0(D^\lambda(v))|\mathcal{F}(v)] = J(v).
\end{aligned} \tag{D.25}
$$

It follows that $\{J(t), \mathcal{F}(t); 0 \le t \le T\}$ is a supermartingale and that $t \mapsto EJ(t)$ is a nonincreasing function. By Fatou's lemma and the right continuity of $Z^0(\cdot)$ and $D^\lambda(\cdot)$ we also have $\lim_{s \downarrow t} EJ(s) = \lim_{s \downarrow t} EZ^0(D^\lambda(s)) \ge EZ^0(D^\lambda(t)) = EJ(t)$. Therefore, the mapping $t \mapsto EJ(t)$ is right continuous; and by Theorem 1.3.13 in Karatzas and Shreve (1991) there is an RCLL supermartingale $J^0(\cdot) = \{J^0(t), \mathcal{F}(t); 0 \le t \le T\}$ such that

$$P[J(t) = J^0(t)] = 1, \quad \forall t \in [0, T]. \tag{D.26}$$

In the spirit of the first part of Theorem D.7, we now extend (D.26) to the statement

$$P[J(v) = J^0(v)] = 1, \quad \forall v \in \mathcal{S}. \tag{D.27}$$

Fix $v \in \mathcal{S}$ and let $\{v_n\}_{n=1}^\infty$ be a decreasing sequence of stopping times in $\mathcal{S}_v$ with $v = \lim_{n \to \infty} v_n$ a.s. and with values in $\{T\} \cup \mathbf{D}$, where $\mathbf{D}$ is the set of dyadic rationals. For any given $A \in \mathcal{F}(v)$, we have from (D.25)

$$\int_A J(v_n)\, dP = \int_A E[J(v_n)|\mathcal{F}(v)]\, dP \le \int_A J(v)\, dP, \quad \forall n \in \mathbb{N},$$

so that $\overline{\lim}_{n \to \infty} \int_A J(v_n)\, dP \le \int_A J(v)\, dP$. On the other hand, Fatou's lemma implies

$$
\begin{aligned}
\overline{\lim}_{n \to \infty} \int_A J(v_n)\, dP &= \overline{\lim}_{n \to \infty} \int_A E[Z^0(D^\lambda(v_n))|\mathcal{F}(v_n)]\, dP \\
&= \overline{\lim}_{n \to \infty} \int_A E[Z^0(D^\lambda(v_n))|\mathcal{F}(v)]\, dP \\
&\ge \int_A E[\lim_{n \to \infty} Z^0(D^\lambda(v_n))|\mathcal{F}(v)]\, dP
\end{aligned}
$$

$$= \int_A E[Z^0(D^\lambda(v))|\mathcal{F}(v)]\, dP = \int_A J(v)\, dP.$$

It follows that

$$\int_A J(v)\, dP = \lim_{n\to\infty} \int_A J(v_n)\, dP.$$

A similar argument, using the optional sampling theorem instead of (D.25), shows that

$$\lim_{n\to\infty} \int_A J^0(v_n)\, dP = \int_A J^0(v)\, dP.$$

Finally, (D.26) implies that $\int_A J(v_n)\, dP = \int_A J^0(v_n)\, dP$ for all n. It follows from these equations that $\int_A J(v)\, dP = \int_A J^0(v)\, dP$. Since $A \in \mathcal{F}(v)$ is arbitrary, (D.27) must hold.

We show next that $J^0(\cdot)$ *dominates* $Z^0(\cdot)$. Consider the RCLL supermartingale $\lambda Z^0(\cdot) + (1 - \lambda)J^0(\cdot)$. Fix $t \in [0,T]$. Then on $\{D^\lambda(t) = t\}$ we have

$$\lambda Z^0(t) + (1 - \lambda)J^0(t) = \lambda Z^0(t) + (1 - \lambda)E[Z^0(t)|\mathcal{F}(t)] = Z^0(t) \geq Y(t)$$

from Theorem D.7. On the other hand, the definition (D.21) of $D^\lambda(t)$ implies that on the event $\{D^\lambda(t) > t\}$ we have

$$\lambda Z^0(t) + (1 - \lambda)J^0(t) \geq \lambda Z^0(t) \geq Y(t).$$

The right continuity of $Z^0(\cdot)$, $J^0(\cdot)$, and $Y(\cdot)$ allows us to conclude that the supermartingale $\lambda Z^0(\cdot) + (1 - \lambda)J^0(\cdot)$ dominates $Y(\cdot)$, and thus dominates $Z^0(\cdot)$ as well. It follows that $J^0(\cdot)$ dominates $Z^0(\cdot)$, and consequently $P[J^0(v) \geq Z^0(v)] = 1, \forall v \in \mathcal{S}$. Combining this with (D.24) and (D.27), we obtain $P[J(v) = Z^0(v)] = 1, \forall v \in \mathcal{S}$. This is (D.23).   □

For fixed $v \in \mathcal{S}$ the family of stopping times $\{D^\lambda(v)\}_{0<\lambda<1}$ is nondecreasing in λ, so we may define the limiting stopping time

$$D_*(v) \triangleq \lim_{\lambda\uparrow 1} D^\lambda(v) \quad \text{a.s.} \tag{D.28}$$

Under the assumption that $Y(\cdot)$ has continuous paths and that

$$E\left[\sup_{0\leq t\leq T} Y(t)\right] < \infty \tag{D.29}$$

holds, we shall show that $D_*(0)$ is optimal (i.e., attains the supremum in (D.1)). As one might expect, this argument relies on the *left continuity* of $Y(\cdot)$. The following example illustrates that in the absence of the assumption (D.29), there need not exist an optimal stopping time.

Example D.11: (*Nonexistence of an optimal stopping time*): Set $T = \infty$ and $Y(t) = \exp\{W(t) - \frac{1}{2}t - \frac{1}{t+1}\}, 0 \leq t < \infty$, where $W(\cdot)$ is a one-dimensional Brownian motion and $\{\mathcal{F}(t)\}$ is the augmentation by null sets of the filtration generated by $W(\cdot)$. According to the "strong law of large numbers," $\lim_{t\to\infty}\frac{W(t)}{t} = 0$ a.s. (e.g., Karatzas and Shreve (1991), Problem 2.9.3 and solution, p. 124), and we have $Y(\infty) = \lim_{t\to\infty} Y(t) = 0$ almost surely; however, the condition (D.29) is *not* satisfied. We show that the Snell envelope of $Y(\cdot)$ is

$$M(t) \stackrel{\triangle}{=} \exp\left\{W(t) - \frac{1}{2}t\right\} = Y(t)\exp\left(\frac{1}{t+1}\right) > Y(t), \quad 0 \le t < \infty.$$

The process $\{M(t), \mathcal{F}(t); 0 \le t < \infty\}$ is a martingale, and when we include the last element $M(\infty) \equiv 0$ we have a continuous supermartingale $M(\cdot) = \{M(t), \mathcal{F}(t); 0 \le t \le \infty\}$ dominating $Y(\cdot)$. According to Theorem D.7, $M(\cdot)$ dominates the Snell envelope $Z^0(\cdot)$. Furthermore, for each $t \in [0,\infty)$ we have from the same theorem that

$$\begin{aligned}
Z^0(t) &= \text{ess sup}_{\tau\in\mathcal{S}_t} E[Y(\tau)|\mathcal{F}(t)] \\
&\ge \text{ess sup}_{t\le s<\infty} E[Y(s)|\mathcal{F}(t)] \\
&= M(t) \cdot \sup_{t\le s<\infty} \exp\left\{-\frac{1}{s+1}\right\} = M(t) \quad \text{a.s.}
\end{aligned}$$

Hence $M(\cdot)$ is the Snell envelope of $Y(\cdot)$; in particular,

$$\sup_{\tau\in\mathcal{S}} EY(\tau) = Z^0(0) = M(0) = 1.$$

There is no stopping time $v\in\mathcal{S}$ satisfying $EY(v) = 1$, because condition (D.19) of Theorem D.9 is satisfied only by $v \equiv \infty$ but $EY(\infty) = 0$. For $\lambda\in(1/e, 1)$, we have

$$D^\lambda(0) \stackrel{\triangle}{=} \inf\{t \ge 0; \quad \lambda M(t) \le Y(t)\} = -\frac{1}{\log\lambda} - 1,$$

and $\lim_{\lambda\uparrow 1} EY(D^\lambda(0)) = \lim_{\lambda\uparrow 1}\lambda = 1$, but the limit $D_*(0) \stackrel{\triangle}{=} \lim_{\lambda\uparrow 1} D^\lambda(0) = \infty$ of these "approximately optimal" stopping times is very clearly "suboptimal," since $EY(D_*(0)) = 0$.

It is of course possible to introduce a deterministic time change in this example, so that T is finite and there still is no optimal stopping time.

Theorem D.12: *Suppose that $Y(\cdot)$ has continuous paths and that the assumption (D.29) holds. Then, for each $v\in\mathcal{S}$, the stopping time $D_*(v)$ defined by (D.28) satisfies*

$$E[Y(D_*(v))|\mathcal{F}(v)] = Z^0(v) = \text{ess sup}_{\tau\in\mathcal{S}_v} E[Y(\tau)|\mathcal{F}(v)] \quad a.s. \tag{D.30}$$

In particular, $D_*(0)$ attains the supremum in (D.1). Furthermore, for any $v\in\mathcal{S}$:

$$D_*(v) = \inf\{t\in[v,T]; \quad Z^0(t) = Y(t)\} \quad a.s. \tag{D.31}$$

PROOF.   Proposition D.10 and the inequality (D.22) imply

$$Z^0(v) = E[Z^0(D^\lambda(v))|\mathcal{F}(v)] \le \frac{1}{\lambda} E[Y(D^\lambda(v))|\mathcal{F}(v)] \quad \text{a.s.}$$

Now, for all $\lambda \in (0,1)$ we have $Y(D^\lambda(v)) \le \bar{Y}$, where

$$\bar{Y} \triangleq \sup_{0 \le t \le T} Y(t), \tag{D.32}$$

so we may use the dominated convergence theorem for conditional expectations, coupled with the left continuity of $Y(\cdot)$, to conclude that

$$Z^0(v) \le \lim_{\lambda \uparrow 1} E[Y(D^\lambda(v))|\mathcal{F}(v)] = E[Y(D_*(v))|\mathcal{F}(v)] \quad \text{a.s.}$$

Because $Z^0(\cdot)$ is a supermartingale dominating $Y(\cdot)$, we have the reverse inequality

$$E[Y(D_*(v))|\mathcal{F}(v)] \le E[Z^0(D_*(v))|\mathcal{F}(v))] \le Z^0(v) \quad \text{a.s.}$$

and (D.30) is proved.

From (D.30) and the supermartingale property, we have

$$EY(D_*(v)) = EZ^0(v) \ge EZ^0(D_*(v)).$$

But $Z^0(\cdot)$ dominates $Y(\cdot)$, so it must be true that $Y(D_*(v)) = Z^0(D_*(v))$ a.s. and

$$D_*(v) \ge \inf\{t \in [v,T]; Z^0(t) = Y(t)\} \quad \text{a.s.}$$

For the reverse inequality, observe first that the definition (D.21) of $D^\lambda(v)$ implies $Y(t) < \lambda Z^0(t) \le Z^0(t)$ for all $t \in (v, D^\lambda(v))$ and $\lambda \in (0,1)$. Hence

$$D_*(v) = \lim_{\lambda \uparrow 1} D^\lambda(v) \le \inf\{t \in (v,T]; \quad Z^0(t) = Y(t)\} \quad \text{a.s.}$$

To conclude further that

$$D_*(v) \le \inf\{t \in [v,T]; \quad Z^0(t) = Y(t)\} \quad \text{a.s.}$$

we must show

$$D_*(v) = v \quad \text{a.s. on } \{Z^0(v) = Y(v)\}.$$

On the set $\{Z^0(v) = Y(v) > 0\}$, the right continuity of $Z^0(\cdot)$ and $Y(\cdot)$ implies $D^\lambda(v) = v$ a.s. for all $\lambda \in (0,1)$, and hence $D_*(v) = v$ a.s. On the other hand, we appeal to the optional sampling theorem

$$E\left[1_{\{Z^0(v)=0\}} Z^0((v+\theta)\wedge T)\right] \le E\left[1_{\{Z^0(v)=0\}} Z^0(v)\right] = 0 \quad \text{a.s.,} \quad \theta \ge 0,$$

to conclude that $\lambda Z^0(t) = 0 \le Y(t)$ for $v \le t \le T$ a.s. on $\{Z^0(v) = Y(v) = 0\}$. Thus, on this set also we have $D_*(v) = v$ a.s.   □

Theorem D.13: *Under the condition (D.29), the RCLL supermartingale $Z^0(\cdot)$ admits the Doob–Meyer decomposition*

$$Z(\cdot) = M(\cdot) - \Lambda(\cdot). \tag{D.33}$$

Here $M(\cdot)$ is a uniformly integrable RCLL martingale with respect to the filtration $\{\mathcal{F}(t)\}_{0 \leq t \leq T}$, and $\Lambda(\cdot)$ is a right-continuous natural nondecreasing, $\{\mathcal{F}(t)\}_{0 \leq t \leq T}$-adapted process with $\Lambda(0) = 0$ and $E\Lambda(T) < \infty$. Moreover, if $Y(\cdot)$ has continuous paths, then $\Lambda(\cdot)$ is continuous and "flat" away from the set $\mathcal{H}(\omega) \overset{\triangle}{=} \{t \in [0, T]; Z^0(t, \omega) = Y(t, \omega)\}$; i.e.,

$$\int_0^T 1_{\{Z^0(t) > Y(t)\}} d\Lambda(t) = 0 \quad a.s. \tag{D.34}$$

PROOF. We first prove the uniform integrability of the family $\{Z^0(v)\}_{v \in \mathcal{S}}$. From the equality (D.17) we have $Z^0(v) \leq E[\bar{Y} | \mathcal{F}(v)]$ in the notation of (D.32), so $EZ^0(v) \leq E\bar{Y} < \infty$ for all $v \in \mathcal{S}$. Given $\varepsilon > 0$, there exists $\delta > 0$ such that

$$A \in \mathcal{F}(T) \text{ and } P(A) < \delta \Rightarrow \int_A \bar{Y} \, dP < \varepsilon.$$

Let $\alpha > \frac{1}{\delta} E\bar{Y}$ be given. Then

$$P\left[Z^0(v) > \alpha\right] \leq \frac{1}{\alpha} E\bar{Y} < \delta,$$

$$\int_{\{Z^0(v) > \alpha\}} Z^0(v) \, dP \leq \int_{\{Z^0(v) > \alpha\}} E[\bar{Y} | \mathcal{F}(v)] \, dP$$

$$= \int_{\{Z^0(v) > \alpha\}} \bar{Y} \, dP < \varepsilon, \quad \forall v \in \mathcal{S}.$$

We show next that if $Y(\cdot)$ has continuous paths, then $Z^0(\cdot)$ is regular; i.e., if $\{v_n\}_{n=1}^\infty$ is a nondecreasing sequence of stopping times in $\mathcal{S}$ and $v = \lim_{n \to \infty} v_n$ a.s., then $EZ^0(v) = \lim_{n \to \infty} EZ^0(v_n)$. The inequality $EZ^0(v) \leq \varliminf_{n \to \infty} EZ^0(v_n)$ follows from the supermartingale property. For the reverse inequality, observe that the sequence of stopping times $\{D_*(v_n)\}_{n=1}^\infty$ satisfies (D.31) and is nondecreasing with $\lim_{n \to \infty} D_*(v_n) \in \mathcal{S}_v$; from (D.30), the dominated convergence theorem, the continuity of $Y(\cdot)$ and its domination by $Z^0(\cdot)$, and the supermartingale property for $Z^0(\cdot)$, we have

$$\varlimsup_{n \to \infty} EZ^0(v_n) = \varlimsup_{n \to \infty} EY(D_*(v_n)) \leq EY\left(\lim_{n \to \infty} D_*(v_n)\right)$$

$$\leq EZ^0\left(\lim_{n \to \infty} D_*(v_n)\right) \leq EZ^0(v).$$

All the assertions of the theorem, except (D.34), now follow from Karatzas and Shreve (1991), Theorems 1.4.10 and 1.4.14. In particular, the martingale $M(\cdot)$ in the Doob–Meyer decomposition (D.33) has a last element $M(T)$ (ibid., Problem 1.3.19 and solution (p. 42)).

For (D.34), define the family of stopping times

$$\rho_t \overset{\triangle}{=} \inf\{s \in [t, T); \Lambda(t) < \Lambda(s)\} \wedge T, \quad \text{for } t \in [0, T). \tag{D.35}$$

Equation (D.30), the domination of $Y(\cdot)$ by $Z^0(\cdot)$, and the supermartingale property for $Z^0(\cdot)$ yield

$$EZ^0(\rho_t) = EY(D_*(\rho_t)) \leq EZ^0(D_*(\rho_t)) \leq EZ^0(\rho_t).$$

According to the optional sampling theorem (*ibid.*, Theorem 1.3.22), the uniformly integrable martingale $M(\cdot)$ in (D.32) satisfies $EM(\rho_t) = EM(D_*(\rho_t))$. Therefore, $E\Lambda(\rho_t) = E\Lambda(D_*(\rho_t))$. But $\Lambda(\cdot)$ is nondecreasing, so $\Lambda(\rho_t) = \Lambda(D_*(\rho_t))$ a.s. The definition of ρ_t shows then that $\rho_t = D_*(\rho_t)$ a.s., and (D.31) now implies $\rho_t(\omega) \in \mathcal{H}(\omega)$ for P-a.e. $\omega \in \Omega$. Letting $\mathbb{Q}$ denote the set of rational numbers, we conclude that for P-a.e. $\omega \in \Omega$, we have

$$\{\rho_q(\omega); q \in [0, T] \cap \mathbb{Q}\} \subseteq \mathcal{H}(\omega). \tag{D.36}$$

We now fix an $\omega \in \Omega$ for which (D.36) holds, for which the mappings $t \mapsto \Lambda(t, \omega)$ and $t \mapsto Y(t, \omega)$ are continuous on $[0, T]$ and for which $t \mapsto Z^0(t, \omega)$ is RCLL. To understand the set on which $\Lambda(\cdot, \omega)$ is "flat," we define

$$J(\omega) \triangleq \{t \in (0, T); \quad \exists \varepsilon > 0 \quad \text{with} \quad \Lambda(t - \varepsilon, \omega) = \Lambda(t + \varepsilon, \omega)\}.$$

It is apparent that the set $J(\omega)$ is open, and thus can be written as a countable union of disjoint open intervals: $J(\omega) = \bigcup_i (\alpha_i(\omega), \beta_i(\omega))$. We are interested in the set

$$\hat{J}(\omega) \triangleq \bigcup_i [\alpha_i(\omega), \beta_i(\omega)) = \{t \in [0, T); \exists \varepsilon > 0 \text{ with } \Lambda(t, \omega) = \Lambda(t + \varepsilon, \omega)\}$$

and in its complement $\hat{J}^c(\omega)$ in $[0, T) \times \Omega$. The function $t \mapsto \Lambda(t, \omega)$ is "flat" on $\hat{J}(\omega)$, in the sense that $\int_0^T 1_{\hat{J}(\omega)}(t) d\Lambda(t, \omega) = \sum_i [\Lambda(\beta_i(\omega), \omega) - \Lambda(\alpha_i(\omega), \omega)] = 0$.

Our task is to show that $\mathcal{H}^c(\omega) \subseteq \hat{J}(\omega)$, *or equivalently, that* $\hat{J}^c(\omega) \subseteq \mathcal{H}(\omega)$. Note that

$$\hat{J}^c(\omega) = \{t \in [0, T); \quad \forall s \in (t, T), \quad \Lambda(t, \omega) < \Lambda(s, \omega)\}.$$

Let $t \in \hat{J}^c(\omega)$ be given. Then there is a strictly decreasing sequence $\{t_n\}_{n=1}^\infty$ such that $\{\Lambda(t_n, \omega)\}_{n=1}^\infty$ is also strictly decreasing and

$$t = \lim_{n \to \infty} t_n, \quad \Lambda(t, \omega) = \lim_{n \to \infty} \Lambda(t_n, \omega).$$

For each n, let q_n be a rational number in (t, t_{n+1}). Then $t \leq \rho_{q_n}(\omega) \leq t_n$ and $t = \lim_{n \to \infty} \rho_{q_n}(\omega)$. From (D.36) we have $Z^0(\rho_{q_n}(\omega), \omega) = Y(\rho_{q_n}(\omega), \omega)$, and letting $n \to \infty$, using the right continuity of $Z^0(\cdot)$ and $Y(\cdot)$, we discover that $t \in \mathcal{H}(\omega)$. $\square$

We have so far constructed an optimal stopping time $D_*(0)$ in (D.30) and a candidate optimal stopping time ρ_0 in (D.35). Another candidate optimal stopping time is

$$\sigma_0 \triangleq \inf\{t \in (0, T]; Z^0(t) = Y(t)\}. \tag{D.37}$$

These three can all be different, as the following simple example shows.

Example D.14:   Consider the deterministic process

$$Y(t) = \begin{cases} |x - 1|, & 0 \leq x \leq 2, \\ 1, & 2 \leq x \leq 3, \\ e^{3-x}, & x \geq 3. \end{cases}$$

The Snell envelope is

$$Z^0(t) = \begin{cases} 1, & 0 \leq x \leq 3, \\ e^{3-x}, & x \geq 3, \end{cases}$$

and $D_*(0) = 0, \sigma_0 = 2, \rho_0 = 3$.

Theorem D.12 asserts the optimality of $D_*(0)$. Under the integrability condition of that theorem, the optimality of σ_0 and ρ_0 is provided by the following corollary.

Corollary D.15:   *Under the condition (D.29) and the continuity of $Y(\cdot)$, the stopping times σ_0 and ρ_0 are optimal:*

$$Z^0(0) = \sup_{\tau \in \mathcal{S}} EY(\tau) = EY(D_*(0)) = EY(\sigma_0) = EY(\rho_0).$$

Moreover, $D_*(0) \leq \sigma_0 \leq \rho_0$ a.s.

PROOF.   It is clear from (D.31) and (D.37) that $D_*(0) \leq \sigma_0$. Furthermore, we have $Z^0(t) > Y(t)$ for $t \in (0, \sigma_0)$, and from (D.34), $\Lambda(\sigma_0) = \Lambda(0) = 0$. Hence, $\sigma_0 \leq \rho_0$.

The optimality of $D_*(0)$ and the supermartingale property give

$$Z^0(0) = \sup_{\tau \in \mathcal{S}} EY(\tau) = EY(D_*(0)) = EZ^0(D_*(0)) \geq EZ^0(\sigma_0) \geq EZ^0(\rho_0).$$

$$(D.38)$$

But from the definition (D.35) of ρ_0, we see that $\Lambda(\rho_0) = 0$ a.s., and taking expectations in (D.33), using the optional sampling theorem, we are led to the equality $EZ^0(\rho_0) = Z^0(0)$. Thus, the inequalities in (D.38) are actually equalities.

From (D.37) and the right continuity of $Z^0(\cdot)$ and $Y(\cdot)$, we have $Z^0(\sigma_0) = Y(\sigma_0)$ a.s. This proves the optimality of $\sigma_0 : Z^0(0) = EZ^0(\sigma_0) = EY(\sigma_0)$. From (D.36) we have $\rho_0 \in \mathcal{H}$ a.s., which means $Z^0(\rho_0) = Y(\rho_0)$ a.s. This proves the optimality of $\rho_0 : Z^0(0) = EZ^0(\rho_0) = EY(\rho_0)$. $\square$

Appendix E
The Clark Formula

We offer in this appendix a brief overview of the formula of J.M.C. Clark (1970) for the stochastic integral representation of Brownian functionals. The approach is that of Bismut (1981), as presented in Rogers and Williams (1987).

Let $W_t(\omega) = \omega(t)$, $0 \leq t \leq T$, be standard, one-dimensional Brownian motion on the canonical space $(\Omega, \mathcal{F}, P), \{\mathcal{F}_t\}_{0 \leq t \leq T}$ with $\Omega = C([0, T])$, the space of continuous functions $\omega : [0, T] \rightarrow \mathbb{R}$, and with P taken to be Wiener measure. Here $\mathcal{F} = \mathcal{F}_T$ is the P-completion of the σ-algebra generated by $\{W_s; \ 0 \leq s \leq T\}$, and $\mathcal{F}_t$ is the σ-algebra generated by $\{W_s; \ 0 \leq s \leq t\}$ and augmented by the P-null sets in $\mathcal{F}$, for $0 \leq t \leq T$.

Let F be a square-integrable *Brownian functional*, i.e., an $\mathcal{F}_T$-measurable mapping from Ω to $\mathbb{R}$ with $EF^2(W) < \infty$. The martingale representation theorem for Brownian functionals (e.g., Karatzas and Shreve (1991), Proposition 3.4.18) states that there exists a unique (up to a.e. equivalence on $[0, T] \times \Omega$) $\{\mathcal{F}_t\}$-progressively measurable process $\quad$ with $E \int_0^T {}_s^2 ds < \infty$ such that

$$F(W) = EF(W) + \int_0^T {}_s dW_s \quad \text{a.s.} \tag{E.1}$$

In the construction of hedging portfolios, it is important to obtain the process $\quad$ explicitly.

Taking conditional expectations on both sides of (E.1) with respect to $\mathcal{F}_t$, we obtain $\Lambda_t \overset{\triangle}{=} E[F(W)|\mathcal{F}_t] = EF(W) + \int_0^t {}_s dW_s$, which we may write in differential notation as

$$d\Lambda_t = d_t\left(E[F(W)|\mathcal{F}_t]\right) = \psi_t dW_t. \tag{E.2}$$

If the differential of the Lévy martingale Λ can be computed, then ψ will be determined. Here is an example of such a computation.

Example E.1: Let $F(\omega) \triangleq \max_{0 \le t \le T} \omega(t)$, and define $M_{s,t} \triangleq \max_{s \le u \le t} W_u$ and $M_t \triangleq M_{0,t}$, so that $F(W) = M_T$. From the reflection principle we have

$$P[M_T > b] = 2P[W_T > b] = 2\left[1 - \Phi\left(\frac{b}{\sqrt{T}}\right)\right], \quad b > 0,$$

and therefore

$$P[M_T \in db] = 2\frac{\partial}{\partial b}\Phi\left(\frac{b}{\sqrt{T}}\right)db = \frac{2}{\sqrt{2\pi T}}e^{-\frac{b^2}{2T}}db, \quad b > 0,$$

where $\Phi(x) \triangleq \frac{1}{\sqrt{2\pi}}\int_{-\infty}^{x} e^{-\frac{1}{2}\xi^2}d\xi$ (e.g., Karatzas and Shreve (1991), §2.8A). It follows easily that $EF^2(W) < \infty$. For $0 \le t \le T$ and $m \ge 0$ we have

$$P[M_T > m|\mathcal{F}_t] = 1_{\{M_t > m\}} + 1_{\{M_t \le m\}}P[M_{t,T} > m|W_t],$$

and thus $P[M_T > m|\mathcal{F}_t]$ depends only on W_t and M_t. For $a < b < m$ and $b > 0$, we obtain

$$P[M_T \in dm|W_t = a, M_t = b] = P[M_{T-t} \in dm - a]$$
$$= \frac{2}{\sqrt{2\pi(T-t)}}e^{-\frac{(m-a)^2}{2(T-t)}}dm.$$

Furthermore,

$$P[M_T = b|W_t = a, M_t = b] = 1 - P[M_T > b|W_t = a, M_t = b]$$
$$= 1 - P[M_{T-t} > b - a] = 2\Phi\left(\frac{b-a}{\sqrt{T-t}}\right) - 1.$$

If follows that

$$E[M_T|W_t = a, M_t = b] = 2b \cdot \Phi\left(\frac{b-a}{\sqrt{T-t}}\right) - b$$
$$+ \int_b^\infty \frac{2m}{\sqrt{2\pi(T-t)}}e^{-\frac{(m-a)^2}{2(T-t)}}dm$$
$$= 2(b-a)\Phi\left(\frac{b-a}{\sqrt{T-t}}\right) - b + 2a$$
$$+ \sqrt{\frac{2(T-t)}{\pi}}\exp\left\{-\frac{(b-a)^2}{2(T-t)}\right\},$$

or equivalently,

$$\Lambda_t \triangleq E[M_T|\mathcal{F}_t] = 2(M_t - W_t)\Phi\left(\frac{M_t - W_t}{\sqrt{T-t}}\right)$$

$$- M_t + 2W_t + \sqrt{\frac{2(T-t)}{\pi}} \exp\left\{-\frac{(M_t - W_t)^2}{2(T-t)}\right\}.$$

A tedious calculation, based on the Itô formula and on the identity $\Phi''(x) = -x\Phi'(x)$, leads to

$$d\Lambda_t = 2\left[1 - \Phi\left(\frac{M_t - W_t}{\sqrt{T-t}}\right)\right] dW_t - \left[1 - 2\Phi\left(\frac{M_t - W_t}{\sqrt{T-t}}\right)\right] dM_t.$$

Because the process M is flat off the set $\{t \in [0,T]; M_t - W_t = 0\}$ and $\Phi(0) = \frac{1}{2}$, the dM_t term on the right-hand side of this last equation drops out, and we have identified the integrand ψ in (E.1) as

$$\psi_t = 2\left[1 - \Phi\left(\frac{M_t - W_t}{\sqrt{T-t}}\right)\right]. \tag{E.3}$$

While many more examples along these lines can be given, the calculations become increasingly laborious and tedious. The *Clark formula* is a theoretical characterization of the integrand in (E.1) that sometimes shortens these computations substantially. As presented here, the derivation of this formula requires the Brownian functional F to satisfy the following three conditions:

$$EF^2(W) < \infty ; \tag{E.4}$$

$$\left\{\begin{array}{c} \text{there is a nonnegative Brownian functional } h \\ \text{satisfying } Eh^2(W) < \infty \text{ and a function} \\ g : [0, \infty) \to [0, \infty) \text{ satisfying } \overline{\lim}_{\varepsilon \downarrow 0}\left(\frac{g(\varepsilon)}{\varepsilon}\right) < \infty, \\ \text{such that with } \|\cdot\| \text{ denoting the supremum norm,} \\ |F(\omega + \varphi) - F(\omega)| \leq h(\omega)g(\|\varphi\|), \forall(\omega, \varphi) \in \Omega^2; \end{array}\right\} \tag{E.5}$$

$$\left\{\begin{array}{c} \text{there is a measurable mapping} \\ \omega \mapsto \partial F(\omega; \cdot) : (\Omega, \mathcal{F}_T) \to (\mathbf{M}, \mathcal{M}), \\ \text{where } \mathbf{M} \text{ is the set of finite Borel measures on} \\ \mathcal{B}([0,T]) \text{ and } \mathcal{M} \text{ is the } \sigma\text{-field generated by} \\ \text{the topology of weak convergence on } \mathbf{M}, \text{that satisfies} \\ \lim_{\varepsilon \to 0} \frac{1}{\varepsilon}[F(\omega + \varepsilon\varphi) - F(\omega)] = \int_0^T \varphi(t)\partial F(\omega; dt), \\ \forall \varphi \in C^1([0,T]), \ P\text{-a.e. } \omega \in \Omega. \end{array}\right\} \tag{E.6}$$

Note that if the mapping ∂F in (E.6) exists, then it must be unique up to P-equivalence.

Theorem E.2: (Clark (1970)): *Under the conditions (E.4)–(E.6), the process ψ of (E.1) is the predictable projection of the (not necessarily adapted) process $\partial F(W; (t,T])$, $0 \leq t \leq T$. Equivalently, for*

Lebesgue-almost-every $t \in [0, T]$, *we have*

$$\psi_t = E[\partial F(W; (t, T)) | \mathcal{F}_t] \quad a.s. \tag{E.7}$$

PROOF. Let X be a bounded, continuous $\{\mathcal{F}_t\}$-adapted process; define $\varphi_t = \int_0^t X_s ds$ for $0 \le t \le T$; and introduce the exponential martingales and the probability measures

$$Z_t^\varepsilon \overset{\triangle}{=} \exp\left[\varepsilon \int_0^t X_s \, dW_s - \frac{\varepsilon^2}{2} \int_0^t X_s^2 ds\right] , \quad Z^\varepsilon \equiv Z_T^\varepsilon,$$

$$P^\varepsilon(A) \overset{\triangle}{=} E[Z^\varepsilon \cdot 1_A], \quad A \in \mathcal{F}_T,$$

for $\varepsilon \in \mathbb{R}$. Then $\{W_t - \varepsilon\varphi_t, \mathcal{F}_t; 0 \le t \le T\}$ is a P^ε-Brownian motion, according to the Girsanov theorem (Karatzas and Shreve (1991), §3.5), and we thus have the *invariance principle*

$$EF(W) = E^\varepsilon F(W - \varepsilon\varphi) = E[Z^\varepsilon F(W - \varepsilon\varphi)], \quad \forall \varepsilon \in \mathbb{R}. \tag{E.8}$$

A stochastic calculus of variations (known as the *Malliavin calculus*) can be based on (E.8). In the spirit of this calculus, we first rewrite (E.8) and then we differentiate with respect to ε. From (E.8) we have for $\varepsilon \ne 0$ that

$$\frac{1}{\varepsilon} E[F(W) - F(W - \varepsilon\varphi)] + E\left[\frac{Z^\varepsilon - 1}{\varepsilon}\{F(W) - F(W - \varepsilon\varphi)\}\right]$$

$$= E\left[F(W)\frac{Z^\varepsilon - 1}{\varepsilon}\right]. \tag{E.9}$$

Using the representation $Z_t^\varepsilon = 1 + \varepsilon \int_0^t X_s Z_s^\varepsilon dW_s$, one can show first that $\lim_{\varepsilon \downarrow 0} E \int_0^T (Z_s^\varepsilon - 1)^2 ds = 0$, and then that $\frac{Z_t^\varepsilon - 1}{\varepsilon} \xrightarrow[\varepsilon \downarrow 0]{L^2} \int_0^t X_s dW_s$ for every $t \in [0, T]$. Upon letting $\varepsilon \downarrow 0$ in (E.9), we obtain from these considerations, (E.5), (E.6), and the dominated convergence theorem that

$$E \int_0^T \varphi_s \partial F(W; ds) = E\left[F(W) \cdot \int_0^T X_s dW_s\right]. \tag{E.10}$$

The left-hand side of (E.10) is equal to

$$E \int_0^T \int_0^T 1_{\{0 \le t < s \le T\}} X_t \, dt \, \partial F(W; ds) = E \int_0^T X_t \, \partial F(W; (t, T]) \, dt$$

(Fubini's theorem), whereas the right-hand side equals (see (E.1))

$$E\left[\left\{EF + \int_0^T \psi_s \, dW_s\right\} \cdot \int_0^T X_s \, dW_s\right] = E \int_0^T X_t \psi_t \, dt.$$

Consequently,

$$E \int_0^T X_t \, \partial F(W; (t, T]) \, dt = E \int_0^T X_t \psi_t \, dt$$

holds for every bounded, continuous, and $\{\mathcal{F}_t\}$-adapted process X, and (E.7) follows.   □

Example E.3: We return to Example E.1, where $F(\omega) = \max_{0 \leq t \leq T} \omega(t)$. This maximum is attained at a unique number $\eta(\omega) \in [0, T]$ (Karatzas and Shreve (1991), Remark 8.16, p. 102), so $F(\omega) = \omega(\eta(\omega))$, for a.e. $\omega \in \Omega$. Conditions (E.4) and (E.5) are clearly satisfied, and as for (E.6), we have

$$\lim_{\varepsilon \downarrow 0} \frac{1}{\varepsilon}[F(\omega + \varepsilon\varphi) - F(\omega)] = \varphi(\eta(\omega)), \quad \forall \varphi \in C^1([0, T]),$$

for P-a.e. $\omega \in \Omega$. Thus $\partial F(\omega; (t, T]) = 1_{\{\eta(\omega) > t\}}$, and in the notation of Example E.1 we have

$$\begin{aligned}
\psi_t = E[\partial F(W; (t, T])|\mathcal{F}_t] &= P[\eta > t|\mathcal{F}_t] = P[M_{t,T} > M_t|\mathcal{F}_t] \\
&= P[M_{T-t} > b]|_{b=M_t-W_t} \\
&= 2\left[1 - \Phi\left(\frac{M_t - W_t}{\sqrt{T-t}}\right)\right] \quad \text{a.s.}
\end{aligned}$$

We have recovered the expression (E.3).

Example E.4: With $\sigma > 0$, take $G(\omega) \stackrel{\triangle}{=} e^{\sigma F(\omega)} = \exp\{\sigma \cdot \max_{0 \leq t \leq T} \omega(t)\}$. Then G satisfies (E.4) and (E.5) with $h = G$, $g(\varepsilon) = e^{\sigma\varepsilon} - 1$. Furthermore, $\partial G(\omega; (t, T]) = \sigma G(\omega) 1_{\{\eta(\omega) > t\}}$ for P-a.e. $\omega \in \Omega$, where $\eta(\omega)$ is as in Example E.3. In the notation of Example E.1, it follows that

$$\begin{aligned}
E[\partial G(W; (t, T])|\mathcal{F}_t] &= \sigma E[\exp(\sigma M_T) 1_{\{\eta(W) > t\}}|\mathcal{F}_t] \\
&= \sigma E[\exp(\sigma M_{t,T}) 1_{\{M_{t,T} > M_t\}}|\mathcal{F}_t] \\
&= \sigma E[\exp(\sigma(M_{T-t} + a)) 1_{\{M_{T-t} > b-a\}}]\Big|_{\substack{a=W_t \\ b=M_t}} \\
&= \frac{2\sigma e^{\sigma a}}{\sqrt{2\pi(T-t)}} \int_{b-a}^{\infty} e^{-\frac{m^2}{2(T-t)} + \sigma m} dm \Big|_{\substack{a=W_t \\ b=M_t}} \\
&= 2\sigma e^{\frac{\sigma^2}{2}(T-t) + \sigma W_t}\left[1 - \Phi\left(\frac{M_t - W_t}{\sqrt{T-t}} - \sigma\sqrt{T-t}\right)\right].
\end{aligned}$$

According to the Clark formula, we have

$$d\Lambda_t = 2\sigma e^{\frac{\sigma^2}{2}(T-t) + \sigma W_t}\left[1 - \Phi\left(\frac{M_t - W_t}{\sqrt{T-t}} - \sigma\sqrt{T-t}\right)\right] dW_t$$

for $\Lambda_t \stackrel{\triangle}{=} E[G|\mathcal{F}_t]$. This last displayed equation can also be verified directly by first computing

$$\begin{aligned}
\Lambda_t = e^{\sigma M_t}&\left[2\Phi\left(\frac{M_t - W_t}{\sqrt{T-t}}\right) - 1\right] \\
&+ 2e^{\frac{\sigma^2}{2}(T-t) + \sigma W_t}\left[1 - \Phi\left(\frac{M_t - W_t}{\sqrt{T-t}} - \sigma\sqrt{T-t}\right)\right]
\end{aligned}$$

and then applying Itô's formula.

Example E.5: Consider now the *Brownian motion* $\tilde{W}_t \stackrel{\triangle}{=} W_t + \nu t$, $0 \le t \le T$, with drift $\nu \in \mathbb{R}$, and define $\tilde{M}_{s,t} \stackrel{\triangle}{=} \max_{s \le u \le t} \tilde{W}_u$, $\tilde{M}_t \stackrel{\triangle}{=} \tilde{M}_{0,t}$. By analogy with Examples E.3 and E.4, we consider the functionals

$$\tilde{F}(\omega) \stackrel{\triangle}{=} \max_{0 \le t \le T} (\omega(t) + \nu t) \quad \text{and} \quad \tilde{G}(\omega) \stackrel{\triangle}{=} \exp(\sigma \tilde{F}(\omega)), \quad \forall \omega \in \Omega,$$

so that $\tilde{F}(W) = \tilde{M}_T$, $\tilde{G}(W) = \exp(\sigma \tilde{M}_T)$. According to Girsanov's theorem, $\tilde{W}$ is a standard Brownian motion under the measure given by $\tilde{P}(A) = E[\exp(-\nu W_t - \frac{\nu^2}{2}t)1_A] = E[\exp(-\nu \tilde{W}_t + \frac{\nu^2}{2}t)1_A]$ for all $A \in \mathcal{F}_t$ and all $t \in [0, T]$. From the *reflection principle* for Brownian motion, we have

$$\tilde{P}[\tilde{M}_t \in db, \tilde{W}_t \in da] = \frac{2(2b-a)}{\sqrt{2\pi t^3}} \exp\left[-\frac{(2b-a)^2}{2t}\right] da\, db, \; a \le b, \; b \ge 0$$

(cf. Karatzas and Shreve (1991), Proposition 2.8.1), and then

$$P[\tilde{M}_t \in db, \tilde{W}_t \in da] = \exp\left[\nu a - \frac{\nu^2}{2}t\right] \cdot \tilde{P}[\tilde{M}_t \in db, \tilde{W}_t \in da]$$

$$= \frac{2(2b-a)}{\sqrt{2\pi t^3}} e^{2b\nu} \exp\left[-\frac{(2b+\nu t-a)^2}{2t}\right] da\, db,$$

for $a \le b$, $b \ge 0$. Consequently,

$$P[\tilde{M}_t \in db] = 2e^{2b\nu}\left[\frac{1}{\sqrt{2\pi t}}e^{-\frac{(b+\nu t)^2}{2t}} - \nu\left\{1 - \Phi\left(\frac{b+\nu t}{\sqrt{t}}\right)\right\}\right]db, \, b \ge 0,$$

or equivalently,

$$P[\tilde{M}_t > b] = f(t,b) \stackrel{\triangle}{=} 1 - \Phi\left(\frac{b-\nu t}{\sqrt{t}}\right) + e^{2\nu b}\left[1 - \Phi\left(\frac{b+\nu t}{\sqrt{t}}\right)\right], \, b \ge 0 .$$

$$\text{(E.11)}$$

As in Examples E.3 and E.4, we obtain

$$E[\partial \tilde{F}(W;(t,T))|\mathcal{F}_t] = P[\tilde{M}_{T-t} > b]|_{b=\tilde{M}_t - \tilde{W}_t} = f(T-t, \tilde{M}_t - \tilde{W}_t),$$

as well as

$$E[\partial \tilde{G}(W;(t,T))|\mathcal{F}_t] = \sigma E\left[\exp(\sigma(\tilde{M}_{T-t} + \tilde{a}))1_{\{\tilde{M}_{T-t} > \tilde{b}-\tilde{a}\}}\right]\Big|_{\substack{\tilde{a}=\tilde{W}_t \\ \tilde{b}=\tilde{M}_t}}$$

$$= \sigma e^{\sigma \tilde{W}_t} \int_{\tilde{M}_t - \tilde{W}_t}^{\infty} e^{\sigma m} P[\tilde{M}_{T-t} \in dm]$$

$$= \sigma e^{\sigma \tilde{M}_t} f(T-t, \tilde{M}_t - \tilde{W}_t)$$

$$+ \sigma^2 e^{\sigma \tilde{W}_t} \int_{\tilde{M}_t - \tilde{W}_t}^{\infty} f(T-t, \xi)e^{\sigma \xi} d\xi,$$

$$E[\tilde{G}(W)|\mathcal{F}_t] = E\left[e^{\sigma \tilde{M}_t}1_{\{\tilde{M}_{t,T} \le \tilde{M}_t\}} + e^{\sigma \tilde{M}_{t,T}}1_{\{\tilde{M}_{t,T} > \tilde{M}_t\}} \mid \mathcal{F}_t\right]$$

$$= e^{\sigma \tilde{M}_t} P\left[\tilde{M}_{T-t} \le \tilde{b} - \tilde{a}\right]\Big|_{\substack{\tilde{a}=\tilde{W}_t \\ \tilde{b}=\tilde{M}_t}}$$

$$+ E\left[\exp(\sigma(\tilde{M}_{T-t} + \tilde{a}))1_{\{\tilde{M}_{T-t} > \tilde{b} - \tilde{a}\}}\right]\Big|_{\substack{\tilde{a} = \tilde{W}_t \\ \tilde{b} = \tilde{M}_t}}$$

$$= e^{\sigma \tilde{M}_t} + \sigma e^{\sigma \tilde{W}_t} \int_{\tilde{M}_t - \tilde{W}_t}^{\infty} f(T - t, \xi) e^{\sigma \xi}\, d\xi.$$

There is also an *extended Clark formula* (E.13) below, which is described as follows. Consider a *smooth Brownian functional*, i.e., a function $F : \Omega \to \mathbb{R}$ of the form $F(\omega) = f(\omega(t_1), \ldots, \omega(t_n))$ for some $n \in \mathbb{N}, (t_1, \ldots, t_n) \in [0, T]^n$, and some element f in the space $C_b^\infty(\mathbb{R}^n)$ of functions with continuous and bounded derivatives of every order. Conditions (E.4)–(E.6) are satisfied by F, with

$$\partial F(\omega; (t, T]) = D_t F(\omega) \triangleq \sum_{j=1}^{n} \frac{\partial}{\partial x_j} f(\omega(t_1), \ldots, \omega(t_n)) 1_{[0, t_j)}(t), \ 0 \le t \le T.$$

For every $p \ge 1$, introduce the norm $\| \cdot \|_{p,1}$ on the space $\mathbf{S}$ of smooth Brownian functionals by the formula

$$\| F \|_{p,1}^p \triangleq E\left[|F(W)|^p + \left(\int_0^T (D_t F(W))^2\, dt\right)^{p/2}\right], \tag{E.12}$$

and denote by $\mathbf{D}_{p,1}$ the Banach space that is the completion of $\mathbf{S}$ under $\| \cdot \|_{p,1}$. As shown by Shigekawa (1980), DF is well-defined by continuity on the entirety of $\mathbf{D}_{p,1}$. We have the following "extended Clark formula."

Theorem E.6: (Ocone (1984); Karatzas, Ocone, and Li (1991)): *For every $F \in \mathbf{D}_{1,1}$ the process ψ in (E.1) is given as the predictable projection of $D_t F$, $0 \le t \le T$. Equivalently, for Lebesgue-almost-every $t \in [0, T]$, we have*

$$\psi_t = E[D_t F | \mathcal{F}_t] \quad a.s. \tag{E.13}$$

References

[1] AASE, K.K. (1993) A jump/diffusion consumption-based capital asset pricing model and the equity premium puzzle. *Mathematical Finance* **3** (2), 65–84.

[2] AASE, K. and ØKSENDAL, B. (1988) Admissible investment strategies in continuous trading. *Stoch. Proc. Appl.* **30**, 291–301.

[3] ADLER, M. and DUMAS, B. (1983) International portfolio choice and corporation finance: a synthesis. *J. Finance* **38**, 925–984.

[4] AKAHORI, J. (1995) Some formulae for a new type of path-dependent option. *Ann. Appl. Probability* **5**, 383–388.

[5] AKIAN, M., MENALDI, J. L., and SULEM, A. (1996) On an investment-consumption model with transaction costs. *SIAM J. Control & Optimization* **34**, 329–364.

[6] ALEXANDER, S.S. (1961) Price movements in speculative markets: trends or random walks. *Industrial Management Rev.* **2**, 7–26. Reprinted in *Cootner (1964)*, 199–218.

[7] ALGOET, P.H. and COVER, T.M. (1988) Asymptotic optimality and asymptotic equipartition properties of log-optimum investment. *Ann. Probability* **16**, 876–898.

[8] ALVAREZ, O. (1994) A singular stochastic control problem in an unbounded domain. Centre de Recherche de Mathématiques de la Décision, Université Paris–Dauphine.

[9] AMENDINGER, J., IMKELLER, P., and SCHWEIZER, M. (1997) Additional logarithmic utility of an insider. Preprint, Technical University, Berlin.

[10] AMIN, K. (1991) On the computation of continuous-time option prices using discrete-time models. *J. Financial & Quantit. Anal.* **26**, 477–496.

[11] AMIN, K. and KHANNA, A. (1994) Convergence of American option values from discrete-to continuous-time financial models. *Mathematical Finance* **4** (4), 289–304.

[12] ANDERSEN, T. (1994) Stochastic autoregressive volatility: a framework for volatility modelling. *Mathematical Finance* **4** (2), 75–102.

[13] ANDERSON, R.W. (1984) *The Industrial Organization of Futures Markets.* D.C. Heath and Co., Lexington, Mass.

[14] ANSEL, J.P. and STRICKER, C. (1992) Lois de martingale, densités, et décomposition de Föllmer-Schweizer. *Ann. Inst. Henri Poincaré* **28** *(Probab. et Statist.)*, 375–392.

[15] ANSEL, J.P. and STRICKER, C. (1994) Couverture des actifs contingents. *Ann. Inst. Henri Poincaré* **30** *(Probab. et Statist.)*, 303–315.

[16] ARAUJO, A. and MONTEIRO, P.K. (1989a) Equilibrium without uniform conditions. *J. Econ. Theory* **48**, 416–427.

[17] ARAUJO, A. and MONTEIRO, P.K. (1989b) General equilibrium with infinitely many goods: The case of separable utilities. Preprint.

[18] ARROW, K. (1952) Le rôle des valeurs boursières pour la repartition la meilleure des risques. *Econometrie*, Colloq. Internat. du CNRS, Paris **40**, 41–47 (with discussion, 47–48). English translation in *Review of Economic Studies* **31** (1964), 1–96.

[19] ARROW, K. (1970) *Essays in the Theory of Risk-Bearing.* North-Holland, Amsterdam.

[20] ARROW, K. (1983) *Collected Papers of Kenneth J. Arrow, Vol. 2: General Equilibrium.* Belknap Press of Harvard University Press, Cambridge, Massachusetts.

[21] ARROW, K. and DEBREU, G. (1954) Existence of equilibrium for a competitive economy. *Econometrica* **22**, 265–290.

[22] ARTZNER, P. and DELBAEN, F. (1989) Term-structure of interest rates: the martingale approach. *Adv. Appl. Mathematics* **10**, 95–129.

[23] ARTZNER, P. and HEATH, D. (1995) Approximate completeness with multiple martingale measures. *Mathematical Finance* **5**, 1–11.

[24] AVELLANEDA, M., LEVY, A., and PARÁS, A. (1995) Pricing and hedging derivative securities in markets with uncertain volatilities. *Appl. Math. Finance* **2**, 73–88.

[25] AVELLANEDA, M. and PARÁS, A. (1994) Dynamic hedging portfolios for derivative securities in the presence of large transaction costs. *Appl. Math. Finance* **1**, 165–194.

[26] AVELLANEDA, M. and PARÁS, A. (1996) Portfolios of derivative securities: the Lagrangian uncertain volatility model. *Appl. Math. Finance* **3**, 23–51.

[27] BACHELIER, L. (1900) Théorie de la spéculation. *Ann. Sci. Ecole Norm. Sup.* **17**, 21–86. Reprinted in *Cootner (1964)*.

[28] BACK, K. (1991) Asset pricing for general processes. *J. Math. Economics* **20**, 371–395.

[29] BACK, K. (1992) Insider trading in continuous-time. *Rev. Financial Studies* **5**, 387–409.

[30] BACK, K. (1993) Time-varying liquidity trading, price-pressure, and volatility. Preprint, Washington University, St. Louis.

[31] BACK, K. and PLISKA, S. (1987) The shadow price of information in continuous-time decision problems. *Stochastics* **22**, 151–186.

[32] BACK, K. and PLISKA, S. (1991) On the fundamental theorem of asset pricing with an infinite state-space. *J. Math. Economics* **20**, 1–18.

[33] BARLES, G., BURDEAU, J., ROMANO, M., and SAMSOEN, N. (1995) Critical stock price near expiration. *Mathematical Finance* **5**, 77–95.

[34] BARLES, G. and SONER, H.M. (1998) Option pricing with transaction costs, and a nonlinear Black-Scholes equation. *Finance & Stochastics*, to appear.

[35] BARONE-ADESI, G. and WHALLEY, R. (1987) Efficient analytic approximation of American option values. *J. Finance* **42**, 301–320.

[36] BARRAQUAND, J. and MARTINEAU, D. (1995) Numerical valuation of high-dimensional multivariate American securities. *J. Financial & Quantit. Analysis* **30**, 383–405.

[37] BARRAQUAND, J. and PUDET, Th. (1996) Pricing of American path-dependent contingent claims. *Mathematical Finance* **6**, 17–51.

[38] BARRON, E.N. (1990) The Bellman equation for control of the running max of a diffusion and applications to look-back options. Preprint, Department of Mathematical Sciences, Loyola University, Chicago.

[39] BARRON, E.N. and JENSEN, R. (1990) A stochastic control approach to the pricing of options. *Math. Operations Research* **15**, 49–79.

[40] BARRON, E.N. and JENSEN, R. (1991) Total risk-aversion and the pricing of options. *Appl. Math. Optimization* **23**, 51–76.

[41] BAŞAK, S. (1993) *General equilibrium continuous-time asset pricing in the presence of portfolio insurers and non-price-taking investors.* Ph.D. dissertation, Graduate School of Industrial Administration, Carnegie-Mellon University.

[42] BAŞAK, S. (1995) A general equilibrium model of portfolio insurance. *Rev. Financial Studies* **8**, 1059–1090.

[43] BAŞAK, S. (1996a) General equilibrium effects of portfolio insurance. Preprint, Wharton School, University of Pennsylvania.

[44] BAŞAK, S. (1996b) Dynamic consumption-portfolio choice and asset pricing with non-price-taking agents. Preprint, Wharton School, University of Pennsylvania.

[45] BAŞAK, S. (1996c) A model of dynamic equilibrium asset pricing with extraneous risk. Preprint, Wharton School, University of Pennsylvania.

[46] BAŞAK, S. and CUOCO, D. (1996) An equilibrium model with restricted stock market participation. *Rev. Financial Studies* **11**, to appear.

[47] BATES, D.S. (1988) Pricing options under jump-diffusion processes. Preprint, Wharton School of Business, University of Pennsylvania.

[48] BATES, D.S. (1992) Jumps and stochastic volatility: exchange rate processes implicit in foreign currency options. Preprint, Wharton School of Business, University of Pennsylvania.

[49] BAXTER, M.W. and RENNIE, A. (1996) *Financial Calculus: An Introduction to Derivative Pricing.* Cambridge University Press, Cambridge, U.K.

[50] BEAGHTON, P.J. (1988) Boundary integral valuation of American options. Preprint, Salomon Brothers, New York.

[51] BECKERS, S. (1980) The constant elasticity of variance model and its implications for option-pricing. *J. Finance* **35**, 661–673.

[52] BECKERS, S. (1983) Variances of security prices returns based on high, low and closing prices. *J. Business* **56**, 97–112.

[53] BEINERT, M. and TRAUTMANN, S. (1991) Jump-diffusion models of stock returns: a statistical investigation. *Statistische Hefte* **32**, 269–280.

[54] BENSAID, B., LESNE, J.-P., PAGÈS, H., and SCHEINKMAN, J. (1992) Derivative asset pricing with transaction costs. *Mathematical Finance* **2**, 63–86.

[55] BENSOUSSAN, A. (1984) On the theory of option pricing. *Acta Appl. Math.* **2**, 139–158.

[56] BENSOUSSAN, A., CROUHY, M., and GALLAI, D. (1994) Stochastic equity volatility and the capital structure of the firm. *Phil. Trans. Roy. Soc. London, Ser. A* **347**, 531–541.

[57] BENSOUSSAN, A., CROUHY, M., and GALLAI, D. (1994/95) Stochastic equity volatility related to the leverage effect: I, Equity volatility behaviour; and II, Valuation of European equity options and warrants. *Appl. Math. Finance* **1**, 63–85; and **2**, 43–59.

[58] BERGMAN, Y. (1985) Time preference and capital asset pricing models. *J. Financial Economics* **14**, 145–159.

[59] BERGMAN, Y. (1995) Option pricing with differential interest rates. *Rev. Financial Studies* **8**, 475–500.

[60] BERNSTEIN, P. (1992) *Capital Ideas.* Macmillan, New York.

[61] BERTSEKAS, D. and SHREVE, S.E. (1978) *Stochastic Optimal Control: The Discrete-Time Case.* Academic Press, New York. Republished by Athena Scientific, Cambridge, MA, 1996.

[62] BEWLEY, T. (1972) Existence of equilibria in economies with infinitely many commodities. *J. Economic Theory* **4**, 514–540.

[63] BEWLEY, T. (1986) Stationary monetary equilibrium with a continuum of independently fluctuating consumers. In *Contributions to Mathematical Economics: Essays in Honor of Gérard Debreu* (W. Hildenbrand and A. Mas-Colell, eds.), 79–102. North-Holland, Amsterdam.

[64] BHATTACHARYA, M. (1983) Transactions data tests of efficiency of the Chicago Board Options Exchange. *J. Financial Economics* **12**, 161–185.

[65] BICK, A. and WILLINGER, W. (1994) Dynamic spanning without probabilities. *Stoch. Process Appl.* **50**, 349–374.

[66] BISMUT, J. M. (1973) Conjugate convex functions in optimal stochastic control, *J. Math. Anal. Appl.* **44**, 384–404.

[67] BISMUT, J. M. (1975) Growth and optimal intertemporal allocations of risks. *J. Economic Theory* **10**, 239–287.

[68] BISMUT, J. M. (1981) Mécanique Aléatoire. *Lecture Notes in Mathematics* **866.** Springer-Verlag, Berlin.

[69] BISMUT, J. M. and SKALLI, B. (1977), Temps d'arrêt, théorie générale de processus et processus de Markov. *Z. Wahrscheinlichkeitstheorie verw. Gebiete* **39**, 301–313.

[70] BISIÈRE, Ch. (1997) *La Structure par Terme des Taux d'Interêt.* Presses Universitaires de France, Paris.

[71] BJÖRK, Th. (1997) Interest Rate Theory. *Lecture Notes in Mathematics* **1656**, 53–122. Springer-Verlag, Berlin.

[72] BJÖRK, Th., DI MASI, G., KABANOV, Yu., and RUNGGALDIER, W. (1997) Towards a general theory of bond-markets. *Finance & Stochastics* **1**, 141–174.

[73] Björk, Th., Kabanov, Yu., and Runggaldier, W. (1997) Bond-market structure in the presence of marked point processes. *Mathematical Finance* **7**, 211–239.

[74] Black, F. (1975) Fact and fantasy in the use of options. *Financial Anal. J.* **31**, 36–41, 61–72.

[75] Black, F. (1976a) The pricing of commodity contracts, *J. Financial Econom.* **3**, 167–179.

[76] Black, F. (1976b) Studies of the stock price volatility changes. In *Proc. 1976 Meeting of the Amer. Stat. Assoc.* (Business and Economic Statistics Section), 177–181.

[77] Black, F. (1986) Noise. *J. Finance* **41**, 529–543.

[78] Black, F. and Perold, A. (1992) Theory of constant proportion portfolio insurance. *J. Economic Dynamics & Control* **16**, 403–426.

[79] Black, F. and Scholes, M. (1973) The pricing of options and corporate liabilities. *J. Political Economy,* **81**, 637–659.

[80] Black, F., Derman, E., and Toy, W. (1990) A one-factor model of interest rates and its application to treasury bond options. *Fin. Anal. Journal* **46**, 33–39.

[81] Blattberg, R. and Gonedes, N. (1974) A comparison of stable and Student distributions as statistical models of stock prices. *J. Business* **47**, 244–280.

[82] Boness, A.J. (1964) Some evidence of the profitability of trading put and call options. In *Cootner (1964)*, 475–496.

[83] Border, K.C. (1985) *Fixed point theorems with applications to economics and game theory.* Cambridge University Press.

[84] Bouaziz, L., Briys, E., and Crouhy, M. (1994) The pricing of forward-starting Asian options. *J. Banking & Finance* **18**, 823–839.

[85] Boykov, Y. (1996) Inference consistency in case of countably additive measures. Ph.D. dissertation, School of Oper. Res. Indus. Eng., Cornell University.

[86] Boyle, P. (1977) Options: a Monte-Carlo approach. *J. Financial Economics* **4**, 323–338.

[87] Boyle, P. (1988) A lattice framework for option-pricing with two state-variables. *J. Financial & Quantit. Analysis* **23**, 1–12.

[88] Boyle, P., Broadie, M., and Glasserman, P. (1997) Monte-Carlo methods for security pricing. *J. Econom. Dynamics & Control,* **21**, 1267–1322.

[89] Boyle, P., Evnine, J., and Gibbs, S. (1989) Numerical evaluation of multivariate contingent claims. *Rev. Financial Studies* **2**, 241–250.

[90] Boyle, P. and Vorst, T. (1992) Option replication in discrete time with transaction costs. *J. Finance* **47**, 272–293.

[91] Brace, A., Gatarek, D., and Musiela, M. (1997) The market model of interest-rate dynamics. *Mathematical Finance* **7**, 127–154.

[92] Breeden, D. (1979) An intertemporal asset pricing model with stochastic consumption and investment opportunities. *J. Financial Economics* **7**, 265–296.

[93] Breeden, D., Gibbons, M., and Litzenberger, R. (1989) Empirical tests of the consumption-oriented CAPM. *J. Finance* **44**, 231–262.

[94] BREIMAN, L. (1961) Optimal gambling systems for favorable games. *Fourth Berkeley Symposium on Math. Statist. and Probability* **1**, 65–78.

[95] BRENNAN, M.J. (1971) Capital market equilibrium with divergent borrowing and lending rates. *J. Financ. & Quantit. Analysis* **6**, 1197–1205.

[96] BRENNAN, M.J. and SCHWARTZ, E. (1977) The valuation of the American put option. *J. Finance* **32**, 449–462.

[97] BRENNAN, M.J. and SCHWARTZ, E.S. (1979) A continuous-time approach to the pricing of bonds. *J. Banking & Finance* **3**, 133–155.

[98] BRENNAN, M.J. and SCHWARTZ, E.S. (1982) An equilibrium model of bond pricing and a test of market efficiency. *J. Financial & Quantit. Analysis* **17**, 301–329.

[99] BRENNAN, M.J. and SCHWARTZ, E.S. (1988) Time-invariant portfolio insurance strategies. *J. Finance* **43**, 283–299.

[100] BRENNAN, M.J. and SCHWARTZ, E.S. (1989) Portfolio insurance and financial market equilibrium. *J. Business* **62**, 455–472.

[101] BROADIE, M., CVITANIĆ, J., and SONER, H.M. (1998) On the cost of super-replication with transaction costs. *Rev. Financial Studies* **11**, 59–79.

[102] BROADIE, M. and DETEMPLE, J. (1995) American capped call options on dividend-paying assets. *Rev. Financial Studies* **8**, 161–191.

[103] BROADIE, M. and DETEMPLE, J. (1996) American option valuation: new bounds, approximations, and a comparison of existing methods. *Rev. Financial Studies* **9**, 1211–1250.

[104] BROADIE, M. and DETEMPLE, J. (1997) The valuation of American options on multiple assets. *Mathematical Finance* **7**, 241–286.

[105] BROADIE, M. and GLASSERMAN, P. (1997) Pricing American-style securities using simulation. *J. Econom. Dynamics & Control*, **21**, 1323–1352.

[106] BROADIE, M., GLASSERMAN, P., and KOU, S.G. (1996) Connecting discrete and continuous path-dependent options. *Finance & Stochastics*, to appear.

[107] BROADIE, M., GLASSERMAN, P., and KOU, S.G. (1997) A continuity correction for discrete barrier options. *Mathematcal Finance* **7**, 325–350.

[108] BROCK, W.A. and MAGILL, M.J.P. (1979) Dynamics under uncertainty. *Econometrica* **47**, 843–868.

[109] BROWNE, S. (1995) Optimal investment policies for a firm with random risk process: exponential utility and minimization of the probability of ruin. *Math. Operations Research* **20**, 937–958.

[110] BUNCH, D. and JOHNSON, H. (1992) A simple and numerically efficient valuation method for American puts, using a modified Geske–Johnson approach. *J. Finance* **47**, 809–816.

[111] CADENILLAS, A. (1992) *Contributions to the Stochastic Version of Pontryagin's Maximum Principle*. Doctoral dissertation, Department of Statistics, Columbia University.

[112] CADENILLAS, A. and KARATZAS, I. (1995) The stochastic maximum principle for linear, convex optimal control with random coefficients. *SIAM J. Control & Optimization* **33**, 590–624.

[113] CADENILLAS, A. and PLISKA, S.R. (1997) Optimal trading of a security when there are taxes and transaction costs. Preprint, University of Alberta, Edmonton.

[114] CAMPBELL, J.Y, LO, A.W., and MACKINLAY, A.C. (1997) *The Econometrics of Financial Markets*. Princeton University Press.

[115] CARR, P. (1993) Bibliography on *Exotic Options*. Preprint.

[116] CARR, P. and FAGUET, D. (1994) Fast accurate valuation of American options. Preprint, Graduate School of Management, Cornell University.

[117] CARR, P. (1998) Randomization and the American put, *Rev. Financial Studies*, to appear.

[118] CARR, P. and JARROW, R. (1990) The stop-loss, start-gain paradox and option valuation: a new decomposition into intrinsic and time-value. *Rev. Financial Studies* **3**, 469–492.

[119] CARR, P., JARROW, R., and MYNENI, R. (1992) Alternative characterizations of American put-options. *Mathematical Finance* **2** (2), 87–106.

[120] CARVERHILL, A. and CLEWLOW, L. (1990) Average-rate options. *Risk* **3**, 25–29.

[121] CARVERHILL, A. and WEBBER, N. (1990) American options: theory and numerical analysis. In *Options: Recent Advances in Theory and Practice*. Manchester University Press.

[122] CHAN, Y.K. (1992) Term structure as a second-order dynamical system and pricing of derivative securities. Bear Stearns & Co., New York.

[123] CHATELAIN, M. and STRICKER, C. (1992) A characterization of complete security markets on a Brownian filtration. Preprint.

[124] CHATELAIN, M. and STRICKER, C. (1994) On component-wise and vector stochastic integration. *Mathematical Finance* **4**, 57–65.

[125] CHESNEY, M., ELLIOT, R.J., and GIBSON, R. (1993) Analytical solutions for the pricing of American bond and yield options. *Mathematical Finance* **3**, 277–294.

[126] CHESNEY, M. and SCOTT, L.O. (1989) Pricing European options: a comparison of the simplified Black–Scholes model and a random-variance model. *J. Financial & Quantit. Analysis* **24**, 267–284.

[127] CHESNEY, M., JEANBLANC-PICQUÉ, M., and YOR, M. (1997) Brownian excursions and Parisian barrier options. *Adv. Appl. Probability* **29**, 165–184.

[128] CHEW, S.H. and EPSTEIN, L. (1991) Recursive utility under uncertainty. In *Equilibrium Theory in Infinite-Dimensional Spaces* (A. Khan and N. Yannelis, eds.). Springer-Verlag, New York.

[129] CHRISTIE, A.A. (1982) The stochastic behavior of common stock variance. *J. Financial Econom.* **10**, 407–432.

[130] CHUNG, K.L. (1974) *Probability Theory*. Academic Press, New York.

[131] CLARK, J.M.C. (1970) The representation of functionals of Brownian motion as stochastic integrals. *Ann. Math. Statistics* **41**, 1282–1295. Correction 1778.

[132] COLWELL, D.B., ELLIOTT, R.J., and KOPP, P.E. (1991) Martingale representation and hedging policies. *Stoch. Processes Appl.* **38**, 335–345.

[133] CONSTANTINIDES, G. M. (1979) Multiperiod consumption and investment behavior with convex transaction costs. *Management Sci.* **25**, 1127–1137.

[134] CONSTANTINIDES, G. (1982) Intertemporal asset pricing with heterogeneous consumers and without demand aggregation. *J. Business* **55**, 253–267.

[135] CONSTANTINIDES, G.M. (1986) Capital market equilibrium with transactions costs. *J. Political Economy* **94**, 842–862.

[136] CONSTANTINIDES, G.M. (1990) Habit formation: a resolution of the equity premium puzzle. *J. Political Economy* **98**, 519–543.

[137] CONSTANTINIDES, G.M. (1993) Option pricing bounds with transaction costs. Preprint, Graduate School of Business, University of Chicago.

[138] CONSTANTINIDES, G.M. (1997) Transaction costs and the volatility implied by option prices. Preprint, Graduate School of Business, University of Chicago.

[139] CONSTANTINIDES, G.M. and ZARIPHOPOULOU, Th. (1997) Bounds on option prices in an intertemporal economy with proportional transaction costs and general preferences. Preprint.

[140] CONZE, A. and VISWANATHAN, R. (1991) European path-dependent options: the case of geometric averages. *J. Finance* **46**, 1893–1907.

[141] COOTNER, P.H. (ed.) (1964) *The Random Character of Stock Market Prices.* MIT Press, Cambridge, MA.

[142] CORNELL, B. (1981) The consumption-based asset pricing model. *J. Financial Economics* **9**, 103–108.

[143] COVER, T.M. (1984) An algorithm for maximizing expected log-investment return. *IEEE Trans. Inform. Theory* **30**, 369–373.

[144] COVER, T. (1991) Universal portfolios. *Mathematical Finance* **11**, 1–29.

[145] COX, J.C. and HUANG, C.F. (1989) Optimal consumption and portfolio policies when asset prices follow a diffusion process. *J. Economic Theory* **49**, 33–83.

[146] COX, J.C. and HUANG, C.F. (1991) A variational problem arising in financial economics. *J. Math. Economics* **20**, 465–487.

[147] COX, J.C., INGERSOLL, J.E., and ROSS, S. (1981) The relation between forward prices and futures prices. *J. Financial Economics* **9**, 321–346.

[148] COX, J.C., INGERSOLL, J.E., and ROSS, S. (1985a) An intertemporal general equilibrium model of asset prices. *Econometrica*, **53**, 363–384.

[149] COX, J.C., INGERSOLL, J.E., and ROSS, S. (1985b) A theory of the term structure of interest rates. *Econometrica*, **53**, 385–407.

[150] COX, J.C. and ROSS, S.A. (1976) The valuation of options for alternative stochastic processes. *J. Financial Economics* **3**, 145–166.

[151] COX, J.C., ROSS, S., and RUBINSTEIN, M. (1979) Option pricing: a simplified approach. *J. Financial Economics* **7**, 229–263.

[152] COX, J.C. and RUBINSTEIN, M. (1985) *Options Markets.* Prentice-Hall, Englewood Cliffs, N.J.

[153] CUOCO, D. (1997) Optimal consumption and equilibrium prices with portfolio constraints and stochastic income. *J. Econom. Theory* **72**, 33–73.

[154] CUOCO, D. and CVITANIĆ, J. (1998) Optimal consumption choices for a "large investor." *J. Econom. Dynamics & Control,* **22**, 401–436.

[155] CUOCO, D. and HE, H. (1993) Dynamic aggregation and computation of equilibria in finite-dimensional economies with incomplete financial markets. Preprint, Wharton School, University of Pennsylvania.

[156] CUOCO, D. and HE, H. (1994) Dynamic equilibrium in infinite-dimensional economies with incomplete financial markets. Preprint, Wharton School, University of Pennsylvania.

[157] CUOCO, D. and LIU, H. (1997) Optimal consumption of a divisible durable good. Preprint, Wharton School, University of Pennsylvania.

[158] CUTLAND, N.J., KOPP, P.E., and WILLINGER, W. (1991) A non-standard approach to option pricing. *Mathematical Finance* **1** (4), 1–38.

[159] CUTLAND, N.J., KOPP, P.E., and WILLINGER, W. (1993a) From discrete to continuous financial models: new convergence results for option pricing. *Mathematical Finance* **3** (2), 101–123.

[160] CUTLAND, N.J., KOPP, P.E., and WILLINGER, W. (1993b) A nonstandard treatment of options driven by Poisson processes. *Stochastics* **42**, 115–133.

[161] CVITANIĆ, J. (1997) Optimal trading under constraints. *Lecture Notes in Mathematics* **1656**, 123–190. Springer-Verlag, Berlin.

[162] CVITANIĆ, J. and KARATZAS, I. (1992) Convex duality in convex portfolio optimization. *Annals Appl. Probability* **2**, 767–818.

[163] CVITANIĆ, J. and KARATZAS, I. (1993) Hedging contingent claims with constrained portfolios. *Annals Appl. Probability* **3**, 652–681.

[164] CVITANIĆ, J. and KARATZAS, I. (1995) On portfolio optimization under "drawdown" constraints. In *Mathematical Finance* (M. Davis, D. Duffie, W. Fleming, and S. Shreve, eds.) *IMA Vol.* **65**, 35–46. Springer-Verlag, New York.

[165] CVITANIĆ, J. and KARATZAS, I. (1996) Hedging and portfolio optimization under transaction costs: a martingale approach. *Mathematical Finance* **6**, 133–165.

[166] CVITANIĆ, J. and MA, J. (1996) Hedging options for a large investor, and forward-backward SDEs. *Annals Appl. Probability* **6**, 370–398.

[167] CVITANIĆ, J., PHAM, H., and TOUZI, N. (1997) Super-replication in stochastic volatility models under portfolio constraints. *Adv. Appl. Probability*, to appear.

[168] CVITANIĆ, J., PHAM, H., and TOUZI, N. (1998) A closed-form solution to problems of super-replication under transaction costs. *Finance & Stochastics*, to appear.

[169] DALANG, R.C., MORTON, A., and WILLINGER, W. (1990) Equivalent martingale measures and no-arbitrage in stochastic security market models. *Stochastics* **29**, 185–201.

[170] DANA, R.A. (1993a) Existence and uniqueness of equilibria when preferences are additively separable. *Econometrica* **61**, 953–957.

[171] DANA, R.A. (1993b) Existence, uniqueness and determinacy of Arrow–Debreu equilibria in finance models. *J. Math. Economics* **22**, 563–579.

[172] DANA, R.A. and PONTIER, M. (1992) On existence of an Arrow–Radner equilibrium in the case of complete markets: a remark. *Math. Operations Research* **17**, 148–163.

[173] DASSIOS, A. (1995) The distribution of the quantile of a Brownian motion with drift, and the pricing of related path-dependent options. *Ann. Appl. Probability* **5**, 389–398.

[174] DAVIS, M.H.A. (1994) Option pricing in incomplete markets. Preprint, Imperial College, London.

[175] DAVIS, M.H.A. (1996) The Margrabe formula. Research Note, Tokyo-Mitsubishi International, London.

[176] DAVIS, M.H.A. and CLARK, J.M.C. (1994) A note on super-replicating strategies. *Phil. Trans. Royal Soc. London, Ser. A* **347**, 485–494.

[177] DAVIS, M.H.A. and NORMAN, A. (1990) Portfolio selection with transaction costs. *Math. Operations Research* **15**, 676–713.

[178] DAVIS, M.H.A. and PANAS, V.G. (1994) The writing price of a European contingent claim under proportional transaction costs. *Comp. & Appl. Mathematics* **13**, 115–157.

[179] DAVIS, M.H.A., PANAS, V.G., and ZARIPHOPOULOU, T. (1993) European option pricing with transaction costs. *SIAM J. Control & Optimization* **31**, 470–493.

[180] DAVIS, M.H.A. and ZARIPHOPOULOU, T. (1995) American options and transaction fees. In *Mathematical Finance* (M. Davis, D. Duffie, W. Fleming and S. Shreve, eds.) *IMA Vol.* **65**, 47–62. Springer-Verlag, New York.

[181] DEBREU, G. (1959) *Theory of Value.* J. Wiley & Sons, New York.

[182] DEBREU, G. (1982) Existence of competitive equilibrium. In *Handbook of Mathematical Economics, Vol. II* (K. Arrow and M. Intriligator, eds.), North-Holland, Amsterdam.

[183] DEBREU, G. (1983) *Mathematical Economics: Twenty Papers of G. Debreu.* Cambridge University Press.

[184] DE FINETTI, B. (1937) La prévision: ses lois logiques, ses sources subjectives. *Ann. Inst. Henri Poincaré* **7**, 1–68.

[185] DE FINETTI, B. (1974) *Theory of Probability.* J. Wiley & Sons, New York.

[186] DELBAEN, F. (1992) Representing martingale measures when asset prices are continuous and bounded. *Mathematical Finance* **2**(2), 107–130.

[187] DELBAEN, F., MONAT, P., SCHACHERMAYER, W., SCHWEIZER, M., and STRICKER, C. (1997) Weighted-norm inequalities and hedging in incomplete markets. *Finance & Stochastics* **1**, 181–227.

[188] DELBAEN, F. and SCHACHERMAYER, W. (1994a) A general version of the fundamental theorem of asset pricing. *Mathematische Annalen* **300**, 463–520.

[189] DELBAEN, F. and SCHACHERMAYER, W. (1994b) Arbitrage and free-lunch with bounded risk, for unbounded continuous processes. *Mathematical Finance* **4** (4), 343–348.

[190] DELBAEN, F. and SCHACHERMAYER, W. (1995a) The no-arbitrage property under a change of numéraire. *Stochastics* **53**, 213–226.

[191] DELBAEN, F. and SCHACHERMAYER, W. (1995b) Arbitrage possibilities in Bessel processes and their relations to local martingales. *Probab. Theory & Rel. Fields* **102**, 357–366.

[192] DELBAEN, F. and SCHACHERMAYER, W. (1995c) The existence of absolutely continuous local martingale measures. *Ann. Appl. Probability* **5**, 926–945.

[193] DELBAEN, F. and SCHACHERMAYER, W. (1996a) The variance-optimal martingale measure for continuous processes. *Bernoulli* **2**, 81–105.

[194] DELBAEN, F. and SCHACHERMAYER, W. (1996b) Attainable claims with pth moments. *Ann. Inst. Henri Poincaré* **32** (*Probab. and Statist.*), 743–763.

[195] DELBAEN, F. and SCHACHERMAYER, W. (1997a) The Banach space of workable contingent claims in arbitrage theory. *Ann. Inst. Henri Poincaré* **33** (*Probab. et Statist.*), 113–144.

[196] DELBAEN, F. and SCHACHERMAYER, W. (1997b) The fundemantal theorem of asset-pricing for unbounded stochastic processes. Preprint.

[197] DELBAEN, F. and SCHACHERMAYER, W. (1997c) Non-arbitrage and the fundemantal theorem of asset-pricing: summary of main results. *Proc. Symp. Appl. Mathematics*, to appear.

[198] DEMARZO, P. and SKIADAS, C. (1996) Aggregation, determinacy, and informational efficiency for a class of economies with asymmetric information. *Journal of Economic Theory*, to appear.

[199] DEMARZO, P. and SKIADAS, C. (1997) On the uniqueness of fully informative rational expectations equilibria. *Economic Theory*, to appear.

[200] DENGLER, H. (1993) *Poisson approximations to continuous security market models*. Doctoral dissertation, Department of Mathematics, Cornell University.

[201] DERMODY, J.C. and ROCKAFELLAR, R.T. (1991) Cash-stream valuation in the presence of transaction costs and taxes. *Mathematical Finance* **1**(1), 31–54.

[202] DERMODY, J.C. and ROCKAFELLAR, R.T. (1995) Tax basis and nonlinearity in cash-stream valuation. *Mathematical Finance* **5**, 97–119.

[203] DETEMPLE, J. (1986a) A general equilibrium model of asset-pricing with partial or heterogeneous information. *Finance* **7**, 183–201.

[204] DETEMPLE, J. (1986b) Asset pricing in a production economy with incomplete information. *J. Finance* **XLI**, 383–390.

[205] DETEMPLE, J. and MURTHY, S. (1993) Intertemporal asset pricing with heterogeneous beliefs. *J. Econom. Theory* **62**, 294–320.

[206] DETEMPLE, J. and MURTHY, S. (1996) Equilibrium asset prices and no-arbitrage with portfolio constraints. Preprint.

[207] DETEMPLE, J. and ZAPATERO, F. (1991) Asset prices in an exchange economy with habit formation. *Econometrica* **59**, 1633–1657.

[208] DETEMPLE, J. and ZAPATERO, F. (1992) Optimal consumption-portfolio policies with habit formation. *Mathematical Finance* **2**, 251–274.

[209] DEWYNNE, J.N., WHALLEY, A.E., and WILMOTT, P. (1994) Path-dependent options and transaction costs. *Phil. Trans. Roy Soc. Lond. A* **347**, 517–529.

[210] DIXIT, A. and PINDYCK, R.S. (1994) *Investment under Uncertainty*. Princeton University Press.

[211] DOOB, J.L. (1953) *Stochastic Processes*. J. Wiley & Sons, New York.

[212] DOTHAN, M. (1978) On the term structure of interest rates. *J. Fin. Economics* **7**, 229–264.

[213] DOTHAN, M. (1990) *Prices in Financial Markets*. Oxford University Press, New York.

[214] DRITSCHEL, M. and PROTTER, P. (1997) Complete markets with discontinuous security prices. Preprint.

[215] DUBINS, L. and SAVAGE, L. (1965) *How to Gamble if You Must: Inequalities for Stochastic Processes*. McGraw-Hill, New York. Reprinted by Dover Publications, New York, 1976.

[216] DUBOFSKY, D.A. (1992) *Options and Financial Futures: Valuation and Uses*. McGraw-Hill, New York.

[217] DUFFIE, D. (1986) Stochastic equilibria: existence, spanning number, and the "no expected gain from trade" hypothesis. *Econometrica* **54**, 1164–1184.

[218] DUFFIE, D. (1987) Stochastic equilibria with incomplete financial markets. *J. Economic Theory* **41**, 405–416.

[219] DUFFIE, D. (1988) *Security Markets: Stochastic Models*. Academic Press, Orlando.

[220] DUFFIE, D. (1989) *Futures Markets*. Prentice-Hall, Englewood Cliffs, NJ.

[221] DUFFIE, D. (1992) *Dynamic Asset Pricing Theory.* Princeton University Press.

[222] DUFFIE, D. and EPSTEIN, L. (Appendix C with C. SKIADAS) (1992a) Stochastic differential utility. *Econometrica* **60**, 353–394.

[223] DUFFIE, D. and EPSTEIN, L. (1992b) Asset pricing with stochastic differential utility. *Rev. Economic Studies* **5**, 411–436.

[224] DUFFIE, D., FLEMING, W., SONER, H.M., and ZARIPHOPOULOU, T. (1997) Hedging in incomplete markets with HARA utility. *J. Economic Dynamics & Control* **21**, 753–782.

[225] DUFFIE, D., GEANAKOPLOS, J., MAS-COLELL, A., and McLENNAN, A. (1994) Stationary Markov equilibria. *Econometrica* **62**, 745–781.

[226] DUFFIE, D., GEOFFARD, P.Y., and SKIADAS, C. (1994) Efficient and equilibrium allocations with stochastic differential utility. *J. Math. Economics* **23**, 133–146.

[227] DUFFIE, D. and GLYNN, P. (1995) Efficient Monte Carlo simulation of security prices. *Annals Appl. Probability* **5**, 897–905.

[228] DUFFIE, D. and HARRISON, J.M. (1993) Arbitrage pricing of Russian options and perpetual look-back options. *Annals Appl. Probability* **3**, 641–573.

[229] DUFFIE, D. and HUANG, C.F. (1985) Implementing Arrow–Debreu equilibria by continuous trading of few long-lived securities. *Econometrica* **53**, 1337–1356.

[230] DUFFIE, D. and HUANG, C.F. (1986) Multi-period security markets with differential information. *J. Math. Econom.* **15**, 283–303.

[231] DUFFIE, D. and HUANG, C.F. (1987) Stochastic production-exchange equilibria. Research paper, Graduate School of Business, Stanford University.

[232] DUFFIE, D. and KAN, R. (1994) Multi-factor term structure models. *Phil. Trans. Roy. Soc. London, Ser. A* **347**, 577–586.

[233] DUFFIE, D. and KAN, R. (1996) A yield-factor model of interest rates. *Mathematical Finance* **6**, 379–406.

[234] DUFFIE, D. and LIONS, P.-L. (1990) Partial differential equation solutions of stochastic differential utility. *J. Math. Economics* **21**, 577–606.

[235] DUFFIE, D. and PROTTER, P. (1992) From discrete- to continuous-time finance: weak convergence of the financial gain process. *Mathematical Finance* **2** (1), 1–15.

[236] DUFFIE, D. and RICHARDSON, H.R. (1991) Mean-variance hedging in continuous time. *Ann. Appl. Probability* **1**, 1–15.

[237] DUFFIE, D. and SHAFER, W. (1985) Equilibrium in incomplete markets. I: A basic model of generic existence. *J. Math. Economics* **14**, 285–300.

[238] DUFFIE, D. and SHAFER, W. (1986) Equilibrium in incomplete markets. II: Generic existence in stochastic economies.*J. Math. Economics* **15**, 199–216.

[239] DUFFIE, D. and SKIADAS, C. (1994) Continuous-time security pricing: a utility gradient approach. *J. Math. Economics* **23**, 107–132.

[240] DUFFIE, D. and STANTON, R. (1992) Pricing continuously resettled contingent claims. *J. Economic Dynamics & Control* **16**, 561–573.

[241] DUFFIE, D. and ZAME, W. (1989) The consumption-based capital asset pricing model. *Econometrica* **57**, 1279–1297.

[242] DUFFIE, D. and ZARIPHOPOULOU, T. (1993) Optimal investment with undiversifiable income. *Mathematical Finance* **3**, 135–148.

[243] DUMAS, B. (1989) Two-person dynamic equilibrium in the capital market. *Review of Financial Studies* **2**, 157–188.

[244] DUMAS, B. and LUCIANO, E. (1989) An exact solution to a dynamic portfolio choice problem under transaction costs. *J. Finance* **46**, 577–595.

[245] DUNFORD, N. and SCHWARTZ, J.T. (1957) *Linear Operators. Part I: General Theory.* Wiley-Interscience, New York.

[246] DUNN, K. and SINGLETON, K. (1986) Modelling the term structure of interest rates under nonseparable utility and durability of goods. *J. Financial Economics* **14**, 27–55.

[247] DUPIRE, B. (1993/94) Model art; pricing with a smile. *Risk* **6**, 118–124; **7**, 18–20.

[248] DYBVIG, P.H. and HUANG, C.F. (1988) Nonnegative wealth, absence of arbitrage, and feasible consumption plans. *Rev. Financial Studies* **1**, 377–401.

[249] DYBVIG, P. (1993) Occasional ratcheting: optimal dynamic consumption and investment given intolerance for any decline in standard of living. Preprint, Olin School of Business, Washington University.

[250] DYBVIG, P. (1997) Bond and bond option pricing based on the current term structure. In *Mathematics of Derivative Securities*, M.A.H. Dempster and S.R. Pliska, eds., Cambridge University Press.

[251] EBERLEIN, E. (1992) On modelling questions in security valuation. *Mathematical Finance* **2** (1), 17–32.

[252] EBERLEIN, E. and JACOD, J. (1997) On the range of option prices. *Finance & Stochastics* **1**, 131–140.

[253] EBERLEIN, E. and KELLER, U. (1995) Hyperbolic distributions in finance. *Bernoulli* **1**, 281–299.

[254] EDIRISINGHE, C., NAIK, V., and UPPAL, R. (1993) Optimal replication of options with transaction costs and trading restrictions. *J. Financial & Quantit. Analysis* **28**, 117–138.

[255] EDWARDS, F.R. and MA, C.W. (1992) *Futures and Options.* McGraw-Hill, New York.

[256] EKELAND, I. and TEMAM, R. (1976) *Convex Analysis and Variational Problems.* North Holland, Amsterdam and American Elsevier, New York.

[257] EL KAROUI, N. (1981) Les aspects probabilistes du contrôle stochastique. *Lecture Notes in Mathematics* **876**, 73–238. Springer-Verlag, Berlin.

[258] EL KAROUI, N. and JEANBLANC-PICQUÉ, M. (1997) Optimization of consumption with labor income, *Finance & Stochastics*, to appear.

[259] EL KAROUI, N., JEANBLANC-PICQUÉ, M., and VISWANATHAN, R. (1992) On the robustness of the Black–Scholes equation. *Lecture Notes on Control & Information Sciences* **177**, 224–247. Springer-Verlag, Berlin.

[260] EL KAROUI, N., JEANBLANC-PICQUÉ, M., and SHREVE, S.E. (1998) On the robustness of the Black-Scholes equation. *Mathematical Finance*, **8**, 93–126.

[261] EL KAROUI, N. and KARATZAS, I. (1995) The optimal stopping problem for a general American put-option. In *Mathematical Finance* (M. Davis, D. Duffie, W. Fleming, and S. Shreve, eds.) *IMA Vol. 65*, 77–88. Springer-Verlag, New York.

[262] EL KAROUI, N., MYNENI, R., and VISWANATHAN, R. (1992) Arbitrage pricing and hedging of interest rate claims with state variables, I (theory) and II (applications). Preprint.

[263] EL KAROUI, N., PENG, S., and QUENEZ, M.C (1997) Backward stochastic differential equations in finance. *Mathematical Finance* **7**, 1–71.

[264] EL KAROUI, N. and QUENEZ, M.C. (1991) Programmation dynamique et évaluation des actifs contingents en marché incomplet. *C. R. Acad. Sci. Paris, Série I* **313**, 851–854.

[265] EL KAROUI, N. and QUENEZ, M.C. (1995) Dynamic programming and pricing of contingent claims in an incomplete market. *SIAM J. Control & Optimization* **33**, 29–66.

[266] EL KAROUI, N. and QUENEZ, M.C. (1997) Nonlinear pricing theory and backward stochastic differential equations. *Lecture Notes in Mathematics* **1656**, 191–246. Springer-Verlag, Berlin.

[267] EL KAROUI, N. and ROCHET, J.C. (1989) A pricing formula for options on coupon bonds. Preprint.

[268] ELLERMAN, D.P. (1984) Arbitrage theory: a mathematical introduction. *SIAM Review* **26**, 241–261.

[269] ELLIOTT, R.J. (1982) *Stochastic Calculus and Applications.* Springer-Verlag, New York.

[270] ELLIOTT, R.J., GEMAN, H., and KORKIE, B.M. (1997) Portfolio optimization and contingent claim pricing with differential information. *Stochastics* **60**, 185–204.

[271] ELLIOTT, R.J. and KOPP, P.E. (1990) Option pricing and hedge portfolios for Poisson processes. *Stoch. Anal. Appl.* **9**, 429–444.

[272] ELLIOTT, R.J., MYNENI, R., and VISWANATHAN, R. (1990) A theorem of El Karoui & Karatzas applied to the American option. Preprint, University of Alberta, Edmonton.

[273] EPSTEIN, L. (1990) Behavior under risk: recent developments in theory and applications. Preprint, Department of Economics, University of Toronto.

[274] EPSTEIN, L. and ZIN, S. (1989) Substitution, risk aversion and the temporal behavior of asset returns: a theoretical framework. *Econometrica* **57**, 937–969.

[275] FAKEEV, A.G. (1970) Optimal stopping rules for processes with continuous parameter. *Theory Probab. Appl.* **15**, 324–331.

[276] FAKEEV, A.G. (1971) Optimal stopping of a Markov process. *Theory Probab. Appl.* **16**, 694–696.

[277] FAMA, E. (1965) The behavior of stock-market prices. *J. Business* **38**, 34–105.

[278] FIGLEWSKI, S. (1989) Options arbitrage in imperfect markets. *J. Finance* **44**, 1289–1311.

[279] FIGLEWSKI, S. and GAO, B. (1997) The adaptive mesh method: a new approach to efficient option pricing. Preprint, New York University.

[280] FITZPATRICK, B. and FLEMING, W. (1991) Numerical methods for optimal investment-consumption models. *Math. Operations Research.* **16**, 823–841.

[281] FLEMING, W., GROSSMAN, S., VILA J.-L., and ZARIPHOPOULOU, T. (1990) Optimal portfolio rebalancing with transaction costs. Preprint, Division of Applied Math., Brown University.

[282] FLEMING, W.H. and RISHEL, R.W. (1975) *Deterministic and Stochastic Optimal Control.* Springer-Verlag, New York.

[283] FLEMING, W.H. and SONER, H.M. (1993) *Controlled Markov Processes and Viscosity Solutions.* Springer-Verlag, New York.

[284] FLEMING, W. and ZARIPHOPOULOU, T. (1991) An optimal investment/consumption model with borrowing. *Math. Operations Research* **16**, 802–822.

[285] FOLDES, L. (1978a) Martingale conditions for optimal saving—discrete time. *J. Math. Economics* **5**, 83–96.

[286] FOLDES, L. (1978b) Optimal saving and risk in continuous time. *Rev. Economic Studies* **45**, 39–65.

[287] FOLDES, L. (1990) Conditions for optimality in the infinite-horizon portfolio-cum-saving problem with semimartingale investments. *Stochastics* **29**, 133–170.

[288] FOLDES, L. (1991a) Certainty equivalence in the continuous-time portfolio-cum-saving model. In *Applied Stochastic Analysis* (M.H.A. Davis and R.J. Elliott, eds.), 343–387. Gordon and Breach, London.

[289] FOLDES, L. (1991b) Optimal sure portfolio plans. *Math. Finance* **12**, 15–55.

[290] FOLDES, L. (1992) Existence and uniqueness of an optimum in the infinite-horizon portfolio-cum-saving model with semimartingale investments. *Stochastics* **41**, 241–267.

[291] FÖLLMER, H. (1991) Probabilistic aspects of options. Preprint, University of Bonn.

[292] FÖLLMER, H. and KABANOV, Yu. (1998) Optional decompositions and Lagrange multipliers. *Finance & Stochastics*, **2**, 69–81.

[293] FÖLLMER, H. and KRAMKOV, D. (1997) Optional decompositions under constraints. *Probab. Theory & Rel. Fields* **109**, 1–25.

[294] FÖLLMER, H. and LEUKERT, P. (1998) Quantile hedging. Preprint, Humbolt University, Berlin.

[295] FÖLLMER, H. and SCHWEIZER, M. (1991) Hedging of contingent claims under incomplete information. In *Applied Stochastic Analysis* (M.H.A. Davis and R.J. Elliott, eds.), *Stochastics Monographs* **5**, 389–414. Gordon & Breach, New York.

[296] FÖLLMER, H. and SCHWEIZER, M. (1993) A microeconomic approach to diffusion models for stock prices. *Math. Finance* **3**, 1–23.

[297] FÖLLMER, H. and SONDERMANN, D. (1986) Hedging of non-redundant contingent claims. In *Contributions to Mathematical Economics: Essays in Honor of G. Debreu* (W. Hildenbrand and A. MasColell, eds.), 205–223. North-Holland, Amsterdam.

[298] FREY, R. (1998) Perfect option hedging for a large trader. *Finance & Stochastics*, **2**, 115–141.

[299] FREY, R. and SIN, A. (1997) Bounds on European option prices under stochastic volatility. Preprint, Statistical Laboratory, University of Cambridge.

[300] FREY, R. and STREMME, A. (1997) Market volatility and feedback effects from dynamic hedging. *Mathematical Finance* **7**, 351–374.

[301] FRIEDMAN, A. (1964) *Partial Differential Equations of Parabolic Type.* Prentice-Hall, Englewood Cliffs, N.J.

[302] FRITELLI, M. and LAKNER, P. (1994) Almost sure characterization of martingales. *Stochastics* **49**, 181–190.

[303] FRITELLI, M. and LAKNER, P. (1995) Arbitrage and free-lunch in a general financial market model: the fundamental theorem of asset pricing. In *IMA*

Vol. 65: Mathematical Finance (M. Davis, D. Duffie, W. Fleming and S. Shreve, eds.), 89–92. Springer-Verlag, New York.

[304] FUJITA, T. (1997) On the price of α-percentile options. Working Paper 24, Faculty of Commerce, Hitotsubashi University, Tokyo.

[305] GAO, B., HUANG, J., and SUBRAHMANYAM, M. (1996) An analytic approach to the valuation of American path-dependent options. Preprint, Stern School, New York University.

[306] GARMAN, M. and KLASS, M.J. (1980) On the estimation of security price volatilities from historical data. *J. Business* **53**, 97–112.

[307] GEMAN, H. and YOR, M. (1992) Quelques relations entre processus de Bessel, options asiatiques, et fonctions confluents hypergéometriques. *C.R. Acad. Sci. Paris* **314**, Série I, 471–474.

[308] GEMAN, H. and YOR, M. (1993) Bessel processes, Asian options, and perpetuities. *Mathematical Finance* **3**, 349–375.

[309] GEMAN, H. and YOR, M. (1996) Pricing and hedging double-barrier options: a probabilistic approach. *Mathematical Finance* **6**, 365–378.

[310] GESKE, R. (1979) The valuation of compound options. *J. Financial Economics* **7**, 63–81.

[311] GESKE, R. and JOHNSON, H. (1984) The American put-option valued analytically. *Journal of Finance* **39**, 1511–1524.

[312] GILSTER, J. and LEE, W. (1984) The effects of transaction costs and different borrowing and lending rates on the option pricing problem: A note. *J. Finance* **39**, 1215–1222.

[313] GOLDMAN, M.B., SOSIN, H.B., and GATTO, M.A. (1979) Path-dependent options. *J. Finance* **34**, 1111–1127.

[314] GRABBE, J.O. (1983) The pricing of call and put options on foreign exchange. *J. International Money Finance* **2**, 239–253.

[315] GRANNAN, E.R. and SWINDLE, G.H. (1996) Minimizing transaction costs of option-hedging strategies. *Mathematical Finance* **6**, 341–364.

[316] GROSSMAN, S. and LAROQUE, G. (1990) Asset pricing and optimal portfolio choice in the presence of illiquid durable consumption goods. *Econometrica* **58**, 25–51.

[317] GROSSMAN, S. and ZHOU, Z. (1993) Optimal investment strategies for controlling drawdowns. *Math. Finance* **3**, 241–276.

[318] GROSSMAN, S. and ZHOU, Z. (1996) Equilibrium analysis of portfolio insurance. *J. Finance* **LI**, 1379–1403.

[319] HAGERMAN, R. (1978) More evidence on the distribution of security returns. *J. Finance* **33**, 1213–1220.

[320] HAKANSSON, N. (1970) Optimal investment and consumption strategies under risk for a class of utility functions. *Econometrica* **38**, 587–607.

[321] HANSEN, L. and SINGLETON, K. (1982) Generalized instrumental variables estimation of nonlinear rational expectations models. *Econometrica* **50**, 1269–1282.

[322] HANSEN, L. and SINGLETON, K. (1983) Stochastic consumption, risk aversion, and the temporal behavior of asset returns. *J. Political Economy* **91**, 249–262.

[323] HARRISON, J.M. and KREPS, D.M. (1979) Martingales and arbitrage in multiperiod security markets. *J. Econom. Theory,* **20**, 381–408.

[324] HARRISON, J.M. and PLISKA, S.R. (1981) Martingales and stochastic integrals in the theory of continuous trading. *Stochastic Processes & Appl.* **11**, 215–260.

[325] HARRISON, J.M. and PLISKA, S.R. (1983) A stochastic calculus model of continuous trading: complete markets. *Stochastic Processes & Appl.* **15**, 313–316.

[326] HART, O. (1975) On the optimality of equilibrium when the market structure is incomplete. *J. Economic Theory* **11**, 418–443.

[327] HE, H. (1990) Convergence from discrete to continuous-time contingent claim prices. *Rev. Financial Studies* **3**, 523–546.

[328] HE, H. (1991) Optimal consumption/portfolio policies: a convergence from discrete to continuous-time models. *J. Econom. Theory* **55**, 340–363.

[329] HE, H. and HUANG, C.F. (1991) Efficient consumption-portfolio policies. Haas School of Business, University of California at Berkeley.

[330] HE, H. and HUANG, C.F. (1994) Consumption-portfolio policies: an inverse optimal problem. *J. Econ. Theory* **62**, 257–293.

[331] HE, H. and PAGÈS, H. (1993) Labor income, borrowing constraints, and equilibrium asset prices: a duality approach. *Economic Theory* **3**, 663–696.

[332] HE, H. and PEARSON, N. (1991a) Consumption and portfolio policies with incomplete markets and short-sale constraints: The infinite-dimensional case. *J. Economic Theory* **54**, 259–304.

[333] HE, H. and PEARSON, N.D. (1991b) Consumption and portfolio with incomplete markets and short-sale constraints: The finite-dimensional case. *Mathematical Finance* **1** (3), 1–10.

[334] HEATH, D. (1993) A continuous-time version of Kulldorff's result. Unpublished manuscript, personal communication.

[335] HEATH, D. and JARROW, R. (1987) Arbitrage, continuous trading, and margin requirements. *J. Finance* **5**, 1129–1142.

[336] HEATH, D., JARROW, R., and MORTON, A. (1992) Bond pricing and the term structure of interest rates: a new methodology for contingent claims valuation. *Econometrica* **60**, 77–105.

[337] HEATH, D., JARROW, R., and MORTON, A. (1996) Bond pricing and the term structure of interest rates: A discrete time approximation. *Fin. Quant. Anal.* **25**, 419–440.

[338] HEATH, D. and SUDDERTH, W. (1978) On finitely-additive priors, coherence, and extended admissibility. *Ann. Statistics* **6**, 333–345.

[339] HENROTTE, P. (1993) Transaction costs and duplication strategies. Preprint, Graduate School of Business, Stanford University.

[340] HESTON, S.L. (1990) Sticky consumption, optimal investment, and equilibrium prices. Preprint, Yale School of Organization and Management.

[341] HESTON, S.L. (1993) A closed-form solution for options with stochastic volatility, with application to bond and currency options. *Review of Financial Studies* **6**, 327–343.

[342] HINDY, A. and HUANG, C.F. (1989) On intertemporal preferences with a continuous time dimension II: the case of uncertainty. *J. Math. Economics*, to appear.

[343] HINDY, A. and HUANG, C.F. (1992) Intertemporal preferences for uncertain consumption: a continuous time approach. *Econometrica* **60**, 781–801.

[344] HINDY, A. and HUANG, C.F. (1993) Optimal consumption and portfolio rules with durability and local substitution. *Econometrica* **61**, 85–122.

[345] HINDY, A., HUANG, C.F., and KREPS, D. (1992) On intertemporal preferences with a continuous time dimension: the case of certainty. *J. Math. Economics* **21**, 401–440.

[346] HINDY, A., HUANG, C.F., and ZHU, H. (1993) Numerical analysis of a free-boundary singular control problem in financial economics. Preprint, Graduate School of Business, Stanford University.

[347] HO, T. and LEE, S. (1986) Term-structure movements and pricing interest rate contingent claims. *Journal of Finance* **41**, 1011–1029.

[348] HOBSON, D. (1998) Volatility misspecification, option pricing and super-replication via coupling, *Ann. Appl. Probability* **8**, 193–205.

[349] HOBSON, D. and ROGERS, L.C.G. (1998) Complete models with stochastic volatility. *Mathematical Finance* **8**, 27–48.

[350] HODGES, S.D. and NEUBERGER, A. (1989) Optimal replication of contingent claims under transaction costs. *Rev. Futures Markets* **8**, 222–239.

[351] HOFMANN, N., PLATEN, E., and SCHWEIZER, M. (1992) Option pricing under incompleteness and stochastic volatility. *Mathematical Finance* **2**, 153–188.

[352] HOGAN, M. (1993) Problems in certain two-factor term structure models. *Ann. Appl. Probability* **3**, 576–581.

[353] HOGGARD, T., WHALLEY, A., and WILMOTT, P. (1994) Hedging option portfolios in the presence of transaction costs. *Adv. Futures & Options Research* **7**, 21–35.

[354] HSU, D.A., MILLER, R., and WICHERN, D. (1974) On the stable-Paretian behavior of stock-market prices. *J. Amer. Statist. Assoc.* **69**, 108–113.

[355] HUANG, C.F. (1987) An intertemporal general equilibrium asset pricing model: the case of diffusion information. *Econometrica,* **55**, 117–142.

[356] HUANG, C.F. and LITZENBERGER, R. (1988) *Foundations for Financial Economics.* North Holland, Amsterdam.

[357] HUANG, C.F. and PAGÈS, H. (1992) Optimal consumption and portfolio policies with an infinite horizon: existence and convergence. *Ann. Appl. Probability* **2**, 36–64.

[358] HULL, J. (1993) *Options, Futures, and other Derivative Securities.* Second Edition. Prentice-Hall, NJ.

[359] HULL, J. and WHITE, A. (1987) The pricing of options on assets with stochastic volatilities, *J. Finance* **42**, 281–300.

[360] HULL, J. and WHITE, A. (1988a) An analysis of the bias in option pricing caused by stochastic volatility *Advances in Futures and Options Research* **3**, 29–61.

[361] HULL, J. and WHITE, A. (1988b) The use of the control-variate technique in option-pricing. *J. Financial & Quantit. Analysis* **23**, 237–251.

[362] HULL, J. and WHITE, A. (1990) Pricing interest rate derivative securities. *Rev. Financial Studies* **3**, 573–592.

[363] INGERSOLL, J.E. JR. (1987) *Theory of Financial Decision Making.* Rowman & Littlefield, Totowa, N.J.

[364] JACKA, S. (1984) Optimal consumption of an investment. *Stochastics* **13**, 45–60.

[365] JACKA, S.D. (1991) Optimal stopping and the American put. *Mathematical Finance* **1** (2), 1–14.

[366] JACKA, S.D. (1992) A martingale representation result and an application to incomplete financial markets. *Mathematical Finance* **2** (4), 239–250.

[367] JAILLET, P., LAMBERTON, D., and LAPEYRE, B. (1990) Variational inequalities and the pricing of American options. *Acta Appl. Math.* **21**, 263–289.

[368] JAMSHIDIAN, F. (1990) The preference-free determination of bond and option prices from the spot interest rate. *Adv. Futures & Options Research* **4**, 51–67.

[369] JAMSHIDIAN, F. (1991) Asymptotically optimal portfolios. *Math. Finance* **3**, 131–150.

[370] JAMSHIDIAN, F. (1992) An analysis of American options. *Rev. Futures Markets* **11**, 72–80.

[371] JAMSHIDIAN, F. (1997a) A note on the analytical valuation of double-barrier options. Working Paper, Sakura Global Capital, London.

[372] JAMSHIDIAN, F. (1997b) LIBOR and swap market models and measures. *Finance & Stochastics* **1**, 261–291.

[373] JARROW, R. (1988) *Finance Theory*. Prentice-Hall, Englewood Cliffs, NJ.

[374] JARROW, R.A. and MADAN, D.B. (1991a) A characterization of complete security markets on a Brownian filtration. *Mathematical Finance* **1** (3), 31–43.

[375] JARROW, R.A. and MADAN, D.B. (1991b) Valuing and hedging contingent claims on semimartingales. Graduate School of Management, Cornell University.

[376] JARROW, R.A. and MADAN, D.B. (1991c) Option pricing using the term structure of interest rates to hedge systematic discontinuities in asset returns. Preprint.

[377] JARROW, R.A. and OLDFIELD, G.S. (1981) Forward contracts and futures contracts. *J. Financial Economics* **9**, 373–382.

[378] JARROW, R.A. and ROSENFELD, E. (1984) Jump risks and the intertemporal capital asset pricing model. *J. Business* **57**, 337–351.

[379] JARROW, R.A. and TURNBULL, S. (1995) *Derivative Securities*. Southwestern Publishers, Cleveland.

[380] JEANBLANC-PICQUÉ, M. and PONTIER, M. (1990) Optimal portfolio for a small investor in a market model with discountinuous prices. *Appl. Math. & Optimization* **22**, 287–310.

[381] JENSEN, M. (1972) Capital markets: theory and evidence. *Bell J. Econ. and Management Science* **3**, 357–398.

[382] JOHNSON, H. (1983) An analytic approximation for the American put price. *J. Financial Quant. Analysis* **18**, 141–148.

[383] JOHNSON, H. and SHANNO, D. (1987) Option pricing when the variance is changing. *J. Financ. & Quantit. Analysis* **22**, 143–151.

[384] JOUINI, E. (1997) Market imperfections, equilibrium and arbitrage. *Lecture Notes in Mathematics* **1656**, 247–307. Springer-Verlag, Berlin.

[385] JOUINI, E. and KALLAL, H. (1995a) Arbitrage in security markets with short-sale constraints. *Math. Finance* **5**, 197–232.

[386] JOUINI, E. and KALLAL, H. (1995b) Martingales and arbitrage in security markets with transaction costs. *J. Econ. Theory* **66**, 178–197.

[387] JONES, E.P. (1984) Option arbitrage and strategy with large price changes. *J. Financial Economics* **13**, 91–113.

[388] KABANOV, Yu. (1997) Hedging and liquidation under transaction costs. Preprint.

[389] KABANOV, Yu. and KRAMKOV, D.O. (1994a) No-arbitrage and equivalent martingale measures: an equivalent proof of the Harrison–Pliska theorem. *Theory Probab. Appl.* **39**, 523–527.

[390] KABANOV, Yu. and KRAMKOV, D.O. (1994b) Large financial markets: asymptotic arbitrage and contiguity. *Theory Probab. Appl.* **39**, 222–229.

[391] KABANOV, Yu. and SAFARIAN, M.M. (1997) On Leland's strategy of option pricing with transaction costs. *Finance & Stochastics* **1**, 239–250.

[392] KAN, R. (1991) Structure of Pareto optima when agents have stochastic recursive prefences. Graduate School of Business, Stanford University.

[393] KARATZAS, I. (1988) On the pricing of American options. *Appl. Math. Optimization* **17**, 37–60.

[394] KARATZAS, I. (1989) Optimization problems in the theory of continuous trading. *SIAM J. Control & Optimization* **27**, 1221–1259.

[395] KARATZAS, I. (1996) *Lectures on the Mathematics of Finance.* CRM Monographs **8**, American Mathematical Society.

[396] KARATZAS, I. (1997) Adaptive control of a diffusion to a goal, and a parabolic Monge–Ampère equation. *Asian J. Math.* **1**, 324–341.

[397] KARATZAS, I. and KOU, S.G. (1996) On the pricing of contingent claims with constrained portfolios. *Ann. Appl. Probability* **6**, 321–369.

[398] KARATZAS, I. and KOU, S.G. (1998) Hedging American contingent claims with constrained portfolios. *Finance and Stochastics*, **3**, 215–258.

[399] KARATZAS, I., LAKNER, P., LEHOCZKY, J.P., and SHREVE, S.E. (1991) Dynamic equilibrium in a multi-agent economy: construction and uniqueness. In *Stochastic Analysis: Liber Amicorum for Moshe Zakai* (E. Meyer-Wolf, A. Schwartz and O. Zeitouni, eds.), 245–272. Academic Press.

[400] KARATZAS, I., LEHOCZKY, J.P., SETHI, S.P., and SHREVE, S.E. (1986) Explicit solution of a general consumption/investment problem. *Math. Operations Research* **11**, 261–294.

[401] KARATZAS, I., LEHOCZKY, J.P., and SHREVE, S.E. (1987) Optimal portfolio and consumption decisions for a "small investor" on a finite horizon. *SIAM J. Control & Optimization* **25**, 1557–1586.

[402] KARATZAS, I., LEHOCZKY, J.P., and SHREVE, S.E. (1990) Existence and uniqueness of multi-agent equilibrium in a stochastic, dynamic consumption/investment model. *Math. Operations Research* **15**, 80–128.

[403] KARATZAS, I., LEHOCZKY, J.P., and SHREVE, S.E. (1991) Equilibrium models with singular asset prices. *Mathematical Finance* **1**, 11–29.

[404] KARATZAS, I., LEHOCZKY, J.P., SHREVE, S.E., and XU, G.L. (1991) Martingale and duality methods for utility maximization in an incomplete market. *SIAM J. Control & Optimization* **29**, 702–730.

[405] KARATZAS, I., OCONE, D.L., and LI, J. (1991) An extension of Clark's formula. *Stochastics* **37**, 127–131.

[406] KARATZAS, I. and SHREVE, S.E. (1991) *Brownian Motion and Stochastic Calculus.* Second Edition, Springer-Verlag, New York.

[407] KARATZAS, I., SHUBIK, M., and SUDDERTH, W.D. (1994) Construction of stationary Markov equilibria in a strategic market game. *Math. Operations Research* **19**, 975–1006.

[408] KARATZAS, I., SHUBIK, M., and SUDDERTH, W.D. (1997) A strategic market game with secured lending. *J. Math. Economics* **28**, 207–247.

[409] KARATZAS, I. and ZHAO, X. (1998) Bayesian adaptive portfolio optimization. Preprint, Columbia University.

[410] KEMNA, A.G.Z. and VORST, C.F. (1990) A pricing method for options based on average asset-values. *Journal of Banking & Finance* **14**, 113–129.

[411] KENDALL, M.G. (1953) The analysis of economic time-series—Part I: Prices. *J. Royal Statist. Soc.* **96, Part I**, 11–25. Reprinted in Cootner (1964), 85–99.

[412] KHANNA, A. and KULLDORFF, H. (1998) A generalization of the mutual fund theorem. *Finance and Stochastics*, to appear.

[413] KIM, I.J. (1990) The analytic valuation of American options. *Rev. Financial Studies* **3**, 547–572.

[414] KIND, P., LIPTSER, R.S., and RUNGGALDIER, W.J. (1991) Diffusion approximation in past-dependent models and applications to option pricing. *Ann. Appl. Probability* **1**, 379–405.

[415] KLOEDEN, P. and PLATEN, E. (1992) *Numerical Solution of Stochastic Differential Equations.* Springer-Verlag, New York.

[416] KNASTER, B., KURATOWSKI, K., and MAZURKIEWICZ, S. (1929) Ein Beweis des Fixpunktsatzes für n-dimensionale Simplexe. *Fundamenta Mathematica* **14**, 132–137.

[417] KON, S.J. (1984) Models of stock returns—a comparison. *J. Finance* **39**, 147–165.

[418] KORN, R. (1992) Option pricing in a model with higher interest rate for borrowing than for lending. Preprint, University of Mainz.

[419] KORN, R. (1998) Portfolio optimization with strictly positive transaction costs and impulse control. *Finance & Stochastics* **2**, 85–114.

[420] KORN, R. and TRAUTMANN, S. (1995) Continuous-time portfolio optimization under terminal wealth constraints. *Zeitschrift Oper. Research—Math. Methods of Operations Research* **42**, 69–92.

[421] KRAMKOV, D. (1996a) On the closure of a family of martingale measures and an optional decomposition of a supermartingale. *Theory Probab. & Appl.* **41** (4).

[422] KRAMKOV, D. (1996b) Optional decomposition of supermartingales and hedging contingent claims in incomplete security markets. *Probab. Theory & Rel. Fields* **105**, 459–479.

[423] KRAMKOV, D. and SCHACHERMAYER, W. (1998) The asymptotic elasticity of utility functions and optimal investment in incomplete markets. Preprint, University of Vienna.

[424] KRAMKOV, D. and VISHNYAKOV, A.N. (1994) Closed form representations for the minimal hedging portfolios of American type contingent claims. Preprint, University of Bonn.

[425] KREPS, D. (1981) Arbitrage and equilibrium in economies with infinitely-many commodities. *J. Math. Economics* **8**, 15–35.

[426] KREPS, D. (1982) Multiperiod securities and the efficient allocation of risk: a comment on the Black–Scholes option pricing model. In *The Economics of Uncertainty and Information* (J. McCall, ed.). University of Chicago Press.

[427] KREPS, D. and PORTEUS, E. (1978) Temporal resolution of uncertainty and dynamic choice theory. *Econometrica* **46**, 185–200.

[428] KULLDORFF, M. (1993) Optimal control of favorable games with a time-limit. *SIAM J. Control & Optimization* **31**, 52–69.

[429] KUNIMOTO, N. and IKEDA, M. (1992) Pricing options with curved boundaries. *Mathematical Finance* **2**, 275–298.

[430] KUSHNER, H. and DUPUIS, P. (1992) *Numerical Methods for Stochastic Control Problems in Continuous Time.* Springer-Verlag, New York.

[431] KUSUOKA, S. (1992) A remark on arbitrage and martingale measures. *Publ. RIMS Kyôto Univ.*, to appear.

[432] KUSUOKA, S. (1995) Limit theorems on option replication with transaction costs. *Ann. Appl. Proba.* **5**, 198–221.

[433] KUWANA, Y. (1993) *Optimal Consumption/Investment Decisions with Partial Observations.* Doctoral dissertation, Department of Statistics, Stanford University.

[434] KUWANA, Y. (1995) Certainty-equivalence and logarithmic utilities in consumption/investment problems. *Mathematical Finance* **5**, 297–310.

[435] KYLE, A. (1985) Continuous auctions and insider trading. *Econometrica* **53**, 1315–1335.

[436] LAKNER, P. (1989) *Consumption/investment and equilibrium in the presence of several commodities.* Doctoral dissertation, Department of Statistics, Columbia University, New York.

[437] LAKNER, P. (1993) Martingale measures for a class of right-continuous processes. *Mathematical Finance* **3** (1), 43–53.

[438] LAKNER, P. (1995) Utility maximization with partial information. *Stoch. Processes & Appl.* **56**, 247–273.

[439] LAKNER, P. and SLUD, E. (1991) Optimal consumption by a bond investor: the case of random interest rate adapted to a point process. *SIAM J. Control & Optimization* **29**, 638–655.

[440] LAMBERTON, D. (1993) Convergence of the critical price in the approximation of American options. *Mathematical Finance* **3** (2), 179–190.

[441] LAMBERTON, D. (1995a) Error estimates for binomial approximation of American options. Preprint.

[442] LAMBERTON, D. (1995b) Critical price for an American option near maturity. In *Seminar in Stochastic Analysis, Random Fields & Applications.* (E. Bolthausen, M. Dozzi, and F. Russo, eds.), 353–358. Birkhäuser, Basel.

[443] LAMBERTON, D. and LAPEYRE, B. (1991) *Introduction au Calcul Stochastique Appliqué à la Finance.* Editions S.M.A.I. Paris.

[444] LAMBERTON, D. and LAPEYRE, B. (1993) Hedging index-options with a few assets. *Mathematical Finance* **3**, 25–42.

[445] LAZRAK, A. (1996) A probabilistic restriction of optimality: an application to the effect of horizon on the risk-aversion of a "small" agent. Preprint, University of Toulouse.

[446] LAZRAK, A. (1997a) Option pricing and portfolio optimization in the stochastic volatility model. Preprint, University of Toulouse.

[447] LAZRAK, A. (1997b) General equilibrium foundation of the stochastic volatility model: a theoretical investigation and an example. Preprint, University of Toulouse.

[448] LEHOCZKY, J. (1997) Simulation methods for option pricing. In *Mathematics of Derivative Securities*, M.A.H. Dempster and S.R. Pliska, eds., Cambridge University Press, Cambridge, U.K.

[449] LEHOCZKY, J., SETHI, S., and SHREVE, S. (1983) Optimal consumption and investment policies allowing consumption constraints and bankruptcy. *Math. Operations Research* **8**, 613–636.

[450] LEHOCZKY, J.P., SETHI, S.P., and SHREVE, S.E. (1985) A martingale formulation for optimal consumption/investment decision making. In *Optimal Control Theory and Economic Analysis* **2** (G. Feichtinger, ed.), 135–153. North-Holland, Amsterdam.

[451] LELAND, H. (1985) Option pricing and replication with transaction costs, *J. Finance* **40**, 1283–1301.

[452] LEROY, S. (1973) Risk-aversion and the martingale property of stock prices. *Intern'l. Economic Review* **14**, 436–446.

[453] LEROY, S.F. (1989) Efficient capital markets and martingales. *J. Economic Literature* **27**, 1583–1621.

[454] LEVENTAL, S. and SKOROHOD, A.V. (1995) A necessary and sufficient condition for absence of arbitrage with tame portfolios. *Ann. Appl. Probability* **5**, 906–925.

[455] LEVENTAL, S. and SKOROHOD, A.V. (1997) On the possibility of hedging options in the presence of transaction costs. *Ann. Appl. Probability* **7**, 410–443.

[456] LINTNER, J. (1965) The valuation of risky assets and the selection of risky investment in stock portfolios and capital budgets. *Rev. Economics and Statistics* **47**, 13–37.

[457] LITTERMAN, R. and SCHEINKMAN, J. (1988) Common factors affecting bond returns. Research paper, Goldman Sachs Financial Strategies Group.

[458] LITTLE, T. (1998) The early exercise boundary for the American put-option as the solution of a Volterra equation. Preprint, Columbia University and Morgan-Stanley, New York.

[459] LO, A.W. (1991) Long-term memory in stock-market prices. *Econometrica* **59**, 1279–1313.

[460] LUCAS, R. (1978) Asset prices in an exchange economy. *Econometrica,* **46**, 1429–1445.

[461] LYONS, T. (1995) Uncertain volatility and risk-free synthesis of securities. *Appl. Math. Finance* **2**, 117–133.

[462] MA, C. (1991) Valuation of derivative securities with mixed Poisson–Brownian information and with recursive utility. Department of Economics, University of Toronto.

[463] MACBETH, J.D. and MERVILLE, L.J. (1979) An empirical examination of the Black–Scholes call option pricing model. *J. Finance* **34**, 1173–1186.

[464] MACMILLAN. L. (1986) Analytic approximation for the American put option. *Adv. Futures & Options Research* **1**, 119–139.

[465] MADAN, D. (1988) Risk measurement in semimartingale models with multiple consumption goods. *J. Economic Theory* **44**, 398–412.

[466] MADAN, D. and MILNE, F. (1991) Option pricing with V.G. martingale components. *Mathematical Finance* **1** (4), 39–56.

[467] MADAN, D. and MILNE, F. (1993) Contingent claims valued and hedged by pricing and investing in a basis. *Mathematical Finance* **3** (4), 223–245.

[468] MADAN, D., MILNE, F., and SHEFRIN, H. (1989) A multinomial option pricing model and its Brownian and Poisson limits. *Rev. Financial Studies* **2**, 251–265.

[469] MADAN, D. and SENETA, E. (1990) The V.G. model for share returns. *J. Business* **63**, 511–524.

[470] MAGILL, M.J.P. (1976) The preferability of investment through a mutual fund. *J. Economic Theory* **13**, 265–271.

[471] MAGILL, M.J.P. (1981) An equilibrium existence theorem. *Math. Anal. Appl.* **84**, 162–169.

[472] MAGILL, M.J.P. and CONSTANTINIDES, G.M. (1976) Portfolio selection with transaction costs. *J. Economic Theory* **13**, 245–263.

[473] MAGILL, M.J.P. and SHAFER, W. (1984) Allocation of aggregate and individual risks through futures and insurance markets. Unpublished manuscript, University of Southern California.

[474] MAGILL, M. and QUINZII, M. (1996) *Theory of Incomplete Markets*. MIT Press.

[475] MALKIEL, B. (1996) *A Random Walk Down Wall Street*. Norton, New York.

[476] MALLIARIS, A.G. (1983) Itô's calculus in financial decision-making. *SIAM Review* **25**, 481–496.

[477] MALLIARIS, A.G. and BROCK, W.A. (1982) *Stochastic Methods in Economics and Finance*. North Holland, Amsterdam.

[478] MARGRABE, W. (1976) A theory of forward and futures prices. Preprint, Wharton School, University of Pennsylvania, Philadelphia.

[479] MARGRABE, W. (1978) The value of an option to exchange one asset for another. *J. Finance* **33**, 177–186.

[480] MARKOWITZ, H. (1952) Portfolio selection. *J. Finance* **8**, 77–91.

[481] MARKOWITZ, H. (1959) *Portfolio Selection: Efficient Diversification of Investment*. J. Wiley & Sons, New York. Second Edition by Blackwell, Oxford, 1991.

[482] MAS-COLELL, A. (1986) The price equilibrium existence problem in topological vector lattices. *Econometrica* **54**, 1039–1053.

[483] MAS-COLELL, A. and ZAME, W. (1991) Equilibrium theory in infinite dimensional spaces. In *Handbook of Mathematical Economics* (W. Hildenbrand & H. Sonnenschein, eds.) **4**. North-Holland, Amsterdam.

[484] MASTROENI, L. and MATZEU, M. (1996) An integro-differential parabolic variational inequality connected with the problem of American option pricing. *Zeitschrift für Analysis und ihre Anwendungen*, to appear.

[485] MCKEAN, H.P. JR. (1965) A free-boundary problem for the heat equation arising from a problem in mathematical economics. *Industr. Manag. Rev.*, **6**, 32–39. Appendix to *Samuelson (1965a)*.

[486] MCMANUS, D. (1984) Incomplete markets: Generic existence of equilibrium and optimality properties in an economy with futures markets. Unpublished manuscript, University of Pennsylvania.

[487] MEHRA, R. and PRESCOTT, E. (1985) The equity premium: a puzzle. *J. Monetary Econ.* **15**, 145–161.

[488] MERCURIO, F. (1996) *Claim Pricing and Hedging under Market Imperfections*. Doctoral dissertation, Erasmus University, Rotterdam.

[489] MERCURIO, F. and RUNGGALDIER, W. (1993) Option pricing for jump diffusions: approximations and their interpretation. *Mathematical Finance* **3** (2), 191–200.

[490] MERRICK, J.J. (1990) *Financial Futures Markets: Structure, Pricing, and Practice*. Harper & Row, New York.

[491] MERTON, R.C. (1969) Lifetime portfolio selection under uncertainty: the continuous-time case. *Rev. Econom. Statist.* **51**, 247–257.

[492] MERTON, R.C. (1971) Optimum consumption and portfolio rules in a continuous-time model. *J. Econom. Theory* **3**, 373–413. Erratum: *ibid.* **6** (1973), 213–214.

[493] MERTON, R.C. (1973a) Theory of rational option pricing. *Bell J. Econom. Manag. Sci.*, **4**, 141–183.

[494] MERTON, R.C. (1973b) An intertemporal capital asset pricing model. *Econometrica*, **41**, 867–888.

[495] MERTON, R.C. (1976) Option pricing when underlying stock returns are discontinuous. *J. Financial Economics* **3**, 125–144.

[496] MERTON, R.C. (1989) Optimal portfolio rules in continuous-time when the nonnegativity constraint on consumption is binding. Working Paper, Harvard Business School.

[497] MERTON, R.C. (1990) *Continuous-Time Finance*. Basil Blackwell, Oxford and Cambridge.

[498] MERTON, R.C. and SAMUELSON, P.A. (1974) Fallacy of log-normal approximation to optimal portfolio decision-making over many periods. *J. Financial Economics* **1**, 67–94.

[499] MEYER, G.H. and VAN DER HOEK, J. (1995) The evaluation of American options with the method of lines. Preprint, University of Adelaide.

[500] MILTERSEN, K.R. (1994) An arbitrage theory for the term-structure of interest rates. *Ann. Appl. Probability* **4**, 953–967.

[501] MILTERSEN, K.R., SANDMANN, K., and SONDERMANN, D. (1997) Closed-form solutions for term structure derivatives with log-normal interest rates. *J. Finance* **52**, 409–430.

[502] MIRRLEES, J. (1974) Optimal accumulation under uncertainty: the case of stationary returns to investment. In *Allocation under Uncertainty: Equilibrium and Optimality* (J. Drèze, ed.), 36–50. J. Wiley & Sons, New York.

[503] MIURA, R. (1992) A note on look-back options based on order statistics. *Hitotsubashi J. Commerce & Management* **27**, 15–28.

[504] MORTON, A. and PLISKA, S. (1995) Optimal portfolio management with fixed transaction costs. *Mathematical Finance* **5**, 337–356.

[505] MOSSIN, J. (1966) Equilibrium in a capital asset market. *Econometrica* **34**, 768–783.

[506] MOSSIN, J. (1968) Optimal multiperiod portfolio policies. *J. Business* **41**, 215–229.

[507] MULINACCI, S. and PRATELLI, M. (1995) Functional convergence of Snell envelopes: application to American option approximations. Preprint.

[508] MÜLLER, S. (1985) Arbitrage Pricing of Contingent Claims. *Lecture Notes in Economics and Mathematical Systems* **254**. Springer-Verlag, New York.

[509] MÜLLER, S. (1989) On complete securities markets and the martingale property on securities prices. *Economic Letters* **31**, 37–41.

[510] MUSIELA, M. and RUTKOWSKI, M. (1997b) *Mathematical Methods in Financial Modelling.* Springer-Verlag, New York.

[511] MUSIELA, M. and RUTKOWSKI, M. (1997a) Continuous-time term structure models: Forward measure approach. *Finance Stochastics* **1**, 261–292.

[512] MYNENI, R. (1992a) The pricing of the American option. *Ann. Appl. Probability* **2**, 1–23.

[513] MYNENI, R. (1992b) Continuous-time relationships between futures and forward prices. Preprint.

[514] NAIK, V. and LEE, M. (1990) General equilibrium pricing of options on the market portfolio with discontinuous returns. *Rev. Financial Studies* **3**, 493–521.

[515] NAIK, V. and UPPAL, R. (1994) Leverage constraints and the optimal hedging of stock- and bond-options. *J. Financ. & Quantit. Analysis* **29**, 199–222.

[516] NEGISHI, T. (1960) Welfare economics and existence of an equilibrium for a competitive economy. *Metroeconomica* **12**, 92–97.

[517] NELSON, D. and RAMASWAMY, K. (1990) Simple binomial processes and diffusion approximations in finance models. *Rev. Financial Studies* **3**, 393–430.

[518] NEVEU, J. (1975) *Discrete-Parameter Martingales.* English translation, North-Holland, Amsterdam and American Elsevier, New York.

[519] OCONE, D.L. (1984) Malliavin's calculus and stochastic integral representations of functionals of diffusion processes. *Stochastics* **12**, 161–185.

[520] OCONE, D. and KARATZAS, I. (1991) A generalized Clark representation formula, with application to optimal portfolios. *Stochastics* **34**, 187–220.

[521] OFFICER, R. (1972) The distribution of stock returns. *J. Amer. Statist. Assoc.* **67**, 807–812.

[522] OMBERG, E. (1987) The valuation of American puts with exponential exercise policies. *Adv. Futures Option Research* **2**, 117–142.

[523] OSBORNE, M.F.M. (1959) Brownian motion in the stock market. *Operations Research* **7**, 145–173. Reprinted in *Cootner (1964)*, 100–128.

[524] PAGÈS, H. (1987) Optimal consumption and portfolio policies when markets are incomplete. MIT mimeo, Massachusetts Institute of Technology.

[525] PAGÈS, H. (1989) *Three Essays in Optimal Consumption.* Doctoral dissertation, Massachusetts Institute of Technology, Cambridge, Mass.

[526] PANAS, V.G. (1993) *Option Pricing with Transaction Costs.* Doctoral dissertation, Imperial College, London.

[527] PAPANICOLAOU, G. and SIRCAR, R. (1996) General Black–Scholes models accounting for increased market volatility from hedging strategies. Preprint, Stanford University.

[528] PARKINSON, M. (1977) Option pricing: the American put. *J. Business* **50**, 21–36.

[529] PARKINSON, M. (1980) The extreme-value method for estimating the variance of the rate of return. *J. Business* **53**, 61–65.

[530] PARTHASARATHY, K.R. (1967) *Probability Measures on Metric Spaces.* Academic Press, New York.

[531] PASKOV, S. and TRAUB, J.F. (1995) Faster valuation of financial derivatives. *J. Portfolio Management*, 113–120.

[532] PELSSER, A. (1997) Pricing double-barrier options using analytical inversion of Laplace transforms. Preprint, Erasmus University, Rotterdam.

[533] PHAM, H. (1995) Optimal stopping, free-boundary, and American option in a jump-diffusion model. *Appl. Mathematics & Optimization*, to appear.

[534] PHAM, H., RHEINLÄNDER, T., and SCHWEIZER, M. (1998) Mean-variance hedging for continuous processes: new proofs and examples. *Finance & Stochastics* **2**, 173–198.

[535] PHAM, H. and TOUZI, N. (1996) Equilibrium state-prices in a stochastic volatility model. *Mathematical Finance* **6**, 215–236.

[536] PIKOVSKY, I. and KARATZAS, I. (1996) Anticipative portfolio optimization. *Adv. Appl. Probab.* **28**, 1095–1122.

[537] PIKOVSKY, I. (1998) Market equilibrium with differential information. *Finance & Stochastics,* to appear.

[538] PLATEN, E. and SCHWEIZER, M. (1994) On smile and skewness. Research Report, Australian National University.

[539] PLATEN, E. and SCHWEIZER, M. (1998) On feedback effects from hedging derivatives. *Mathematical Finance* **8**, 67–84.

[540] PLISKA, S.R. (1986) A stochastic calculus model of continuous trading: optimal portfolios. *Math. Operations Research* **11**, 371–382.

[541] PLISKA, S.R. (1997) *Introduction to Mathematical Finance: Discrete-Time Models*. Basil Blackwell, Oxford and Cambridge.

[542] PLISKA, S. and SELBY, M. (1994) On a free boundary problem that arises in portfolio management. *Phil. Trans. Roy. Soc. London, Ser. A* **347**, 555–561.

[543] PRESCOTT, E. and MEHRA, R. (1980) Recursive competitive equilibrium: the case of homogeneous households. *Econometrica* **48**, 1365–1379.

[544] PRESMAN, E. and SETHI, S. (1991) Risk-aversion behavior in consumption/investment problems. *Mathematical Finance* **11**, 101–124.

[545] PRESMAN, E. and SETHI, S. (1996) Distribution of bankruptcy time in a consumption/portfolio problem. *J. Econom. Dynamics Control* **20**, 471–477.

[546] PROTTER, P. (1990) *Stochastic Integration and Differential Equations: a unified approach*. Springer-Verlag, New York.

[547] RADNER, R. (1972) Existence of equilibrium of plans, prices, and price expectations in a sequence of markets. *Econometrica* **40**, 289–303.

[548] REIMER, M. and SANDMANN, K. (1995) A discrete time approach for European and American barrier options. Preprint, University of Bonn.

[549] RENAULT, E. and TOUZI, N. (1996) Option hedging and implied volatilities in a stochastic volatility model. *Mathematical Finance* **6**, 279–302.

[550] RICHARD, S.F. (1975) Optimal consumption, portfolio, and life insurance rules for an uncertain lived individual in a continuous-time model. *J. Financial Economics* **2**, 187–203.

[551] RICHARD, S.F. (1978) An arbitrage model for the term structure of interest rates. *J. Financial Economics* **6**, 33–57.

[552] RICHARD, S.F. (1979) A generalized capital asset pricing model. In *Portfolio Theory: 25 Years after* (E.J. Elton and M.J. Gruber, eds.) *TIMS Studies in Management Sciences* **11**, 215–232.

[553] RICHARD, S.F. and SUNDARESAN, M. (1981) A continuous-time equilibrium model of forward and future prices in a multigood economy. *J. Financial Economics* **9**, 347–371.

[554] RICHARDSON, H. (1989) A minimum-variance result in continuous trading portfolio optimization. *Management Science* **35**, 1045–1055.

[555] ROCKAFELLAR, R.T. (1970) *Convex Analysis.* Princeton University Press, Princeton, NJ.

[556] ROGERS, L.C.G. (1995a) Equivalent martingale measures and no-arbitrage. *Stochastics* **51**, 41–50.

[557] ROGERS, L.C.G. (1995b) Which model of term-structure of interest rates should one use? In *Mathematical Finance* (M. Davis, D. Duffie, W. Fleming, and S. Shreve, eds.) *IMA Vol.* **65**, 93–116. Springer-Verlag, New York.

[558] ROGERS, L.C.G. (1997) The potential approach to term-structure of interest rates and foreign-exchange options. *Mathematical Finance* **7**, 157–176.

[559] ROGERS, L.C.G. and SATCHELL, S.E. (1991) Estimating variance from high, low and closing prices. *Ann. Appl. Probability* **1**, 504–512.

[560] ROGERS, L.C.G. and SHI, Z. (1995) The value of an Asian option. *J. Appl. Probability* **32**, 1077–1088.

[561] ROGERS, L.C.G. and STAPLETON, E.J. (1998) Fast accurate binomial pricing. *Finance & Stochastics* **2**, 3–18.

[562] ROGERS, L.C.G. and WILLIAMS, D. (1987) *Diffusions, Markov Processes and Martingales, Vol. 2: Itô Calculus.* J. Wiley & Sons, Chichester, G. Britain, and New York, USA.

[563] ROGERS, L.C.G. and ZANE, O. (1997) Valuing moving-barrier options. *J. Computational Finance* **1**, 5–13.

[564] ROMANO, M. and TOUZI, N. (1997) Contingent claims and market completeness in a stochastic volatility model. *Mathematical Finance* **7**, 399–412.

[565] ROSS, S.A. (1976) The arbitrage theory of capital asset pricing. *J. Econom. Theory* **13**, 341–360.

[566] ROTHSCHILD, M. and STIGLITZ, J.E. (1971) Increasing risk II. Its economic consequences. *J. Economic Theory* **3**, 66–84.

[567] RUBINSTEIN, M. (1974) An aggregation theory for securities markets. *J. Financial Economics* **1**, 225–244.

[568] RUBINSTEIN, M. (1983) Displaced diffusion option-pricing. *J. Finance* **40**, 455–480.

[569] RUBINSTEIN, M. (1985) Nonparametric tests of alternative option pricing models using all reported trades and quotes on the 30 most active CBOE option classes from August 23, 1976, through August 31, 1978. *J. Finance* **60**, 455–480.

[570] RUBINSTEIN, M. (1991) *Exotic Options.* Preprint, Haas School of Business, University of California, Berkeley.

[571] SAMUELSON, P.A. (1965a) Rational theory of warrant pricing. *Industr. Manag. Rev.* **6**, 13–31.

[572] SAMUELSON, P.A. (1965b) Proof that properly anticipated prices fluctuate randomly. *Industr. Manag. Rev.* **6**, 41–50.

[573] SAMUELSON, P.A. (1969) Lifetime portfolio selection by dynamic stochastic programming. *Rev. Econom. Statist.* **51**, 239–246.

[574] SAMUELSON, P.A. (1973) Mathematics of speculative prices. *SIAM Review* **15**, 1–39.

[575] SAMUELSON, P.A. (1979) Why we should not make mean-log of wealth big, though years to act are long. *J. Banking Finance* **3**, 305–307.

[576] SAMUELSON, P.A. and MERTON, R.C. (1969) A complete model of warrant-pricing that maximizes utility. *Industr. Mangmt. Review* **10**, 17–46.

[577] SANDMANN, K. and SONDERMANN, D. (1993) A term-structure model for pricing interest-rate derivatives. *Rev. Futures Markets* **12**, 392–423.

[578] SCHACHERMAYER, W. (1992) A Hilbert-space proof of the fundamental theorem of asset-pricing in finite discrete time. *Insurance: Mathematics & Economics* **11**, 249–257.

[579] SCHACHERMAYER, W. (1993) A counterexample to several problems in mathematical finance. *Mathematical Finance* **3** (2), 217–229.

[580] SCHACHERMAYER, W. (1994) Martingale measures for discrete-time processes with infinite horizon. *Mathematical Finance* **4** (1), 25–56.

[581] SCHÄL, M. (1974) A selection theorem for optimization problems. *Arch. Math.* **25**, 219–224.

[582] SCHÄL, M. (1975) Conditions for optimality in dynamic programming, and for the limit of n-stage optimal policies to be optimal. *Z. Wahrscheinlichkeitstheorie und verw. Gebiete (Probability Theory and Related Fields)* **32**, 179–196.

[583] SCHÄL, M. (1994) On quadratic cost criteria for option hedging. *Math. Operations Research* **19**, 121–131.

[584] SCHOENMAKERS, J.G.M. and HEEMINK, A.W. (1997) Fast valuation of financial derivatives. *J. Comp. Finance* **1**, 47–62.

[585] SCHRODER, M. (1989) Computing the constant elasticity of variance option-pricing formula. *J. Finance* **44**, 211–219.

[586] SCHRODER, M. and SKIADAS, C. (1997) Optimal consumption and portfolio selection with stochastic differential utility. Preprint, Kellogg School, Northwestern University.

[587] SCHWEIZER, M. (1988) *Hedging of options in a general semimartingale model.* Doctoral dissertation #8615, ETH-Zentrum, Zürich, Switzerland.

[588] SCHWEIZER, M. (1990) Risk-minimality and orthogonality of martingales. *Stochastics* **30**, 123–131.

[589] SCHWEIZER, M. (1991) Option hedging for semimartingales. *Stoch. Processes Appl.* **37**, 339–363.

[590] SCHWEIZER, M. (1992a) Martingale densities for general asset prices. *J. Math. Economics* **21**, 123–131.

[591] SCHWEIZER, M. (1992b) Mean-variance hedging for general claims. *Annals Appl. Probability* **2**, 171–179.

[592] SCHWEIZER, M. (1993) *Approximating random variables by stochastic integrals, and applications in financial mathematics.* Habilitationsschrift, University of Göttingen.

[593] SCHWEIZER, M. (1994) Approximating random variables by stochastic integrals. *Ann. Probability* **22**, 1536–1575.

[594] SCHWEIZER, M. (1995a) On the minimal martingale measure and the Föllmer–Schweizer decomposition. *Stoch. Anal. Appl.* **13**, 573–599.

[595] SCHWEIZER, M. (1995b) Variance-optimal hedging in discrete time. *Math. Operations Research* **20**, 1–32.

[596] SCHWEIZER, M. (1996) Approximation pricing and the variance-optimal martingale measure. *Ann. Probability* **24**, 206–236.

[597] SCOTT, L.O. (1987) Option pricing when the variance changes randomly: theory, estimation, and an application. *J. Financ. & Quantit. Analysis* **22**, 419–438.

[598] SCOTT, L.O. (1997) Pricing stock options in a jump-diffusion model with stochastic volatility and interest rates: applications of Fourier inversion methods. *Mathematical Finance* **7**, 413–426.

[599] SELBY, M. and HODGES, S. (1987) On the valuation of compound options. *Management Science* **33**, 347–355.

[600] SETHI, S. (1997) *Optimal Consumption and Investment with Bankruptcy*, Kluwer, Boston.

[601] SETHI, S. and TAKSAR, M. (1988) A note on Merton's "Optimal consumption and portfolio rules in a continuous-time model." *J. Econom. Theory* **46**, 395–401.

[602] SETHI, S. and TAKSAR, M. (1992) Infinite-horizon investment consumption model with a nonterminal bankruptcy. *J. Optimization Theory Appl.* **74**, 333–346.

[603] SETHI, S., TAKSAR, M., and PRESMAN, E. (1992) Explicit solution of a general consumption/portfolio problem with subsistence consumption and bankruptcy. *J. Economic Dynamics and Control* **16**, 747–768.

[604] SHARPE, W.F. (1964) Capital asset prices: a theory of market equilibrium under conditions of risk. *J. Finance* **19**, 425–442.

[605] SHEPP, L. and SHIRYAEV, A.N. (1993) The Russian option: reduced regret. *Annals Appl. Probability* **3**, 631–640.

[606] SHEPP, L. and SHIRYAEV, A.N. (1994) A new look at the Russian option. *Theory Probab. Appl.* **39**, 103–119.

[607] SHIGEKAWA, I. (1980) Derivatives of Wiener functionals, and absolute continuity of induced measures. *J. Math. Kyôto Univ.* **20**, 263–289.

[608] SHIRAKAWA, H. (1990) Optimal dividend and portfolio decisions with Poisson and diffusion-type return processes. Preprint IHSS 90–20, Tokyo Institute of Technology.

[609] SHIRAKAWA, H. (1991) Interest rate option pricing with Poisson–Gaussian forward rate curve processes. *Mathematical Finance* **1** (4), 77–94.

[610] SHIRAKAWA, H. and KONNO, H. (1995) Pricing of options under proportional transaction costs. Preprint, Tokyo Institute of Technology.

[611] SHIRYAEV, A.N. (1978) *Optimal Stopping Rules.* Springer-Verlag, New York.

[612] SHREVE, S.E. and SONER, H.M. (1994) Optimal investment and consumption with transaction costs. *Ann. Appl. Probab.* **4**, 609–692.

[613] SHREVE, S.E., SONER, H.M., and XU, G.-L. (1991) Optimal investment and consumption with two bonds and transaction costs. *Math. Finance* **13**, 53–84.

[614] SHREVE, S.E. and XU, G.L. (1992) A duality method for optimal consumption and investment under short-selling prohibition: I, General market coefficients; and II, Constant market coefficients. *Ann. Appl. Probability* **2**, 87–112 and 314–328.

[615] SIN, C. (1996) Strictly local martingales and hedge ratios on stochastic volatility models. Ph.D. dissertation, School of Oper. Res. Indus. Eng., Cornell University.

[616] SINGLETON, K.J. (1987) Specification and estimation of intertemporal asset pricing models. In *Handbook of Monetary Economics* (B. Friedman and F. Hahn, eds.) North-Holland, Amsterdam.

[617] SMITH, C.W. JR. (1976) Option pricing: a review. *J. Financial Econom.* **3**, 3–51.

[618] SONER, H.M., SHREVE, S.E., and CVITANIĆ, J. (1995) There is no nontrivial hedging portfolio for option pricing with transaction costs. *Ann. Appl. Probab.* **5**, 327–355.

[619] SPIVAK, G. (1998) *Maximizing the probability of perfect hedge.* Doctoral dissertation, Columbia University.

[620] SPRENKLE, C.M. (1961) Warrant prices as indicators of expectations and preferences. *Yale Economic Essays* **1**, 178–231. Reprinted in Cootner (1964), 412–474.

[621] STEIN, E.M. and STEIN, J.C. (1991) Stock-price distributions with stochastic volatility. *Rev. Financial Studies* **4**, 727–752.

[622] STRICKER, C. (1990) Arbitrage et lois de martingale. *Ann. Inst. Henri Poincaré* **26**, 451–460.

[623] STROOCK, D.W. and VARADHAN, S.R.S. (1979) *Multidimensional Diffusion Processes.* Springer-Verlag, Berlin.

[624] SUDDERTH, W. (1994) Coherent Inference and Prediction in Statistics. In *Logic, Methodology and Philosophy of Science* **IX** (D. Prawitz, B. Skyrms and D. Westerstahl, eds.), 833–844. Elsevier Science, New York.

[625] SUNDARESAN, S. (1989) Intertemporally dependent preferences and the volatility of consumption and wealth. *Rev. Financial Studies* **2**, 73–89.

[626] SUNDARESAN, S. (1997) *Fixed Income Markets and their Derivatives.* South-Western College Publishing, Cincinnati, Ohio.

[627] SUTCLIFFE, C.M.S. (1993) *Stock Index Futures.* Chapman & Hall, London.

[628] TAKSAR, M., KLASS, M.J., and ASSAF, D. (1988) A diffusion model for optimal portfolio selection in the presence of brokerage fees. *Math. Operations Research* **13**, 277–294.

[629] TALAY, D. and TUBARO, L. (1997) Book on Numerical Methods for Stochastic and Partial Differential Equations. In preparation.

[630] TAQQU, M. and WILLINGER, W. (1987) The analysis of finite security markets using martingales. *Adv. Appl. Probab.* **19**, 1–25.

[631] TAYLOR, S. (1994) Modelling stochastic volatility: a review and comparative study. *Mathematical Finance* **4** (2), 171–179.

[632] THORP, E.O. (1971) Portfolio choice and the Kelly criterion. In *Stochastic Models in Finance.* (W.T. Ziemba and R.G. Vickson, eds.), 599–619. Academic Press, New York.

[633] TILLY, J. (1993) Valuing American options in a path-simulation model. *Trans. Soc. Actuaries* **45**, 83–104.

[634] TOBIN, J. (1958) Liquidity preference as behavior towards risk. *Rev. Economic Studies* **25**, 68–85.

[635] TURNBULL, S.M. and WAKEMAN, L.M. (1991) A quick algorithm for pricing European average options. *J. Fin. Quant. Analysis* **26**, 377–389.

[636] UZAWA, H. (1968) Time preference, the consumption function and optimal asset holdings. In *Value, Capital and Growth: Papers in Honor of Sir John Hicks* (J. Wolfe, ed.) 485–504. Aldine, Chicago.

[637] VAN MOERBEKE, P. (1976) On optimal stopping and free boundary problems. *Arch. Rational Mech. Anal.* **60**, 101–148.

[638] VASICEK, O. (1977) An equilibrium characterization of the term structure. *J. Financial Economics* **5**, 177–188.

[639] VILA, J.L. and ZARIPHOPOULOU, T. (1991) Optimal consumption and portfolio choice with borrowing constraints. Preprint, Department of Mathematics, University of Wisconsin.

[640] WALD, A. (1936) Über einige Gleichungsysteme der mathematischen Ökonomie. *Zeitschrift für Nationalökonomie* **7**, 637–670. English translation in *Econometrica* **19**, 368–403.

[641] WALRAS, L. (1874/77) *Eléments d'économie politique pure.* Fourth edition, L. Corbaz, Lausanne. English translation in *Elements of Pure Economics* (W. Jeffé, ed.) Allen & Unwin, London.

[642] WEERASINGHE, A. (1996) Singular optimal strategies for investment with transaction costs. Preprint, Iowa State University.

[643] WIGGINS, J.B. (1987) Option values under stochastic volatility: theory and empirical estimates. *J. Financial Economics* **19**, 351–372.

[644] WILLINGER, W. and TAQQU, M. (1988) Pathwise approximations of processes based on the fine structure of their filtrations. *Lecture Notes in Mathematics* **1321**, 542–559. Springer-Verlag, Berlin.

[645] WILLINGER, W. and TAQQU, M. (1991) Towards a convergence theory for continuous stochastic securities market models. *Mathematical Finance* **1** (1), 55–99.

[646] WILMOTT, P., DEWYNNE, J.N., and HOWISON, S. (1993) *Option Pricing: Mathematical Models and Computation.* Oxford Financial Press, Oxford.

[647] WILMOTT, P., DEWYNNE, J.N., and HOWISON, S. (1995) *The Mathematics of Financial Derivatives: A Student Edition,* Cambridge University Press.

[648] WYSTUP, U. (1998) *Valuation of exotic options under short-selling constraints as a singular stochastic control problem.* Ph.D. dissertation, Dept. Math. Sci., Carnegie-Mellon University.

[649] XU, G.L. (1989) Zero investment in a high-yield asset can be optimal. *Math. Operations Research* **14**, 457–461.

[650] XU, G.L. (1990) *A duality method for optimal consumption and investment under short-selling prohibition.* Doctoral dissertation, Department of Mathematics, Carnegie-Mellon University.

[651] XUE, X.H. (1984) On optimal stopping of continuous-parameter stochastic processes. *Proceedings of the China–Japan Symposium on Statistics* (Beijing, China, 6–12 November 1984), 371–373.

[652] XUE, X.X. (1992) Martingale representation for a class of processes with independent increments, and its applications. *Lecture Notes in Control and Information Sciences* **177**, 279–311. Springer-Verlag, Berlin.

[653] YOR, M. (1995) The distribution of Brownian quantiles. *J. Appl. Probability* **32**, 405–416.

[654] ZARIPHOPOULOU, Th. (1989) *Optimal investment/consumption models with constraints.* Doctoral dissertation, Division of Applied Mathematics, Brown University, Providence, RI.

[655] ZARIPHOPOULOU, Th. (1992) Investment/consumption model with transaction costs and Markov-chain parameters. *SIAM J. Control Optimization* **30**, 613–636.

[656] ZHANG, P. (1997) *Exotic Options,* World Scientific, Singapore.

[657] ZHANG, X. (1993) Options américaines et modèles de diffusion avec sauts. *C. R. Acad. Sci. Paris, Sér. I (Math.)* **317**, 857–862.

Index

Applications of Mathematics

(continued from page ii)

FINANCE AND STOCHASTICS

Editor: **D. SONDERMANN**, University of Bonn, Germany
e-mail: sonderma@addi.or.uni-bonn.de

Co-Editors: **P. ARTZNER**, University of Strasbourg, France; **D. DUFFIE**, Stanford University, CA; **W. RUNGGALDIER**, Université degli Studi di Padova, Italy; **M. SCHWEIZER**, Technische Universität Berlin, Germany; **A.N. SHIRYAEV**, Steklov Mathematics Institute, Moscow, Russia and **S.E. SHREVE**, Carnegie Mellon University University, Pittsburgh, PA

Advisory Board: **F. DELBAEN**, Swiss Federal Institute, Zürich, Switzerland; **P. EMBRECHTS**, Swiss Federal Institute, Zürich, Switzerland; **H. FÖLLMER**, Humboldt University, Berlin, Germany and **Y. KABANOV**, Université de Franche–Comté, Besançon, France

Associate Editors: **Y. AÏT-SAHALIA**, The University of Chicago, IL; **T. BJÖRK**, Stockholm School of Economics, Sweden; **P.H. DYBVIG**, Washington University, St. Louis, MO; **N. EL KAROUI**, Université P. & M. Curie, Paris, France; **R.J. ELLIOTT**, University of Alberta, Edmonton; **D. HEATH**, Cornell University, Ithaca, NY; **F. JAMSHIDIAN**, Sakura Global Capital, London, UK; **I. KARATZAS**, Columbia University, New York, NY; **U. KÜCHLER**, Humboldt University, Berlin, Germany; **S. KUSUOKA**, Tokyo University; **Yu.A. KUTOYANTS**, Université du Maine, Le Mans, France; **R. LIPTSER**, Tel Aviv University, Israel ; **M. MUSIELA**, The University of New South Wales, Sydney, Australia; **J.A. NIELSEN**, Aarhus University; **B. ØKSENDAL**, University of Oslo, Norway; **E. PLATEN**, The Australian National University, Canberra; **M. RUTKOWSKI**, Politechnika, Warsaw; **W. SCHACHERMAYER**, University of Vienna; **J. SCHEINKMAN**, The University of Chicago, IL; **W.M. SCHMIDT**, Deutsche Bank, Frankfurt/Main; **D. TALAY**, INRIA Centre de Sophia Antipolis, Valbonne; **T. VORST**, Erasmus University Rotterdam; **M. YOR**, Université P. & M. Curie, Paris, France

Assistant Editor: **F. OERTEL**, University of Bonn

Finance and Stochastics serves as a publication platform for both theoretical and applied financial economists using advanced stochastic methods, and researchers in stochastics motivated by and interested in applications in finance and insurance. The purpose of *Finance and Stochastics* is to provide a high standard publication forum for research in all areas of finance based on stochastic methods, and on specific topics in mathematics (in particular probability theory, statistics and stochastic analysis) motivated by the analysis of problems in finance.

Finance and Stochastics encompasses–but is not limited to–the following fields: Theory and analysis of financial markets, continuous time finance, derivatives research, insurance in relation to finance, portfolio selection, credit and market risks, term structure models, statistical and empirical financial studies, numerical and stochastic solution techniques for problems in finance, intertemporal economics, uncertainty and information in relation to finance.

Finance and Stochastics also publishes surveys on financial topics of general interest if they clearly picture and illuminate the basic ideas and techniques at work, the interrelationship of different approaches and the central questions which remain open. Special issues may be devoted to specific topics in rapidly growing research areas.

Visit the Editor's homepage:
http://addi.or.uni-bonn.de:1048/finasto.html

ISSN 0949-2984 Title No. 780
ISSN (Electronic Edition) 1432-1122

To order a sample copy or for subscription information:

In North America:
- Call: 1-800-SPRINGER or 212-460-1500
- Fax: 201-348-4505
- Write: Springer-Verlag, 175 Fifth Avenue, New York, NY 10010
- E-mail: journals@ springer-ny.com
- Visit our website: http:// www.springer-ny.com

Outside North America:
- Call: +49 (30) 827 870
- Fax: +49/30/ 8 27 87-448
- Write: Springer-Verlag, P.O. Box 3113 40, D-10643, Berlin, Germany
- E-mail: subscriptions@ springer.de
- Visit our website: http://www.springer.de

Now available online
http://link.springer-ny.com

Springer

http://www.springer-ny.com